Nanopores

Samir M. Iqbal • Rashid Bashir
Editors

Nanopores

Sensing and Fundamental Biological Interactions

Springer

Editors
Samir M. Iqbal
Department of Electrical Engineering
Nanotechnology Research and
Teaching Facility
Joint Graduate Studies Committee
of the Bioengineering Program
University of Texas at Arlington
and University of Texas Southwestern
Medical Center at Dallas
TX 76019, USA
smiqbal@uta.edu

Rashid Bashir
Department of Electrical and Computer
Engineering
Department of Bioengineering
Micro and Nanotechnology Lab
University of Illinois at Urbana-Champaign
Champaign, IL 61820, USA
rbashir@illinois.edu

ISBN 978-1-4419-8251-3 e-ISBN 978-1-4419-8252-0
DOI 10.1007/978-1-4419-8252-0
Springer New York Dordrecht Heidelberg London

Library of Congress Control Number: 2011929039

Printed on acid-free paper

Springer is part of Springer Science+Business Media (www.springer.com)

Preface

Dear Colleagues,

We are delighted to present this book to you, which is a compilation of knowledge from leaders and experts in the field of nanopore sensors. We believe that this book will provide a good source of review articles for researchers and graduate students interested and working in the general areas of nanobiotechnology and especially on nanopore sensors. Over the years, nanopores have rightly emerged as promising tools in three areas of scientific investigations and applications. Firstly, these sensors have been used to study interesting biophysics at the single molecule level and these studies have provided useful insights into mechanisms like folding/unfolding of DNA molecules, unzipping of double-stranded DNA, examination of single molecule DNA-protein interactions, and elegant force spectroscopy measurements of biomolecule, all utilizing electronic or optical means. Secondly, these single molecule sensors can potentially be used in applications which could be loosely termed as 'early diagnostics', such as detection of important attributes of individual biomolecules such as methylation, identification of specific short strands such as miRNA, and discriminating molecules of different lengths, etc. And thirdly, nanopores sensors can provide a path towards the holy grail of direct sequencing of single molecules of DNA. And certainly, tremendous strides have been made towards this grand challenge.

Nanopore sensors can be made from either biological components or from solid-state materials. The use of alpha hemolysin, reported in 1996 by Kasianowicz et al. introduced the use of α-hemolysin protein channels for characterization of single molecules of DNA. Within a few years, a number of groups reported solid-state nanopores as possible robust replacements of lipid bilayers and protein channels. Since then a lot more has been done in moving forward both the solid-state nanopores and the proteinous nanochannels. As you will see in the book, the advancements have focused at many dimensions to improve overall nanopore systems. The endeavors have focused on novel fabrication approaches, interesting measurement techniques, variety in data analysis, ever more detailed modeling to elucidate molecule level interactions, equipping nanopores with reproducible chemistries, and much more. Evanescent optical methods in solid-state nanochannels provide exquisite signal to noise ratio for optical detection of sequencing by synthesis.

More recently, graphene nanopores have been shown to provide the possibility of atomically thin membranes and electrodes. Similarly, the elegant approaches of α-hemolysin based sequencing-by-synthesis, and use of new channels such as the connector protein of the phi29 bacterial phage raise interesting possibilities for characterization of double-stranded DNA. All these have brought close to reality the goal of skimming direct sequencing information from a single DNA molecule.

In summary, the field of nanopore sensors is moving fast. We expect great advancements and contributions in the near future towards realization of these sensors for biophysical, biological, clinical, and many more sensing applications. This book certainly serves as a stepping stone toward many exciting discoveries at the interface of synthetic materials and biology.

Acknowledgments The foremost gratitude is due to our families, who have been supportive for us to tackle this challenge. And then we thank the contributors. This work would not have been possible without their expertise, and their diligent and timely support. Many colleagues and students have helped in editing and proof-reading the work contained in this book, and we are in debt for their gracious share in spite of their busy schedules; notably W. Asghar, J. Azhar, M. J. Bari, J. A. Billo, M. Y. Danawala, A. H. Faraz, K. H. U. Rahman, S. Shoeb, and B. M. K. Venkatesan.

Arlington, TX Samir M. Iqbal
Champaign, IL Rashid Bashir

Contents

Contributors

Bala Murali Venkatesan Department of Electrical and Computer Engineering, Micro and Nanotechnology Lab, University of Illinois at Urbana-Champaign, Champaign, IL 61820, USA
bvenkate@illinois.edu

Rashid Bashir Department of Electrical and Computer Engineering, Department of Bioengineering, Micro and Nanotechnology Lab, University of Illinois at Urbana-Champaign, Champaign, IL 61820, USA
rbashir@illinois.edu

Adam R. Hall Kavli Institute of Nanoscience, Delft University of Technology, Lorentzweg 1, Delft, CJ 2628, The Netherlands

Joint School of Nanoscience and Nanoengineering, University of North Carolina at Greensboro, Greensboro, NC 27401, USA
adam.hall@uncg.edu

Cees Dekker Kavli Institute of Nanoscience, Delft University of Technology, Lorentzweg 1, Delft, CJ 2628, The Netherlands
c.dekker@tudelft.nl

Li-Qun Gu Department of Biological Engineering and Dalton Cardiovascular Research Center, University of Missouri, Columbia, MO 65211, USA
gul@missouri.edu

Farzin Haque Nanobiomedical Center, Vontz Center for Molecular Studies, University of Cincinnati, ML #0508, 3125 Eden Avenue, Room 2308, Cincinnati, OH 45267, USA
haquefn@ucmail.uc.edu

Peixuan Guo Nanobiomedical Center, Vontz Center for Molecular Studies, University of Cincinnati, ML #0508, 3125 Eden Avenue, Room 2308, Cincinnati, OH 45267, USA
guop@purdue.edu; guopn@ucmail.uc.edu

Waseem Asghar Department of Electrical Engineering, Nanotechnology Research and Teaching Facility, University of Texas at Arlington, Arlington, TX 76019, USA
waseem.asghar@mavs.uta.edu

Joseph A. Billo Department of Electrical Engineering, Nanotechnology Research and Teaching Facility, University of Texas at Arlington, Arlington, TX 76019, USA
joseph.billo@mavs.uta.edu

Samir M. Iqbal Department of Electrical Engineering, Nanotechnology Research and Teaching Facility, Joint Graduate Studies Committee of the Bioengineering Program, University of Texas at Arlington and University of Texas Southwestern Medical Center at Dallas, TX 76019, USA
smiqbal@uta.edu

Bradley Ledden Microelectronics-Photonics, University of Arkansas, 248 Physics Bldg., Fayetteville, AR 72701, USA
bledden@uark.edu

Daniel Fologea Department of Biological Sciences, University of Arkansas, SCEN601, Fayetteville, AR 72701, USA
dfologea@uark.edu

David S. Talaga Department of Chemistry and Biochemistry, Montclair State University, Montclair, NJ 07043, USA
talagad@mail.montclair.edu

Jiali Li Department of Physics, University of Arkansas, 226 Physics Bldg., Fayetteville, AR 72701, USA
jialili@uark.edu

Maria E. Gracheva Department of Physics, Clarkson University, 8 Clarkson Avenue, Potsdam, NY 13699, USA
gracheva@clarkson.edu

Amandine Leroux Department of Electrical Engineering and Computer Sciences, Institut Montefiore, Universite de Liège, Liège, Belgium

Beckman Institute, University of Illinois at Urbana-Champaign, Champaign, IL 61820, USA
A.Leroux@ulg.ac.be

Jacques Destiné Department of Electrical Engineering and Computer Sciences, Institut Montefiore, Universite de Liège, Liège, Belgium

Beckman Institute, University of Illinois at Urbana-Champaign, Champaign, IL 61820, USA
destine@montefiore.ulg.ac.be

Jean-Pierre Leburton Department of Electrical and Computer Engineering and Department of Physics, Beckman Institute, University of Illinois at Urbana-Champaign, Urbana, IL 61801, USA
jleburto@uiuc.edu

Xinsheng Sean Ling Department of Physics, Brown University, 182 Hope Street, Providence, RI 02912, USA
xsling@brown.edu

Erik C. Yusko Department of Biomedical Engineering, University of Michigan, 1101 Beal Avenue, Lurie Biomedical Engineering Building, Room 2493, Ann Arbor, MI 48109-2099, USA
ecyus@umich.edu

Yazan N. Billeh Department of Biomedical Engineering, University of Michigan, 1101 Beal Avenue, Lurie Biomedical Engineering Building, Room 2493, Ann Arbor, MI 48109-2099, USA
ynbilleh@umich.edu

Jerry Yang Department of Chemistry and Biochemistry, University of California, San Diego, 9500 Gilman Drive, MC 0358, 6100C Pacific Hall, La Jolla, CA 92093-0358, USA
jerryyang@ucsd.edu

Michael Mayer Department of Biomedical Engineering, University of Michigan, 1101 Beal Avenue, Lurie Biomedical Engineering Building, Room 2174, Ann Arbor, MI 48109-2099, USA
mimayer@umich.edu

Meni Wanunu Department of Physics and Astronomy, University of Pennsylvania, 209 S 33rd St., David Rittenhouse Labs, 1N19, Philadelphia, PA 19104, USA
wanunu@sas.upenn.edu

Allison Squires Department of Biomedical Engineering, Boston University, 44 Cummington St., Boston, MA 02215, USA
asquires@bu.edu

Amit Meller Department of Biomedical Engineering, Boston University, 44 Cummington St., Boston, MA 02215, USA
ameller@bu.edu

Gustavo Stolovitzky IBM Thomas J. Watson Research Center, 1101 Kitchawan Road, Yorktown Heights, NY 10598, USA
gustavo@us.ibm.com

Hongbo Peng IBM Thomas J. Watson Research Center, 1101 Kitchawan Road, Yorktown Heights, NY 10598, USA
pengho@us.ibm.com

Binquan Luan IBM Thomas J. Watson Research Center,
1101 Kitchawan Road, Yorktown Heights, NY 10598, USA
bluan@us.ibm.com

Gregory Timp Stinson-Remick Hall, University of Notre Dame,
Notre Dame, IN 46556, USA
gtimp@nd.edu

Utkur Mirsaidov University of Illinois at Urbana-Champaign,
Beckman Institute, 405 N Mathews Ave, Urbana, IL 61801, USA
uzphoton@uiuc.edu

Winston Timp School of Medicine, Johns Hopkins University,
855 N. Wolfe St., Rangos 570, Baltimore, MD 21205, USA
wtimp1@jhmi.edu

Jiwook Shim Department of Electrical Engineering, University of
Notre Dame, B20 Stinson-Remick Hall, Notre Dame, IN 46556, USA
Jiwook.Shim.5@nd.edu

Deqiang Wang Department of Electrical Engineering, Unversity of
Notre Dame, B20 Stinson-Remick, Notre Dame, IN 46556, USA
dwang2@nd.edu

Valentin Dimitrov University of Illinois at Urbana-Champaign,
Beckman Institute, 405 N Mathews Ave, Urbana, IL 61801, USA
dimitrov@uiuc.edu

Jan Scrimgeour School of Physics, Petit Institute for Bioengineering and
Bioscience, Molecular Science and Engineering Building, Georgia Institute
of Technology, 901 Atlantic Dr. Atlanta, GA 30332-0400
jan.scrimgeour@physics.gatech.edu

Chunchen Lin Department of Electrical Engineering, University of
Notre Dame, B20 Stinson-Remick Hall, Notre Dame, IN 46556, USA
clin7@nd.edu

Jeffrey Comer Department of Physics, Beckman Institute
for Advanced Science and Technology, University of Illinois
at Urbana-Champaign, 1110 W Green Street Urbana, Urbana, IL 61801, USA
jcomer2@illinois.edu

Anthony H. Ho Department of Physics, University of Illinois
at Urbana-Champaign, Beckman Institute for Advanced Science
and Technology, 405 N Mathews Ave, Urbana, IL 61801, USA
hho7@illinois.edu

Xueqing Zou Physics Department, University of Illinois at Urbana-Champaign,
Beckman Institute, Room 3159, 405 N Mathews Ave, Urbana, IL 61801, USA
xqzou@ks.uiuc.edu

Aleksei Aksimentiev Department of Physics, Beckman Institute for Advanced Science and Technology, University of Illinois at Urbana-Champaign, 1110 W Green Street Urbana, Urbana, IL 61801, USA
aksiment@illinois.edu

Klaus Schulten University of Illinois at Urbana-Champaign, Beckman Institute 285 for Advanced Science and Technology, 405 N Mathews Ave, Urbana, IL 61801, USA
kschulte@ks.uiuc.edu

Xiyun Guan Department of Chemistry and Biochemistry, The University of Texas at Arlington, 700 Planetarium Place, Arlington, TX 76019-0065, USA
xguan@uta.edu

Ranulu Samanthi S. de Zoysa Department of Chemistry and Biochemistry, The University of Texas at Arlington, 700 Planetarium Place, Arlington, TX 76019-0065, USA
samanthi.dezoysa@mavs.uta.edu

Dilani A. Jayawardhana Department of Chemistry and Biochemistry, The University of Texas at Arlington, 700 Planetarium Place, Arlington, TX 76019-0065, USA
dilani.jayawardhana@mavs.uta.edu

Qitao Zhao Department of Chemistry and Biochemistry, The University of Texas at Arlington, 700 Planetarium Place, Arlington, TX 76019-0065, USA
qzhao@uta.edu

Chapter 1
Solid-State Nanopore Sensors for Nucleic Acid Analysis

Bala Murali Venkatesan and Rashid Bashir

Abstract Solid-state nanopores are *nm* sized apertures formed in thin synthetic membranes. These single molecule sensors have been used in a variety of biophysical and diagnostic applications and serve as a potential candidate in the development of cost-effective, next generation DNA sequencing technologies, critical to furthering our understanding of inheritance, individuality, disease and evolution. The versatility of solid-state nanopore technology allows for both interfacing with biological systems at the nano-scale as well as large scale VLSI integration promising reliable, affordable, mass producible biosensors with single molecule sensing capabilities. In addition, this technology allows for truly parallel, high throughput DNA and protein analysis through the development of nanopore and micropore arrays in ultra-thin synthetic membranes. This chapter is focused on the development of solid-state nanopore sensors in synthetic membranes and the potential benefits and challenges associated with this technology. Biological nanopores, primarily *α-hemolysin* and the *phi29* connector are also reviewed. We conclude with a detailed discussion on chemically modified solid-state nanopores. These surface functionalized nanopore sensors combine the stability and versatility of solid-state nanopores with the sensitivity and selectivity of biological nanopore systems and may play an important role in drug screening and medical diagnostics.

B.M. Venkatesan (✉)
Department of Electrical and Computer Engineering, Micro and Nanotechnology Lab, University of Illinois at Urbana-Champaign, Champaign, IL 61820, USA
e-mail: bvenkate@illinois.edu

R. Bashir (✉)
Department of Electrical and Computer Engineering, Department of Bioengineering, Micro and Nanotechnology Lab, University of Illinois at Urbana-Champaign, Champaign, IL 61820, USA
e-mail: rbashir@illinois.edu

S.M. Iqbal and R. Bashir (eds.), *Nanopores: Sensing and Fundamental Biological Interactions*, DOI 10.1007/978-1-4419-8252-0_1,

Keywords α-hemolysin • phi29 • Nanopores in Al_2O_3 Membranes • Surface Charges in Nanopores • Surface Enhanced DNA Transport

1.1 Introduction

Our quest to better understand the origins of inheritance, individuality, disease and evolution has led to some of the greatest scientific and technological discoveries in history. This journey has taken us from the macro-scale world that we live in, to the micro-scale environments in which cells and bacteria operate, down to the nano-scale in which single molecules of DNA, RNA and proteins exist. The discovery of microorganisms in the late 1600s by Anton Van Leeuwenhoek laid the foundation for modern microbiology and bacteriology, making possible the discovery of processes such as cell division and cell differentiation. Gregor Mendel's discovery of the laws of inheritance through observations involving the physical traits passed on between generations of pea plants, laid the foundation for modern day genetics. Latter work by Oswald Avery confirmed that this genetic information is carried by deoxyribonucleic acid or DNA.

The discovery of DNA as the blue prints of life in all living organisms is of fundamental importance in medicine and biology. DNA contains the instruction set that is used to encode RNA and proteins, the machinery that drives all cellular activity. Chemically, DNA consists of two long polymers composed of simple sub-units called nucleotides arranged in a double helix structure. Each nucleotide contains a sugar-phosphate backbone attached to one of four types of molecules called bases, specifically Adenine, Thymine, Cytosine and Guanine. It is the sequence of these four bases along the DNA backbone that encodes the genetic information that defines the various characteristics of an organism. Due to the vast information content of DNA and its importance in regulating cellular behavior, widespread research is focused on the development of technologies applicable to DNA analysis.

Gel electrophoresis is the most commonly used tool in DNA analysis and is the work-horse of conventional DNA Sanger sequencing platforms [73]. In gel electrophoresis, charged biopolymers are electrically driven through a 'gel', consisting of a heterogeneous, three-dimensional matrix of pores ranging in diameter from a few nanometers up to hundreds of nanometers. The gel is composed of either polyacrylamide or agarose depending on the specific weight and composition of the analyte of interest. Electrophoresis of DNA is made possible by the charged nature of this polymer in solution. The isoelectric point of the phosphate group on the DNA backbone is ~1 resulting in a single negative charge per nucleotide under most experimental conditions, including physiological pH. In solution, this charge is partially shielded due to counterion condensation according to Manning theory [57]. During gel electrophoresis, strong interactions between DNA and the pore network result in the fine separation of even relatively similar molecular species, based on the enhanced mobility of shorter molecules in the gel. This technique finds application in DNA and protein purification, DNA length separation assays and in DNA sequencing [9]. The limitation of this technique however, is that it involves the simultaneous processing of millions or billions of molecules in order to provide a macroscale optical readout. This in turn translates to large analyte volumes, increased preparation time and high cost. In addition, the output is averaged over a population of molecules and is less sensitive to subtle structural variations amongst molecules. Single molecule sensing methods could help overcome these limitations. Single molecule sensing methods employ highly sensitive optical and electrical technologies to interrogate and analyze individual molecules thereby reducing required analyte volumes and cost. One such technology that finds application in single molecule DNA analysis with potential application to next generation DNA sequencing, is the use of nanopores.

Nanopores are *nm* sized apertures embedded in biological membranes or fabricated in solid-state membranes. Though the passage of biomolecules and ions through nanopores is commonplace in biology, it is only recently that researchers have been able to successfully drive single biomolecules such as DNA through proteinaceous and solid-state nanopores in-vitro [22, 43]. Solid-state nanopore platforms are capable of resistively sensing individual biomolecules including DNA, RNA and small proteins. The concept of resistive particle sensing in solutions was first pioneered by Coulter in the early 1950s [21]. This work led to the development of the Coulter Counter, now a commonly used device for obtaining complete blood cell counts. The principle governing the operation of the Coulter counter is relatively simple. An aperture, slightly larger than the analyte of interest separates two chambers filled with conductive electrolyte. Electrodes immersed in each chamber are used to apply an electric potential, creating a current of ions through the aperture. As the analyte of interest passes through the aperture, the ionic current is partially blocked and this perturbation is sensed electrically revealing useful information about the particle.

Nanopore based single molecule sensors use this exact same principle, only at the nano-scale where the size of this aperture is comparable to the 2.2 nm cross sectional diameter of an individual dsDNA molecule. Briefly, a silicon support containing a single nanopore of diameter comparable to the diameter of an individual DNA molecule is fabricated (Fig. 1.1a) and then inserted into a flow cell containing two chambers filled with conductive electrolyte (Fig. 1.1b). An electrode is immersed in each chamber as shown, and a potential is applied across the nanopore chip, resulting in an ionic current through the pore corresponding to the open pore current. Target DNA molecules are next inserted into the *cis* chamber of the fluidic setup. Two-terminal electrophoresis is used to drive the negatively charged DNA molecule through the nanopore (Fig. 1.1c), resulting in transient current blockades as seen in Fig. 1.1d. These electrical signatures are then analyzed, revealing useful information about the translocating molecule [89, 90]. This technique has been used to study various biophysical phenomena at the single molecule level including the label-free

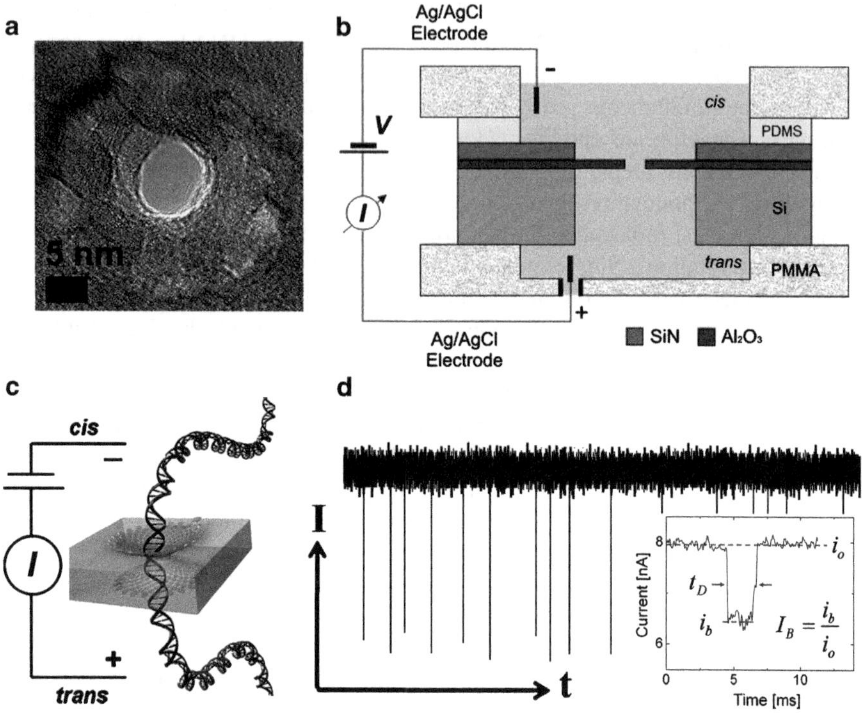

Fig. 1.1 (**a**) Transmission Electron Microscope (TEM) image of a 7 nm Al_2O_3 nanopore (**b**) Nanopore fluidic setup showing Si nanopore chip encapsulated between two chambers of a PMMA flow cell filled with conductive electrolyte (**c**) Schematic of DNA transport through a nanopore (**d**) Characteristic downward current blockades seen during the transport of individual DNA molecules through the pore (Inset) Current signature corresponding to a single DNA translocation event showing, blockage ratio I_B, and translocation time, t_D, specific to the translocation event

detection of single nucleotide polymorphisms (SNP's) with application to cancer diagnostics [36], stretching transitions in individual dsDNA [31], unzipping kinetics of hairpin DNA molecules [59], DNA sizing and sieving [85], and the detection of DNA-protein complexes essential in DNA repair [80].

1.2 Nanopores in Biology

The biological cell contains various types of nanopores and nanochannels that regulate the flow of ions and molecules into and out of the cell. These nanopores play a vital role in cellular processes such as intercellular communication and signaling between subcellular structures. Examples include gated, selective ion channels that connect the cell cytosol to the cell exterior; nuclear membrane pores that control the passage of biomolecules such as messenger RNA (mRNA) from the cell nucleus into the cytosol; proteins that are secreted across pores in the membranes of cell organelles; and viruses, which dump their genomes into cells via pores that insert into the cell membrane [22]. An example of a biological nanopore that is frequently used in-vitro is the *α-hemolysin* channel.

1.2.1 α-Hemolysin

α-hemolysin is a naturally occurring biological protein complex extracted from the bacterium *Staphylococcus aureus* that when inserted into a lipid bilayer membrane, forms a ~1.5 nm diameter pore allowing the passage of ions and ssDNA. In vivo, bacterium *Staphylococcus aureus* secretes alpha-hemolysin monomers that bind to the outer membrane of host cells. These monomers self assemble into seven subunit oligomers to form a water-filled transmembrane channel that facilitates the uncontrolled permeation of water, ions, and small organic molecules in and out of the host cell. The resulting discharge of vital molecules from the host cell, osmotic swelling and a loss in ionic gradient can result in irreversible cell damage and eventually death of the host cell (apoptosis). Apoptosis induced by *α-hemolysin* insertion in various cell types including rabbit erythrocytes, human erythrocytes, monocytes, and lymphocytes has been reported [4, 40].

In-vitro studies of DNA transport through biological pores have traditionally focused on *α-hemolysin* as the transmembrane channel of choice. The structure of the heptameric *α-hemolysin* pore embedded in a lipid bilayer is shown in Fig. 1.2a.

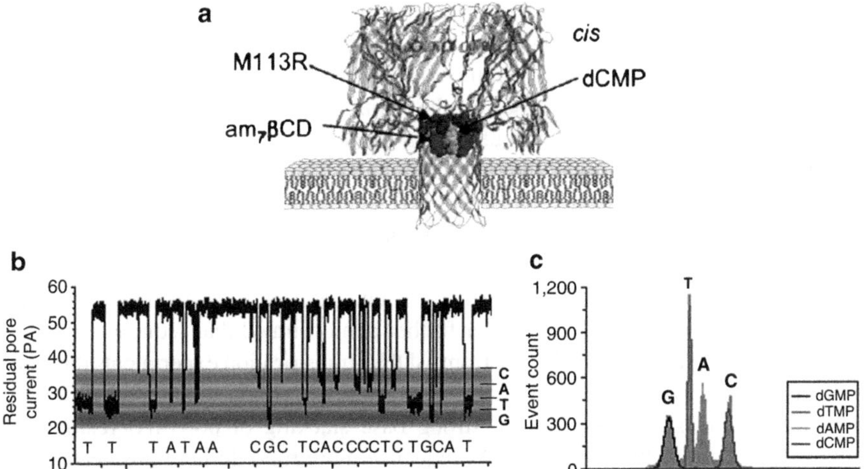

Fig. 1.2 Single nucleotide detection using a mutant *α-hemolysin* nanopore [18], reprinted with permission. (**a**) Mutant *α-hemolysin* protein channel modified with a cyclodextrin adapter embedded in a lipid bilayer (**b**) Characteristic current recordings from a modified *α-hemolysin* channel, showing different blockade levels for each base (**c**) Current blockade histogram confirms single base discrimination ability of modified *α-hemolysin* pore

The total channel length is 10 nm and is comprised of a 5 nm vestibule that protrudes into the *cis* compartment and a 5 nm transmembrane domain embedded in the lipid bilayer. Between pH 7–9, *α-hemolysin* forms a relatively stable and reproducible non-gating channel with less than 2% variation in open pore current under temperature stabilized conditions [96]. The comparable inner channel diameter of *α-hemolysin* to ssDNA (diameter ~1.3 nm) suggests that less than one Debye length (~3 Å in 1 M KCl) separates the translocating biomolecule from the amino acid residues in the pore. Although dsDNA is too large to translocate through *α-hemolysin*, up to a 10 bp fragment can reside in the vestibule. This makes *α-hemolysin* a very powerful tool for examining biomolecular interactions and the binding affinities of individual molecules at the single molecule level.

In a landmark study, Kasianowicz et al. demonstrated the ability to electrically detect individual ssDNA and ssRNA molecules using *α-hemolysin* nanopores embedded in planar phospholipid bilayers [43]. A plethora of studies have since followed elucidating the biophysics of single molecule transport through proteinaceous *α-hemolysin*. For example, Meller et al. examined the effects of polymer length on translocation velocity [61]. Polymers longer than the pore length were seen to translocate at constant speed however shorter polymers exhibited a length dependent velocity. Studies by Mathe et al. revealed that *α-hemolysin* nanopore sensors are sensitive enough to differentiate between 3′ and 5′ threading of ssDNA in the pore with 5′ threading resulting in a twofold increase in translocation times relative to 3′ threading, attributed to the tilt reorientation of bases towards the 5′ end of the molecule [58]. Brun et al. showed that biomolecule flux through proteinaceous *α-hemolysin* is highly dependent on the applied voltage with the capture rate of ssDNA [60, 64] and small polyelectrolytes [8], following a simple Van't Hoff-Arrhenius relationship. Kasianowicz and co-workers further showed that the asymmetric structure of *α-hemolysin* promotes biomolecule entry from the *cis* side (side with the vestibule) as opposed to the *trans* side [32]. Reduced biomolecule flux from the *trans* side was attributed to a combination of factors; (1) the high entropic barrier associated with the highly confined geometry of the β barrel on the *trans* side and (2) electrostatic repulsion of DNA by the negatively charged asparatic acid residues located on the *trans* side. The unzipping of hair-pin DNA structures using *α-hemolysin* was observed by Vercoutere et al. for sufficiently short hairpins. Vercoutere demonstrated the ability to discriminate between 3 and 8 bp long hairpins with single base resolution [92]. Meller further demonstrated that sequence specific information could be derived directly from ssDNA translocation experiments through biological *α-hemolysin* [60]. Poly(dA) and Poly(dC) strands exhibited different translocation times attributed to the strong base stacking of poly(dA) relative to poly(dC), thereby making the poly(dA) sequence more rigid during translocation.

α-hemolysin nanopores also hold tremendous value in the field of DNA sequencing. Stoddart recently demonstrated the ability to resolve individual nucleotides located in homopolymeric and heteropolymeric ssDNA immobilized in biological *α-hemolysin* [83]. Mitchell et al. showed that chemical labels attached to bases could be used to resolve individual bases in a translocating DNA strand [62]. Interestingly blockage durations and amplitudes could be tuned by varying the chemistry, charge and size of these chemical tags, suggesting the possibility of base discrimination based on peptide labeling with application to DNA sequencing. Another novel nanopore-based sequencing approach was proposed by Cockcroft et al. that exploited the selective, base-by-base activity of DNA Polymerase [19]. By anchoring a DNA/DNA-Polymerase complex in the nanopore, the authors were able to electrically monitor single nucleotide primer extension events. Primer extensions were controlled by providing each nucleotide set sequentially and temporally extracting sequence information [19]. The Bayley group recently demonstrated the ability to continuously resolve indigenous single nucleotides (dAMP, dCMP, dGMP, dTMP) through nanopore based resistive current measurements [18]. Remarkably, individual bases could be discriminated based on current blockade levels without any prior base labeling or chemical modification as shown in Fig. 1.2b, c. Base selectivity was achieved by modifying the mutant α-*hemolysin* pore with a cyclodextrin adapter ($am_6amPDP_1\beta CD$) covalently bound within the β barrel of the transmembrane domain, thereby constricting the nanopore channel while enhancing the chemical specificity of the sensor. Raw bases were read with over 99% confidence under optimal operating conditions. By integrating this base identification platform with a highly processive exonuclease (chemical attachment or genetic fusion of exonuclease), a single molecule sequencing by digestion approach may be feasible.

1.2.2 Bacteriophage phi29 Connector

Another biological nanopore that is receiving much interest of late is the connector protein from the bacteriophage *phi29* DNA packaging motor. In bacteriophage *phi29*, linear dsDNA is packaged into a viral capsid, an entropically unfavorable process that requires the hydrolysis of ATP. During packaging, the linear DNA passes through a narrow channel of inner diameter ~3.6 nm, termed the connector. The connector is comprised of 12 GP10 protein subunits that readily self assemble in solution to form a stable, repeatable do-decameric structure [95]. As the crystal structure of this biological nanopore channel has been resolved [100], explicit site engineering is possible. Wendell et al. modified the *phi29* connector protein to include hydrophilic sites on the crown and base, thereby allowing its integration into liposomes [95]. The insertion of the *phi29* connector into preformed lipid bilayers was achieved through vesicle fusion resulting in steady, repeatable pore conductance. Pore conductance was comparatively approximately five times higher than that observed in α-*hemolysin* under similar conditions and did not show any voltage gating effects. *Phi29* has one distinct advantage over α-*hemolysin*. The larger barrel allows the translocation of dsDNA and a variety of proteins that are simply too large to pass through α-*hemolysin*. Thus the *phi29* connector system allows the experimentalist to examine a broader spectrum of biomolecule interactions at the single molecule level. Wendell et al. demonstrated the successful detection of 5.5 kbp dsDNA and 35 bp dsDNA using this novel biological nanopore platform.

1.3 Solid-State Nanopores

With advances in microfabrication technologies, focus has shifted to the solid-state domain with numerous groups studying biomolecule transport through solid-state nanopores. Solid-state nanopores exhibit superior chemical, thermal, and mechanical stability over their biological counterparts, and can be fabricated using conventional semiconductor processes, thereby facilitating mass fabrication and size tunability. In addition, solid-state nanopores are functional over a wide range of pH and do not exhibit any voltage gating, whilst allowing for integration with electrical contacts and optical probes.

1.3.1 Fabrication of Single Nanopores

There are four primary techniques available for the fabrication of solid-state nanopores in thin Si_3N_4, SiO_2, Al_2O_3 or polymer membranes; surface tension driven oxide reflow, ion milling, track-etch method and electron beam based decompositional sputtering. Other lithography-free techniques for creating individual nanopores include focused ion beam (FIB) techniques coupled with ion-beam sculpting to achieve pore sizes as low as 10 nm [54], and laser ablation methods capable of achieving sub 100 nm pore diameters [99, 101].

1.3.1.1 Electron Beam Induced Oxide Reflow

The oxide reflow technique involves the use of e-beam lithography to pattern large 40–100 nm holes in micromachined silicon membranes. These pores are subsequently oxidized and shrunk to the sub-10 nm range using a TEM. The TEM shrinking process, discovered by Storm [84] uses a high energy electron beam to locally fluidize the oxide surface in the vicinity of the nanopore causing the oxide to reflow in the direction that minimizes interstitial surface energy. For nanopores with diameter, $d < t$, where t is the membrane thickness, nanopore shrinking was repeatedly observed. Figure 1.3 illustrates this electron beam induced shrinking process. Schenkel et al. attributed this shrinking phenomenon to the build up of a low-Z hydrocarbon layer in the nanopore during electron-beam irradiation [74]. Electron Energy Loss Spectra (EELS) from the localized nanopore region however revealed the presence of only Si and O and the absence of C, [84] thereby confirming that oxide reflow is indeed the mechanism responsible for nanopore shrinking.

The temporal contraction of a SiO_2 nanopore formed in an oxidized free standing Si membrane through electron beam induced oxide reflow processes is shown in Fig. 1.3a–d. The temporal contraction of an Al_2O_3 nanopore is illustrated in Fig. 1.3e–h [89]. The formation of SiO_2 nanopores in oxidized free standing Si membranes presents some inherent limitations. Pores were oxidized (wet oxidation)

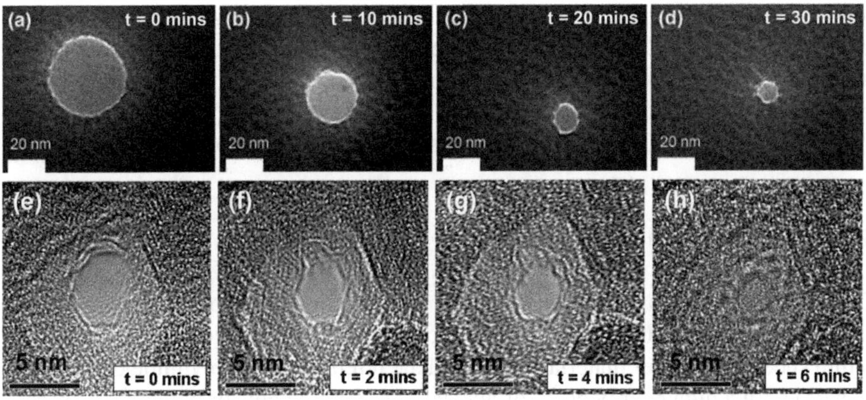

Fig. 1.3 TEM based nanopore contraction process: (**a–d**) Sequence of phase contrast images obtained during TEM based temporal contraction of a SiO_2 nanopore. Pore contraction was observed over a period of 30 mins from an initial pore size of ~40 nm to a final pore size of ~8 nm. (**e–h**) TEM based contraction of an Al_2O_3 nanopore [89], reprinted with permission. A tightly focused electron beam is used to initially sputter a pore of diameter ~4 nm. Exposure to a defocused electron beam results in material reflow into the pore, resulting in pore contraction. Final pore size is approximately 1 nm

at 900°C for 31 min resulting in a SiO_2 layer of thickness between 800 and 900 Å. This oxide layer acts as an insulating barrier, shielding the electrolyte solution from the underlying conductive Si surface. Tilted SEM images of these thermally grown silicon dioxide membranes illustrate high levels of compressive stress and membrane buckling post oxidation [89]. The result is extremely fragile, highly stressed membranes that frequently rupture during DNA translocation experiments. Fitch et al. studied the intrinsic stress and strain in thin films of SiO_2 prepared by the thermal oxidation of crystalline silicon and concluded that large intrinsic stress gradients exist in the layers of SiO_2 at the Si/SiO_2 interface [25]. The residual intrinsic interfacial stress was calculated at ~460 MPa and was independent of the growth temperature and thermal annealing processes, attributed to mismatches in the molar volumes of Si and SiO_2 at the interface. This highlights the need for low stress, mechanically stable membranes for nanopore formation.

1.3.1.2 Ion Beam Sculpting

The ion-beam sculpting process first developed by Li et al. [53], uses an energetic beam of Ar^+ ions to form nanopores with dimensions as low as 1.5 nm in thin Si_3N_4 membranes. Contrary to what one would expect, a 3 keV Ar^+ ion beam rastered continuously over a Si_3N_4 sample at room temperature resulted in ion beam assisted diffusion of atoms into the pore region resulting in nanopore contraction. The flow of matter to the developing nanopore showed temperature dependence with a transition between pore opening and pore closing consistently observed at ~5 °C under the ion-beam conditions used. Pore expansion was attributed to ion sputter erosion at the pore edge, dominant at low temperatures and high ion flux. Pore closure was attributed to the reflow of a stressed viscous surface layer into the nanopore. The reduced viscosity and/or enhanced stress in this layer caused relaxation, thereby filling the nanopore. Feedback control was used to precisely control nanopore size.

1.3.1.3 Track-Etch Method

Conical nanopores are typically formed in μm thick polymer films using the track-etch method [10, 75, 76]. The fabrication process involves first bombarding a thin sheet of polymer material (polyethylene terephthalate, polyimide or polycarbonate) with a high energy beam of nuclear fission fragments or with a high energy ion beam from a MeV accelerator at normal or near normal incidence angle. The irradiated polymer membrane is then placed between two chambers of a conductivity cell and etched chemically from one side. Chemical-etching of the damage track is done in a strong alkaline solution (pH $\approx$ 13) with high chlorine content at elevated temperatures (~50 °C) using a solution such as sodium hypochlorite (NaOCl) [77]. The other compartment of the conductivity cell is filled with 1 M potassium iodide (KI) solution as a stopping medium for the OCl^- ions of the etchant. As soon as the etchant completely penetrates the polymer film, iodide ions reduce OCl^- to Cl^- ions thereby halting the etch process. The result is a tapered, individual conical nanopore with pore diameter as low as ~10 nm in the polymer membrane.

1.3.1.4 Electron Beam Induced Sputtering

Electron beam induced sputtering offers a rapid and reliable method to prototype *nm* sized pores in the TEM. This method involves the use of a focused convergent electron beam with sufficiently high current density to decompositionally sputter *nm* sized pores in thin oxide or nitride membranes (thickness $\leq$ 60 nm). An added benefit of this method is that it allows the operator to inspect pore size during fabrication and avoids the need for electron beam lithography involving e-beam resists and reactive ion etching (RIE) pattern transfer steps. Kim et al. used high-resolution TEM to study nanopore formation kinetics in Si_3N_4. Nanopore formation was a balance between two competing processes: (a) material sputtering and (b) surface-tension-induced shrinking [46, 48]. Nanopores, 4–8 nm in diameter were directly drilled using a JEOL 2010F field emission TEM with an accelerating voltage of 200 keV and a beam current density of 10^8–10^9 e nm^{-2}. Nanopore contraction was achieved by slightly defocusing the e-beam, effectively reducing the beam intensity to ~10^6 e nm^{-2}. TEM tomography was used to map the three-dimensional structure of these solid-state nanopores. It was observed that the sidewalls of the sputtered pores were angled (approximately 65° to the horizontal), attributed to the intensity distribution of the e-beam around its focal point. Post-drilling, pores formed an 'hourglass' structure with pore width being determined by the width of the narrowest constriction [46–48]. Similarly, Heng et al. used a focused convergent electron beam to form nanopores in ultra-thin 10 nm Si_3N_4 membranes. The nanopore structure resembled a double cone structure with a cone angle of 10° [30]. Smeets et al. observed a cone angle of 45° for nanopores sputtered in composite SiO_2/SiN/SiO_2 membranes [81]. In all cases, nanopores formed directly through electron beam induced sputtering exhibited the ability to contract under a defocused electron beam.

1.3.2 *Fabrication of Nanopore Arrays*

Multiple methods exist for the formation of nanopore arrays. The track-etch method is one which has been used to produce commercially viable nanopore arrays with diameters as low as ~10 nm and packing densities as high as 6×10^8 pores/cm^2 [28]. Nanopore arrays can also be fabricated through an anodization process of thin aluminum films. In one such process that we have previously explored, aluminum foil is first anodized in a 0.3 M oxalic acid solution at 5 °C at a constant applied voltage of 40 V for 20 h [63]. The anodized aluminum is then etched in an aqueous mixture of phosphoric/chromic acid at 60 °C. Any remaining Al in the pore region is dissolved using a saturated $HgCl_2$ solution. We used this process to produce Anodized Aluminum Oxide (AAO) membranes with a highly ordered network of nanopores (diameters = 75 nm, center-to-center distance = 105 nm), as shown in Fig. 1.4a. These nanopores can be further reduced in size through atomic layer deposition (ALD). Figure 1.4b shows an array of nanopores with final diameters of 15 ± 1 nm formed using a combination of anodization and ALD processes.

Nanopore arrays can also be formed using a serial write process using ion or electron beams. Figure 1.4c, d show arrays of ~200 nm diameter nanopores with 1 μm pitch, formed in 45 ± 5 nm thick, free standing Al_2O_3 membrane using a FIB tool employing a high energy Ga^+ beam. Arrays were formed in these studies using a FEI DB235 FIB system at an accelerating potential of 30 keV and 10 pA beam current. This process can be used to achieve pore diameters as low as 30 nm by controlling the ion dose, accelerating potential and beam current. Nanopore arrays with pore sizes as low as ~20 nm have also been formed in SiO_2 using electron beam lithography processes [70]. In addition, Kim et al. demonstrated that nanopore arrays could be fabricated by stepping a focused electron probe formed in a TEM over the sample surface [47]. Though the process was time intensive, arrays with pore diameters as low as ~5 nm could be produced this way.

1.3.3 *Nanopore Formation in Thin Al_2O_3 Membranes*

Recently, we demonstrated the fabrication of highly sensitive, mechanically stable nanopore sensors in Al_2O_3 membranes formed via electron beam based

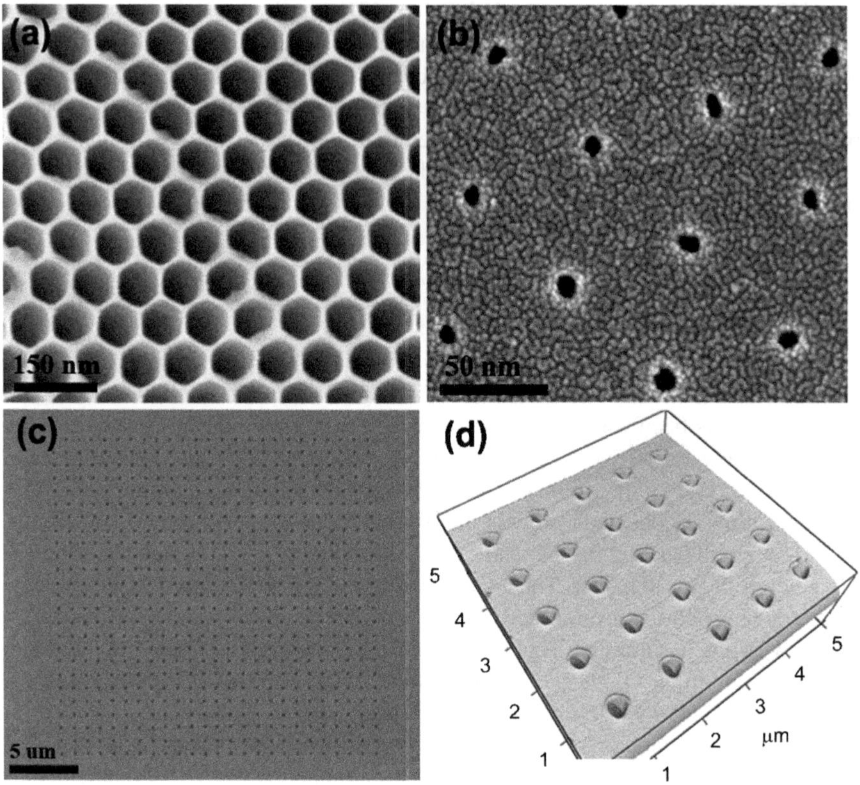

Fig. 1.4 (**a**) SEM image of an AAO nanopore array consisting of 75 nm diameter pores (**b**) SEM of AAO nanopores shrunk using the ALD process to a final pore size of 15 nm [63], reprinted with permission (**c**) SEM of nanopore array formed through FIB milling to form 200 nm pores in a 45 ± 5 nm thick free standing Al_2O_3 membrane (**d**) Atomic force microscope scan of a 5×5 Al_2O_3 nanopore array consisting of 200 nm nanopores, formed using the FIB tool

decompositional sputtering processes [89, 90]. Al_2O_3 membranes were formed through a combination of ALD and standard microfabrication processes resulting in low stress, 45–60 nm thick free standing membranes. In addition, these membranes were mechanically more robust than their SiO_2 counterparts as compressive thermal stress was reduced during fabrication. Nanopores formed in Al_2O_3 membranes exhibited state of the art noise performance and allowed for true VLSI device level integration due to the rapid, low temperature fabrication process developed.

1.3.3.1 Nanopore Nucleation and Expansion Kinetics

SiO_2 and various metal halides/oxides including CaF_2, AlF_3, Al_2O_3, MgO and TiO_x exhibit unique sputtering characteristics under an intense electron beam. The formation of trenches and slots in these material systems has been previously demonstrated [3, 13, 72]. The sputtering process in Al_2O_3 is attributed to the Coulomb explosion displacement of atoms based on the Knotek-Feibelman electron-stimulated desorption mechanism. This decompositional sputtering mechanism was exploited to form nanopores ranging in diameter from 2 to 30 nm [89, 90]. Figure 1.5 shows the Gaussian/Lorentzian intensity profiles of various focused electron probes used in the pore formation process. Probe diameters of 2.3, 2.7, 3.2 and 3.9 nm full width at half maximum (FWHM) were investigated, corresponding to beam current-densities of 2.6×10^6 A/m^2, 4.2×10^6 A/m^2, 6.1×10^6 A/m^2 and 1.2×10^7 A/m^2 respectively. The inset of Fig. 1.5a is a TEM image of a 3.2 nm probe, light areas indicating regions of maximum electron intensity located at the center of the probe and darker areas indicating less intense regions located at the tail of the probe. Larger probe sizes exhibited higher peak intensities and a broader Gaussian/Lorentzian profile and were well suited to form large nanopores with diameters in the range of 10–30 nm, applicable for single molecule protein analysis and the detection of "large" analytes. Smaller probes (2.7 and 3.2 nm) exhibited lower peak intensity and a narrower profile, ideal for the high precision fabrication of 2–10 nm pores in Al_2O_3, well suited for ssDNA, dsDNA and RNA analysis.

Three stages were identified during nanopore formation in Al_2O_3, I, Pore Nucleation, II, Rapid Expansion and III, Controlled Growth as shown in Fig. 1.5b. A critical beam current density in excess of 2.6×10^6 A/m^2 was required for nanopore nucleation in Al_2O_3 membranes. This is in good agreement with threshold current densities extracted by Salisbury et al. in experiments involving electron beam sputtered anodized alumina [72]. Below this threshold, topographical damage corresponding to the cleaving of Al-O bonds (bond dissociation energy of 513 kJ/mol) [67], was observed but electron momentum was insufficient to induce an embryonic nanopore structure. Pore contraction mechanisms were also seen to dominate at low beam current densities, possibly due to surface tension driven oxide reflow, generation/recombination of closely spaced Frenkel pairs [71] and mass transport of mobile atoms into the nucleation site. This is consistent with the nanopore contraction phenomenon observed previously in SiO_2 [11, 84], and Al_2O_3 systems [89]. The sputter rate transition observed at the boundary of the Rapid Expansion and Controlled Growth stage was attributed to electron beam induced crystallization and metallization of the nanopore region.

1.3.3.2 Electron Beam Induced Crystallization

Structural phase transformations in the membrane material around the pore region was observed during electron beam induced decompositional sputtering of Al_2O_3 [89, 90]. Discrete spot reflections of α, γ, δ and κ phase Al_2O_3 were initially identified, confirming the formation of nanocrystalline clusters of preferred phases. In α-Al_2O_3, Al^{3+} cations are octahedrally coordinated with average Al-O bond lengths of 1.92 Å [6]. However, γ-Al_2O_3 typically exhibits a cubic defect-spinel type structure with average Al-O bond lengths of 1.89 Å [5]. The presence of multiple heterogeneous phases with varying bond lengths and co-ordinations,

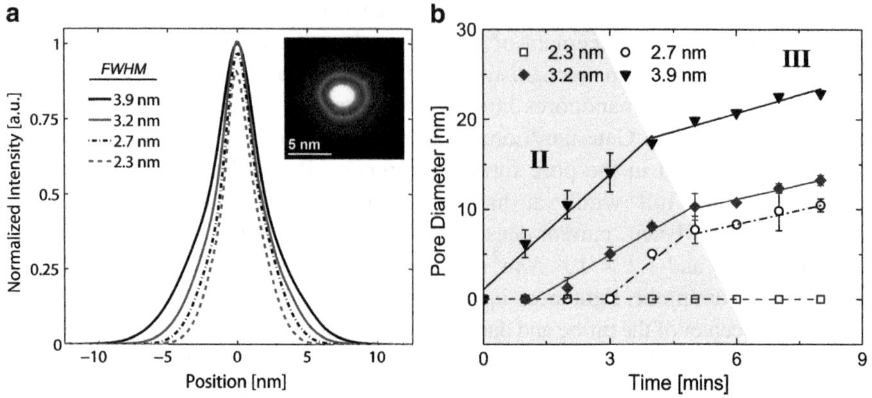

Fig. 1.5 Al_2O_3 nanopore formation kinetics [90], reprinted with permission (**a**) Intensity profiles of the focused electron probes used during nanopore formation in Al_2O_3 thin films. Intensity is normalized with respect to peak intensity of the 3.9 nm *FWHM* probe (Inset) TEM image of a 3.2 nm probe showing spatial intensity distribution (**b**) Nanopore sputtering kinetics illustrating pore diameter vs. time for the various probe sizes examined. Distinct expansion rates were observed delineating the three stages of pore formation, I Pore Nucleation (not shown), II Rapid Expansion and III Controlled Growth

confirms that an irregular density of exposed Al-O groups exists at the pore surface which in turn corresponds to an irregular surface charge distribution in a hydrated Al_2O_3 nanopore. This irregular charge distribution is expected to strongly impact DNA translocation kinetics. Prolonged exposure to the electron beam resulted in a polycrystalline structure with preferred α and/or γ phases only [90]. These results provide evidence that the surface charge in the nanopore can be engineered based on critical electron doses. Electron beam induced crystallization was not observed in Si_3N_4 and SiO_2 systems [89].

1.3.3.3 Variations in the Nanopore Stoichiometry

Wu et al. demonstrated that the electron beam inadvertently modifies the stoichiometry of the nanopore [97, 98]. Nanopores formed in ~60 nm thick SiO_2/SiN/SiO_2 membranes through electron beam sputtering, induced the formation of Si rich clusters in the vicinity of the nanopore as confirmed through electron energy loss spectroscopy (EELS) and energy filtered TEM (EFTEM). O and N were preferentially sputtered away in this case. Electron beam induced sputtering also has a dramatic impact on the composition of Al_2O_3 based nanopores. EELS analysis confirmed that Al-rich regions were formed at and near the pore edge due to the preferential desorption of O [90]. Compositional variations were calculated by the *k-factor* method [24], and revealed that the O to Al ratio in the local nanopore region decreased from 1.5 before pore formation to ~0.6 after pore formation. This result confirms that the sputtering process preferentially desorbs oxygen, leaving behind Al-rich nanocrystals resulting in a partially metalized nanopore. Similar phenomenon has been observed in hole drilling experiments conducted in Na-β Al_2O_3 [24]. Coupled with studies by Berger et al. demonstrating the formation of continuous Al regions and "plugs" in e-beam irradiated metal β-aluminas, [3] this work provides a unique method to potentially form nano-scale metallic contacts within a nanopore for bio-sensing applications. Simulation work by Lagerqvist et al. demonstrated the ability to achieve single nucleotide resolution by employing a nanopore sensor with embedded transverse sensing electrodes, with potential application to nanopore-based DNA sequencing [50]. These results could help enable the possible realization of such a structure. Local nanopore stoichiometry is also very important when chemically modifying or functionalizing a nanopore with various biomolecules or organosilanes. The packing density of these molecules in the lumen and barrel of the nanopore are dependent on the density of Aluminol surface groups in the nanopore. Thus a thorough oxidation in an O_2 plasma is required before any surface functionalization steps.

1.3.4 Ionic Conduction Through Solid-State Nanopores

1.3.4.1 Nanopore Conductance

The conductance of the nanopore can be measured in monovalent electrolyte, typically KCl, by placing the nanopore between two electrically isolated, fluidic reservoirs. Typically, high salt solutions well in excess of physiological conditions are used (~1 M KCl, 10 mM Tris–HCl, pH 7.5) to obtain sufficiently high baseline current levels that can be monitored using a Pico ammeter. Faradaic Ag/AgCl electrodes are placed in each reservoir allowing for a localized redox based exchange reaction to occur at each electrode. After a 30 s O_2 plasma treatment, immediate wetting and ionic conduction through the nanopore is observed. Linear current-voltage characteristics are typically observed for nanopores in SiO_2, Si_3N_4 and Al_2O_3 membranes formed using TEM based decompositional sputtering processes [30, 81, 89]. The linear current-voltage characteristics of a ~11 nm pore in 1 M, 100 and 10 mM KCl electrolyte are shown in Fig. 1.6a. To further probe the performance of Al_2O_3 nanopores in electrolyte, the conductance, G, of 11 different nanopores of varying diameter (4–16 nm) were measured in 1 M KCl, as shown in Fig. 1.6b. Two geometric models were proposed to fit G [46, 48, 81]. The first model assumed a symmetric double cone structure with cone angle, α [81]. Assuming $\alpha = 30°$, [46, 48] an upper conductance bound can be derived (solid black curve of Fig. 1.6b). The second model assumed a purely cylindrical channel of length, $L_{pore} = 60$ nm with a cross sectional diameter equal to the pore diameter, d_{pore} (solid gray curve of Fig. 1.6b). This model provided a lower bound for the measured pore conductance. Applying a least squares fit to the measured data (black dashed curve of Fig. 1.6b), an effective length of $h_{eff} \approx 26.5$ nm and cone angle of $\alpha \approx 24°$ were extracted for Al_2O_3 pores formed via decompositional sputtering. The effects of surface charge were neglected in these

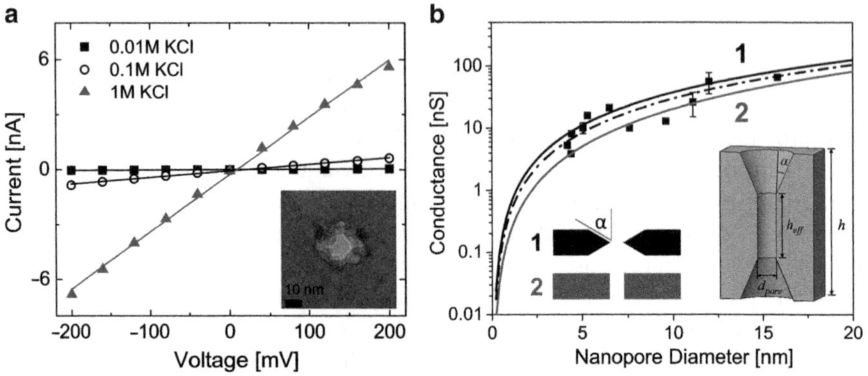

Fig. 1.6 Electrical characterization of Al_2O_3 nanopores [89], reprinted with permission. (**a**) Current-Voltage characteristics of a 11 nm diameter pore measured in 10 mM KCl, 100 mM KCl and 1 M KCl. Linear current-voltage characteristics suggest pore geometry is symmetric in the z-direction (**b**) Pore conductance of 11 nanopores ranging in diameter from 4 to 16 nm. Black and gray solid lines represent conductance models for geometries 1 (double cone) and 2 (cylindrical), shown in the inset. Black dashed line is a least squares fit to the measured data, used to extract pore parameters $h_{eff} \approx 26.5$ nm and $\alpha \approx 24°$. (Right Inset) Predicted cylindrical, double cone geometry of an Al_2O_3 pore from conductance measurements and energy filtered TEM imaging

models as the Debye screening length given by κ^{-1} (where $K^2 = 2e^2 n_{KCL}/k_B T \varepsilon\varepsilon_0$ in 1 M KCl) $<< d_{pore}$. At these high salt concentrations, charge carriers in the bulk were expected to dominate current flow. Electro osmotic flows resulting from counterion condensation on the charged pore surface should be negligible. These results confirmed that a symmetric, double cone nanopore structure was formed through decompositional sputtering. TEM tomograms taken by Wu et al. and Kim et al. on SiO_2/SiN/SiO_2 and Si_3N_4 nanopores further confirmed the double cone, symmetric structure of a TEM sputtered solid-state nanopore [46, 48, 98].

Asymmetric current-voltage characteristics have been reported in synthetic nanopores formed in polyethylene terephthalate (PET) polymer membranes [76, 78]. The track etch method used to form these nanopores produced an asymmetric, conical geometry resulting in a structure that significantly rectifies the ionic current. This conically shaped, highly charged nanopore is cation selective, exhibiting diode like behavior in fluid with a preferential direction for the cation flow from the narrow entrance toward the wide opening of the pore. Siwy et al. further demonstrated some of the novel characteristics of this architecture by pumping ions against a concentration gradient using a fluctuating electric force applied across the membrane in the form of an AC voltage signal [76].

1.3.4.2 Surface Charge Effects

The effects of surface charge on pore conductance were investigated by Ho et al. using nanopores formed in 10 nm thick Si_3N_4 membranes [33]. At low electrolyte concentrations ($\leq$10 mM), pore conductivity was found to be much larger than bulk approximations calculated using pore geometry. This conductance deviation was attributed to Debye layer overlap in the pore (Debye length is comparable to or larger than the pore radius). Multiscale simulations of ion transport through these pores coupled with experimental results suggested the presence of fixed negative charges on the pore wall, and a reduction of the ion mobility due to fixed charge and ion proximity to the pore wall.

Consistent with Ho's results, Smeets et al. [81] found that in ~10 nm diameter SiO_2 pores, the negative surface charge lining the pore walls dominates pore conductivity at salt concentration below ~0.1 M. Interestingly, a variable surface charge density in the pore was extracted as a function of electrolyte concentration in these experiments. pH dependent studies performed by Wanunu et al. in Si_3N_4 nanopores did not reveal any significant changes in the conductance with varying pH [93]. Nanopores formed in PET membranes also exhibited a net negatively charged surface at neutral and slightly basic pH's due to the deprotonation of carboxylate groups on the pore walls. The average density of carboxylate groups was estimated to be 1.5 groups/nm^2 [78]. In contrast to SiO_2, TiO_2 only presented a slightly negatively charged surface at neutral or slightly basic pH. Unlike the variable surface charge observed by Smeets et al. in SiO_2 pores, TiO_2 nanopore conductance saturated at very low electrolyte concentrations [65, 68]. Nam et al. extracted a surface charge density of ~0.005 mC/m^2 in TiO_2 pores, significantly lower than the charge density observed in SiO_2 pores which is estimated at between ~25 and 50 mC/m^2 [65]. Nam suggested that this low charge density may be responsible for the saturation in ionic conductivity at lower KCl concentrations than expected.

In contrast to the aforementioned systems, Al_2O_3 nanopores are expected to be positively charged at neutral or slightly basic pH. The formation of hetero-phase crystalline domains (in particular γ and α-phases) of varying bond lengths and co-ordinations during electron beam irradiation impart a non-uniform charge density in the pore. α-Al_2O_3 and γ-Al_2O_3 both exhibit different points-of-zero-charge (pzc's), estimated at pH 9.1 and pH 8.5 in monovalent salt solution [1, 88]. In addition, the Zeta potentials of these materials measured in pH 7.5 electrolyte are ~50 mV and ~25 mV respectively, [7, 27] and thus these charged nanocrystalline domains are expected to interact with ions and charged polymers in the pore differently [90].

1.3.4.3 Manipulating Surface Charge in Nanopores

As mentioned, precise control over phase transformations in Al_2O_3 nanopore systems by varying electron dose provides a novel method to engineer surface charge at the nanopore/fluid interface via the electron beam. This is very interesting from a materials perspective. Another approach for tuning surface charge in the nanopore is through electrical mechanisms. Nam et al. embedded a TiN gate electrode directly in the nanopore and showed current rectification by applying potentials to the gate electrode [65]. This gating behavior was only observed at very low salt concentrations ($<10^{-3}$ M) where the effects of surface charge are dominant and Debye layer overlap in the nanopore is indeed expected. P-type unipolar behavior was observed suggesting that K^+ ions are the majority carriers in these TiO_2 gated nanopores. Studies by Kalman et al. focused on integrating a Au electrode into a conical nanopore [42]. By modulating the electric potential applied to the gate, one alters the distribution of ions in the overlapping Debye layer in the pore and thus the potential distribution in the pore. Using this approach Kalman et al. were able to manipulate the current through the device from the rectifying behavior synonymous with conical nanopores to a near linear type behavior as seen in structurally symmetric nanopores. The mechanism for this change in transport behavior was accredited to the enhancement of concentration polarization induced by the gate. The manipulation of surface charge through the chemical modification of solid-state nanopores will be discussed in subsequent sections.

1.3.5 Noise in Solid-State Nanopores

Electrical noise in ionic current measurements involving solid-state nanopores limits the utility of these systems in wide spread nucleic acid based diagnostics. Two dominant sources of noise have been documented in the literature; a low frequency current fluctuation with *1/f* characteristics (flicker noise) and a high-frequency background noise component associated with the relatively high capacitance of the insulating membrane on the support chip (dielectric noise) [16, 35, 79, 80, 82, 86, 87]. Minimizing these respective noise components is integral to improving the sensitivity and signal to noise ratio of nanopore sensors.

1.3.5.1 *1/f* Noise in Solid-State Nanopores

1/f noise has been observed in many physical and biological systems. *1/f* noise is present in the form of fluctuations in the voltages or currents of semiconductors, the voltage across nerve membranes and synthetic membranes and in the resistance of aqueous ionic solutions [45]. The power spectrum, denoted by *S(f)*, is proportional to the reciprocal of the frequency in a narrow bandwidth as illustrated in eq. 1.1.

$$S(f) = \frac{constant}{|f|^{\alpha}} \quad where\ 0 < \alpha < 2 \tag{1.1}$$

Hoogerheide et al. studied the *1/f* noise characteristics of Si_3N_4 nanopores as a function of pH and electrolyte ionic strength and concluded that *1/f* noise originates from surface charge fluctuations at the nanopore surface [35]. The model presented was based on protonization of surface functional groups and was sensitive to as few as tens of active surface groups in the nanopore. In contrast, Smeets et al. concluded that low frequency noise was predominantly due to the number of charge carriers in the nanopore thereby following Hooge's phenomenological relation rather than on the surface charge fluctuations and composition of the nanopore surface [79, 80, 82]. Surface modifications however have been shown to significantly improve the 1/*f* noise characteristics of nanopores. Chen et al. used an atomic layer deposition process to coat Si_3N_4 nanopores with Al_2O_3 and saw significant reductions in 1/*f* noise [14, 15]. Tabard-Cossa et al. demonstrated a significant reduction in 1/*f* noise by treating nanopore chips with piranha solution [86]. It is therefore likely that 1/*f* noise in nanopores is a combination of the two mechanisms described previously; that is, fluctuations in the total number of charge carriers in the nanopore coupled with a fluctuation in their mobilities due to trapping at surface sites. As expected, 1/*f* noise was observed in our experiments involving nanopores sputtered in Al_2O_3 thin films and were comparable in magnitude to noise levels observed in biological nanopores. By addressing the surface properties of solid-state nanopores, through either chemical surface treatment or material choice, improved noise performance may be achieved.

1.3.5.2 Dielectric Noise

Dielectric noise in nanopores is associated with the capacitance of the nanopore chip and scales linearly with frequency. Nanopores are typically fabricated in dielectric thin film such as SiO_2 or Si_3N_4, anchored on a conductive Si substrate. These dielectric materials are typically lossy and have a dissipation factor, *D*, associated with them. Smeets et al. extracted a dissipation factor of 0.27 ± 0.07 for Si_3N_4 pores, strongly deviating from $D = 0$ for an ideal capacitor [80, 82]. The dielectric noise can be reduced by minimizing the capacitance of the substrate. To achieve this, the thickness of the Si substrate can be increased or the fluidic contact area on the chip can be minimized. Tabard-Cossa et al. selectively patterned PDMS on Si_3N_4 chips to reduce the fluidic contact area and thereby minimized dielectric noise [86].

Using various micro-fabrication processes and PDMS fluidic isolation techniques, we achieved noise performance that surpasses the state-of-the-art in Si_3N_4 technology as reported by Tabard-Cossa et al. [86]. Noise reduction was attributed to a decrease in device capacitance, measured at 20 ± 5 pF [89] as compared to device capacitance in Si_3N_4 structures, which was measured in excess of 300 pF [80, 82]. Noise performance was further optimized using the three structures illustrated in Fig. 1.7. Architecture 1 consisted of 45 ± 5 nm thick Al_2O_3 membranes formed on a low resistivity Si substrate. Architecture 2 consisted of a 500 nm thick SiN passivation layer added on top of the Al_2O_3 layer with a 30 μm square opening forming the membrane region. PDMS gaskets with 300 μm openings were bonded to the chip to further decrease device capacitance. Architecture 3 consisted of a ~1.5 μm thick SiN passivation layer added on top of the Al_2O_3 layer with a 5 μm circular opening containing the Al_2O_3 membrane area. This stack was formed on a high resistivity Si wafer ($\rho = 10{,}000$ Ω-cm), effective in reducing leakage currents through the substrate. PDMS gaskets containing a 300 μm opening were bonded to the front side of the nanopore chip. These structures helped us incrementally reduce device capacitance and thereby reduce dielectric noise. Impedance spectra (magnitude and phase) were obtained for each architecture as shown in Fig. 1.7d, e, and fitted to an equivalent RC circuit. A capacitance of 1 nF, 300 pF and 20 pF was extracted for architectures 1, 2 and 3 respectively, corresponding to peak-to-peak noise values of approximately 1.2 nA, 400 pA and 200 pA at the 100 kHz bandwidth setting on the Axopatch 200B measurement platform, as shown in Fig. 1.7f. Capacitance minimization has proven to be an effective method in decreasing high frequency dielectric noise, improving signal-to-noise ratio and enhancing the overall sensitivity of these nanopore sensors in DNA translocation experiments.

1.3.6 DNA Translocation Through Solid-State Nanopores

The first demonstrations of DNA translocation through a solid-state nanopore were shown by Li et al. [53]. Deep current blockades were observed as dsDNA was electrophoretically driven through nanopores formed in thin Si_3N_4 membranes using the ion beam sculpting process described earlier. Further studies confirmed the dependence of dsDNA transport kinetics on bias voltage, DNA length and DNA conformation [52]. Li et al. further showed that by reducing the bias voltage by a factor of two, the dwell time of the DNA molecule in the nanopore could be approximately doubled [52]. Multiple configurations of the translocating molecule in the nanopore were also observed in these experiments attributed to dsDNA folding, a phenomenon observed primarily in large nanopores. Smaller ~3 nm pores however, were shown to restrict the passage of folded molecules and promoted only the linear passage of unfolded molecules. Heng et al. demonstrated that by reducing nanopore diameter to below that of dsDNA, the electrophoretic separation of ssDNA from dsDNA could be achieved using a solid-state nanopore [29]. Narrow ~2 nm pores were seen to block the passage of dsDNA and allowed the passage of only ssDNA. Only by applying very high fields was dsDNA permeation through these narrow pores indeed possible attributed to stretching transitions that occur in dsDNA at forces exceeding 60 pN. Comer et al. further demonstrated that very narrow <1.6 nm diameter synthetic nanopores could be effectively used to unzip hairpin DNA [20]. Different modes of hairpin DNA transport were observed in these experiments, the first mode referring to the unzipping of the double helix structure to form ssDNA and the second mode referring to the stretching/distortion of the double helix itself.

Chang et al. studied the effect of buffer concentration on DNA translocation dynamics [11]. Current enhancements were observed in large SiO_2 nanopores at low salt concentrations (100 mM KCl) as opposed to the typical blockades that were observed at higher salt concentrations. A more rigorous study by Smeets et al. and Chang et al. suggested that these current enhancements are due to counterion condensation on the DNA backbone, thereby locally increasing the concentration of

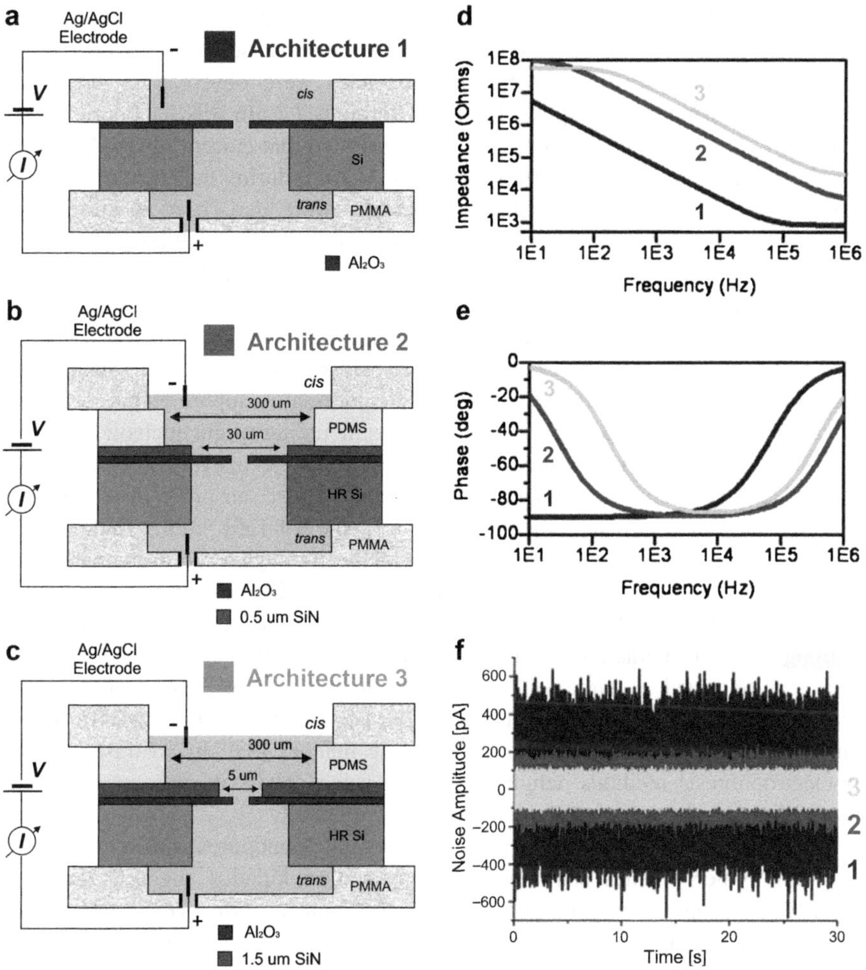

Fig. 1.7 Dielectric noise reduction in solid-state Al_2O_3 nanopores (**a**) Architecture 1 consists of a 45 ± 5 nm thick Al_2O_3 membranes formed on a low resistivity Si substrate. (**b**) Architecture 2 consists of a 500 nm thick SiN passivation layer added on top of the Al_2O_3 film with a 30 μm square opening forming the membrane region. PDMS gaskets with 300 μm openings were bonded to the chip to further decrease device capacitance. (**c**) Architecture 3 consists of a 1.5 μm thick SiN passivation layer added on top of the Al_2O_3 film with a patterned 5 μm circular opening constituting the Al_2O_3 membrane area. This stack was formed on a high resistivity Si wafer ($\rho = 10{,}000\Omega$-cm) (**d** and **e**) Impedance magnitude and phase spectra for the various architectures investigated. By fitting to an equivalent RC circuit, capacitances of 1 nF, 300 pF and 20 pF were extracted for architectures 1, 2 and 3 respectively (**f**) Normalized current traces in 1 M KCl, 10 mM Tris, pH 7.5 buffer at an applied voltage of 200 mV for architectures 1, 2 and 3. Capacitance reduction decreases the peak-to-peak noise from 1.2 nA (architecture 1), to 400 pA (architecture 2), to 200 pA (architecture 3)

counterions in the pore during DNA transport [12, 81]. Current enhancements were observed at concentrations below 0.4 M, a phenomenon that seems localized to only large SiO_2 nanopores. Distinct current blockades were observed in 100 mM salt solution during the transport of λ-DNA through 2.2 nm SiN nanopores [23]. In biological *α-hemolysin*, Benner et al. demonstrated that current blockades were still observed at low salt concentrations (300 mM KCl) during the entry of dsDNA into the lumen of the pore [2]. Current blockades were also observed during the transport of dsDNA through Al_2O_3 nanopores in 100 mM KCl salt [90].

Polymer velocity in the nanopore is also a key topic of interest. Translocation velocities of up to ~30 bases/μs have been reported at relatively low bias voltages in Si_3N_4 nanopores [26]. Chen et al. observed similar translocation velocities in large Al_2O_3 coated Si_3N_4 nanopores estimated at ~27 bases/μs [14, 15]. Such high translocation velocities limit the utility of conventional nanopore technologies in high end DNA sensing and analysis applications including single nucleotide detection. Fologea et al. demonstrated that by increasing electrolyte viscosity using glycerol and by decreasing temperature and bias voltage, an order of magnitude reduction in translocation velocity could be achieved [26]. Remarkably, even with these improvements, the translocation velocities through a solid-state nanopore are still more than an order of magnitude faster than that observed in biological *α-hemolysin* [14, 15]. Lubensky accredited the slow translocation rates in *α-hemolysin* to strong polymer interactions with the pore walls [56]. High translocation velocities were also observed in large ~10 nm SiO_2 nanopores [85]. Despite these high velocities, Storm et al. showed that it is indeed possible to size long dsDNA using solid-state nanopores, in a rapid and label free manner [85]. In contrast to bulk gel-electrophoresis methods, length separation using solid-state nanopores allows each molecule to be screened and interrogated individually.

The kinetics of DNA transport through solid-state nanopores is also of interest from a polymer physics stand point. Translocation kinetics suggests that the majority of events in larger nanopores are fast translocation events, where the dwell time, t_D, is significantly less than the characteristic relaxation time or *Zimm* time [85]. The *Zimm* time, t_Z, is an upper bound on the time taken by a polymer to reach an entropically and sterically favored state. For events where $t_D < t_Z$, the molecule was said to exhibit a *frozen* polymer configuration during transport, hindered by only the hydrodynamic drag on the part of the molecule outside the pore [85]. The effects of specific polymer-pore interactions were not accounted for in these studies. Wanunu et al. discussed the importance of surface interactions on dsDNA transport through Si_3N_4 nanopores [94]. Studies performed using small 2.7–5 nm pores revealed an order of magnitude increase in dwell times as pore diameter was decreased from 5 to 2.7 nm. In addition, strong temperature dependence was observed confirming that surface interactions play an important role in polymer transport. Surface interactions were also seen to play an important role in the transport of dsDNA through small ~5 nm Al_2O_3 nanopores [89]. These interactions were characterized by a monoexponential decay in dwell time histograms with time constants consistent with timescales observed in Si_3N_4 systems [89].

1.3.6.1 Surface Enhanced DNA Transport Through Al_2O_3 Nanopores

More and more evidence is emerging supporting the notion that nanopore surface interactions play an integral role in determining the dynamics of DNA transport. Nanopore surface characteristics including stoichiometry, morphology, surface charge density, charge polarity, cone angle and *rms* roughness are all expected to factor into this argument. In fact, we recently reported that surface interactions can help enhance the detection capabilities of solid-state nanopore sensors [90]. In experiments involving the electrophoretic transport of 5 kbp dsDNA through 7 nm diameter nanocrystalline Al_2O_3 nanopores, mean dwell-times at 100 mV yielded a translocation velocity of ~1.4 nucleotides/μs, more than an order of magnitude slower than dsDNA transport through Si_3N_4 nanopores (~30 nucleotides/μs) at similar biases [26], but an order of magnitude faster than single stranded DNA translocation through α-hemolysin [14, 15].

Figure 1.8 illustrates the voltage dependent transport of dsDNA through a 7 nm nanocrystalline Al_2O_3 nanopore. Two distinct timescales are observed in the translocation time histograms of Fig. 1.8, summarized in Fig. 1.8d [90]. The shorter timescale exhibited strong voltage dependence and was associated with fast polymer transport through the nanopore with minimal DNA-nanopore interactions. Such fast translocations are indeed probable in large 7 nm pores via translocation through the central pore region where the effects of surface binding sites and surface charge are significantly screened. Fast translocation events were not observed in smaller ~5 nm Al_2O_3 nanopores suggesting that pore size and Debye layer thickness indeed play an important role in regulating the velocity of DNA transport [89]. The longer timescale observed was associated with DNA translocations involving significant interactions with the nanopore. The origins of these interactions are hydrophobic and/or electrostatic in nature and are dependent partially on the material properties of the pore (stoichiometry, morphology and surface roughness). As previously discussed, materials analysis confirmed the formation of hetero-phase crystalline domains (in particular γ and α phase Al_2O_3) of varying bond lengths and coordinations in the nanopore region, resulting in non-uniform distributions of exposed Al-O groups at the pore surface. In a hydrated nanopore, these surface sites react with adsorbed water to form protonated hydroxyl groups at pH 7.5, resulting in a net positive, non-homogeneous surface charge density across the pore. These positively charged nanocrystalline domains are expected to interact strongly with anionic DNA. In fact, modeling results by Kejian et al. confirmed that polymer translocation velocities in a solid-state nanopore are heavily dependent on zeta potential and surface charge [44]. Alterations to pore stoichiometry due to the preferential desorption of O and the aggregation of Al is also expected to result in a distribution of equilibrium constants (pK's) for the protonizable chemical sites across the pore. The resulting electrostatic interactions/binding between the non-homogeneous, net positively charged nanopore surface and anionic DNA is one factor contributing to the *slow* translocation velocities observed in experiments involving Al_2O_3 nanopores. This strong electrostatic binding was not reported in SiO_2 and Si_3N_4, likely as these systems exhibit a net

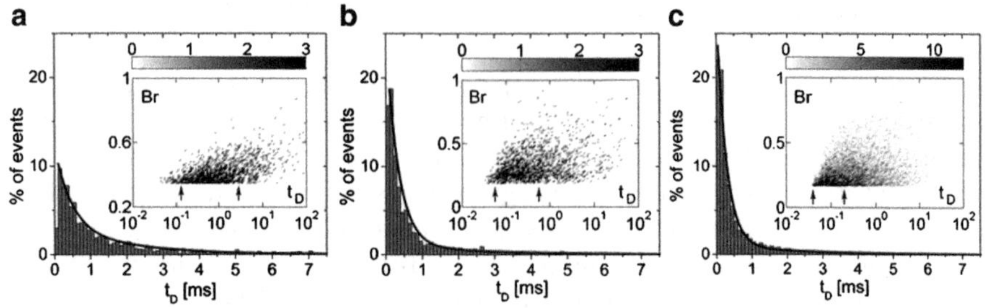

d

Voltage	100 mV	300 mV	500 mV
I (Open Pore) [nA]	0.23	0.7	1.2
$\mu(t_D)$ [ms]	3.5 ± 10.8	1.3 ± 4.3	0.8 ± 1.9
t_1 [ms]	0.48 ± 0.05	0.35 ± 0.01	0.28 ± 0.02
t_2 [ms]	2.0 ± 0.7	2.9 ± 1.3	2.6 ± 2.2
$\mu(B_r)$ [%]	0.42	0.32	0.28
n [events]	1421	1954	5351
R [Hz]	4.7	6.5	17.9

Fig. 1.8 Dwell time histograms for the transport of 5 kbp dsDNA through a 7 nm nanocrystalline Al_2O_3 nanopore in 100 mM KCl, 10 mM Tris, pH 7.5 buffer at applied voltages of (**a**) 100 mV (**b**) 300 mV and (**c**) 500 mV [90], reprinted with permission. Each distribution is fitted with a bi-exponential function (*black line*) with two time constants, t_1 and t_2, indicating two distinct types of polymer transport, fast translocation governed by polymer hydrodynamics and slow translocation regulated by polymer-pore interactions. (Insets) Blockage Ratio (B_r) vs. Dwell time (t_D) [ms] event density plots at each voltage. Grayscale bar represents number of events. At higher voltages, a greater percentage of events exhibit fast translocation dynamics bounded by the arrows in the insets. Translocation events exhibit clear voltage dependence (**d**) Summary of results from the electrical sensing of 5 kbp dsDNA through 7 nm Al_2O_3 nanopores. t_D Dwell time (time biomolecule resides in the pore); t_1 Time constant corresponding to fast translocation; t_2 Time constant corresponding to slow translocation; B_r Blockage Ratio (percentage of open pore current that is blocked during DNA translocation); *n* Biomolecule Flux (total number of events during 5 min of recording); *R* Capture Rate (average number of translocation events per second)

negatively charged surface at pH 7.5 resulting from the deprotonation of surface silanol groups [35]. Furthermore, a comparison of the surface charge density of Si_3N_4 and γ-Al_2O_3 surfaces at pH 7.5 (in monovalent salt solution at concentration 1×10^{-4} M) revealed a charge density that is approximately six times higher in γ-Al_2O_3 (50 mC/m^2) than in Si_3N_4 (8 mC/m^2) systems [1, 81]. Thus, polymer-pore interactions involving electrostatic binding events are expected to be more pronounced in Al_2O_3 nanopores.

The importance of polymer-pore interactions are explored further in the following section through nanopore based chemical modification. Chemical modification and surface functionalization of nanopores has helped usher in the next generation of nanopore sensors relying on polymer-pore interactions to achieve selective and facilitated transport of various biomolecules through individual nanopores and nanopore arrays.

1.3.7 Chemically Modified Solid-State Nanopore Sensors

Chemical modification of the surface of solid-state nanopores offers a method to tailor the physical and chemical properties of the nanopore. Selective transport through functionalized solid-state nanopore arrays was previously demonstrated by varying nanopore size (thereby restricting biomolecule passage based on molecular weight) [38], nanopore surface charge [17], and nanopore polarity [39] (to achieve the selective transport of either hydrophobic or hydrophilic molecules). More recently, focus has shifted to the attachment of specific recognition sequences or tethered receptors in the nanopore for target specific molecular recognition. In drug screening and medicine, such a technique provides a means for label-free, real-time kinetic analysis of biomolecular interactions at the single molecule level including protein-protein, protein-DNA and receptor-ligand interactions. In fact, Lee et al. demonstrated that enantiomeric drug separations could be achieved using an antibody functionalized nanoporous array [51]. Functionalized nanopore channels can also help elucidate the mechanisms driving biological processes, including cell signaling and regulation. Jovanovic-Talisman demonstrated that functionalized polycarbonate nanoporous arrays can reproduce the selectivity of nuclear pore complexes (NPCs), an essential component in the trafficking of specific macromolecules between the cell nucleus and cytoplasm [41]. Proteins referred to as phenylalanine-glycine (FG)-nucleoporins typically line the walls of NPCs and facilitate the transient binding and passing of transport factors and their cargo-bound complexes, while restricting the passage of proteins that fail to specifically bind to FG-nucleoporins [69]. Using nanopore channels of the correct dimensions coated with FG-nucleoporins, Jovanovic-Talisman et al. were able to reproduce key features of nucleocytoplasmic transport, selectively discriminating against control proteins in favor of transport factors and transport factor cargo complexes. Kohli demonstrated that selective permeation through synthetic nanoporous membranes could be achieved using DNA hybridization as the selective transport mechanism [49]. In this specific example, a gold nanoporous array was functionalized using hairpin DNA with a thiol substituent at the $5'$ end allowing it to be covalently attached to the inside walls of the array. The analyte of interest was 18 base long ssDNA which was either a perfect complement to the loop of the hairpin or contained a single base mismatch. Using optical absorbance methods, Kohli demonstrated that single nucleotide polymorphisms could be detected using this chemically modified nanopore platform under optimal conditions.

Various strategies have been implemented to chemically modify solid-state nanopores. Covalent attachment chemistries are generally preferred due to the stability and high packing density of self assembled monolayers (SAM's) on well prepared sufaces. A very common SAM preparation involves the reaction of molecules with a sulfhydrl termination group (−SH) with Au surfaces to form S-Au attachments to the surface. An extensive review on the formation of SAM's on Au surfaces of varying curvatures is provided by Love et al. [55]. In many cases however, the surface of the nanopore may be an insulating oxide or nitride (SiO_2, Si_3N_4, Al_2O_3). In these cases a covalent attachment chemistry specific to this insulating surface is required. Liquid phase silane based chemistries are the most commonly used technique to functionalize individual nanopores in such insulating membranes [66, 93]. While these surface chemistries have been characterized in detail on planar surfaces, questions still remain as to the exact packing density, molecular orientation and thickness of SAM's in a highly confined environment that is a nanopore. In addition, nanopores formed via TEM decompositional sputtering processes typically exhibit high surface roughness, high surface curvature and a non-stoichiometric material composition due to selective material sputtering as observed in SiO_2 coated Si_3N_4 nanopores [97, 98], further complicating the nanopore functionalization process. In these cases it is vital to thoroughly oxidize the surface through an extensive O_2 plasma treatment or a liquid based treatment in 1:3 H_2O_2:H_2SO_4. Using such a process, Wanunu et al. showed a change in the pH response of Si_3N_4 nanopore functionalized with various amine terminated silane chemistries in comparison to non-functionalized Si_3N_4 nanopores [93]. Ionic conductance measurements were used to monitor in-situ the formation of the SAM in the nanopore and to calculate the thickness of the molecular layer directly attached to the internal surface of the nanopore. The calculated values suggested the upright orientation of the attached molecules on the nanopore surface. Note, in this specific example, the entire membrane containing the nanopore was functionalized with the silane chemistry.

For certain applications however, it may be desirable to functionalize only the nanopore region itself. For example, in applications where the analyte of interest is present only at very low concentrations, a functionalized membrane may reduce the detection limits of the nanopore due to delocalized binding events on the membrane surface between immobilized receptors and the target species, without yielding detectable changes in the output signal [37]. In addition, receptors immobilized on the membrane may themselves modulate the conductance of the nanopore, even in the absence of the target species. Hofler et al. showed via coarse-grained molecular dynamics simulations that DNA anchored on the membrane surface can electrically gate the nanopore if bound sufficiently close to the pore opening [34]. A localized nanopore functionalization process was explored by Nilsson et al. and involved the localized deposition of a tetraethylorthosilicate (TEOS) based oxide ring around the nanopore [66]. A focused ion beam was used to decompose the TEOS precursor near the Si nanopore surface, thereby reducing the diameter of the pore to a final diameter of between 25 and 30 nm. DNA probes were immobilized in the nanopore via a silane based chemistry thereby introducing local chemical functionality at the entrance of the nanopore without functionalizing the remainder of the Si membrane. SAM coatings may also help to reduce the speed of polymer translocation through nanopores. Kim et al. derivatized Al_2O_3 nanopore surfaces with aminopropyltriethoxysilane (APTES) resulting in a positively charged surface in pH 6.0 buffer, attractive to anionic dsDNA [46, 48]. The resulting strong electrostatic polymer-pore interactions enabled the detection of short dsDNA molecules that are typically under the detection limits of conventional solid-state nanopore sensors. In addition to SAM coatings, highly functional lipid bilayer coatings on nanopores are also possible [91], permitting the potential integration with sensitive biological nanopore channels.

Perhaps the most striking example of nanopore functionalization impacting the sensing capabilities of single nanopore channels involves the use of oxidized Si nanopores functionalized with hairpin-loop, probe DNAs to selectively transport short "target" (complementary) ssDNAs under an applied electrical field [36]. Iqbal et al. functionalized SiO_2 nanopores with 20 base long hairpin-loop DNA probes containing a 6 bp stem forming region. Mismatches were introduced in the target ssDNA at points complementary to the stem forming region of the immobilized probes. Higher flux and smaller translocation times were observed during the passage of Perfectly Complementary (PC) DNA as opposed to single base mismatch (1MM) DNA. This was attributed to an attractive potential between the immobilized probe and the target PC-DNA resulting in hairpin unzipping and facilitated transport of the PC sequence. In the case of 1MM however, a repulsive potential was proposed between the target sequence and the probe resulting in probe-target interaction without unzipping of the hairpin. Reduced biomolecule flux and increased translocation times were also consistently observed in the cases of two and three base mismatch DNAs (2MM and 3MM respectively) relative to PC-DNA. These studies confirm that it is indeed possible to impart chemical selectivity in single solid-state nanopores and this selectivity can be electrically monitored through translocation signatures at the single-molecule level. Such devices could help further unravel the physics of selective and facilitated transport of biomolecules through nanoscale channels and could play an important role in medical diagnostics.

1.4 Conclusions

Solid-state nanopore sensors are highly versatile platforms for the rapid, label-free detection and analysis of single molecules, with potential application to next generation DNA sequencing. The versatility of this technology allows for both interfacing with biological systems at the nano-scale as well as large scale VLSI integration promising reliable, affordable, mass producible biosensors with single molecule sensing capabilities. This technology may also serve as a base to provide further insight into the mechanisms driving biological processes, including cell signaling and regulation through gated, selective ion channels, protein secretion across cellular membranes and viral infection by phages. The applications for solid-state nanopore technology are diverse. Point-of-care diagnostic devices employing solid-state nanopores can be used to detect and monitor infectious diseases e.g. influenza, an effective tool in public health strategies. In defense, solid-state nanopores can be used for the rapid detection of high priority agents such as Bacillus anthracis (anthrax) at very low concentrations. In drug screening and medical applications, solid-state nanopores provide a means for label-free, real-time kinetic analysis of biomolecular interactions at the single molecule level including protein-protein, protein-DNA and receptor-ligand interactions. This technology finds broad application in bio-nanotechnology.

Acknowledgments We thank the staff at *Micro and Nanotechnology Lab and Frederick Seitz Materials Research Lab*, University of Illinois at Urbana-Champaign for their assistance. We acknowledge the funding from the National Institutes of Health through the NIH Roadmap for Medical Research Nanomedicine Development Center (PN2 EY 018230) and NIH R21 EB007472.

References

1. Alami-Younssi, S., Larbot, A., Persin, M., Sarrazin, J., & Cot, L. (1995). Rejection of mineral salts on a gamma alumina nanofiltration membrane Application to environmental process. *Journal of Membrane Science, 102*, 123–129.
2. Benner, Seico, Chen, Roger J. A., Wilson, Noah A., Abu-Shumays, Robin, Hurt, Nicholas, Lieberman, Kate R., et al. (2007). Sequence-specific detection of individual DNA polymerase complexes in real time using a nanopore. *Nat Nano, 2*(11), 718–724.
3. Berger, S. D., Salisbury, I. G., Milne, R. H., Imeson, D., & Humphreys, C. J. (1987). Electron energy-loss spectroscopy studies of nanometre-scale structures in alumina produced by intense electron-beam irradiation. *Philosophical Magazine Part B, 55*(3), 341–358.
4. Bhakdi, S., & Tranum-Jensen, J. (1991). Alpha-toxin of Staphylococcus aureus. *Microbiol. Mol. Biol. Rev., 55*(4), 733–751.
5. Bouchet, Danièle, & Colliex, Christian (2003). Experimental study of ELNES at grain boundaries in alumina: intergranular radiation damage effects on Al-L23 and O-K edges. *Ultramicroscopy, 96*(2), 139–152.
6. Bourdillon, A. J., El-mashri, S. M., & Forty, A. J. (1984). Application of TEM extended electron energy loss fine structure to the study of aluminium oxide films. *Philosophical Magazine A, 49*(3), 341–352.
7. Bowen, Paul, Carry, Claude, Luxembourg, David, & Hofmann, Heinrich (2005). Colloidal processing and sintering of nanosized transition aluminas. *Powder Technology, 157*(1–3), 100–107.
8. Brun, L., Pastoriza-Gallego, M., Oukhaled, G., Mathe, J., Bacri, L., Auvray, L., et al. (2008). Dynamics of polyelectrolyte transport through a protein channel as a function of applied voltage. *Physical Review Letters, 100*(15).
9. Cantor, C. R., & Schimmel, P. R. (1980). *Biophysical Chemistry, Part II: Techniques for the Study of Biological Structure and Function*. New York: W.H. Freeman.
10. Chad Harrell, C., Siwy, Z. S., & Martin, C. R. (2006). Conical Nanopore Membranes: Controlling the Nanopore Shape13. *Small, 2*(2), 194–198.
11. Chang, H., Kosari, F., Andreadakis, G., Alam, M. A., Vasmatzis, G., & Bashir, R. (2004). DNA-Mediated Fluctuations in Ionic Current through Silicon Oxide Nanopore Channels. *Nano Lett., 4*(8), 1551–1556.
12. Chang, H., Venkatesan, B. M., Iqbal, S., Andreadakis, G., Kosari, F., Vasmatzis, G., et al. (2006). DNA counterion current and saturation examined by a MEMS-based solid state nanopore sensor. *Biomedical Microdevices, 8*(3), 263–269.
13. Chen, G. S., Boothroyd, C. B., & Humphreys, C. J. (1998). Electron-beam-induced damage in amorphous SiO2 and the direct fabrication of silicon nanostructures. *Philosophical Magazine A, 78*, 491–506.
14. Chen, P., Gu, J., Brandin, E., Kim, Y. R., Wang, Q., & Branton, D. (2004). Probing Single DNA Molecule Transport Using Fabricated Nanopores. *Nano Letters, 4*(11), 2293–2298.
15. Chen, P., Mitsui, T., Farmer, D.B., Golovchenko, J., Gordon, R.G., & Branton, D. (2004). Atomic Layer Deposition to Fine-Tune the Surface Properties and Diameters of Fabricated Nanopores. *Nano Lett., 4*(7), 1333–1337.
16. Chen, Peng, & Gillis, Kevin D. (2000). The Noise of Membrane Capacitance Measurements in the Whole-Cell Recording Configuration. *Biophysical Journal, 79*(4), 2162–2170.

17. Chun, Kyoung-Yong, & Stroeve, Pieter (2002). Protein Transport in Nanoporous Membranes Modified with Self-Assembled Monolayers of Functionalized Thiols. *Langmuir, 18*(12), 4653–4658.
18. Clarke, J., Wu, H. C., Jayasinghe, L., Patel, A., Reid, S., & Bayley, H. (2009). Continuous base identification for single-molecule nanopore DNA sequencing. *Nature Nanotechnology, 4*(4), 265–270.
19. Cockroft, Scott L., Chu, John, Amorin, Manuel, & Ghadiri, M. Reza (2008). A Single-Molecule Nanopore Device Detects DNA Polymerase Activity with Single-Nucleotide Resolution. *Journal of the American Chemical Society, 130*(3), 818–820.
20. Comer, J., Dimitrov, V., Zhao, Q., Timp, G., & Aksimentiev, A. (2009). Microscopic mechanics of hairpin DNA translocation through synthetic nanopores. *Biophys J, 96*(2), 593–608.
21. Coulter, W. H. (1953). Means for counting particles suspended in a fluid, United States Patent 2656508.
22. Dekker, Cees (2007). Solid-state nanopores. *Nat Nano, 2*(4), 209–215.
23. Dimitrov, V., Mirsaidov, U., Mansfield, W., Miner, J., Klemens, F., Cirelli, R., et al. (2009). Nanopores in solid-state membranes engineered for single molecule detection. *Nanotechnology, 21*(6), 065502.
24. Egerton, R. F. (1996). *Electron Energy-Loss Spectroscopy in the Electron Microscope* (Second Edition ed.). New York: Plenum Press.
25. Fitch, J. T., Bjorkman, C. H., Lucovsky, G., Pollak, F. H., & Yin, X. (1989). *Intrinsic stress and stress gradients at the SiO_2/Si interface in structures prepared by thermal oxidation of Si and subjected to rapid thermal annealing*. Paper presented at the Proceedings of the 16th annual conference on the physics and chemistry of semiconductor interfaces, Bozeman, Montana, USA.
26. Fologea, Daniel, Uplinger, James, Thomas, Brian, McNabb, David S., & Li, Jiali (2005). Slowing DNA Translocation in a Solid-State Nanopore. *Nano Letters, 5*(9), 1734–1737.
27. Franks, George V., & Meagher, Laurence (2003). The isoelectric points of sapphire crystals and alpha-alumina powder. *Colloids and Surfaces A: Physicochemical and Engineering Aspects, 214*(1–3), 99–110.
28. Gyurcsányi, R. E. (2008). Chemically-modified nanopores for sensing. *TrAC Trends in Analytical Chemistry, 27*(7), 627–639.
29. Heng, J. B., Aksimentiev, A., Ho, C., Marks, P., Grinkova, Y. V., Sligar, S., et al. (2006). The electromechanics of DNA in a synthetic nanopore. *Biophysical Journal, 90*(3), 1098–1106.
30. Heng, J. B., Ho, C., Kim, T., Timp, R., Aksimentiev, A., Grinkova, Y. V., et al. (2004). Sizing DNA Using a Nanometer-Diameter Pore. *Biophys. J., 87*(4), 2905–2911.
31. Heng, J. B., Aksimentiev, A., Ho, C., Marks, P., Grinkova, Y.V., Sligar, S., et al. (2005). Stretching DNA Using the Electric Field in a Synthetic Nanopore. *Nano Lett., 5*(10), 1883–1888.
32. Henrickson, Sarah E., Misakian, Martin, Robertson, Baldwin, & Kasianowicz, John J. (2000). Driven DNA Transport into an Asymmetric Nanometer-Scale Pore. *Physical Review Letters, 85*(14), 3057.
33. Ho, Chuen, Qiao, Rui, Heng, Jiunn B., Chatterjee, Aveek, Timp, Rolf J., Aluru, Narayana R., et al. (2005). Electrolytic transport through a synthetic nanometer-diameter pore. *Proceedings of the National Academy of Sciences of the United States of America, 102*(30), 10445–10450.
34. Höfler, Lajos, & Gyurcsányi, Róbert E. (2008). Coarse Grained Molecular Dynamics Simulation of Electromechanically-Gated DNA Modified Conical Nanopores. *Electroanalysis, 20*(3), 301–307.
35. Hoogerheide, D. P., Garaj, S., & Golovchenko, J. A. (2009). Probing Surface Charge Fluctuations with Solid-State Nanopores. *Physical Review Letters, 102*(25).

36. Iqbal, Samir M., Akin, Demir, & Bashir, Rashid (2007). Solid-state nanopore channels with DNA selectivity. *Nat Nano, 2*(4), 243–248.
37. Jagerszki, Gyula, Gyurcsanyi, Robert E., Hofler, Lajos, & Pretsch, Erno (2007). Hybridization-Modulated Ion Fluxes through Peptide-Nucleic-Acid- Functionalized Gold Nanotubes. A New Approach to Quantitative Label-Free DNA Analysis. *Nano Letters, 7* (6), 1609–1612.
38. Jirage, Kshama B., Hulteen, John C., & Martin, Charles R. (1997). Nanotubule-Based Molecular-Filtration Membranes. *Science, 278*(5338), 655–658.
39. Jirage, Kshama B., Hulteen, John C., & Martin, Charles R. (1999). Effect of Thiol Chemisorption on the Transport Properties of Gold Nanotubule Membranes. *Analytical Chemistry, 71*(21), 4913–4918.
40. Jonas, D., Walev, I., Berger, T., Liebetrau, M., Palmer, M., & Bhakdi, S. (1994). Novel path to apoptosis: small transmembrane pores created by staphylococcal alpha-toxin in T lymphocytes evoke internucleosomal DNA degradation. *Infect. Immun., 62*(4), 1304–1312.
41. Jovanovic-Talisman, T., Tetenbaum-Novatt, J., McKenney, A. S., Zilman, A., Peters, R., Rout, M. P., et al. (2009). Artificial nanopores that mimic the transport selectivity of the nuclear pore complex. *Nature, 457*(7232), 1023–1027.
42. Kalman, Eric B., Sudre, Olivier, & Siwy, Zuzanna S. (2009). Control of Ionic Transport through an Ionic Transistor based on Gated Single Conical Nanopores. *Biophysical Journal, 96*(3, Supplement 1), 648a-648a.
43. Kasianowicz, J. J., Brandin, E., Branton, D., & Deamer, D. W. (1996). Characterization of individual polynucleotide molecules using a membrane channel. *Proc. Natl Acad. Sci. USA, 93*(24), 13770.
44. Kejian, Ding, Weimin, Sun, Haiyan, Zhang, Xianglei, Peng, & Honggang, Hu (2009). Dependence of zeta potential on polyelectrolyte moving through a solid-state nanopore. *Applied Physics Letters, 94*(1), 014101–014103.
45. Keshner, M. S. (1982). 1/f noise. *Proceedings of the IEEE, 70*(3), 212–218.
46. Kim, M. J., McNally, B., Murata, K., & Meller, A. (2007). Characteristics of solid-state nanometre pores fabricated using a transmission electron microscope. *Nanotechnology 20*, 205302
47. Kim, M. J., Wanunu, M., Bell, D. C., & A. Meller (2006). Rapid Fabrication of Uniformly Sized Nanopores and Nanopore Arrays for Parallel DNA Analysis. *Advanced Materials, 18*(23), 3149–3153.
48. Kim, Y. R., Min, J., Lee, I. H., Kim, S., Kim, A. G., Kim, K., et al. (2007). Nanopore sensor for fast label-free detection of short double-stranded DNAs. *Biosensors and Bioelectronics, 22*(12), 2926–2931.
49. Kohli, Punit, Harrell, C. Chad, Cao, Zehui, Gasparac, Rahela, Tan, Weihong, & Martin, Charles R. (2004). DNA-Functionalized Nanotube Membranes with Single-Base Mismatch Selectivity. *Science, 305*(5686), 984–986.
50. Lagerqvist, Johan, Zwolak, Michael, & Di Ventra, Massimiliano (2006). Fast DNA Sequencing via Transverse Electronic Transport. *Nano Letters, 6*(4), 779–782.
51. Lee, Sang Bok, Mitchell, David T., Trofin, Lacramioara, Nevanen, Tarja K., Soderlund, Hans, & Martin, Charles R. (2002). Antibody-Based Bio-Nanotube Membranes for Enantiomeric Drug Separations. *Science, 296*(5576), 2198–2200.
52. Li, Jiali, Gershow, Marc, Stein, Derek, Brandin, Eric, & Golovchenko, J. A. (2003). DNA molecules and configurations in a solid-state nanopore microscope. *Nat Mater, 2*(9), 611–615.
53. Li, Jiali, Stein, Derek, McMullan, Ciaran, Branton, Daniel, Aziz, Michael J., & Golovchenko, Jene A. (2001). Ion-beam sculpting at nanometre length scales. *Nature, 412*(6843), 166–169.
54. Lo, Chih Jen, Aref, Thomas, & Bezryadin, Alexey (2006). Fabrication of symmetric sub-5 nm nanopores using focused ion and electron beams. *Nanotechnology, 17*(13), 3264–3267.
55. Love, J. Christopher, Estroff, Lara A., Kriebel, Jennah K., Nuzzo, Ralph G., & Whitesides, George M. (2005). Self-Assembled Monolayers of Thiolates on Metals as a Form of Nanotechnology. *Chemical Reviews, 105*(4), 1103–1170.

56. Lubensky, D. K., & Nelson, D. R. (1999). Driven Polymer Translocation Through a Narrow Pore. *Biophysical Journal, 77*(4), 1824–1838.
57. Manning, G. S. (1993). A Condensed Counterion Theory for Polarization of Polyelectrolyte Solutions in High Fields. *Journal of Chemical Physics, 99*(1), 477–486.
58. Mathe, Jerome, Aksimentiev, Aleksei, Nelson, David R., Schulten, Klaus, & Meller, Amit (2005). Orientation discrimination of single-stranded DNA inside the α-hemolysin membrane channel. *Proceedings of the National Academy of Sciences of the United States of America, 102*(35), 12377–12382.
59. McNally, Ben, Wanunu, Meni, & Meller, Amit (2008). Electromechanical Unzipping of Individual DNA Molecules Using Synthetic Sub-2 nm Pores. *Nano Lett.*
60. Meller, A., & Branton, D. (2002). Single molecule measurements of DNA transport through a nanopore. *Electrophoresis, 23*(16), 2583–2591.
61. Meller, A., Nivon, L., & Branton, D. (2001). Voltage-Driven DNA Translocations through a Nanopore. *Physical Review Letters, 86*(15), 3435.
62. Mitchell, Nick, & Howorka, Stefan (2008). Chemical Tags Facilitate the Sensing of Individual DNA Strands with Nanopores13. *Angewandte Chemie International Edition, 47*(30), 5565–5568.
63. Moon, Jeong-Mi, Akin, Demir, Xuan, Yi, Ye, Peide, Guo, Peixuan, & Bashir, Rashid (2009). Capture and alignment of *phi29* viral particles in sub-40 nanometer porous alumina membranes. *Biomedical Microdevices, 11*(1), 135–142.
64. Nakane, J., Akeson, M., & Marziali, A. (2002). Evaluation of nanopores as candidates for electronic analyte detection. *Electrophoresis, 23*(16), 2592–2601.
65. Nam, Sung-Wook, Rooks, Michael J., Kim, Ki-Bum, & Rossnagel, Stephen M. (2009). Ionic Field Effect Transistors with Sub-10 nm Multiple Nanopores. *Nano Letters, 9*(5), 2044–2048.
66. Nilsson, J., Lee, J. R. I., Ratto, T. V., & Létant, S. E. (2006). Localized Functionalization of Single Nanopores. *Advanced Materials, 18*(4), 427–431.
67. O'Keeffe, M., & Stuart, J. A. (2002). Bond energies in solid oxides. *Inorganic Chemistry, 22*(1), 177–179.
68. Parks, George A. (2002). The Isoelectric Points of Solid Oxides, Solid Hydroxides, and Aqueous Hydroxo Complex Systems. *Chemical Reviews, 65*(2), 177–198.
69. Peters, Reiner (2005). Translocation Through the Nuclear Pore Complex: Selectivity and Speed by Reduction-of-Dimensionality. *Traffic, 6*(5), 421–427.
70. Petrossian, L., Wilk, S. J., Joshi, P., Hihath, S., Goodnick, S. M., & Thornton, T. J. (2007). Fabrication of Cylindrical Nanopores and Nanopore Arrays in Silicon-On-Insulator Substrates. *Journal of Microelectromechanical Systems, 16*(6), 1419–1428.
71. Pivin, Jean Claude (1983). An overview of ion sputtering physics and practical implications. *Journal of Materials Science, 18*(5), 1267–1290.
72. Salisbury, I. G., Timsit, R. S., Berger, S. D., & Humphreys, C. J. (1984). Nanometer scale electron beam lithography in inorganic materials. *Applied Physics Letters, 45*(12), 1289–1291.
73. Sanger, F., Nicklen, S., & Coulson, A. R. (1977). DNA sequencing with chain-terminating inhibitors. *Proceedings of the National Academy of Sciences of the United States of America, 74*(12), 5463–5467.
74. Schenkel, T., Radmilovic, V., Stach, E. A., Park, S. J., & Persaud, A. (2003). Formation of a few nanometer wide holes in membranes with a dual beam focused ion beam system. *Journal of Vacuum Science & Technology B: Microelectronics and Nanometer Structures, 21*(6), 2720–2723.
75. Siwy, Z., Dobrev, D., Neumann, R., Trautmann, C., & Voss, K. (2003). Electro-responsive asymmetric nanopores in polyimide with stable ion-current signal. *Applied Physics A: Materials Science & Processing, 76*(5), 781–785.
76. Siwy, Z., & Fulinski, A. (2004). A nanodevice for rectification and pumping ions. *American Journal of Physics, 72*(5), 567–574.

77. Siwy, Z., Heins, E., Harrell, C. Chad, Kohli, P., & Martin, C.R. (2004). Conical-Nanotube Ion-Current Rectifiers: The Role of Surface Charge. *Journal of the American Chemical Society, 126*(35), 10850–10851.
78. Siwy, Z. S. (2006). Ion-Current Rectification in Nanopores and Nanotubes with Broken Symmetry. *Advanced Functional Materials, 16*(6), 735–746.
79. Smeets, R. M. M., Dekker, N. H., & Dekker, C. (2009). Low-frequency noise in solid-state nanopores. *Nanotechnology, 20*(9), 095501.
80. Smeets, R. M. M., Keyser, U. F., Dekker, N. H., & Dekker, C. (2008). Noise in solid-state nanopores. *Proceedings of the National Academy of Sciences, 105*(2), 417–421.
81. Smeets, R. M. M., Keyser, U. F., Krapf, D., Wu, M. Y., Dekker, N. H., & Dekker, C. (2006). Salt Dependence of Ion Transport and DNA Translocation through Solid-State Nanopores. *Nano Lett., 6*(1), 89–95.
82. Smeets, R. M. M., Kowalczyk, S. W., Hall, A. R., Dekker, N. H., & Dekker, C. (2008). Translocation of RecA-Coated Double-Stranded DNA through Solid-State Nanopores. *Nano Letters*.
83. Stoddart, David, Heron, Andrew J., Mikhailova, Ellina, Maglia, Giovanni, & Bayley, Hagan (2009). Single-nucleotide discrimination in immobilized DNA oligonucleotides with a biological nanopore. *Proceedings of the National Academy of Sciences, 106*(19), 7702–7707.
84. Storm, A. J., Chen, J. H., Ling, X. S., Zandbergen, H. W., & Dekker, C. (2003). Fabrication of solid-state nanopores with single-nanometre precision. *Nat Mater, 2*(8), 537–540.
85. Storm, A. J., Storm, C., Chen, J., Zandbergen, H., Joanny, J.-F., & Dekker, C. (2005). Fast DNA Translocation through a Solid-State Nanopore. *Nano Lett., 5*(7), 1193–1197.
86. Tabard-Cossa, Vincent, Trivedi, Dhruti, Wiggin, Matthew, Jetha, Nahid N., & Marziali, Andre (2007). Noise analysis and reduction in solid-state nanopores. *Nanotechnology, 18*(30), 305505.
87. Uram, Jeffrey D., Ke, Kevin, & Mayer, Michael (2008). Noise and Bandwidth of Current Recordings from Submicrometer Pores and Nanopores. *ACS Nano, 2*(5), 857–872.
88. Veeramasuneni, S., Yalamanchili, M. R., & Miller, J. D. (1996). Measurement of Interaction Forces between Silica and [alpha]-Alumina by Atomic Force Microscopy. *Journal of Colloid and Interface Science, 184*(2), 594–600.
89. Venkatesan, B. M., Dorvel, B., Yemenicioglu, S., Watkins, N., Petrov, I., & Bashir, R. (2009). Highly Sensitive, Mechanically Stable Nanopore Sensors for DNA Analysis. *Advanced Materials, 21*(27), 2771–2776.
90. Venkatesan, B. M., Shah, A. B., Zuo, J. M., & Bashir, R. (2010). DNA Sensing Using Nanocrystalline Surface-Enhanced Al_2O_3 Nanopore Sensors. *Advanced Functional Materials, 20*(8), 1266–1275.
91. Venkatesan, B. M., Polans, J., Comer, J., Sridhar, S., Wendell, D., Aksimentiev, A., & Bashir, R. (2011). Lipid Bilayer Coated Al_2O_3 Nanopore Sensors: Towards a Hybrid Biological Solid-State Nanopore. *Biomedical Microdevices*, DOI: 10.1007/s10544-011-9537-3.
92. Vercoutere, Wenonah, Winters-Hilt, Stephen, Olsen, Hugh, Deamer, David, Haussler, David, & Akeson, Mark (2001). Rapid discrimination among individual DNA hairpin molecules at single-nucleotide resolution using an ion channel. *Nat Biotech, 19*(3), 248–252.
93. Wanunu, Meni, & Meller, Amit (2007). Chemically Modified Solid-State Nanopores. *Nano Letters, 7*(6), 1580–1585.
94. Wanunu, Meni, Sutin, Jason, McNally, Ben, Chow, Andrew, & Meller, Amit (2008). DNA Translocation Governed by Interactions with Solid State Nanopores. *Biophys. J.*, biophysj.108.140475.
95. Wendell, David, Jing, Peng, Geng, Jia, Subramaniam, Varuni, Lee, Tae Jin, Montemagno, Carlo, et al. (2009). Translocation of double-stranded DNA through membrane-adapted phi29 motor protein nanopores. *Nat Nano, 4*(11), 765–772.
96. Winters-Hilt, Stephen (2007). The alpha-Hemolysin nanopore transduction detector - single-molecule binding studies and immunological screening of antibodies and aptamers. *BMC Bioinformatics, 8*(Suppl 7), S9.

97. Wu, M. Y., Krapf, D., Zandbergen, M., Zandbergen, H. W., & Batson, P. E. (2005). Formation of nanopores in a SiN/SiO_2 membrane with an electron beam. *Applied Physics Letters, 87*(11), 113106–113103.
98. Wu, M. Y., Smeets, R. M. M., Zandbergen, M., Ziese, U., Krapf, D., Batson, P. E., et al. (2009). Control of Shape and Material Composition of Solid-State Nanopores. *Nano Letters, 9*(1), 479–484.
99. Wu, Shanshan, Park, Sang Ryul, & Ling, Xinsheng Sean (2006). Lithography-Free Formation of Nanopores in Plastic Membranes Using Laser Heating. *Nano Letters, 6*(11), 2571–2576.
100. Xiang, Ye, Morais, Marc C., Battisti, Anthony J., Grimes, Shelley, Jardine, Paul J., Anderson, Dwight L., et al. (2006). Structural changes of bacteriophage [phi]29 upon DNA packaging and release. *EMBO J, 25*(21), 5229–5239.
101. Yu, Minrui, Kim, Hyun-Seok, & Blick, Robert H. (2009). Laser drilling of nano-pores in sandwiched thin glass membranes. *Opt. Express, 17*(12), 10044–10049.

Chapter 2
Molecular Detection and Force Spectroscopy in Solid-State Nanopores with Integrated Optical Tweezers

Adam R. Hall and Cees Dekker

Abstract We describe how individual biopolymer molecules can be captured, detected and manipulated inside a solid-state nanopore using an integrated optical tweezer system. The combination of nanopore and tweezer technologies offers measurement capabilities like size and length discrimination similar to translocation experiments along with arbitrary position control and the ability to perform direct force spectroscopy. We discuss the experimental setup and measurements on two different types of molecules (bare DNA and protein-coated DNA), and we describe a model for the force on a charged molecule in a nanopore.

Keywords Force Spectroscopy • Integrated Optical Tweezers • Size and length discrimination • Protein-Coated DNA Molecules

2.1 Introduction

Solid-state nanopores show great promise as a high-throughput means by which to detect and analyze biopolymers on an individual basis [1, 2]. A commonly used method towards this end involves the use of a thin, insulating membrane with a single fabricated pore in it (typically prepared by either a transmission electron microscope [3] or a focused ion beam [4]) that acts as a barrier between two supplies of ionic solution. The application of a voltage across the membrane sets up an electric field that is strongly focused inside the pore and can be used to transport charged molecules from one side to the other electrophoretically. Upon doing so,

A.R. Hall (✉)
Kavli Institute of Nanoscience, Delft University of Technology, Lorentzweg 1,
Delft, CJ 2628, The Netherlands
and
Joint School of Nanoscience and Nanoengineering, University of North Carolina
at Greensboro, Greensboro, NC 27401, USA
e-mail: adam.hall@uncg.edu

S.M. Iqbal and R. Bashir (eds.), *Nanopores: Sensing and Fundamental Biological Interactions*, DOI 10.1007/978-1-4419-8252-0_2,

the presence of the molecule can be detected by monitoring its effect on the trans-membrane current, which is measured by a patch-clamp amplifier.

During electrophoretic translocation, molecules typically pass through the nanopore at high speed. A 48 kbp double-strand DNA (dsDNA) molecule, for instance, traverses the pore in only about 1 ms under an applied voltage of 120 mV [5]. While this presents advantages for some applications (high throughput rapid screening to name one), it poses a challenge for others. For example, accurate detection of sub-molecular structure like local protein position or even nucleotide sequence – an especially promising goal in the field – would require very high bandwidth to achieve. One potential way to address this is to adjust experimental variables like temperature and viscosity in order to slow down the translocation speed. However, this places inherent limitations on solvent conditions and thus on what can be measured. A more elegant solution would be a mechanism by which to "hold on" to a given molecule and control its position relative to the pore. Such a situation would offer the ability to slow, halt, or even reverse translocation arbitrarily.

Here, we describe a technique that yields this level of control: solid-state nanopores with integrated optical tweezers [6]. We detail how the experiments are performed, describe how the resultant measurements are interpreted, and discuss a model developed along the way that gives insight into the dominant forces involved in nanopore translocation.

2.2 Experimental Methods

Solid-state nanopores are fabricated using a method that has been described elsewhere [7]. Briefly, common microfabrication techniques are used to produce a 20-nm thin, 5 μm wide, free-standing window of silicon nitride supported in a silicon chip. This chip is then mounted in a transmission electron microscope (TEM) and the highly-focused electron beam is used to locally ablate the surface, effectively "drilling" a single hole through the membrane (Fig. 2.1a, inset). The properties of the resultant nanopore (diameter, shape) can be controlled to some degree by adjusting the beam intensity, beam size, and the exposure time [8]. After pore formation, the entire chip is stored in a solution of 50% ethanol in water. The mixture was chosen in order to keep the membrane clean and hydrated, but to allow fluid to wet the interior of the small pore more easily by reducing the surface tension.

Prior to use the chip is cleaned with water, acetone and ethanol, then dried under nitrogen flow and exposed to an oxygen plasma for 30 s in order to create a hydrophilic surface. Directly following this treatment, the chip is introduced with measurement solution (KCl solution with 10 mM tris-HCl at pH 8.0). The sample cell (Fig. 2.1a) is composed of a sandwich-type structure with a poly(methyl methacrylate) (PMMA) flow cell below the chip and a Perspex flow cell above. This allows independent exchange of solvent to both sides of the membrane while still permitting the optical path necessary for optical tweezer integration (Fig. 2.1b). The assembled flow cell is positioned on a three-dimensional piezo stage above a 60× water immersion objective, which acts as the focusing lens of both the optical trapping laser (4 W, $\lambda = 1{,}064$ nm) and a second position-reading laser (5 mW, $\lambda = 650$ nm) the light of

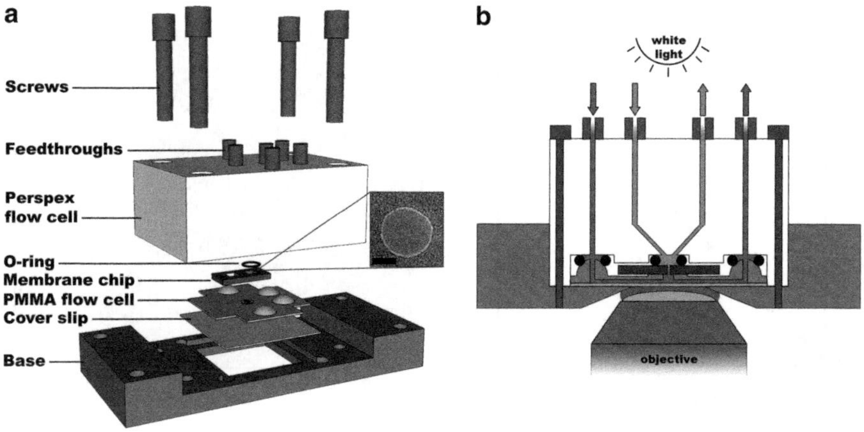

Fig. 2.1 The experimental setup. (**a**) Diagram of the nanopore flow cell, detailing each component. Inset: TEM image of a typical nanopore (scale bar 10 nm). (**b**) Side view of the assembled flow cell, showing the liquid feedthroughs to both the top (*inner*) and bottom (*outer*) of the nanopore chip, and the optical path for transmitted light to the objective

which is reflected onto a position-sensitive detector (PSD). The objective is used additionally as a means for optical feedback using a CCD camera. Electrical contact is made to measurement solutions with agarose salt bridges and Pt connecting wires using small, enclosed reservoirs of 1 M KCl containing 100 mM ferro/ferri cyanide as a redox agent. Voltage application and ionic current measurements are performed using a patch-clamp amplifier (Axopatch 200B). This experimental setup is described in greater detail elsewhere [9].

A given nanopore is tested for linear I–V characteristics (Fig. 2.2a) and low electrical noise (<20 pA RMS) before further use. In the event of unacceptable properties, the pore can be cleaned briefly (~5 min) with 1 M NaOH and/or rinsed, plasma treated and remounted for additional testing. Poor qualities are often the result of bubbles [10] and can also be remedied by exchanging the solution with water and then ethanol before exchanging back to measurement buffer. If all treatments fail, the pore is abandoned and a new chip is used. Nanopores can sometimes be unsuitable for measurement even immediately after fabrication due to unknown reasons. Using the ethanol/water storage conditions, we find that 80% of pores are fit for measurement.

Following identification of a useful nanopore, its precise location on the membrane is determined. This is done by rastering the high-power trapping laser across the membrane surface and monitoring the trans-pore current at each step. As the laser focus comes near the nanopore, it locally heats the solution, increasing its mobility and thus the measured ionic current. The resulting 2D array displays a (nominally gaussian) spike [11] at the location of the nanopore. The membrane is then positioned such that the trapping laser is co-local with the nanopore and moved about 10 μm above the focal plane. Following this, microbeads (2 μm polystyrene) with target molecules attached to them [6] are introduced to the bottom chamber at very low concentration (0.001% w/v). Once a single bead is trapped in the optical

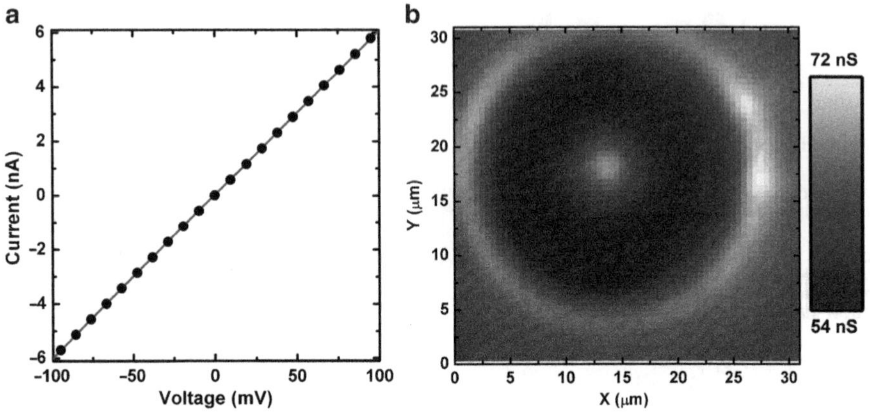

Fig. 2.2 (**a**) Typical I–V measurement on a 8 nm nanopore, demonstrating linear response. (**b**) Measured conductance through the same pore as the trapping laser is rastered across the membrane surface. Local heating increases the ionic mobility when the laser beam is at the pore opening, making its location detectable as the bright peak near $(x, y) = (13, 18\ \mu\text{m})$

tweezer focus, clean measurement solution is flowed into the chamber to prevent additional beads from entering the trap during subsequent measurement. Since the target molecule is mechanically attached to the bead, the membrane-bead separation must be minimized in order to ensure that the molecule is affected by the applied electric field. Therefore, after rinsing, the membrane is lowered to within a short distance (generally 1–3 μm) of the trapped bead; best results are found with a separation roughly equal to the molecular radius of gyration.

Once positioned correctly, the application of a voltage with appropriate polarity across the membrane will attract molecules toward the nanopore. The size of the necessary voltage varies somewhat, but generally falls between 50 and 150 mV. Too low voltage does not create a strong enough field to pull a molecule into the pore in a practical amount of time, while too high voltage creates the risk of capturing multiple molecules or pulling the bead out of the optical tweezer. Each system and molecule will require finesse to identify the optimal conditions. The capture of a molecule in the nanopore is manifested as a sudden change in the measured trans-membrane ionic current together with a (roughly) simultaneous change in the PSD signal, indicating bead motion away from its initial position (Fig. 2.3). When this occurs, the applied voltage is immediately reduced to 10–20 mV; enough to keep the captured molecule from diffusing out of the pore, but too weak to introduce others. At this point, the bead position can be changed relative to the nanopore (by moving the membrane in z) without losing the molecule.

Force spectroscopy is performed by increasing the applied voltage in a step-wise fashion while continuously monitoring the bead position via the PSD. Each increase in voltage creates an increased electrophoretic force on the molecule, and thus pulls the microbead farther from its initial position and closer to the membrane. A measurement of the position can be done using a detector scheme in which the PSD signal scales directly with the bead position [12], but the results we describe here utilize a calibration curve taken directly prior to incremental voltage stepping [9, 13].

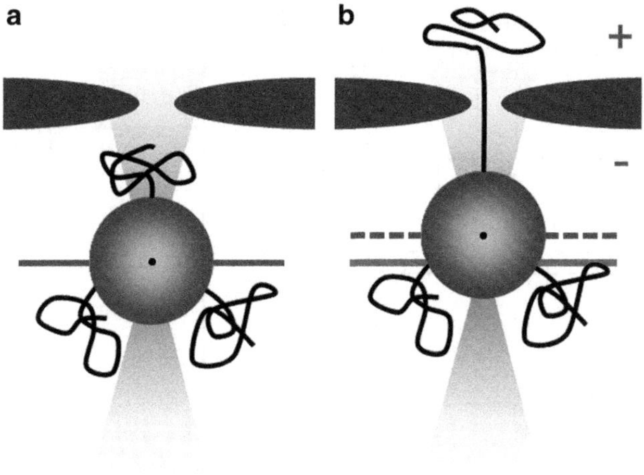

Fig. 2.3 Capturing scheme. (**a**) A microbead (center indicated by *dot*) with target molecules attached is delivered near the nanopore opening with an optical tweezer. (**b**) Upon application of an appropriate voltage, a molecule is pulled in to the pore and held there statically by the optically trapped bead. This action both blocks the nanopore current and displaces the bead from its initial position relative to the membrane (*horizontal lines*)

2.3 DNA Detection

dsDNA (λ-phage, 48.5 kb) is end-labeled through hybridization to a biotin-linked primer with the complimentary 12-base overhang. The resulting material is then attached to streptavidin-coated microbeads with a 1-h incubation at 37°C under buffer conditions of 1 M KCl and 10 mM Tris-HCl (pH 8.0). The concentration of DNA is controlled relative to the beads in order to attach only 1–10 molecules per bead on average. This acts to further reduce the chance of having multiple molecules enter the pore simultaneously.

Captures are performed by applying a voltage across the nanopore membrane while monitoring nanopore current (conductance) and bead position (PSD signal). In order to confirm that target molecules are indeed the cause of such signal changes, a range of applied voltages can be used, as shown in Fig. 2.4a. In this case, the solvent contains 1 M KCl. As increasing positive voltage is applied, dsDNA capture events are eventually observed in both the conductance and the PSD signal (*I, II*). However, by reversing the polarity (*III*), an opposing force is applied which ejects the molecules from the nanopore. Thus, when a positive voltage is applied (*IV*), the original open-pore conductance is again measured until another molecule is eventually captured (*V*) when one returns to negative voltages again.

A closer look at a typical dsDNA capture event is shown in Fig. 2.4b. Two aspects of this measurement bare further mentioning here. First, the amount of conductance change measured (referred to as conductance blockade in high ionic strength conditions [14]) is found to be very consistent at ~1 nS. This is in

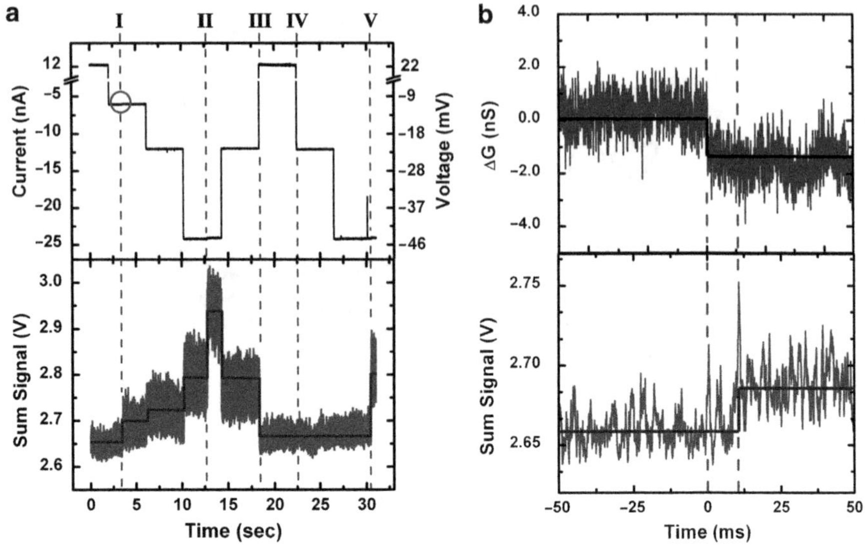

Fig. 2.4 Typical dsDNA captures. (**a**) Simultaneous measurement of nanopore current (*top*) and PSD voltage on the position-sensitive detector (which measures the bead position, *bottom*) under varying applied voltage to the *trans* chamber. See text for descriptions. (**b**) Close-up of the dsDNA capture event circled in (**a**). The *dashed lines* mark the difference in time between the conductance drop (*top*) and the change in bead position (*bottom*)

agreement with the (unfolded) event depth measured in translocation experiments for the same molecule [15] (Fig. 2.5a) and is thus similarly indicative of the size sensitivity of nanopore measurements. Second, although the molecule is statically captured inside the nanopore so the measured conductance does not return to a baseline value as in translocation, temporal information can also be garnered from these optical tweezer measurements. When a molecule is captured in the nanopore, its presence immediately affects the trans-membrane conductance, but there is a delay in the bead position change, because the entire molecule must be threaded through the pore before it is pulled taut and the force is transmitted to the trapped bead. This delay time (Fig. 2.4b, red lines) is a measure of the molecular contour length in the same way that the dwell time is in free translocation [5]. In fact, measurements of the delay consistently match the dwell times for the same length of molecule quantitatively (Fig. 2.5b).

Once a single dsDNA is statically held inside the nanopore, the piezo stage can be used to change the bead-membrane separation and control the position of the molecule relative to the nanopore. An example of this is shown in Fig. 2.6, where the bead is slowly moved away from the membrane surface to a distance greater than the contour length of the molecule (16.5 μm). Here, the measured pore conductance changes suddenly back to the original open-pore level, indicating that the opening is no longer occluded and that the dsDNA has been removed. This occurs at a total distance of 15 μm from the membrane surface, in reasonable agreement with the contour length. The slight difference is due to settling in the system over the long measurement time

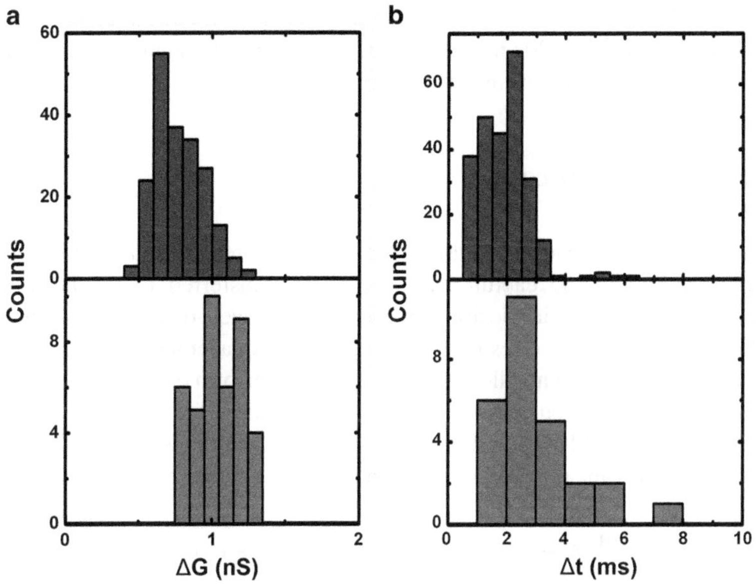

Fig. 2.5 Comparison of translocation to static capturing of dsDNA. (**a**) Conductance change and (**b**) dwell time measurements compared between free translocation (*above*) and static capturing (*below*) [6]

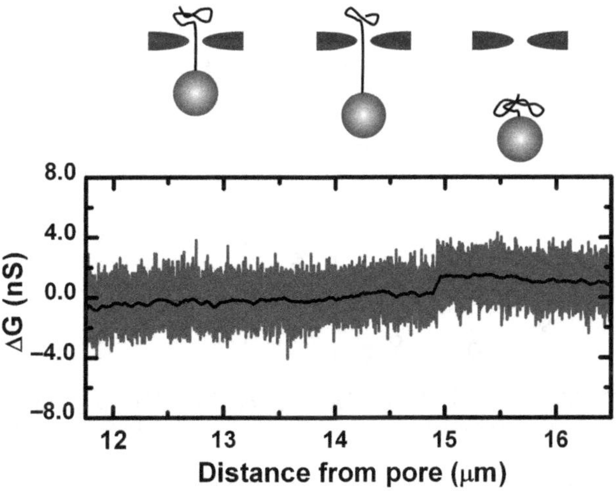

Fig. 2.6 Retraction of captured dsDNA. Measured nanopore conductance as a bead with a captured molecule attached is moved away from the pore using the optical tweezer. At 15 μm, the conductance increases suddenly by ~1 nS, corresponding to the removal of an individual dsDNA from the pore

(largely a result of laser heating of the membrane and optics). The presented retraction is performed at a speed of 20 nm/s; more than five orders of magnitude slower than a common free translocation event (~8 nm/μs).

2.4 Force Spectroscopy

The optically trapped bead can be used as more than merely a handle, however. The net forces acting on the captured molecule are transferred to the bead itself, displacing it from its initial position in the potential well of the optical tweezer. Therefore, changes in force result in relative bead displacements that are measurable through the PSD signal, allowing the system to perform one-dimensional force curves on a single-molecule basis. A typical force curve for dsDNA is shown in Fig. 2.7a (taken in 1 M KCl and at a starting bead-membrane distance of 4 μm), demonstrating the linear relation between applied voltage and measured force.

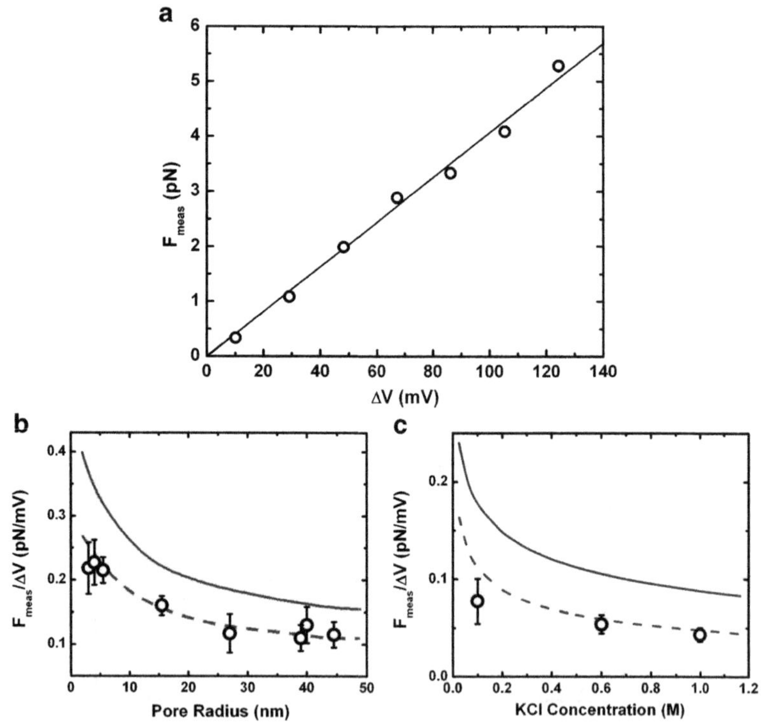

Fig. 2.7 Nanopore force measurements on dsDNA. (**a**) A typical dsDNA force curve (nanopore radius 12 nm, 1 M salt conditions) showing the linear dependence of measured force on applied voltage (*gray line* is a linear fit to the data). The observed relationship between the slope of the force curve and both nanopore radius (**b**) and salt concentration (**c**). The *solid gray line* in each represents the theoretical relation (2) and the *dashed line* represents the same relation calculated with a reduction of dsDNA charge of 50%

Phenomenologically, one can interpret this force as the electrophoretic pulling force $F = Q_{eff}\Delta V$ on the molecule, where Q_{eff} is the effective molecular line charge density and ΔV is the applied voltage [6]. Comparison of the slope of the 1D force curve to the known line charge density of dsDNA (2 e^-/bp) would in this view yield a measure of the charge reduction caused by screening in the ionic solution.

Further measurements, however, prove this model to be too simple. The dsDNA force curves, while always linear, are found to vary in slope when performed under different experimental conditions. Particularly, the slope is found to depend on nanopore diameter [16]; smaller pores are found to yield higher levels of force per unit voltage when all other conditions are kept constant (Fig. 2.7b). Concurrently, the ionic strength of the measurement solution is also found to affect the slope [13], with lower molarity salt resulting in greater measured force per unit voltage (Fig. 2.7c). As pure counterion screening should not be changed by shifting in these conditions, the purely electrophoretic model is shown to be incomplete; there must be forces that are unaccounted for in the current description. A revised model [16–19] that captures the results well reveals at least one of those additional forces: electroosmotic drag.

2.5 Modeling: Electrophoresis and Electroosmotic Shear

Charged ions in an electric field experience an electrostatic force. When these ions are in solution, their motion due to this force tends to drag along surrounding liquid molecules. Near a charged surface, the density of counterions is much higher than that of coions, creating a unidirectional net fluid flow in a process called electroosmosis. This effect is especially pertinent to nanopore measurements: because of the small scale, the entire volume contained inside the pore is effectively near a charged surface (the walls of the pore). The introduction of a charged molecule like dsDNA into that confined space will further modify the fluid flow profile. Fluid motion creates a drag force on the captured molecule, affecting the net force acting on it, and it must therefore be accounted for in a model describing the system. The forces used in the present model are detailed in Fig. 2.8a. The net force created by the applied voltage is in fact the electrophoretic force reduced by the electroosmotic drag on the molecule, and it is this combination that is opposed by the restoring force of the optical tweezer.

Ghosal [17] and independently van Dorp et al. [16] showed that the Poisson-Boltzmann relation can be combined with the Stokes equation such that it describes the force on a captured molecule in a nanopore:

$$F_{meas} = \frac{2\pi\varepsilon(\Phi(a) - \Phi(R))}{\ln(R/a)}\Delta V, \tag{2.1}$$

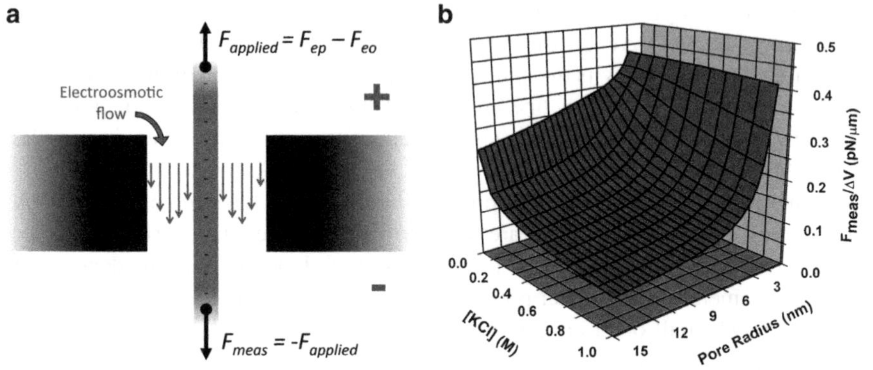

Fig. 2.8 Net electrical forces on a captured dsDNA. (**a**) Diagram of forces on dsDNA (*center*) in a nanopore (*black*), showing that measured force F_{meas} is balanced by a combination of the electrophoretic force (F_{ep}) and the drag force caused by electroosmotic flow (F_{eo}). Flow direction and qualitative magnitude are indicated by the vertical arrows. (**b**) Calculated force curve slope for dsDNA under various ionic concentrations and nanopore radii [2]

where ε is the dielectric constant of water, a and R are radii of the molecule and the nanopore, respectively, and $\Phi(a)$ and $\Phi(R)$ are the surface potentials of the molecule and the nanopore, respectively [17]. A $ln^{-1}(R/a)$ size dependence is therefore predicted, explaining why the measured force is affected by nanopore dimension as observed. Furthermore, the surface potentials are dependent on the distribution of ions surrounding those surfaces; a higher concentration of counterions causes greater screening of surface charge, thereby reducing the surface potential. This is correlated to the overall ion concentration, indicating the source of the observed force dependence on the measurement solution salt concentration.

Comparison of this description to experimental results is achieved by creating a computer model wherein the nanopore is represented by an annulus and the captured molecule by a cylinder, each of uniform charge density. For a designated salt concentration, this allows the ion distribution between the two entities to be calculated numerically. The distribution can then be used to solve the Stokes equation and describe the fluid flow profile, which can subsequently be used to yield the viscous drag acting on the molecule at a given voltage [16]. This can be combined with pure electrophoresis ($F_{EP}=Q\Delta V$) to result in a total prediction of the measured force per unit voltage on a captured molecule for any combination of pore diameter and salt concentration (Fig. 2.8b).

The trends of the size and salt dependences are captured well by the model (Fig. 2.7b, c), but the values do not agree quantitatively; the measured force is apparently overestimated by ~50%. This quantitative difference, while troublesome, has a physical explanation that becomes clearer by comparing these dsDNA measurements to ones on a second type of molecule with different physical properties.

2.6 Measurements on Protein-Coated DNA Molecules

Experiments are also performed by forming a second type of bead-conjugated target molecules: RecA-coated dsDNA [13]. RecA is a protein with an important role in homologous recombination and dsDNA break repair [20]. For the present experiments it was chosen for two main reasons. First, it is able to bind cooperatively along an entire molecule of dsDNA [21], resulting in an uninterrupted coating of protein that can be formed under non-dissociable conditions (upon using ATP-γS). Second, the attachment of this protein coating changes the physical properties of the molecule, the most relevant changes being an increase in diameter (from 2.2 nm for bare dsDNA to ~7 nm for RecA coated dsDNA [22]) and the overall charge (RecA monomers carry a net negative charge).

When nanopore capture experiments are performed on this nucleoprotein filament, the measured change in conductance upon insertion of a molecule is found to differ significantly from that of bare dsDNA under the same experimental conditions. Examples of typical conductance blockades for both kinds of molecule taken in 1 M KCl are shown in Fig. 2.9a. At 1 M KCl, the RecA-dsDNA is found to cause a much larger conductance blockade of about 7 nS, similar to translocation measurements on the same type of molecule [23]. This also agrees qualitatively with the description of how the presence of a molecule affects the measured nanopore conductance [14], wherein the conductance blockade ΔG can be expressed as

$$\Delta G = \frac{1}{L_{pore}} \left(-\pi a^2 (\mu_{cat} + \mu_{an}) n_{tot} e + \mu_{counter} q_l^* \right). \tag{2.2}$$

Here, L_{pore} represents the length of the pore, μ_{cat} and μ_{an} are the electrophoretic mobilities of the cation and anion, respectively, n_{tot} is the total number density of ions in solution, e is the elementary charge, $\mu_{counter}$ is the mobility of counterions near the surface of the molecule (which we adopt to equal the bulk mobility), and q_l^* is the net charge per unit length of carriers surrounding the molecule, which to first order equals the effective line charge density Q_{eff} of the molecule [13]. This expression shows that when the ionic concentration is high (n_{tot} is large), the first term, which scales with the square of the diameter a of the molecule, is dominant, and thus the conductance change is essentially a measure of molecular size. Since RecA-coated dsDNA has a larger diameter than bare dsDNA, it makes sense that the conductance change for the former is greater than the latter.

A more complete comparison can be made by examining the conductance change for each molecule under various salt concentrations (Fig. 2.9). In both cases, measurements indicate a transition from conductance blockades at high ionic strength to conductance enhancement at low ionic strength. A fit to the data [13] for bare dsDNA yields an apparent charge of -0.22 nC/m which compares reasonably with translocation measurements [14] on the same molecule which yielded

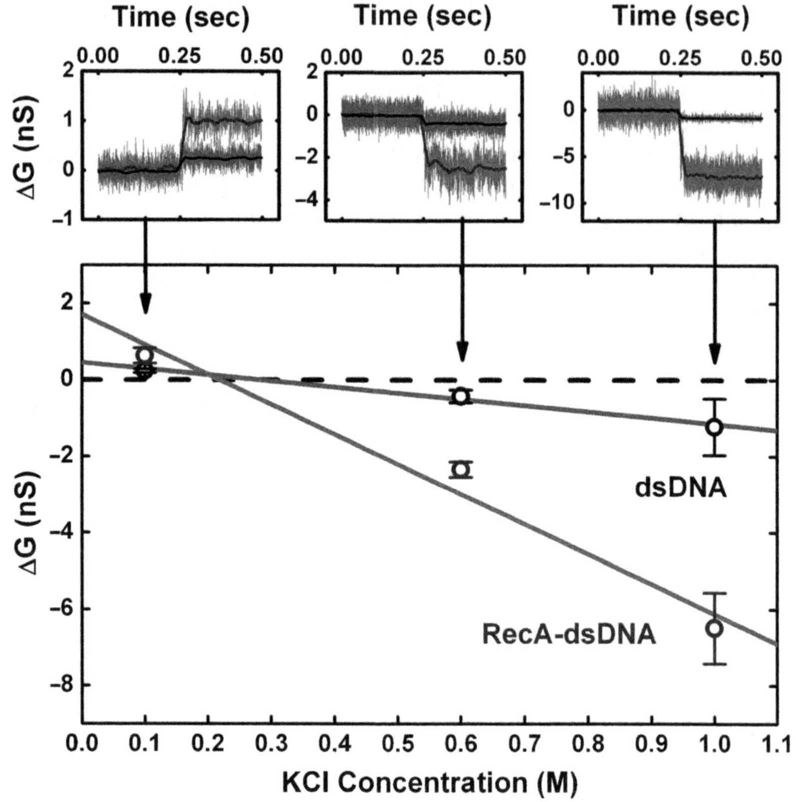

Fig. 2.9 Conductance change during static capture. Measured conductance change upon molecular capture for both bare (*dark points*) and RecA-coated (*light points*) dsDNA under various salt concentrations (*top*). The relation is found to be linear for each (*bottom*) and can be described well by fitting to (2) (*solid lines*) [13]

-0.38 nC/m. As with the force spectroscopy measurements, these analyses show that a reduction from the known line charge density of bare dsDNA (-1.1 nC/m) is necessary to have quantitative agreement between model and experiment. A fit to the conductance changes measured for RecA-coated dsDNA, meanwhile, yields an apparent line charge density of -1.6 ± 0.6 nC/m. When compared with the theoretical nucleoprotein filament charge of -1.8 nC/m (calculated by simple charge summation of the RecA amino acids with the underlying dsDNA [13]), this result indicates that a charge reduction is *not* necessary to achieve quantitative agreement for RecA-dsDNA.

Force spectroscopy is also performed on the protein-coated molecule under multiple ionic strengths (Fig. 2.10a). Several things can be noted from these results. First, the slopes of RecA-dsDNA force curves can be easily distinguished from those of bare dsDNA under all examined conditions. This points toward the ability

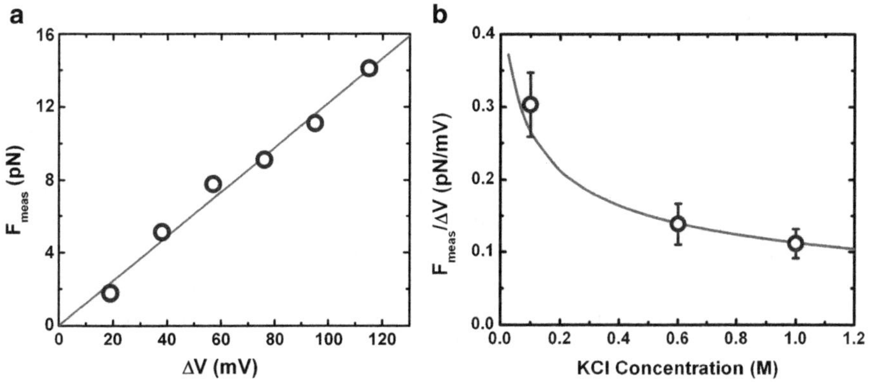

Fig. 2.10 Nanopore force measurements on RecA-dsDNA. (**a**) A typical force curve on RecA-coated dsDNA, taken at 1 M KCl. (**b**) Measured relationship between RecA-dsDNA force curve slope and salt concentration. The *solid line* is the calculated relation with no fitting parameters

of the measurement technique to identify different molecules with a mechanism separate from that of conductance blockade. Second, it is apparent that the force per unit voltage acting on the protein-coated molecule is greater as ionic strength decreases. As with bare dsDNA measurements, this fits qualitatively with the model, which anticipates that lower salt concentration reduces the opposing electroosmotic drag force. Third, and perhaps most importantly, the experimental relation between measured force and salt concentration is described well by the electrophoretic/electroosmotic model without any additional fitting (Fig. 2.10b). As with the conductance change measurements, this suggests that no charge reduction is needed for a RecA-dsDNA molecule.

These results taken in total are suggestive of the source of the deviation between the model and experiments for bare dsDNA. One potential cause is the presence of stationary charges on the surface of the nanopore. However, as the same nanopores were often used to measure both types of molecule presented here, this would produce a systematic variation. Since no adjustment is needed for the fit to the nucleoprotein filament data, this source can be excluded. Instead, a more likely explanation is a reduction in the mobility of counterions close to the molecular surface. Although the RecA-dsDNA molecule has higher total charge than its uncoated counterpart, its larger diameter results in a lower surface charge density (-0.08 C/m^2 compared to -0.14 C/m^2 for bare dsDNA) [13]. Non-linear charge accumulation will therefore cause counterions to interact more strongly with the bare dsDNA surface, reducing their mobility to a larger degree than for the protein-coated molecule. The current understanding of charge screening layers in solution cannot account for such short-range effects. Future experiments could be used to confirm this hypothesis by examining molecules of different surface charge density, such as dsDNA coated with proteins of different charge.

2.7 Conclusions

The combination of optical tweezers with solid-state nanopores represents a versatile system for the study of biomolecules. The use of an optical tweezer to deliver bead-conjugated target molecules to the nanopore allows measurements similar to those measured with pure translocation (size and length discrimination) but additionally presents new capabilities unattainable in traditional systems (arbitrary position control, force spectroscopy). We have demonstrated that measurements on conductance change and dwell time with the combination of technologies allow bare dsDNA to be differentiated from RecA-coated dsDNA, comparable to translocations. However, we have also shown that force curves can be measured on both types molecules, demonstrating the ability to differentiate molecules through a second method in the same system. The dependence of measured force curves on both nanopore dimension and ionic concentration can be described well by a model incorporating both electrophoresis and electroosmosis. This provides a foundation with which to understand the results of forthcoming studies.

Future studies will exploit the position control inherent in the optical tweezer system to allow for this measurement technique to be performed on local structures. Simultaneous detection of both conductance and applied force will allow for a multi-faceted approach to detect small features along an individual biomolecule. This opens the door to, for example, epigenetic footprinting at the molecular scale.

Acknowledgments We acknowledge U. Keyser, J. van der Does and D. Krapf for contributions to system development. M. van den Hout performed the calculations shown in Fig. 2.8b. We also wish to thank S. Lemay, U. Keyser, S. van Dorp and S. Kowalczyk for useful discussions. This work was supported financially by NWO, FOM, and the EC project READNA.

References

1. D. Branton, D. W. Deamer, A. Marziali, H. Bayley, S. A. Benner, T. Butler, M. Di Ventra, S. Garaj, A. Hibbs, X. H. Huang, S. B. Jovanovich, P. S. Krstic, S. Lindsay, X. S. S. Ling, C. H. Mastrangelo, A. Meller, J. S. Oliver, Y. V. Pershin, J. M. Ramsey, R. Riehn, G. V. Soni, V. Tabard-Cossa, M. Wanunu, M. Wiggin and J. A. Schloss, Nature Biotechnology **26** (10), 1146–1153 (2008).
2. C. Dekker, Nature Nanotechnology 2 (4), 209–215 (2007).
3. A. J. Storm, J. H. Chen, X. S. Ling, H. W. Zandbergen and C. Dekker, Nature Materials **2** (8), 537–540 (2003).
4. J. Li, D. Stein, C. McMullan, D. Branton, M. J. Aziz and J. A. Golovchenko, Nature **412** (6843), 166–169 (2001).
5. A. J. Storm, C. Storm, J. H. Chen, H. Zandbergen, J. F. Joanny and C. Dekker, Nano Letters **5** (7), 1193–1197 (2005).
6. U. F. Keyser, B. N. Koeleman, S. Van Dorp, D. Krapf, R. M. M. Smeets, S. G. Lemay, N. H. Dekker and C. Dekker, Nature Physics **2** (7), 473–477 (2006).
7. D. Krapf, M. Y. Wu, R. M. M. Smeets, H. W. Zandbergen, C. Dekker and S. G. Lemay, Nano Letters **6** (1), 105–109 (2006).

8. M. van den Hout, A. R. Hall, M. Y. Wu, H. W. Zandbergen, C. Dekker and N. H. Dekker, Nanotechnology **21** (11) 115304 (2010).
9. U. F. Keyser, J. van der Does, C. Dekker and N. H. Dekker, Review of Scientific Instruments **77** (10) 105105 (2006).
10. R. M. M. Smeets, U. F. Keyser, M. Y. Wu, N. H. Dekker and C. Dekker, Physical Review Letters **97** (8) 088101 (2006).
11. U. F. Keyser, D. Krapf, B. N. Koeleman, R. M. M. Smeets, N. H. Dekker and C. Dekker, Nano Letters **5** (11), 2253–2256 (2005).
12. A. Sischka, C. Kleimann, W. Hachmann, M. M. Schafer, I. Seuffert, K. Tonsing and D. Anselmetti, Review of Scientific Instruments **79** (6) 063702 (2008).
13. A. R. Hall, S. van Dorp, S. G. Lemay and C. Dekker, Nano Letters 9 (12), 4441–4445 (2009).
14. R. M. M. Smeets, U. F. Keyser, D. Krapf, M. Y. Wu, N. H. Dekker and C. Dekker, Nano Letters **6** (1), 89–95 (2006).
15. A. J. Storm, J. H. Chen, H. W. Zandbergen and C. Dekker, Phys Rev E Stat Nonlin Soft Matter Phys **71** (5 Pt 1), 051903 (2005).
16. S. van Dorp, U. F. Keyser, N. H. Dekker, C. Dekker and S. G. Lemay, Nature Physics **5** (5), 347–351 (2009).
17. S. Ghosal, Physical Review E **76** (6) 061916 (2007).
18. U. F. Keyser, S. van Dorp and S. G. Lemay, Chemical Society Reviews **39** (3), 939–947 (2010).
19. J. L. Viovy, Reviews of Modern Physics **72** (3), 813–872 (2000).
20. S. C. Kowalczykowski, D. A. Dixon, A. K. Eggleston, S. D. Lauder and W. M. Rehrauer, Microbiological Reviews **58** (3), 401–465 (1994).
21. R. Galletto, I. Amitani, R. J. Baskin and S. C. Kowalczykowski, Nature **443** (7113), 875–878 (2006).
22. Z. C. Chen, H. J. Yang and N. P. Pavletich, Nature **453** (7194), 489-U483 (2008).
23. R. M. M. Smeets, S. W. Kowalczyk, A. R. Hall, N. H. Dekker and C. Dekker, Nano Letters **9** (9), 3089–3095 (2009).

Chapter 3
Single Molecule Detection with an Aptamer-Integrated Nanopore

Li-Qun Gu

Abstract Aptamers, also called "synthetic" antibodies, are a broad class of engineered oligonucleotides that can specifically bind targets with high affinity. Upon binding, these powerful molecules can form complex three-dimensional structures and possess sophisticated functions to inhibit pathogen protein, catalyze chemical reactions, controlling gene expression, and regulate cellular functions. These can be therefore potentially applied as tools for exploring biological systems and medical diagnosis. Nanopore detection technology, on the other hand, can "visually" capture the dynamic binding of a single molecule to a ligand in a nanometer-scaled pore through the discrete changes in conductance upon binding. This ability to track single molecule kinetics has made the nanopore a promising single molecule detector. This chapter will be focused on the use of nanopores as a research tool that can be combined with laboratory nanofabrication, bio-friendly surface engineering and site-directed protein engineering to understand aptamer folding process, the interaction between a single aptamer and its target, and to develop aptamer-encoded nanopore sensors for medical and bio-defense detection.

Keywords G-quadruplex aptamers • Protein detection with aptamer-encoded nanopore • IgE • Ricin • Ion-regulated switching of aptamer structure

---

Li-Qun Gu (✉)
Department of Biological Engineering and Dalton Cardiovascular Research Center, University of Missouri, Columbia, MO 65211, USA
e-mail: gul@missouri.edu

S.M. Iqbal and R. Bashir (eds.), *Nanopores: Sensing and Fundamental Biological Interactions*, DOI 10.1007/978-1-4419-8252-0_3,

3.1 Overview

3.1.1 What Is Aptamers?

DNA and RNA possess far broader applications than simply storing and transferring hereditary information. With the help of molecular engineering, powerful new RNA and DNA molecules can be synthesized, which are capable of forming complex three-dimensional structures, and possess surprisingly sophisticated functions [9, 12, 53, 91, 106] in protein inhibition [9, 53], reaction catalysis, and gene regulation. These qualities make them potential therapeutic agents and tools for exploring biological systems.

Among these engineered nucleic acids, aptamers are a group of short, structured DNA or RNA molecules usually created through directed in vitro evolution, termed SELEX: Systematic Evolution of Ligands by Exponential Enrichment [23, 110]. Aptamers are called "synthetic" antibodies because they are designed specifically to bind broad species of ligands with high affinities, from antibiotics [94, 108, 126], vitamins [123], amino acids [24, 29], to peptides [4], proteins [67, 118] and pathogen target [20, 54, 58]. However, aptamers possess many advantages over antibodies. They are very small but can selectively recognize and bind specific ligands with nano-molar or pico-molar ranges that match or exceed those of their true antibody counterparts [52]. For example, the catalytic thrombin-binding aptamer is only a 15-base sequence (Fig. 3.1a) [114], but binds with nano-molar affinity [10]. In principle, aptamers can be generated against any target using in vitro methods that are independent of animal, and are not limited by physiological conditions. Aptamer-target interaction can be fine tuned by mutation, and all the sequences of current aptamers have already been published. Aptamers are significantly more durable than most protein receptors, yet simpler to synthesize, modify, and immobilize by low cost methods. These advantages render aptamers great promise in bioanalysis [30, 31, 40, 41, 48, 105], diagnostics [13], therapy [121], bio-catalysis and cell modulation [109].

3.1.2 Molecular Folding, Interaction and Biosensing

A highly specific, high-affinity interaction with its target is the most essential aptamer property, which has ignited a variety of aptamer researches and applications.

Aptamers are much smaller, yet more flexible in structure. They must fold into special structures upon binding with targets under specific conditions [38]. For instance, aptamers for thrombin (Fig. 3.1a) and HIV-1 Integrase should fold from free chains into remarkably stable G-quartet forms in the presence of metal ions [16, 20, 56, 57, 90, 114, 115]. More significantly, the G-quartet is common in areas of the human genome such as telomeric DNA [86, 89]. Thus, stabilization of the G-quartet by interactions with small chemical reagents or ions could inhibit the function of telomerase that is active in cancer cells [32]. Understanding the precise

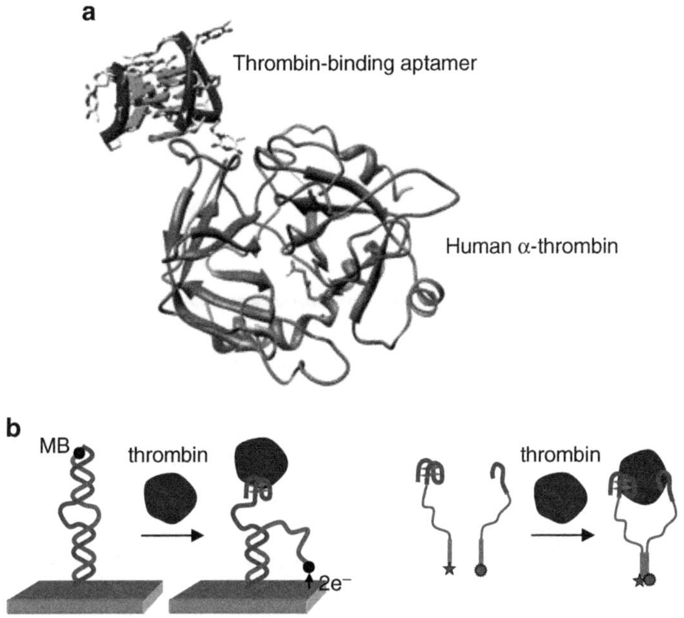

Fig. 3.1 Targeting human α-thrombin with a DNA aptamer and exemplified biosensors employing aptamer-target interactions. (**a**) Crystal structure of the complex of thrombin-binding aptamer (TBA) and its target thrombin (PDB: 1HAP). TBA is a 15 bases DNA oligonucleotide (GGTTGGTGTGGTTGG) that folds into a G-quartet structure upon binding with thrombin. (**b**) TBA-based biosensors for the detection of thrombin (*left*) electrochemical approach [124] and (*right*) Molecular beacon fluorescence technique [41]

folding process is therefore extremely important. This brings us to question, what is the folding duration; what are the intermediate folding states, and do they exist in discrete step or continuously; what is the reversibility and stability; and what are the coordinating steps made by cations during G-quartet formation?

There are many significant processes involved in aptamer-target interactions that need to be elucidated. For instance, what are the potential contributions of association and dissociation to aptamers' high affinities? A fuller understanding of the aptamer-target interaction could provide meaningful insight into the subtleties of diverse nucleic acid-protein interactions in living cells. More significantly, it should benefit to molecular recognition-based rational design and its applications such as biosensor construction (Fig. 3.1b).

3.1.3 Single Molecule Detection and Nanopore

Nanopore is an emerging single molecule detection technique in which the individual binding events of target molecules to a site within the lumen of a nanometer-scaled pore can be electrically "visualized" through discrete changes in ion

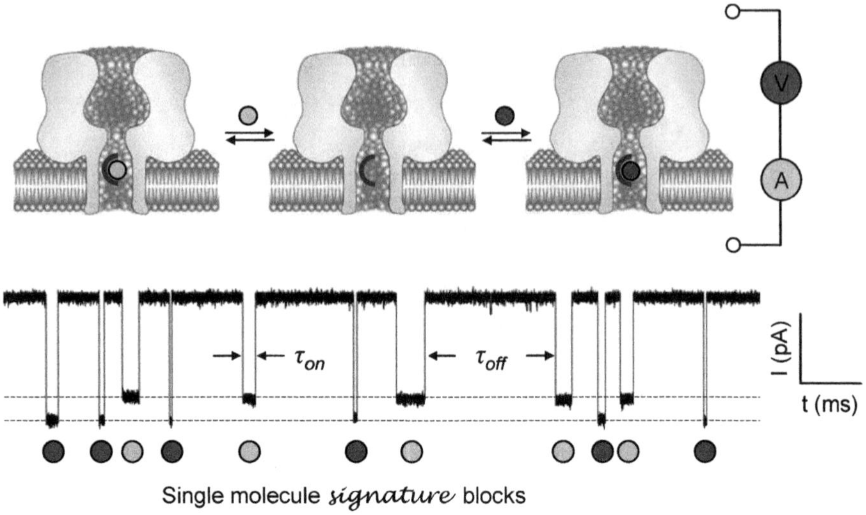

Fig. 3.2 Single molecule detection with a protein nanopore sensor. The nanopore sensor can be constructed by forming a lipid bilayer membrane that insulates solutions on both sides. A protein nanopore embedded in the bilayer constitutes the only path for ionic flow across the membrane. The voltage applied between the two solutions will drive a pico-Ampere ion current through the pore, recorded using an amplifier. The protein pore has been engineered with a recognizing probe in the lumen. Without a bound target, the pore remains open. When a single molecule binds to the probe, the pore current is blocked, but when the molecule is released, the ionic current resumes. Repeated cycles of binding and release of individual target molecules to and from the pore can result in a string of binary (on/off) signature blocks. Different targets in the mixture (represented by *filled* and *unfilled* balls) may competitively bind to the same pore, but give rise to different block amplitude and duration, which can be used for target discrimination. The block duration, τ_{off}, is target-specific, and is used to evaluate the dissociation rate constant k_{off}: $k_{off} = 1/\tau_{off}$. The frequency of block occurrence f can be known from the inter-block duration τ_{on}: $f = 1/\tau_{on}$. f is a measure of target quantity because it is proportional to the target concentration ([T]), using the association rate constant k_{on} as the coefficient, $f = k_{on}$[T]

conductance (Fig. 3.2). From this principle, bio-nanopores [5] and artificial nanopores [42, 51, 69, 71, 93, 102] have been engineered as potential biosensors for substrates including pharmaceuticals [33], secondary cellular messengers [15], metal ions [11, 63], nucleic acids [47], and proteins [82, 93, 103]. Other goals of nanopore use are the understanding of polymer dynamics and transportation [7, 8], and the rapid sequencing of DNA [3, 21, 27, 61, 62, 70, 79, 81, 84, 112, 113]. Invariably, these applications take advantage of the potent capabilities of nanopores in dissecting single-molecule kinetics – a property deduced from the principle that any change in conformation during a molecular process will affect a molecule's occupancy within the nanopore, thereby altering the conductance. In addition to ease-of-use, one of the most outstanding features of nanopores is their ability to directly measure transition rate constants between states in a kinetic path [34–37, 46, 73]. Furthermore, detection using nanopore requires no molecular labels, such as fluorescents or microbeads, which could otherwise perturb molecule function.

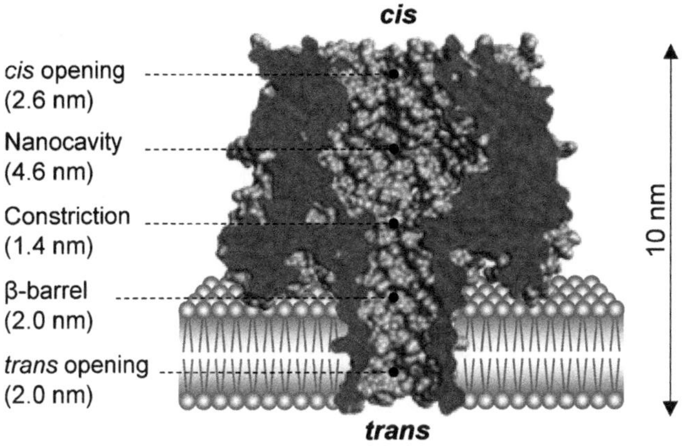

Fig. 3.3 Molecular graph of the heptameric α-hemolysin pore in the lipid bilayer. The diameters in various positions lining the pore lumen were provided [104]

3.1.4 Selection of Nanopores

Through our research, the nanopore is proven a powerful technique in single molecule detections of aptamer molecular processes and interaction with targets. Since detection relies upon a change in occupancy of a molecule in the nanopore, the selection of a nanopore dimensionally matched with the target is the first priority. Protein nanopores are most promising due to their capacity for site-directed modification by protein chemistry and genetic engineering. We employed the protein nanopore assembled by the bacterial toxin α-hemolysin (Fig. 3.3) to study the G-quartet aptamer folding process (Fig. 3.4). This research was followed by the development of portable and stable ion channel chip device for the uses in real-time biosensing (Fig. 3.5).

On the other hand, α-hemolysin may not be suitable to the study of all the aptamer-target interactions due to the small unchangeable pore size, making the pore unable to accommodate the aptamer/target complex. Therefore we developed a facile glass nanopore that is fabricated on a micro-pipette tip (Fig. 3.6) to investigate aptamer-target interactions (Figs. 3.7 and 3.8), based on the hypothesis that because aptamers are much smaller than their targets, target blockades become much more distinguishable. The long term goal of this research is an aptamer-encoded nanopore single molecule biosensor.

3.2 Understanding Ion-Regulated Folding Process of G-Quadruplex Aptamers

G-quadruplex is a special structure formed by guanine-rich single-stranded DNA or RNA. The quadruplex is built with G-tetrads, a planar assembly of four guanine bases networked via hydrogen bonds. G-tetrads stack one on another, with a cation

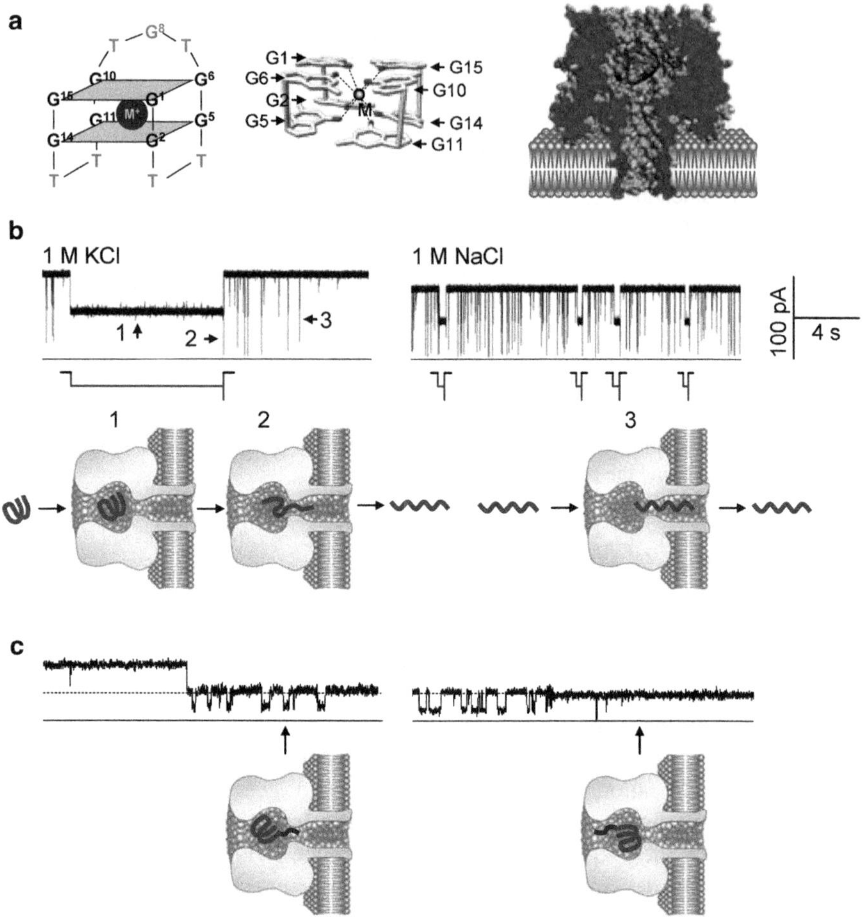

Fig. 3.4 Understanding folding process of single G-quadruplex aptamer using the α-hemolysin-based nanopore sensor. (**a**) The sequence and structure of the G-quadruplex formed by the thrombin-binding aptamer (TBA) (*left*), the two G-tetrad planes coordinated by a metal ion in the TBA G-quadruplex (*middle*), and molecular graph showing the encapsulation of a single TBA G-quadruplex in the nanocavity of the α-hemolysin pore (*right*). (**b**) Current traces showing the interaction of G-quadruplex and the α-hemolysin pore in the presence of K^+ (*left*) and Na^+ (*right*), and models. The signature long blocks with a terminal spike are generated by trapping and unfolding of a G-quadruplex in the pore, and the co-existing short blocks by the translocation of linear form TBA DNA. (**c**) Characteristic blocks produced by tag-TBA (*top*) and the model showing the location and change in position of the G-quadruplex in the cavity (*bottom*) [99, 100]. Reprinted with permission from *J. Phys. Chem. B* 112:8354–9360 (2008) and *Nucleic Acids Research* 37:972–982 (2009)

intervened between adjacent tetrads in coordination with eight carbonyls of guanine bases for stabilization [10, 58]. Native G-quadruplexes frequently occurs at the end of telomeres in the human genome where they participate in gene regulation [2]. G-quadruplexes in the human genome [95, 96] are important drug targets [49, 64, 85] because they are associated with many gene control-related mechanisms [2, 26,

95, 101, 125], such as the function block of telomerase active in cancer cells [89], and serve as targets for drugs in cancer treatment [85]. G-quadruplexes can also be created in vitro through molecular design. Synthetic G-quadruplexes with controllable folding/unfolding properties act as building blocks for assembling nano-structures [19] and nanomachines [1, 72]. Due to the high affinity for target proteins, G-quadruplex aptamers are ideal candidates for biosensor construction [10] and show usefulness as potent pharmaceuticals [107]. Understanding the cation-selective folding/unfolding of the G-quadruplex is very useful because a properly folded quadruplex is necessary for the molecular recognition involved in many quadruplex functions and is beneficial for designing quadruplex applications.

The 15 bases thrombin-binding aptamer (TBA) [10] is a famous one that folds the simplest quadruplex structure, with two parallel intra-molecular G-tetrads coordinated by a metal ion (Fig. 3.3a) [60]. The top tetrad is assembled with guanine 1, 6, 10 and 15, and the bottom one with guanine 2, 5, 11 and 14 [77, 88, 116] (Fig. 3.1b). The TBA's capability in forming G-quadruplex is cation-selective. Upon folding, the TBA quadruplex functions as an ultra-sensitive thrombin detector and an efficient inhibitor of thrombin clotting activity [41, 43, 48].

We demonstrated that the protein pore can be used to learn various issues on the folding of G-quadruplex [99, 100]: how to discriminate a single TBA molecule in the G-quartet form or in linear form? What is the interaction of G-quadruplex with nanopore? What is the approach to the understanding of G-single quadruplex folding/unfolding kinetics? And how do cations regulate the folding process of the G-quadruplex.

3.2.1 Encapsulation of a G-Quadruplex in the Nanopore Nanocavity

TBA was first found to interact with the α-hemolysin pore from the *cis* side in the presence of K^+ or Na^+, by generating the signature blocks that are characterized by a long duration (~15 s in K^+ and 3 s in Na^+) and partial reduction of the pore conductance (Fig. 3.4b traces) [99]. These signature blocks are distinguishable from the ~10^2 μs short blocks caused by translocation of a linear form DNA through the pore (Fig. 3.4b right model), and should be attributed to a single TBA G-quadruplex entering the pore from the cis opening and accommodating inside the nanocavity domain (left model). The 2.1-nm wide G-quadruplex is slightly narrower than the 2.6-nm cis opening of α-hemolysin but larger than the 1.4 nm β-barrel in middle of the pore (Fig. 3.3), making it possible for the G-quadruplex to enter and be trapped in the nanocavity (Fig. 3.4a). Moreover, the widest part of nanocavity is 4.6 nm (Fig. 3.3), therefore upon binding of TBA, the unoccupied space in the nanocavity forms an ion pathway. This is evidenced by the large residual current in the long blocks (Fig. 3.4b traces). The TBA's ability to bind thrombin was further investigated. Because thrombin chelates free TBA G-quadruplex in the solution, and the formed G-quadruplex/thrombin complex is too large to enter the pore, consequently the signature blocks became barely detected. Overall, the signature blocks can be used to distinguish a TBA molecule either in the G-quartet form or linear form.

3.2.2 *Spontaneous Unfolding of the G-Quadruplex in the Nanocavity*

The most significant finding is that the signature block is terminated with a spike-like full block (Fig. 3.4b traces), which is similar to the short events for linear form DNA translocation in conductance and duration [99]. This current profile represents such a molecular process that the encapsulated G-quadruplex spontaneously unfolds into a linear DNA that immediately traverses the β-barrel (Fig. 3.4b left model). The unfolding is likely to occur spontaneously, as the electrical field in the nanocavity, which is much weaker than in the β-barrel, has little impact on the G-quadruplex.

3.2.3 *Interaction of Nanopore with the Trapped G-Quadruplex*

The interaction of nanopore with the trapped G-quadruplex can be revealed using a designed TBA variant, called tag-TBA that contains a GACTAC tag at the $5'$ terminal [99]. When folded, the tag-TBA form a ball (quadruplex)-stick (tag) chimera (Fig. 3.4c). The tag-TBA not only generates the TBA signature blocks, but also produces a series of new Level-2 blocks upon the signature block (Fig. 3.4c left). Since the conductance of Level-2 blocks is similar to that for linear form DNA translocation, they are expected to be caused by the interaction of the linear tag with the β-barrel, therefore verifying that the tag-TBA is accommodated around the bottom of the nanocavity, where it vibrates, repeatedly touching the entry of the β-barrel, producing reversible Level-2 blocks. In addition, the Level-2 blocks can also occasionally disappear (Fig. 3.4c right), marking a movement that the tag-TBA suddenly rotates to another orientation within the nanocavity, leaving the β-barrel untouched by the tag-TBA bottom.

3.2.4 *Analytical Approach to Learning Folding/Unfolding Kinetics*

The G-quadruplex folding/unfolding kinetics can be quantitatively dissected using a mathematic approach [100]. Firstly, the duration of the signature block τ has been proven to be the lifetime of the G-quadruplex in the solution. By comparing τ in different ions, we can understand the role of ion in the G-quadruplex stability. For example, the τ value in 1 M KCl is ~50 times as long as that in 1 M NaCl, suggesting that the K^+-induced G-quadruplex is much more stable than the one induced by Na^+. The lifetime of G-quadruplex τ corresponds to the unfolding rate constant k_u,

$$k_u = 1/\tau \tag{3.1}$$

The equilibrium constant K_f can be inferred by comparing the frequency of short-lived blocks produced by the linear form TBA (f_{TBAL}) and a control DNA (f_{Ctrl})

$$K_f = f_{Ctrl}/f_{TBA_L} - 1 \tag{3.2}$$

This expression is based on the fact that f is proportional to the concentration of linear form DNA in the solution. For example, it was found a lower percent of linear TBA molecules or a higher percent of folded ones in K^+ than in Na^+ (Fig. 3.4a traces). The shift in equilibrium toward the folded structure suggests that K^+ is more capable of the induction of the G-quadruplex formation than Na^+. By comparing the short-lived block occurrences for TBA and the control DNA, we can determine the concentrations of both the folded and linear TBA in the solution. These concentration data is then used to obtain the equilibrium constant. Finally, the folding rate constant, k_f, can be determined using

$$K_f = k_f/k_u \tag{3.3}$$

3.2.5 *Ion-Regulated Folding/Unfolding of G-Quadruplex Aptamer*

Aided by this analytical approach, we were enabled to use nanopore to study distributions and variations of different aptamer conformations as the environmental conditions change [100]. We have uncovered the mechanisms that govern how metal ions regulate the folding and unfolding of the G-quadruplex (Table 3.1).

Table 3.1 Equilibrium, folding and unfolding rate constants and free energy for the TBA G-quadruplex in different cations

M•TBA	r (Å)	K_f (no unit)	k_f (s^{-1})	k_u (s^{-1})	$\Delta G°$ (kcal mol^{-1})
Li^+•TBA	0.60	1.5	0.095	0.065	−0.23
Na^+•TBA	0.95	2.3	6.5	2.9	−0.48
K^+•TBA	1.33	7.0	0.46	0.066	−1.1
NH_4^+•TBA	1.45	4.6	1.1	0.25	−0.89
Cs^+•TBA	1.69	2.8	0.23	0.082	−0.61
Ba^{2+}•TBA	1.35	4.2	0.25	0.061	−0.84
Mg^{2+}•TBA	0.65	–	–	–	–
Ca^{2+}•TBA	0.99	–	–	–	–

The G-quadruplex formation is cation-selective. The selectivity sequence is $K^+ > NH_4^+ \sim Ba^{2+} > Cs^+ \sim Na^+ > Li^+$. Ba^{2+} can form a long-lived G-quadruplex with TBA. However, due to the strong cation-DNA interaction, G-quadruplex was not detected in Mg^{2+} and Ca^{2+}. The high formation capability of the K^+-induced G-quadruplex is contributed largely by the slow unfolding reaction. Interestingly, although the Na^+- and Li^+-quadruplexes feature similar equilibrium properties, they undergo radically different pathways. The Na^+-quadruplex folds and unfolds most rapidly, while the Li^+-quadruplex performs both reactions at the slowest rates. The sensitive nanopore also revealed that the cation-selective formation of the G-quadruplex is correlated with the G-quadruplex volume, which varies with cation species.

3.2.6 Significance and Impacts

The research on single aptamer molecule folding/unfolding properties suggests the nanopore provide a powerful sensitive, non-covalent, label-free approach for single-molecule manipulation, based on the principle that when a molecule folds, unfolds or reacts in the lumen of a pore, different molecular states can characteristically alter the pore conductance that can be identified. The current signatures help to recognize these states and their transitions, making it possible to track single-molecule folding/unfolding or reaction pathway. The nanopore method is also applicable to other quadruplexes and their variants, including a variety of bio-relevant intramolecular quadruplexes, such as the i-motif (quadruplexes formed by cytidine-rich sequences) and chemically-modified quadruplexes with unique functionalities. When combined with site-directed nucleotide substitution, the nanopore method could be used to examine the contribution of each guanine to the quadruplex's folding capability. Nanopores could also be used to determine the force involved in the interaction of G-quadruplex aptamers and their targets, such as thrombin and HIV-1 reverse transcriptase. This would further enhance the understanding of the molecular recognition by aptamer-target complexes. Finally, this research may facilitate the generation of new molecular species with tunable properties for nano-construction and the manufacture of biosensors.

3.3 Single Molecule Biosensing with a Robust Nanopore Biochip

The best-learned and most useful property of aptamers is their ability to specifically bind targets with high affinity, which renders aptamers unique sensors for ultra-sensitive bio-detection. For example, due to the high affinity in nanomolar (nM), a number of aptamer-based techniques for thrombin detection have been proposed, such as sensitive electrochemical technique [124] and molecular beacon fluorescence technique [41] (Fig. 3.1b). As studied in Sect. 2.1, we have identified the dramatic decrease in the number of TBA G-quadruplex signature blocks in the nanopore in the presence of thrombin, compared with TBA alone without thrombin. This sensitivity suggested the possibility of developing aptamer-based nanopore sensors for protein detection. Actually this new generation of single molecule biosensors has a vast breadth of applications because the selection of aptamer species is unlimited.

3.3.1 Progress in Construction of Nanopore Sensors

For real-time applications in the future, such nanopore biosensors should be portable for independent storage and free transportation, while they should be independent, pluggable components that connect to other systems. The device should work in the microarray platform for the high-throughput screenings, in which each element contains a single responsive pore. To create a robust device that functions with single pores, however, one of significant challenges is to improve the fragility of the lipid bilayer membrane in which the protein pore is embedded. There have been several strategies to stabilize the lipid membrane. For example, reducing the size of the aperture over which the lipid bilayer is formed may help to promote membrane stability. Through micro-fabrication, the aperture in silicon can be made as small as several micrometers in diameter [25, 92]. The lipid bilayers can also be covalently tethered to a solid surface to form a solid-supported bilayer [17, 68]. Peterson et al. has reported a sandwiched lipid membrane generated by the painted method [83], in which the bilayer was formed on a pre-cast gel slab and covered with another gel slab for double support [18]. With the sandwich structure, Schmidt et al. improved the lipid membrane stability using an UV-triggered hydrogel in place of agarose [55], and demonstrated an efficient system that conjugates the headgroups of lipids in the membrane to the polyethylene glycol-based hydrogel to achieve extended lifetimes and resistance to mechanical perturbation [75]. The White group and Cremer group generated a long-lived suspended lipid bilayer over a 100 nm glass nanopore (GNP) membrane, and incorporated single protein pores in this membrane as a sensor [120]. This GNP-based bilayer can be fabricated using materials and instruments that are commonly available in most laboratories and do not require microlithographic fabrication techniques. While all of the methods described above form bilayers in the aqueous phase, the Bayley group and Wallace group have pursued protein pore research and applications on a bilayer that is formed in the oil phase, called the droplet interface bilayer (DIB) [6]. The DIB demonstrates superior stability, and many droplets can be assembled into a responsive DIB network that functions as a microbattery [45] and a molecular device that simulates the functions of semiconductor circuits, such as half-wave rectification and full-wave rectification [74].

3.3.2 Portable, Durable, Modular Ion Channel Chip

We have developed a long-lived, portable, modular nanopore chips [98] (Fig. 3.5). The chip is composed of a durable lipid membrane that is sandwiched between two agarose layers, with a single protein pore embedded in the membrane functioning as the sensor element for both biomedical detection and membrane protein research [98]. The ion channel-integrated chip device is highly stable with a lifetime over 1 week [59], compared with several hours for conventional planar bilayers. The ion channel chip is also highly portable. It can be repeatedly disconnected and reconnected to any device, such as an electrophysiology recording instrument. The disconnected chip is capable of independent storage, and can be transported from place to place. The unique portability and durability of the ion channel chip make it an independent, pluggable, modular biosensor that is useful for real-time sensing applications. For example, the chip containing a single pore formed by M113R/T145R, an engineered a-hemolysin that can discriminate various phosphate compounds [15], can detect nano-molar concentrations of the second messenger inositol 1,4,5-triphosphate (IP3) in a mixture with high concentration of ATP (Fig. 3.5c).

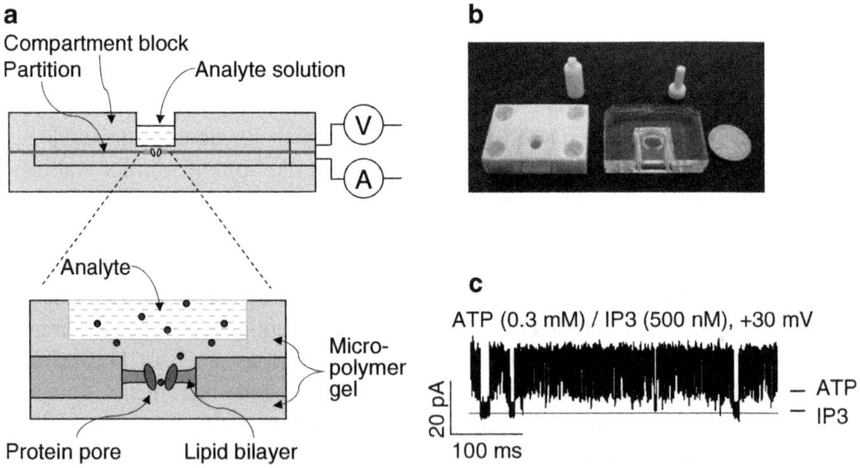

Fig. 3.5 Fabrication, prototype and application of modular ion channel chip. (**a**) Chip assembling. The analyte can be added from the sample cell on the back of either compartment and delivered to the sensor element through agarose layered between the sample cell and the membrane. (**b**) A device prototype. (**c**) Signature blocks by the second messenger IP3 (500 nM) on the chip in a simulated intracellular conditions: 150 mM KCl, 2 mM ATP, 2.3 mM $MgCl_2$, 10 mM Tris, 0.3 mM ATP, pH 7.4 [98]. Reprinted with permission from *Anal.Chem.* 79:2207–2213 (2009)

3.3.3 Significance and Impacts

Any membrane protein can be used in the ion channel chip for single molecule detection, such as screening of enzymes or detection of glucose or neural transmitters. The areas of research where the chip will be of the greatest use include DNA and protein detection in genomics and proteomics, and screening for membrane protein-targeting pharmaceuticals. The chip could also be a valuable tool for investigating the long-term dynamics of membrane proteins, including ion channels, which otherwise are difficult to study by traditional electrophysiology techniques. After improvement, the volume of sample cell on this modular device can be reduced to smaller than 1 μL to significantly reduce reagent consumption, and the chip could be coupled with a micro-fluidic system to rapidly control the sample exchange. This speculation is possible because the protein pore-incorporated lipid membrane has been shown to form across a microfluidic channel [76]. As micropatterned hydrogels have been created [66, 80], the chip is speculated to be able to provide a micro-array in future for high throughput screening with each array element containing a single stochastic sensor. The possibility of indexing each individual ion channel on the array is supported by the ability to quickly transfer ion channels into membrane for single channel assay [44].

3.4 Detection of Single Protein Molecules with Aptamer-Encoded Nanopore

3.4.1 Detection of Binding, but not Translocation

Protein detection is an essential task in most fields including medical diagnosis and biodefense. In the nanopore approach, most protein targets are larger than the size of proteins pores [104]. Such limitations would prevent the pore from accommodating bulky protein targets, therefore in the current stage, the protein pores may not be universally suitable for the detection of large protein target. Alternatively, the stable synthetic nanopores with flexible pore sizes can circumvent these limitations, but almost all past studies on synthetic nanopores have focused on the measurement of single DNA or protein molecule translocation. With the translocation-based detection, any molecules smaller than the nanopore may generate indistinguishable transient pore blocks. This low specificity limits the ability to identify and isolate molecular processes [78]. This may be overcome by coating a layer of probing molecules to the nanopore [50, 69, 87, 103, 117] to force specific targeting. For example, DNA transport was enhanced in nanopores modified with an complementary oligonucleotide probe [50, 69]; a biosensing nanopore coated with antibodies can detect target proteins by measuring the time spent for fully blocking the pore conductance [103]; and recently, a significant step was made toward functionalization at selected points in a solid nanopore [87]. These methods, however, have thus far failed to detect the binding of individual target molecules, and it remains impossible to distinguish between blocks produced by binding and those generated by translocation. Furthermore, high fabrication costs of synthetic nanopores limit studies performed by laboratories with broader research goals.

3.4.2 Integration of Aptamer with Synthetic Nanopores

The integration of sophisticated aptamers and the recent glass nanopore technique [28], led to a novel aptamer-encoded nanopore [22] that may pave an avenue to overcome these challenges. Because aptamers are much smaller than their targets, when they are bound by the target, the target signal is pronounced, allowing one to identify single molecules that are sequentially captured by the immobilized aptamer. Therefore, aptamers outperform antibodies with regard to single-molecule detection in nanopores. With aptamer-encoded nanopore, it become possible to identify the single-molecule binding and release processes in a nanopore, and to diminish the effect of translocation events by adjusting nanopore dimensions and regulating binding kinetics. The benefits of these capabilities include the possibility of using one nanopore to detect multiple targets.

3.4.3 Fabrication and Properties of the Glass Nanopore

The glass nanopore probe used in this research can be fabricated by externally penetrating an enclosed nanocavity in the terminal end of a capillary pipette [28]. The pipette tip is first sealed with a melting process so that a wineglass-shaped nanocavity is formed inside the terminal (Fig. 3.6a left). The tip then is exposed to hydrofluoric acid/ammonium fluoride for external etching, and monitored by the ionic current between solutions inside and outside the pipette (Fig. 3.6a middle). A nanopore is formed once the enclosed nanocavity is perforated (Fig. 3.6a right). The pipette tip is then transferred to an etchant-free solution to determine the pore conductance. The tip can be repeatedly etched until the desired conductance is achieved. Because the nanocavity geometrical profile is uniform, the corresponding pore size can be evaluated from the conductance according to the pore size-conductance correlation [28].

This nanopore can be fashioned to accommodate almost any molecular complex under investigation, and features several distinguishable benefits: ease of fabrication by virtually any laboratory at low cost; precise manipulatable pore size, from one to several hundred nanometers; experimentally verified ability to capture single molecules and perform stochastic sensing; reduced electrical noise; bio-friendly

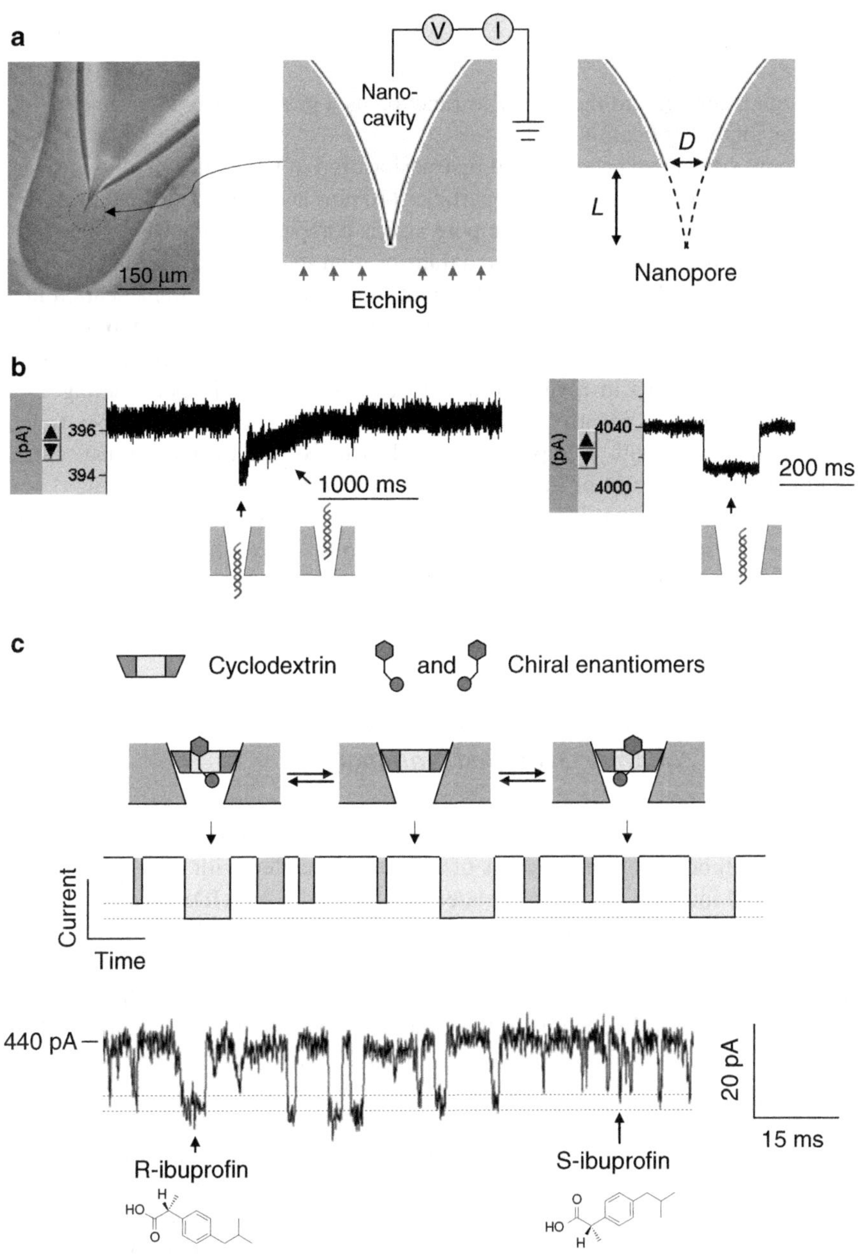

Fig. 3.6 Glass nanopore-terminated probe and single molecule manipulation [28]. (**a**) The glass nanopore is fabricated by sealing the micropipette terminal to enclose a nanocavity, followed by external etching the glass terminal with electrical monitoring to perforate the nanocavity with controllable pore size. (**b**) Current blocks showing translocation of 1 kbp dsDNA through a 2-nm nanopore (*left*) and a 7-nm pore (*right*) in 1 M NaCl (+100 mV). When the pore size is comparable to dsDNA, the DNA translocation speed is slowed down and translocation steps can be revealed from the block type (*left*), which is different from that for DNA translocation in a wider pore (*right*). (**c**) Single molecule discrimination of chiral enantiomers by the cyclodextrin (1.5 nm) trapped in the 1 nm nanopore. The interaction of chiral compounds with cyclodextrin can be separated from their block durations and current amplitudes. The current trace showed the binding of individual enantiomers of ibuprofen to the trapped β-cyclodextrin in the nanopore. The solution contained the mixture of 100 μM R-(−)-ibuprofen and 100 μM S-(+)-ibuprofen. The current block levels by R- and S-ibuprofen were marked with *dash lines*. Reprinted with permission from *Anal. Chem.* 81:80–86 (2009)

surface engineering; and the ability to function as a probe platform for in situ and high throughput applications.

The pore sizes have been verified by translocation of molecules of known sizes, including dsDNA (2 nm), gold nanoparticles (10 nm) and ring-shaped cyclodextrin (1.5 nm) [28]. Particularly when the pore size is comparable to dsDNA, the DNA translocation speed is slowed down and translocation steps can be revealed from the block type (Fig. 3.6b left), which is different from that for DNA translocation in a wider pore (Fig. 3.6b right). Strikingly, the glass nanopore can be fabricated at 1 nm, such that the pore can trap a single cyclodextrin in the lumen. The trapped cyclodextrin functions in a similar way to that in the protein pore: acting as a molecular adapter to identify small chemicals in the mixture [33]. We found the glass nanopore with a trapped β-cyclodextrin is able to discriminate chiral enantiomers on the basis of their characteristic block signatures [28] (Fig. 3.6c). This finding supports that it is possible to observe the signature signal for single molecule binding in the nanopore, a phenomenon that may apply to the aptamer-encoded nanopore for protein detection.

3.4.4 Capture of Single IgE Molecules in the Aptamer-Modified Nanopore

Immunoglobulin E (IgE) is consistently a representative biomarker for clinical detection methods. Abnormal levels of IgE are associated with allergy-mediated disorders and immune deficiency-related diseases, such as AIDS. We first developed aptamer-encoded nanopore for IgE [22]. To test the single molecule detection ability of aptamer-encoded nanopore, the aptamer for IgE was immobilized on the inner surface of nanopore, then low concentration IgE was applied to the nanopore from the wide opening of the pipette, to observe the block signal. Unlike the typical rectangle shape blocks, we can observe a series of stepwise current blocks (Fig. 3.7a). After confirmation with a series control tests, we conclude that the stepwise blocks are associated with single IgE molecules that sequentially bind to the immobilized aptamers in the nanopore, one molecule per block level. It is clear that the stepwise blocks are also distinguished from those produced by protein molecule translocation, as protein molecules traversing a nanopore typically generate short-lived blocks with a duration ranging from 10^2 μs to 10^2 ms [93, 97, 111].

We observed a number of stepwise current blocks, but rarely observed characteristic current increase for releasing IgE from immobilized aptamers [22]. This suggests a long binding duration for IgE•aptamer in hours, which is equivalent to a dissociation rate constant of ~10^{-5} s^{-1}. The slow dissociation rate in the glass nanopore suggests that the confinement in the nanopore could enhance the bonding strength between the target and its aptamer. The apparent association rate constant is evaluated by analyzing the frequency of the stepwise block occurrence was calculated as 1.9×10^5 M^{-1} s^{-1}.

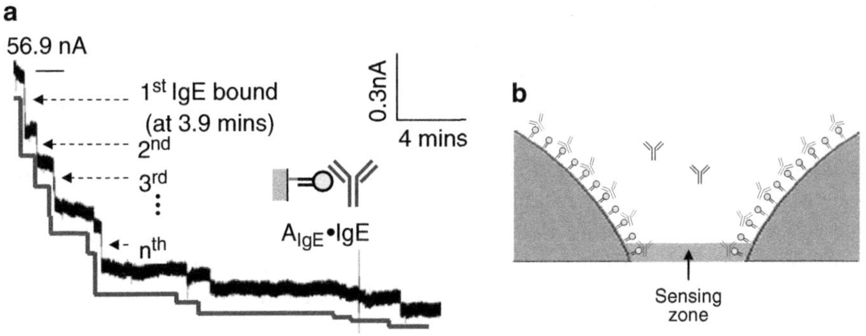

Fig. 3.7 Capture of single IgE molecules by immobilized aptamers in the nanopore. The sequence of IgE aptamer is 5′-GGGGCACGTTTATCCGTCCCTCCTAGTGGCGTGCCCC [122], the voltage was given from the electrode in the pipette and the external bath was grounded. (**a**) Current in a 63-nm A_{IgE}-encoded nanopore at +100 mV with 5 nM IgE in the pipette. Stepwise discrete current blocks are due to individual IgE molecule binding to the immobilized aptamers (**b**) Sensing zone at the narrow opening of the nanopore [22]. Reprinted with permission from *Anal.Chem.* 81:6649–6655 (2009)

The electroosmotic flow can also be identified based on the voltage polarity-dependent blocks, i.e. stepwise block can only observed at positive voltage, but not negative voltage. The voltage polarity switches the direction of the electroosmotic flow, thus enhancing or weakening the binding frequency [28].

According to the number of IgE blocks in a recording, we infer that there is a sensing zone in the nanopore (Fig. 3.7b) [22]. Only single IgE molecules captured by aptamers in the sensing zone generated pronounced signal blocks, whereas molecules outside of the sensing zone yielded signal blocks that are too weak to be detected. Based on the wineglass-shaped profile, we approximated the thickness of sensing zone.

Protein detection with aptamer-encoded nanopore is highly selective, as characteristic stepwise blocks cannot be found if IgG is added to the IgE aptamer-modified nanopore or if a mutant IgE aptamer that has a single nucleotide removed from the recognizing domain (loop). The result of the mutant aptamer further suggests the nanopore's ability in probing aptamer-target interaction [122].

3.4.5 Detection of Bioterrorist Agent Ricin with RNA Aptamer in Nanopore

The aptamer-encoded nanopore is highly programmable and can be versatile to any aptamers. For example, besides DNA, the RNA aptamers targeting high impact substances can use applied. The bioterrorist agent ricin from the castor bean plant is the third-most toxic substance, after plutonium and botulinum toxin, according to the US Environmental Protection Agency and the Center for Defense Information. Ricin has high potential for use as an agent of biological warfare, a weapon of mass destruction (WMD), and a terrorist weapon. Ricin protein consists of the A chain and B chain. The A chain has the toxic enzymic activity (an RNA N-glycosidase) of the protein and the B chain responsible for binding to cells and being taken up by them. The dimension of ricin A protein is ~6.1 nm × 2.4 nm [119] (Fig. 3.8a). The toxicity of ricin is activated only when the A chain meets with B chain. The RNA aptamer targeting ricin A protein has been created by Ellington and co-workers [39]. Recently the ricin A aptamer has been successfully integrated into the chip for ricin detection [65]. However, the detection requires fluorescent-labeling of the ricin protein.

By comparing, aptamer-encoded nanopore provides a novel, label-free method of detecting single ricin molecules (Fig. 3.8a) [22]. Similar to the IgE with its aptamer, the presence of low concentration of ricin can generate a series of stepwise blocks that can be attributed to the capture of single ricin molecules (Fig. 3.8b). Again, no discrete current increase observed for the release of ricin from its aptamer, suggested a long binding time for A_{ricin}•ricin, in concordance with the

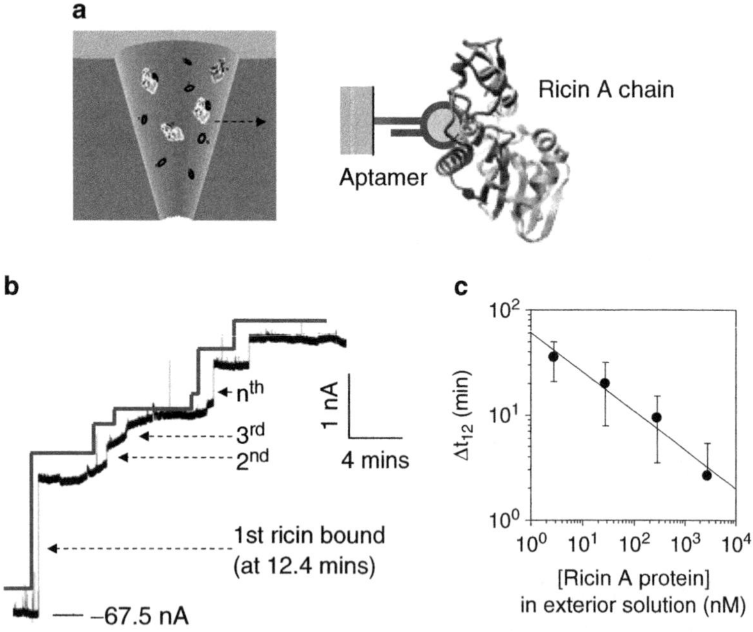

Fig. 3.8 Single-molecule detection of ricin A-chain protein in the external solution using an A_{ricin}-encoded nanopore. (**a**) A model showing the single ricin A-chain molecule binding to the RNA aptamer immobilized in the nanopore. (**b**) Current in a 56-nm A_{Ricin}-encoded nanopore at −100 mV, with 100 nM ricin A-chain protein in the external solution. (**c**) The detection time interval between the first and the second ricin block (t_1-t_2) in various concentrations of ricin A-chain protein [22]. Reprinted with permission from *Anal.Chem.* 81:6649–6655 (2009)

high affinity (7.4 nM) for A_{ricin}•ricin [39]. By determining the interval between the first and second block, $\Delta t_{12} = t_2 - t_1$, and correlating Δt_{12} with the ricin concentration (Fig. 3.8c), we obtained the calibration curve: Δt_{12} decreased from 35 to 2.6 min as the ricin concentration increased from 2.8 nM to 2.8 μM.

3.4.6 *Advantages and Perspectives*

The aptamer-encoded nanopore demonstrates a series of advantages over other synthetic nanopore systems. The first is the selective detection. The aptamer-encoded nanopore can distinguish between transient current blockades caused by non-specific molecules passing through the nanopore and much longer blocks resulting from the target binding. The digital signal produced by individual target molecules (stepwise blocks, Figs. 3.7 and 3.8) distinguishes them from the analog background signal. This is particularly useful in real-time detections, when the background current dynamically drifts or fluctuates with the environment, one is still able to identify discrete single molecule events, thus greatly enhancing the signal/noise ratio. The same important characters also include simple, rapid and label-free detection with high sensitivity.

3.5 Conclusions

The sophisticated aptamers can be integrated with the sensitive nanopore single molecule detectors, to explore life science problems that have great implications in biomedicine, biotechnology and biodefense. These problems range from dynamic structures of aptamers such as folding/unfolding processes triggered by the environmental factors, to the utilization of aptamers for ultrasensitive bio-detection, such as clinical biomarkers and bioterrorist materials (because of their importance to homeland security). The potential of aptamer-encoded nanopore sensors is broad. They can be applied to medical diagnoses, quantifying biomarkers in blood samples, and used for environmental monitoring, detecting pollution or contaminating materials in water. It depends on the aptamer development. If the aptamer is ready, one can use it for just about anything. Using aptamers is also advantageous because they are more durable than most protein receptors, resisting most denaturing and degrading conditions, including immobilization; yet, they are simpler to synthesize, modify, and immobilize using low-cost methods. The affinity and specificity of aptamer-target interactions can also be fine-tuned through rational design or molecular evolution. Having already created a nanopore detector for IgE and ricin, we now aims to devise a nanopore sensor that will use an aptamer to detect anthrax spores – well known as a bioterrorist threat. Scientists already have developed anthrax aptamers [14], paving the way for a nanopore sensor.

Acknowledgments Financial support from NSF CAREER (0546165), National Institutes of Health (GM079613), University of Missouri Research Board and Startup Fund. The author's work was conducted in a facility constructed with support from Research Facilities Improvement Program Grant C06-RR-016489-01 from the National Center for Research Resources, National Institutes of Health.

References

1. Alberti, P. & Mergny, J.L. (2003) DNA duplex-quadruplex exchange as the basis for a nanomolecular machine *Proceedings of the National Academy of Sciences of the United States of America* 100:1569–1573.
2. Arthanari, H. & Bolton, P.H. (2001) Functional and dysfunctional roles of quadruplex DNA in cells *Chemistry & Biology* 8:221–230.
3. Ashkenasy, N., Sanchez-Quesada, J., Bayley, H. & Ghadiri, M.R. (2005) Recognizing a single base in an individual DNA strand: A step toward DNA sequencing in nanopores *Angewandte Chemie-International Edition* 44:1401–1404.
4. Baskerville, S., Zapp, M. & Ellington, A.D. (1999) Anti-Rex aptamers as mimics of the Rex-binding element *Journal of Virology* 73:4962–4971.
5. Bayley, H. & Cremer, P.S. (2001) Stochastic sensors inspired by biology *Nature* 413:226–230.
6. Bayley, H. et al. (2008) Droplet interface bilayers *Mol. Biosyst.* 4:1191–1208.
7. Bayley, H. & Jayasinghe, L. (2004) Functional engineered channels and pores - (Review) *Molecular Membrane Biology* 21:209–220.
8. Bezrukov, S.M., Vodyanoy, I. & Parsegian, V.A. (1994) Counting polymers moving through a single ion channel *Nature* 370:279–281.
9. Bock, C. et al. (2004) Photoaptamer arrays applied to multiplexed proteomic analysis *Proteomics* 4:609–618.
10. Bock, L.C., Griffin, L.C., Latham, J.A., Vermaas, E.H. & Toole, J.J. (1992) Selection of Single-Stranded-DNA Molecules That Bind and Inhibit Human Thrombin *Nature* 355:564–566.
11. Braha, O., Gu, L.Q., Zhou, L., Lu, X.F., Cheley, S. & Bayley, H. (2000) Simultaneous stochastic sensing of divalent metal ions *Nature Biotechnology* 18:1005–1007.
12. Breaker, R.R. (2004) Natural and engineered nucleic acids as tools to explore biology *Nature* 432:838–845.
13. Brody, E.N., Willis, M.C., Smith, J.D., Jayasena, S., Zichi, D. & Gold, L. (1999) The use of aptamers in large arrays for molecular diagnostics *Molecular Diagnosis* 4:381–388.
14. Bruno, J.G. & Kiel, J.L. (1999) In vitro selection of DNA aptamers to anthrax spores with electrochemiluminescence detection *Biosensors & Bioelectronics* 14:457–464.
15. Cheley, S., Gu, L.Q. & Bayley, H. (2002) Stochastic sensing of nanomolar inositol 1,4,5-trisphosphate with an engineered pore *Chem. Biol.* 9:829–838.
16. Chou, S.H., Chin, K.H. & Wang, A.H.J. (2005) DNA aptamers as potential anti-HIV agents *Trends in Biochemical Sciences* 30:231–234.
17. Cornell, B.A. et al. (1997) A biosensor that uses ion-channel switches *Nature* 387:580–583.
18. Costello, R.F., Peterson, I.P., Heptinstall, J., Byrne, N.G. & Miller, L.S. (1998) A robust gel-bilayer channel biosensor *Advanced Materials for Optics and Electronics* 8:47–52.
19. Davis, J.T. & Spada, G.P. (2007) Supramolecular architectures generated by self-assembly of guanosine derivatives *Chemical Society Reviews* 36:296–313.
20. de Soultrait, V.R., Lozach, P.Y., Altmeyer, R., Tarrago-Litvak, L., Litvak, S. & Andreola, M. L. (2002) DNA aptamers derived from HIV-1 RNase H inhibitors are strong anti-integrase agents *Journal of Molecular Biology* 324:195–203.

21. Deamer, D.W. & Branton, D. (2002) Characterization of nucleic acids by nanopore analysis *Acc. Chem. Res.* 35:817–825.
22. Ding, S., Gao, C. & Gu, L.Q. (2009) Capturing Single Molecules of Immunoglobulin and Ricin with an Aptamer-Encoded Glass Nanopore *Anal. Chem.* 81:6649–6655.
23. Ellington, A.D. & Szostak, J.W. (1990) Invitro Selection of Rna Molecules That Bind Specific Ligands *Nature* 346:818–822.
24. Famulok, M. & Szostak, J.W. (1992) Stereospecific Recognition of Tryptophan Agarose by Invitro Selected Rna *Journal of the American Chemical Society* 114:3990–3991.
25. Fertig, N., Blick, R.H. & Behrends, J.C. (2002) Whole cell patch clamp recording performed on a planar glass chip *Biophysical Journal* 82:3056–3062.
26. Fletcher, T.M., Sun, D.K., Salazar, M. & Hurley, L.H. (1998) Effect of DNA secondary structure on human telomerase activity *Biochemistry* 37:5536–5541.
27. Fologea, D., Gershow, M., Ledden, B., McNabb, D.S., Golovchenko, J.A. & Li, J. (2005) Detecting single stranded DNA with a solid state nanopore *Nano Lett.* 5:1905–1909.
28. Gao, C., Ding, S., Tan, Q. & Gu, L.Q. (2009) Method of creating a nanopore-terminated probe for single-molecule enantiomer discrimination *Anal. Chem.* 81:80–86.
29. Geiger, A., Burgstaller, P., vonderEltz, H., Roeder, A. & Famulok, M. (1996) RNA aptamers that bind L-arginine with sub-micromolar dissociation constants and high enantioselectivity *Nucleic Acids Research* 24:1029–1036.
30. German, I., Buchanan, D.D. & Kennedy, R.T. (1998) Aptamers as ligands in affinity probe capillary electrophoresis *Analytical Chemistry* 70:4540–4545.
31. Gokulrangan, G., Unruh, J.R., Holub, D.F., Ingram, B., Johnson, C.K. & Wilson, G.S. (2005) DNA aptamer-based bioanalysis of IgE by fluorescence anisotropy *Analytical Chemistry* 77:1963–1970.
32. Gowan, S.M. et al. (2002) A G-quadruplex-interactive potent small-molecule inhibitor of telomerase exhibiting in vitro and in vivo antitumor activity *Molecular Pharmacology* 61:1154–1162.
33. Gu, L.Q., Braha, O., Conlan, S., Cheley, S. & Bayley, H. (1999) Stochastic sensing of organic analytes by a pore-forming protein containing a molecular adapter *Nature* 398:686–690.
34. Gu, L.Q., Cheley, S. & Bayley, H. (2001) Capture of a single molecule in a nanocavity *Science* 291:636–640.
35. Gu, L.Q., Cheley, S. & Bayley, H. (2003) Electroosmotic enhancement of the binding of a neutral molecule to a transmembrane pore *Proc. Natl. Acad. Sci. U. S. A.* 100:15498–15503.
36. Gu, L.Q., Cheley, S. & Bayley, H. (2005) Dissection of protein-adapter interaction by single-channel recording *Biophysical Journal* 88:657A.
37. Gu, L.Q. et al. (2000) Reversal of charge selectivity in transmembrane protein pores by using noncovalent molecular adapters *Proc. Natl. Acad. Sci. U. S. A.* 97:3959–3964.
38. Hermann, T. & Patel, D.J. (2000) Biochemistry - Adaptive recognition by nucleic acid aptamers *Science* 287:820–825.
39. Hesselberth, J.R., Miller, D., Robertus, J. & Ellington, A.D. (2000) In vitro selection of RNA molecules that inhibit the activity of ricin A-chain *Journal of Biological Chemistry* 275:4937–4942.
40. Heyduk, E. & Heyduk, T. (2005) Nucleic acid-based fluorescence sensors for detecting proteins *Analytical Chemistry* 77:1147–1156.
41. Heyduk, T. & Heyduk, E. (2002) Molecular beacons for detecting DNA binding proteins *Nat. Biotechnol.* 20:171–176.
42. Ho, C. et al. (2005) Electrolytic transport through a synthetic nanometer-diameter pore *Proc. Natl. Acad. Sci. U. S. A.* 102:10445–10450.
43. Ho, H.A. & Leclerc, M. (2004) Optical sensors based on hybrid aptamer/conjugated polymer complexes *Journal of the American Chemical Society* 126:1384–1387.
44. Holden, M.A., Jayasinghe, L., Daltrop, O., Mason, A. & Bayley, H. (2006) Direct transfer of membrane proteins from bacteria to planar bilayers for rapid screening by single-channel recording *Nature Chemical Biology* 2:314–318.

45. Holden, M.A., Needham, D. & Bayley, H. (2007) Functional bionetworks from nanoliter water droplets *J. AM. CHEM. SOC.* 129:8650–8655.
46. Howorka, S. & Bayley, H. (2002) Probing distance and electrical potential within a protein pore with tethered DNA *Biophysical Journal* 83:3202–3210.
47. Howorka, S., Cheley, S. & Bayley, H. (2001) Sequence-specific detection of individual DNA strands using engineered nanopores *Nature Biotechnology* 19:636–639.
48. Huang, C.C., Cao, Z., Chang, H.T. & Tan, W. (2004) Protein-protein interaction studies based on molecular aptamers by affinity capillary electrophoresis *Anal. Chem.* 76:6973–6981.
49. Hurley, L.H. (2002) DNA and its associated processes as targets for cancer therapy *Nature Reviews Cancer* 2:188–200.
50. Iqbal, S.M., Akin, D. & Bashir, R. (2007) Solid-state nanopore channels with DNA selectivity *Nature Nanotechnology* 2:243–248.
51. Ito, T., Sun, L. & Crooks, R.M. (2003) Simultaneous determination of the size and surface charge of individual nanoparticles using a carbon nanotube-based Coulter counter *Anal. Chem.* 75:2399–2406.
52. Jenison, R.D., Gill, S.C., Pardi, A. & Polisky, B. (1994) High-Resolution Molecular Discrimination by Rna *Science* 263:1425–1429.
53. Jenison, R.D., Jennings, S.D., Walker, D.W., Bargatze, R.F. & Parma, D. (1998) Oligonucleotide inhibitors of P-selectin-dependent neutrophil-platelet adhesion *Antisense & Nucleic Acid Drug Development* 8:265–279.
54. Jeon, S.H., Kayhan, B., Ben-Yedidia, T. & Arnon, R. (2004) A DNA aptamer prevents influenza infection by blocking the receptor binding region of the viral hemagglutinin *Journal of Biological Chemistry* 279:48410–48419.
55. Jeon, T.J., Malmstadt, N. & Schmidt, J.J. (2006) Hydrogel-encapsulated lipid membranes *Journal of the American Chemical Society* 128:42–43.
56. Jing, N. et al. (2000) Stability-activity relationships of a family of G-tetrad forming oligonucleotides as potent HIV inhibitors - A basis for anti-HIV drug design *Journal of Biological Chemistry* 275:3421–3430.
57. Jing, N.J. & Hogan, M.E. (1998) Structure-activity of tetrad-forming oligonucleotides as a potent anti-HIV therapeutic drug *Journal of Biological Chemistry* 273:34992–34999.
58. Jing, N.J., Rando, R.F., Pommier, Y. & Hogan, M.E. (1997) Ion selective folding of loop domains in a potent anti-HIV oligonucleotide *Biochemistry* 36:12498–12505.
59. Kang, X.F., Cheley, S., Rice-Ficht, A.C. & Bayley, H. (2007) A storable encapsulated bilayer chip containing a single protein nanopore *Journal of the American Chemical Society* 129:4701–4705.
60. Kankia, B.I. & Marky, L.A. (2001) Folding of the thrombin aptamer into a G-quadruplex with Sr2+: Stability, heat, and hydration *Journal of the American Chemical Society* 123:10799–10804.
61. Karhanek, M., Kemp, J.T., Pourmand, N., Davis, R.W. & Webb, C.D. (2005) Single DNA molecule detection using nanopipettes and nanoparticles *Nano Lett.* 5:403–407.
62. Kasianowicz, J.J., Brandin, E., Branton, D. & Deamer, D.W. (1996) Characterization of individual polynucleotide molecules using a membrane channel *Proc. Natl. Acad. Sci. U. S. A.* 93:13770–13773.
63. Kasianowicz, J.J., Burden, D.L., Han, L.C., Cheley, S. & Bayley, H. (1999) Genetically engineered metal ion binding sites on the outside of a channel's transmembrane beta-barrel *Biophysical Journal* 76:837–845.
64. Kerwin, S.M. (2000) G-quadruplex DNA as a target for drug design *Current Pharmaceutical Design* 6:441–471.
65. Kirby, R. et al. (2004) Aptamer-based sensor arrays for the detection and quantitation of proteins *Anal. Chem.* 76:4066–4075.
66. Klajn, R. et al. (2004) Multicolour micropatterning of thin films of dry gels *Nature Materials* 3:729–735.

67. Klug, S.J., Huttenhofer, A. & Famulok, M. (1999) In vitro selection of RNA aptamers that bind special elongation factor SelB, a protein with multiple RNA-binding sites, reveals one major interaction domain at the carboxyl terminus *Rna-a Publication of the Rna Society* 5:1180–1190.
68. Knoll, W. et al. (2000) Functional tethered lipid bilayers *J Biotechnol* 74:137–58.
69. Kohli, P., Harrell, C.C., Cao, Z.H., Gasparac, R., Tan, W.H. & Martin, C.R. (2004) DNA-functionalized nanotube membranes with single-base mismatch selectivity *Science* 305:984–986.
70. Li, J., Gershow, M., Stein, D., Brandin, E. & Golovchenko, J.A. (2003) DNA molecules and configurations in a solid-state nanopore microscope *Nat. Mater.* 2:611–615.
71. Li, J., Stein, D., McMullan, C., Branton, D., Aziz, M.J. & Golovchenko, J.A. (2001) Ion-beam sculpting at nanometre length scales *Nature* 412:166–169.
72. Li, J.W.J. & Tan, W.H. (2002) A single DNA molecule nanomotor *Nano Letters* 2:315–318.
73. Luchian, T., Shin, S.H. & Bayley, H. (2003) Kinetics of a three-step reaction observed at the single-molecule level *Angewandte Chemie-International Edition* 42:1925–1929.
74. Maglia, G. et al. (2009) Droplet networks with incorporated protein diodes show collective properties *Nat. Nanotechnol.* 4:437–440.
75. Malmstadt, N., Jeon, T.J. & Schmidt, J.J. (2008) Long-lived planar lipid bilayer membranes anchored to an in situ polymerized hydrogel *Adv Mater* 20:84–89.
76. Malmstadt, N., Nash, M.A., Purnell, R.F. & Schmidt, J.J. (2006) Automated formation of lipid-bilayer membranes in a microfluidic device *Nano Letters* 6:1961–1965.
77. Marathias, V.M. & Bolton, P.H. (2000) Structures of the potassium-saturated, 2 : 1, and intermediate, 1 : 1, forms of a quadruplex DNA *Nucleic Acids Research* 28:1969–1977.
78. Martin, C.R. & Siwy, Z.S. (2007) Learning nature's way: Biosensing with synthetic nanopores *Science* 317:331–332.
79. Mathe, J., Visram, H., Viasnoff, V., Rabin, Y. & Meller, A. (2004) Nanopore unzipping of individual DNA hairpin molecules *Biophysical Journal* 87:3205–3212.
80. Mayer, M., Yang, J., Gitlin, I., Gracias, D.H. & Whitesides, G.M. (2004) Micropatterned agarose gels for stamping arrays of proteins and gradients of proteins *Proteomics* 4:2366–2376.
81. Meller, A. (2003) Dynamics of polynucleotide transport through nanometre-scale pores *Journal of Physics-Condensed Matter* 15:R581–R607.
82. Movileanu, L., Howorka, S., Braha, O. & Bayley, H. (2000) Detecting protein analytes that modulate transmembrane movement of a polymer chain within a single protein pore *Nature Biotechnology* 18:1091–1095.
83. Mueller, P., Rudin, D.O., Tien, H.T. & Wescott, W.C. (1962) Reconstitution of cell membrane structure in vitro and its transformation into an excitable system *Nature* 194:979–80.
84. Nakane, J., Wiggin, M. & Marziali, A. (2004) A nanosensor for transmembrane capture and identification of single nucleic Acid molecules *Biophys. J.* 87:615–621.
85. Neidle, S. & Parkinson, G. (2002) Telomere maintenance as a target for anticancer drug discovery *Nature Reviews Drug Discovery* 1:383–393.
86. Neidle, S. & Parkinson, G.N. (2003) The structure of telomeric DNA *Current Opinion in Structural Biology* 13:275–283.
87. Nilsson, J., Lee, J.R.I., Ratto, T.V. & tant, S.E. (2006) Localized functionalization of single nanopores *Adv Mater* 18:427–431.
88. Padmanabhan, K., Padmanabhan, K.P., Ferrara, J.D., Sadler, J.E. & Tulinsky, A. (1993) The Structure of Alpha-Thrombin Inhibited by a 15-Mer Single-Stranded-DNA Aptamer *Journal of Biological Chemistry* 268:17651–17654.
89. Parkinson, G.N., Lee, M.P.H. & Neidle, S. (2002) Crystal structure of parallel quadruplexes from human telomeric DNA *Nature* 417:876–880.
90. Phan, A.T., Kuryavyi, V., Ma, J.B., Faure, A., Andreola, M.L. & Patel, D.J. (2005) An interlocked dimeric parallel-stranded DNA quadruplex: A potent inhibitor of HIV-1 integrase *Proceedings of the National Academy of Sciences of the United States of America* 102:634–639.

91. Rajendran, M. & Ellington, A.D. (2002) Selecting nucleic acids for biosensor applications *Combinatorial Chemistry & High Throughput Screening* 5:263–270.
92. Sackmann, E. (1996) Supported membranes: Scientific and practical applications *Science* 271:43–48.
93. Saleh, O.A. & Sohn, L.L. (2003) Direct detection of antibody-antigen binding using an on-chip artificial pore *Proc. Natl. Acad. Sci. U. S. A.* 100:820–824.
94. Schurer, H. et al. (2001) Aptamers that bind to the antibiotic moenomycin A *Bioorganic & Medicinal Chemistry* 9:2557–2563.
95. Sen, D. & Gilbert, W. (1988) Formation of parallel four-stranded complexes by guanine-rich motifs in DNA and its implications for meiosis *Nature* 334:364–366.
96. Sen, D. & Gilbert, W. (1990) A sodium-potassium switch in the formation of four-stranded G4-DNA *Nature* 344:410–414.
97. Sexton, L.T., Horne, L.P., Sherrill, S.A., Bishop, G.W., Baker, L.A. & Martin, C.R. (2007) Resistive-pulse studies of proteins and protein/antibody complexes using a conical nanotube sensor *Journal of the American Chemical Society* 129:13144–13152.
98. Shim, J.W. & Gu, L.Q. (2007) Stochastic sensing on a modular chip containing a single-ion channel *Anal. Chem.* 79:2207–2213.
99. Shim, J.W. & Gu, L.Q. (2008) Encapsulating a single G-quadruplex aptamer in a protein nanocavity *J Phys Chem B* 112:8354–8360.
100. Shim, J.W., Tan, Q. & Gu, L.Q. (2009) Single-molecule detection of folding and unfolding of a single G-quadruplex aptamer in a nanopore nanocavity *Nucleic Acids Research* 37:972–982.
101. Simonsson, T. (2001) G-quadruplex DNA structures - Variations on a theme *Biological Chemistry* 382:621–628.
102. Siwy, Z. et al. (2002) Rectification and voltage gating of ion currents in a nanofabricated pore *Europhysics Letters* 60:349–355.
103. Siwy, Z., Trofin, L., Kohli, P., Baker, L.A., Trautmann, C. & Martin, C.R. (2005) Protein biosensors based on biofunctionalized conical gold nanotubes *Journal of the American Chemical Society* 127:5000–5001.
104. Song, L.Z., Hobaugh, M.R., Shustak, C., Cheley, S., Bayley, H. & Gouaux, J.E. (1996) Structure of staphylococcal alpha-hemolysin, a heptameric transmembrane pore *Science* 274:1859–1866.
105. Stadtherr, K., Wolf, H. & Lindner, P. (2005) An aptamer-based protein biochip *Analytical Chemistry* 77:3437–3443.
106. Sun, L.Q., Cairns, M.J., Saravolac, E.G., Baker, A. & Gerlach, W.L. (2000) Catalytic nucleic acids: From lab to applications *Pharmacological Reviews* 52:325–347.
107. Tang, Z.W. et al. (2007) Selection of aptamers for molecular recognition and characterization of cancer cells *Anal. Chem.* 79:4900–4907.
108. Tereshko, V., Skripkin, E. & Patel, D.J. (2003) Encapsulating streptomycin within a small 40-mer RNA *Chemistry & Biology* 10:175–187.
109. Tombelli, S., Minunni, M. & Mascini, M. (2005) Analytical applications of aptamers *Biosens. Bioelectron.* 20:2424–2434.
110. Tuerk, C. & Gold, L. (1990) Systematic Evolution of Ligands by Exponential Enrichment - Rna Ligands to Bacteriophage-T4 Dna-Polymerase *Science* 249:505–510.
111. Uram, J.D., Ke, K., Hunt, A.J. & Mayer, M. (2006) Label-free affinity assays by rapid detection of immune complexes in submicrometer pores *Angew. Chem. Int. Ed.* 45:2281–2285.
112. Vercoutere, W., Winters-Hilt, S., Olsen, H., Deamer, D., Haussler, D. & Akeson, M. (2001) Rapid discrimination among individual DNA hairpin molecules at single-nucleotide resolution using an ion channel *Nature Biotechnology* 19:248–252.
113. Wang, H., Dunning, J.E., Huang, A.P., Nyamwanda, J.A. & Branton, D. (2004) DNA heterogeneity and phosphorylation unveiled by single-molecule electrophoresis *Proc. Natl. Acad. Sci. U. S. A.* 101:13472–13477.

114. Wang, K.Y., Krawczyk, S.H., Bischofberger, N., Swaminathan, S. & Bolton, P.H. (1993) The Tertiary Structure of a DNA Aptamer Which Binds to and Inhibits Thrombin Determines Activity *Biochemistry* 32:11285–11292.
115. Wang, K.Y., McCurdy, S., Shea, R.G., Swaminathan, S. & Bolton, P.H. (1993) A DNA Aptamer Which Binds to and Inhibits Thrombin Exhibits a New Structural Motif for DNA *Biochemistry* 32:1899–1904.
116. Wang, K.Y., Swaminathan, S. & Bolton, P.H. (1994) Tertiary Structure Motif of Oxytricha Telomere DNA *Biochemistry* 33:7517–7527.
117. Wanunu, M. & Meller, A. (2007) Chemically modified solid-state nanopores *Nano Letters* 7:1580–1585.
118. Wen, J.D. & Gray, D.M. (2004) Selection of genomic sequences that bind tightly to Ff gene 5 protein: primer-free genomic SELEX *Nucleic Acids Research* 32.
119. Weston, S.A., Tucker, A.D., Thatcher, D.R., Derbyshire, D.J. & Pauptit, R.A. (1994) X-ray structure of recombinant ricin A-chain at 1.8 +à resolution *J. MOL. BIOL.* 244:410–422.
120. White, R.J. et al. (2007) Single ion-channel recordings using glass nanopore membranes *J. Am. Chem. Soc.* 129:11766–11775.
121. White, R.R., Sullenger, B.A. & Rusconi, C.P. (2000) Developing aptamers into therapeutics *Journal of Clinical Investigation* 106:929–934.
122. Wiegand, T.W., Williams, P.B., Dreskin, S.C., Jouvin, M.H., Kinet, J.P. & Tasset, D. (1996) High-affinity oligonucleotide ligands to human IgE inhibit binding to FcΓ receptor I *J. Immunol.* 157:221–230.
123. Wilson, C., Nix, J. & Szostak, J. (1998) Functional requirements for specific ligand recognition by a biotin-binding RNA pseudoknot *Biochemistry* 37:14410–14419.
124. Xiao, Y., Lubin, A.A., Heeger, A.J. & Plaxco, K.W. (2005) Label-free electronic detection of thrombin in blood serum by using an aptamer-based sensor *Angew. Chem. Int. Ed.* 44:5456–5459.
125. Zahler, A.M., Williamson, J.R., Cech, T.R. & Prescott, D.M. (1991) Inhibition of Telomerase by G-Quartet Dna Structures *Nature* 350:718–720.
126. Zimmerman, J.M. & Maher, L.J. (2002) In vivo selection of spectinomycin-binding RNAs *Nucleic Acids Research* 30:5425–5435.

Chapter 4
Membrane-Embedded Channel of Bacteriophage Phi29 DNA-Packaging Motor for Translocation and Sensing of Double-Stranded DNA

Farzin Haque and Peixuan Guo

Abstract Living systems contain a wide variety of nanomachines and highly-ordered macromolecular structures with diverse functions. All linear double-stranded (ds) DNA viruses package their genome into a pre-formed protein shell via an Adenosine-5'-triphosphate (ATP) driven nanomotor. The novel and ingenious design of the bacteriophage phi29 DNA packaging motor with an elegant and elaborate channel has inspired its applications in nanotechnology. The central component of the phi29 motor is a connector composed of 12 copies of the protein gp10, which forms a dodecamer channel acting as a path for the translocation of dsDNA, which enters the protein shell during replication and exits into the host cell during infection. The diameter of the channel is 3.6 nm at the narrow end and 6 nm at its wider end. The connector has been successfully inserted into a lipid bilayer and exhibited robust capability for extremely reliable and precise assessment of the transportation of ions and DNA as revealed by single channel conductance assays. In addition, the bacteriophage phi29 DNA packaging motor exercised a one-way traffic property for dsDNA translocation from N-terminal entrance to C-terminal exit with a valve mechanism in DNA-packaging. These findings have important implications since artificial membrane architecture for DNA packaging motor would allow detailed investigations into discrete mechanisms of motor operation as well as future avenues for therapeutic dsDNA packaging, sampling, and delivery.

Keywords Bacteriophage phi29 • DNA packaging • Nanomotor • Connector • Liposomes • Ion channel • Single channel conductance • Membrane channel • Viral assembly • Stoichiometry quantification • Nanostructure • Bionanotechnology • Nanobiotechnology • Nanomedicine

P. Guo (✉)
Nanobiomedical Center, Vontz Center for Molecular Studies, University of Cincinnati, ML #0508, 3125 Eden Avenue, Room 2308, Cincinnati, OH 45267, USA
e-mail: guop@purdue.edu; guopn@ucmail.uc.edu

S.M. Iqbal and R. Bashir (eds.), *Nanopores: Sensing and Fundamental Biological Interactions*, DOI 10.1007/978-1-4419-8252-0_4,

4.1 Introduction

Viral genomes are enclosed in a protein shell called a capsid. Following biosynthesis, viral structural proteins and the genome interact with each other to form a complete virion through a process referred to as DNA or RNA packaging [1, 2]. All linear double-stranded (ds) DNA or RNA viruses, including dsDNA bacteriophages [1], adenoviruses [3], poxviruses [4], human cytomegaloviruses (HCMV) [5], herpes simplex viruses (HSV) [6], and dsRNA bacteriophages [7], share a common feature in that their genome is packaged into a preformed procapsid. This energetically unfavorable process is accomplished by a packaging motor that harvests energy from ATP [1, 8–10]. One particularly attractive example of DNA-packaging machinery, in which both protein and RNA are involved, is the *Bacillus subtilis* bacteriophage phi29 DNA-packaging motor [9, 11–14].

The bacteriophage phi29 DNA-packaging motor is comprised of a 12-subunit gp10 connector [15, 16], six copies of ATP-binding DNA packaging RNA (pRNA) [11, 17, 18] and an ATPase protein gp16 [19, 20] to provide the chemical energy required for DNA translocation (Fig. 4.1a, b). These purified components can be combined in vitro and assembled into one of the most powerful nanomachines known to date [21]. A central channel, 6.0 nm at the wide end and 3.6 nm at the narrow end (Fig. 4.1c, d), is formed by three long α-helices from each of the 12 gp10 subunits of the connector. The cross-sectional area of the channel is 10 nm^2 at the narrow end and 28 nm^2 at its wider end. The mode of connector insertion and anchoring within the viral capsid is mediated via hydrophobic domains and protein–protein interactions. The portal proteins of many viruses are morphologically similar sharing the common feature of DNA translocation, but exhibit large variations in sequence homologies and molecular weights [22].

Development of a highly sensitive detection system is important in many areas, including, but not limited to pathogen identification, disease diagnosis, environmental monitoring, and national security. A wide range of processes such as transport of DNA, RNA, peptides, proteins, polymers, and pharmaceutical agents have been studied using electrophysiological measurements [23–27]. Using the well studied alpha-hemolysin (α-HL) channel from *Staphylococcus aureus*, the length of single-stranded (ss) DNA or RNA was determined [28]. Subsequently, DNA hairpin molecules have been used to decelerate the DNA translocation rate through the α-HL pore, thereby demonstrating that the nanopore device can distinguish single nucleotide polymorphisms [29]. A hybridization sensor developed by covalent attachment of individual DNA oligonucleotides within the lumen of the pore was used to identify single base pair differences [30]. Detection of base pair stacking and strand orientation within the pore have also been reported [31]. Other protein channels that have been investigated include alamethicin for detection of polyethylene glycol [32] and reengineered MspA protein from *M. smegmatis* for translocation of single-stranded DNA (ssDNA) [33].

Due to limitation of the pore size, to date, most of the single pore studies have focused on the translocation of ssDNA. While other studies have shown the

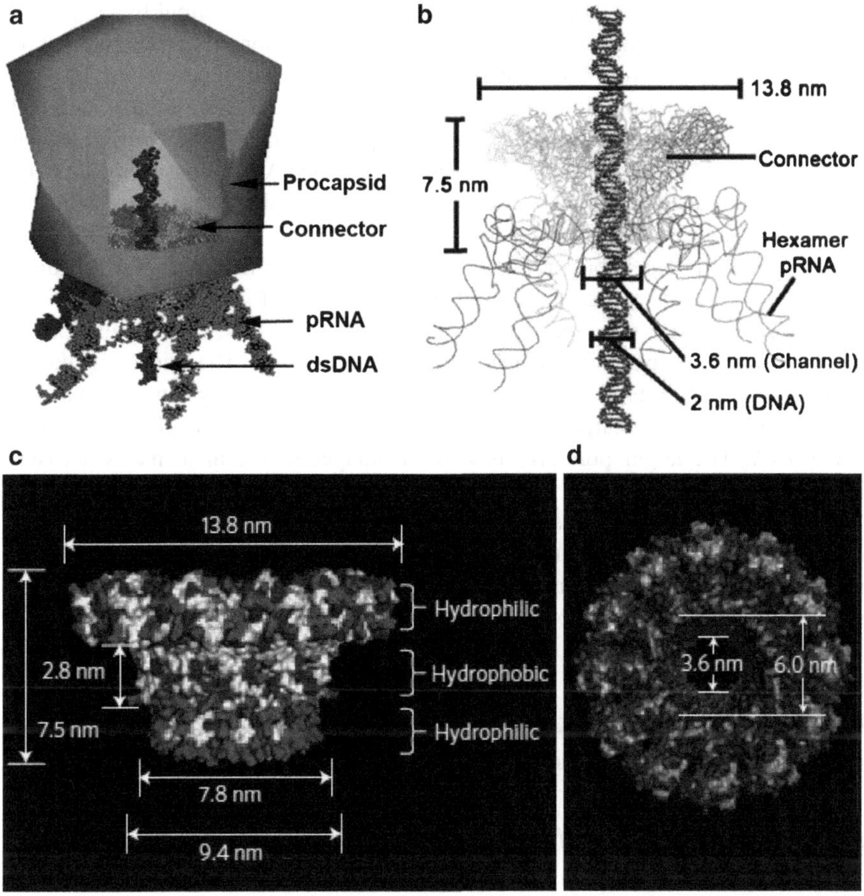

Fig. 4.1 (**a**) The phi29 DNA-packaging motor. (**b**) Illustration of the entire motor showing DNA translocation through the connector. (**c**) Side view of the phi29 connector showing the hydrophobic and hydrophilic layers [15, 16, 40, 41]. (**d**) Top view of the connector showing the diameter of the narrow part and wide part of the channel. Figures reproduced with permissions from: (**a**) Ref. [44], © American Chemical Society; (**b**–**d**) Ref. [42], © Nature Publishing Group

evidence of dsDNA transport [34–36], they have limited sensing applications due to voltage gating and signal fluctuations. For this reason, some researchers have switched to fabricating synthetic pores for potential use in dsDNA sequencing [37–39]. However, generating uniform pore sizes in synthetic nanopores reproducibly within a few nanometers is very challenging and it can also be difficult to reengineer specific pore modifications or conjugations. As a result, the search for alternate nanopores is still ongoing.

In this chapter, we will discuss and demonstrate various utilities of the lipid-embedded channel of phi29 DNA packaging motor. Explicit engineering of the phi29 connector is possible due to its available crystal structure [40, 41].

The connector protein has been reengineered and inserted into lipid bilayers [42]. This is the first instance demonstrating the incorporation of a protein that is neither a membrane protein nor an ion channel into an artificial membrane. The conductance of each pore in the bilayer lipid membrane (BLM) is almost identical and is perfectly linear with respect to the applied voltage. Numerous transient current blockade events in real time induced by dsDNA are consistent with the dimensions of the channel and of dsDNA. Furthermore, the connector channel is stable under a wide range of experimental conditions, including high salt and extreme pH [43]. The robust properties of the connector nanopore have made it possible to develop a simple, reproducible approach for pore quantifications [43]. More recently, we reported that the phi29 connector channel exhibits a one-way traffic property for dsDNA translocation [44]. The most significant advantage of our system, different from other well-studied systems, is that the phi29 connector has a larger channel allowing for the passage of both dsDNA and ssDNA. The larger pore size is also advantageous in that it makes it easier for channel modifications for the insertion or conjugation of chemical groups for enhanced sensing applications. These findings have important implications since artificial membrane architecture for DNA packaging motor would allow detailed investigations into discrete mechanisms of motor operation as well as future avenues for therapeutic dsDNA packaging, sampling, and delivery.

4.2 Reengineering, Expression and Purification of the Phi29 Connector

4.2.1 Reengineering of Phi29 Connector

The construction of the plasmid for the expression of the connector protein and the assembly of the dodecameric connector has been reported previously [41]. The subsequent terminal modifications of the connectors have also been described [40, 45, 46]. Briefly, the modification of one of the plasmids was, for example, achieved by a two-step polymerase chain reaction (PCR). First, the primer pair F1-R1 was used to amplify the GP10 gene. The first PCR product was used as a template for a second step PCR with primer pair F1 and R2, which contained affinity Tags (His-Tag or Strep-Tag) as well as the restriction sites for NdeI and XhoI, respectively. The second PCR product was digested with NdeI/XhoI and ligated into the NdeI/XhoI sites of the vector pET-21a(+) (Novagen) to generate the plasmid.

4.2.2 Expression and Purification of the Connector

Plasmid pETgp10-Cstrep or -Chis was transformed into the *Escherichia coli* strain HMS174 (DE3) for protein expression. A volume of 10 mL of the *E. coli* culture was incubated overnight at 37°C in Luria-Bertani (LB) medium containing 100 μg/mL ampicillin and agitated at 250 rpm. A volume of 5 mL of the culture was inoculated into 500 mL medium and induced with 0.5 mM IPTG when the cell density reached 0.5–0.6 unit at OD600. Cells were harvested 3 h after Isopropyl β-D-1-thiogalactopyranoside (IPTG) induction by centrifugation at 5,000 × *g* for 20 min in a Beckman JS-7.5 rotor and then stored at −70°C before use.

The Strep-tagged connector was purified by affinity chromatography with Strep-Tactin (IBA, St. Louis, MO). Cells were resuspended with Buffer W (15% glycerol, 0.5 M NaCl, 1 mM EDTA, 100 mM Tris–HCl, pH 8.0), and the cleared lysate was loaded onto a Strep-Tactin Sepharose Column and washed with Buffer W. The Strep-tagged connector was eluted by buffer E (15% glycerol, 0.5 M NaCl, 1 mM EDTA, 2.5 mM desthiobiotin, 100 mM Tris–HCl, pH 8.0). The His-tagged connector was purified with Nickel affinity chromatography [46] (Novagen). A plasmid was constructed to express the connector protein by a two-step PCR [41]. The connectors were reengineered at the terminal ends, such as insertion of a His- or Strep-tag at the C-terminus just downstream of a six glycine linker for improved affinity tag flexibility [40, 42, 45, 46]. The plasmid was then transformed into the *E. coli* strain HMS174 (DE3) for protein expression. The His- or Strep-tagged connector was subsequently purified to homogeneity by affinity chromatography (Novagen, IBA) [46].

4.3 Preparation of Lipid Vesicles Containing the Reengineered Connector

Attempts to directly incorporate the connector into planar lipid membranes to serve as a nanopore were unsuccessful. A two step procedure is necessary to enable the insertion of the connector channel into a lipid bilayer with high efficiency: reconstitution of the connector into a liposome, followed by vesicle fusion to insert the connector into a planar lipid membrane.

Analysis of the surface charge of the connector revealed that its central surface region exhibits slight hydrophobicity compared to the two flanking layers at the wider and narrow ends, respectively (Fig. 4.1c) [15, 16]. The C-terminal end of the connector is located at the wider end domain and embedded inside the viral procapsid. The hydrophobic layer is in contact with the shell of the procapsid (Fig. 4.1a).

Lipid bilayers typically contain a hydrophilic-hydrophobic-hydrophilic architecture. To facilitate the lipid membrane insertion, the three hydrophilic-hydrophobic -hydrophilic layers of the connector were made more distinct by replacing the 25 residues at the C-terminal domain of gp10 with a basic peptide. The modified gp10 was purified to homogeneity and the protein self-assembled into the dodecameric structure with similar morphology to the 12-fold symmetric wild type connector, as observed by AFM, TEM and SDS-Page gel (Fig. 4.2a–d). The existence of a native

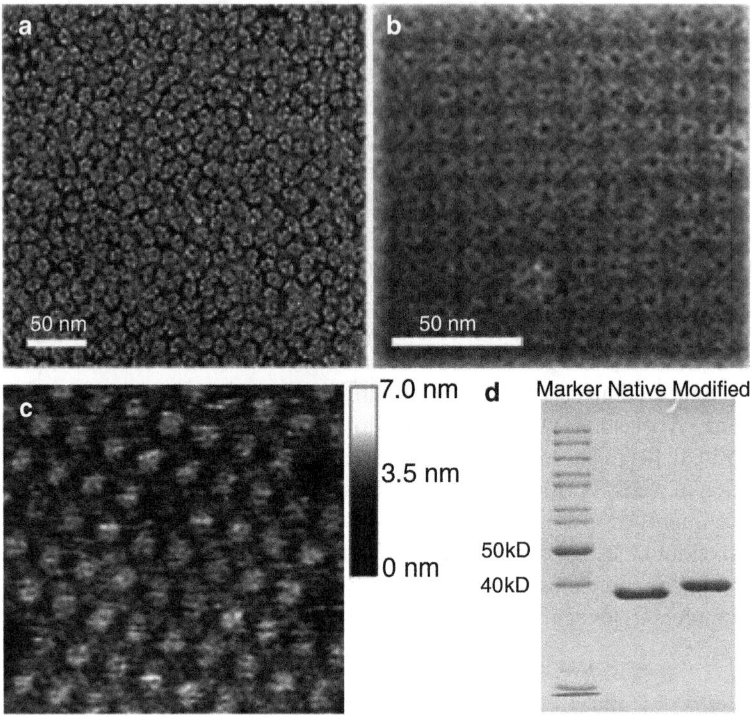

Fig. 4.2 Purified reengineered connector formed arrays, which was observed by EM (**a**, **b**) and AFM (**c**). (**d**) Coomasie-blue stained SDS-gel showing native and reengineered connectors. Figures reproduced with permissions from: (**a**, **b**, **d**) Ref. [42], © Nature Publishing Group; (**c**) Ref. [43], © The Royal Society of Chemistry

and authentic motor configuration was verified through its competency to package the double stranded DNA after incorporation into procapsid (Sect. 6.1, Fig. 4.15) and to assemble the resulting DNA-filled capsid into infectious phi29 virion.

To prepare the connector reconstituted lipid vesicles, lipid stock solutions of 1,2-dioleoyl-*sn*-glycero-3-phosphocholine (DOPC) or 1,2-diphytanoyl-*sn* glycerol-3-phosphocholine (DPhPC) in chloroform (Avanti Polar Lipids) were syringed into a glass vial, dried under a stream of nitrogen, and placed under vacuum overnight to generate a lipid film [42]. After removal of chloroform, the lipids were rehydrated in a buffer containing 200–300 mM sucrose (to create a high osmotic pressure) and co-incubated with the connector. Such incubation provides an opportunity for the hydrophobic layer to interact with the hydrophobic domain of individual lipid molecules by hydrophilic-hydrophobic two-layer alignment. The dehydration–hydration method led to the production of giant liposome up to 50 μm [47]. Small unilamellar vesicles (100 nm or 400 nm) were then generated by extruding through polycarbonate membrane filters for single channel measurements.

The insertion of the connector protein into the lipid membrane was confirmed by fluorescence microscopy, filtration assay, and sedimentation analysis (Fig. 4.3). The presence of the fluorescently labeled connector in the membrane was demonstrated by a clear fluorescent ring around the liposome (Fig. 4.3a-i). The fluorescent ring was very similar to the liposome generated with fluorescent lipids NBD-PE (Fig. 4.3a-ii). No fluorescent ring was observed when the fluorescently-tagged connector was mixed non-specifically with the non-connector inserted liposome (Fig. 4.3a-iii). The free connectors were then removed by filtration using a 0.45 μm filter membrane or by 5–20% sucrose gradient ultracentrifugation (Fig. 4.3b, c). Furthermore, the lipid membrane retained its fluidity after incorporation of the connector protein as demonstrated by FRAP (Fluorescence Recovery After Photobleaching) (Fig. 4.3d).

None of the aforementioned assays can distinguish between the membrane surface attachment of the connector and incorporation into a membrane to form a channel. A direct verification of the incorporation of a membrane channel is by single channel conductance assay, described in the next section.

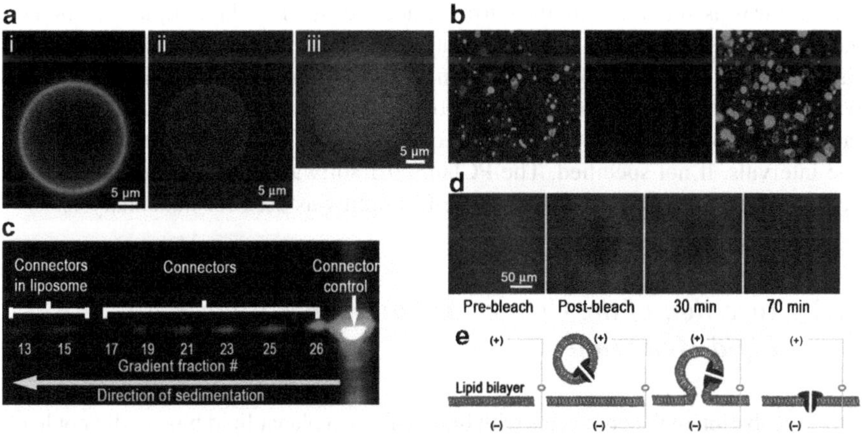

Fig. 4.3 (**a**) Epifluorescence Images of giant liposome (~50 μm) containing the connector: (i) lipid labeled with NBD-PE without connector; (ii) proteoliposomes reconstituted by FITC labeled connectors; (iii) FITC-connector mixed non-specifically with liposomes. (**b**) Membrane filtration isolated most of the free connectors. (**c**) Separation of liposome/FITC-connector complexes by sucrose gradient sedimentation. Free connectors appeared in the top fractions while proteoliposomes remained in the lower fractions. Fractions #1–12 are not shown. (**d**) Fluidity of fluorescent lipid bilayer demonstrated by FRAP (Fluorescence Recovery After Photobleaching) showing that the fluorescence intensity of photobleached area (*black*) was gradually increased over time due to lipid diffusion. (**e**) Schematic showing the insertion of the connector into a planar lipid bilayer via vesicle fusion. Figures reproduced with permissions from: Ref. [42], © Nature Publishing Group

4.4 Characterization of Phi29 Connector Channels by Single Channel Electrophysiological Assays

4.4.1 Experimental Setup

A standard horizontal BLM chamber (Eastern Sci LLC) was utilized in these experiments. A thin Teflon film with an aperture of 100–200 μm (Easter Sci LLC) in diameter was used as a partition to separate the chamber into *cis*- (working volume 250 μL) and *trans*- (working volume 2.5 mL) compartments. For connector insertions, diluted liposome stock solution was directly added to the *cis*- compartment. A final molar ratio of lipid:connector was set to 4,000:1 to 16,000:1 for single channel conduction assays.

The thickness of the BLM is critical for efficient insertion of connector reconstituted liposomes with the BLM. The capacitance of the membrane produced from a voltage ramp is first obtained using the equation $C = I(dV/dt)^{-1}$, where I is the current measured, and $(dV/dt)$is the applied voltage ramp. The thickness can then be expressed as $d = (\varepsilon_0 \cdot \varepsilon_s \cdot A \cdot C^{-1})$, where ε_0 and ε_s are permittivity of free space (8.854×10^{-12} F · m^{-1}) and lipid (~3 F · m^{-1}), respectively. The optimum lipid thickness is ~5 nm, which corresponds to the thickness of a single lipid bilayer.

A pair of Ag/AgCl electrodes connected directly to the head-stage of a current amplifier was used to measure the current traces across the bilayer lipid membrane. The trace was recorded using an Axopatch 200B patch clamp amplifier coupled with the Axon DigiData 1322A or Axon DigiData 1440 analog-digital converter (Axon Instruments). All the voltages reported are those of the *trans*-compartment. Data was low band-pass filtered at a frequency of 1 kHz and acquired at 500 μs time intervals, if not specified. The PClamp 9.1 software (Axon Instruments) was used to collect the data, and the software Clampfit was used for data analysis.

4.4.2 Insertion of the Connector into Planar Bilayer Lipid Membrane

Direct incubation of the connector with liposomes or a planar lipid bilayer did not lead to channel formation in the bilayer membrane (Fig. 4.4a). Connector insertions into the bilayer were only observed upon vesicle fusion of connector reconstituted proteoliposomes with the planar bilayer (see schematic Fig. 4.3e). The channel insertion events were observed as discrete step-wise increase in conductance (each step represents the insertion of a single connector channel) as shown in a continuous current trace (Fig. 4.4b), under either positive or negative transmembrane voltage (Fig. 4.4c). The number of connector channels inserted into the membrane can be obtained by simply counting the steps in the current trace during the course of insertion.

A series of extensive investigations under a wide range of solution conditions revealed that the step size of the current jump per single connector insertion were

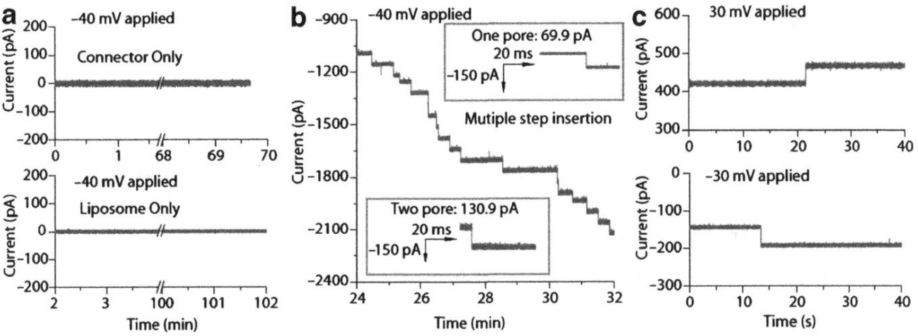

Fig. 4.4 Conductance assays confirm the insertion of the connector into bilayer lipid membranes (BLM). (**a**) BLM with connector only (*top*) or liposomes only (*bottom*). (**b**) Addition of connector-containing proteoliposomes resulted in multiple insertions. Inserts: Insertion of one (*top*) and two connectors simultaneously (*bottom*). (**c**) One connector insertion at positive voltage (*top*) and at negative voltage (*bottom*). Figures reproduced with permissions from: Ref. [42], © Nature Publishing Group

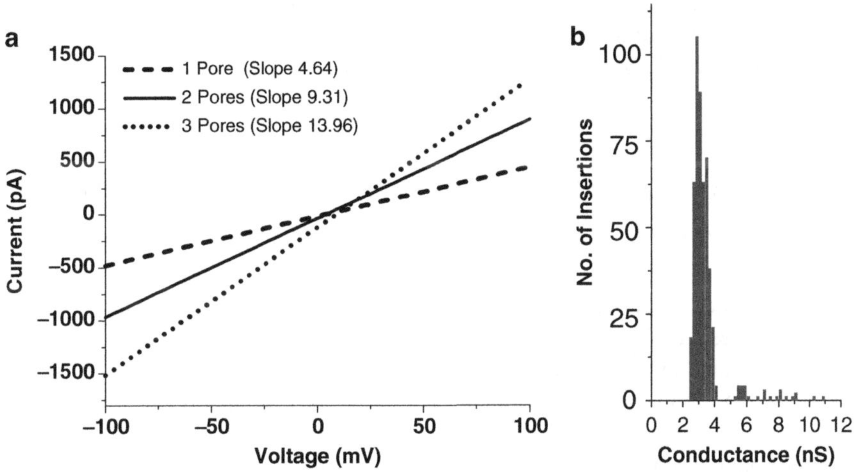

Fig. 4.5 (**a**) Data demonstrating robust properties of connector indicated by strong linear I–V relationship. Increased number of connectors correlates to greater slopes but identical conductance per channel. (**b**) Histogram showing uniform conductance of connector channels in the membrane. Figures reproduced with permissions from: Ref. [43], © The Royal Society of Chemistry

nearly identical (Fig. 4.5b), as demonstrated by a sharp Gaussian distribution. For instance, the insertion of one connector in the BLM results in an increase in the current of ~220 pA (equivalent to ~3 nS) at a potential of −75 mV in a buffer composed of 5 mM Tris (pH 7.9)/1 M NaCl. Occasionally, simultaneous insertions of two and three connectors can be observed in ~5% and 2% of all the connector insertion events, respectively.

4.4.3 Current–Voltage Relationship of Single (or Multiple) Connector Channels

The application of a ramping potential (2.2 mV s^{-1}) from -100 mV to +100 mV to the BLM containing a single (or multiple) connector channel revealed a perfectly linear I–V relationship, thereby demonstrating well-behaved, uniform conductance of the channel. The current traces from one, two and three connector insertions are shown in Fig. 4.5a in the presence of 5 mM Tris (pH 7.8)/2 M NaCl. Increased number of connectors correlated to greater slopes but identical conductance per channel, further demonstrating the uniform size and stability of the phi29 connector channels in the BLM.

In addition to number of connector channels, the conductance also depends on the ion concentration and ion species (Fig. 4.6). From the I–V slope, the conductance in 0.5 M NaCl, 1.0 M NaCl, 1.5 M NaCl and 2.0 M NaCl were determined to be 1.75 $\pm$ 0.11 nS, 3.03 $\pm$ 0.02 nS, 4.07 $\pm$ 0.26 nS, and 4.60 $\pm$ 0.05 nS, respectively, demonstrating the expected increase in conductance with an increase in salt concentration (Fig. 4.6a, c). The conductance values were higher at equal concentrations of KCl compared to NaCl as follows: at 0.5 M KCl, 1.0 M KCl, 1.5 M KCl and 2.0 M KCl, the conductance were 2.70 $\pm$ 0.21 nS, 4.65 $\pm$ 0.11 nS, 6.23 $\pm$ 0.32 nS, and 7.83 $\pm$ 0.20 nS, respectively (Fig. 4.6b, d).

4.4.4 An Analytical Expression for Determining the Number of Connectors Inserted in the Lipid Bilayer

The conductance of the membrane/connector complex, G_m can be calculated using two approaches: (1) The step size of the current jump upon connector insertion under a specific voltage in real time. In this case, the ratio of the measured current to the applied voltage represents the conductance value. (2) The slope of the current trace under a ramp voltage [48]. The second approach gives a more accurate value, since the slope contains numerous data points and also accounts for effects, such as changes in surface potentials of Ag/AgCl electrodes as well as membrane contact potentials.

The conductance of one channel in the BLM, G_c can be obtained readily using the same approaches. Since phi29 connector channels are of uniform size and are stable under a wide range of salt concentrations, a linear relationship is therefore expected between the conductance of the membrane/connector and the number of connectors in the BLM (Fig. 4.6c, d). Indeed a linear relationship is obtained in both NaCl and KCl buffers. Table 4.1 shows the results of calculating single channel conductance using the discrete current steps.

Generally, if G_m and G_c are known, the number of connectors (N) incorporated in a membrane can be deduced as follows [48]:

$$N = \frac{G_m}{G_c} \tag{4.1}$$

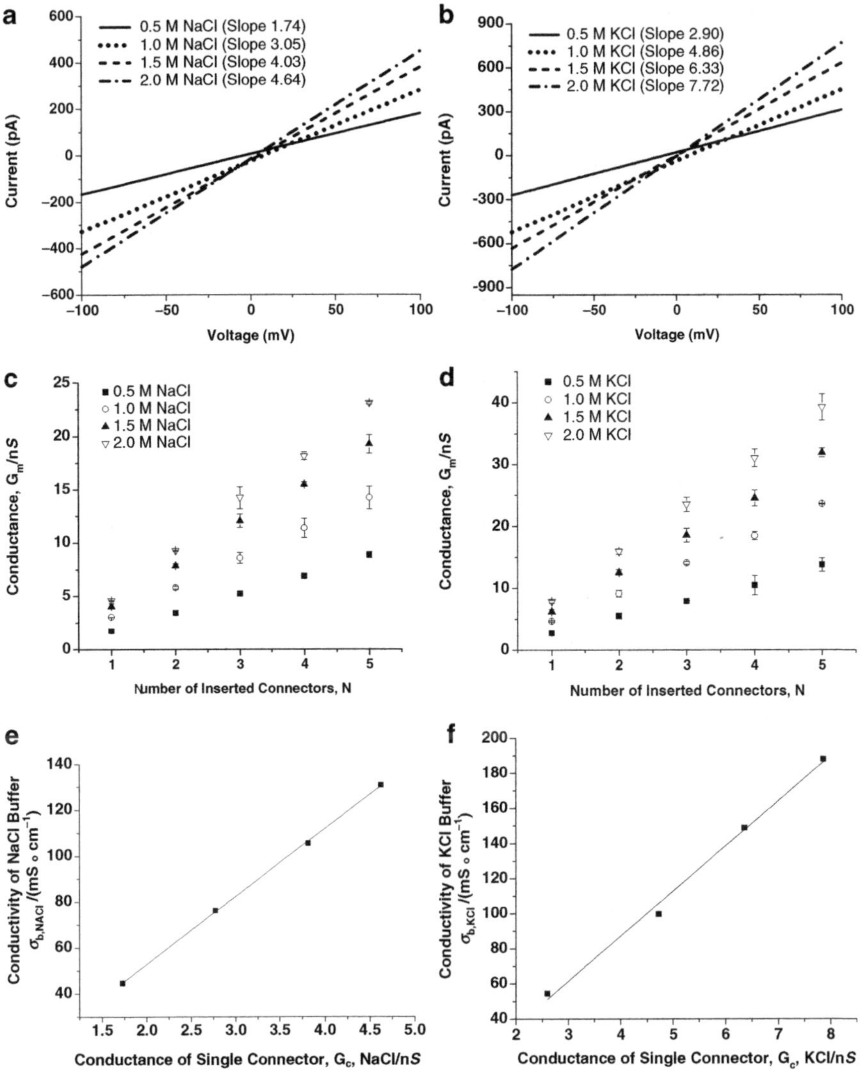

Fig. 4.6 Current–Voltage traces of single connector channel at different concentrations of NaCl with 5 mM Tris, pH 7.8 (**a**) and different concentrations of KCl with 5 mM HEPES, pH 7.8 (**b**). Relationship of measured conductance with number of connectors at different concentrations of NaCl with 5 mM Tris, pH 7.8 (**c**) and different concentrations of KCl with 5 mM HEPES, pH 7.8 (**d**). All the conductance measurements were from more than three individual experiments. Error bars represent standard deviations for the measurements. The regression equation for 0.5 M NaCl is: $G_m/nS = 1.73 \times N + 0.01$; for 1.0 M NaCl is: $G_m/nS = 2.77 \times N + 0.26$; for 1.5 M NaCl is: $G_m/nS = 3.81 \times N + 0.26$; and for 2.0 M NaCl is $G_m/nS = 4.62 \times N - 0.00$. The regression equation for 0.5 M KCl is: $G_m/nS = 2.60 \times N + 0.11$; for 1.0 M KCl: $G_m/nS = 4.73 \times N - 0.10$; for 1.5 M KCl is: $G_m/nS = 6.35 \times N - 0.16$; and for 2.0 M KCl is: $G_m/nS = 7.86 \times N - 0.01$, respectively. Relationship of buffer conductivity versus conductance of single connector at 5 mM Tris with 0.5–2.0 M NaCl (**e**) and at 5 mM HEPES with 0.5–2.0 M KCl (**f**). Figures reproduced with permissions from: Ref. [43], © The Royal Society of Chemistry

Table 4.1 Comparison of conductance induced by single channel measured from discrete current jumps and calculated using empirical equations

No. of Independent experiments	Buffer solutions	Conductance/channel calculated by discrete steps (nS)	Buffer conductivity (mS cm^{-1})	Conductance/channel calculated using the empirical equations (nS)
2	5 mM Tris + 0.5 M NaCl	1.65 ± 0.07 ($n = 62$)	44.6	1.72
6	5 mM Tris + 1 M NaCl	3.04 ± 0.56 ($n = 52$)	76.3	2.79
3	5 mM Tris + 2 M NaCl	4.71 ± 0.18 ($n = 45$)	130.9	4.64
40	TMS + 1 M NaCl	3.14 ± 0.36 ($n = 500$)	80.5	2.94
2	5 mM HEPES + 0.5 M KCl	2.48 ± 0.13 ($n = 10$)	54.4	2.73
4	5 mM HEPES + 1 M KCl	4.73 ± 0.20 ($n = 9$)	99.7	4.49

Note: The conductance in Table was calculated using the ratio of the measured current jump induced by a discrete step to the applied voltage. "(n)" in column three represents the number of insertions

A formula specific to the phi29 system can be derived for calculating the single channel conductance. A plot of solution conductivity, σ_b (measured with a Corning conductivity meter) versus single channel conductance (obtained from the slopes of the regression equations) revealed a linear relationship (Fig. 4.6e, f). Using the regression equations, G_c at NaCl or KCl concentration within the range 0.5–2.0 M can be deduced from buffer conductivity, σ_b.

The NaCl buffer conductivity is given by:

$$\frac{\sigma_{b,NaCl}}{(mS \cdot cm^{-1})} = \left(29.58 \times \frac{G_{c,NaCl}}{nS}\right) - 6.32 \quad (R^2 = 0.9994) \tag{4.2}$$

Thus, $$\frac{G_{c,NaCl}}{nS} = \frac{\left[\frac{\sigma_{b,NaCl}}{(mS \cdot cm^{-1})} + 6.32\right]}{29.58}$$

Similarly, the KCl buffer conductivity can be obtained as follows:

$$\frac{\sigma_{b,KCl}}{(mS \cdot cm^{-1})} = \left(25.69 \times \frac{G_{c,KCl}}{nS}\right) - 15.61 \quad (R^2 = 0.9923) \tag{4.3}$$

Thus, $$\frac{G_{c,KCl}}{nS} = \frac{\left[\frac{\sigma_{b,KCl}}{(mS \cdot cm^{-1})} + 15.61\right]}{25.69}$$

The number of channels in a lipid bilayer can therefore be derived by measuring the bulk solution conductivity and calculating the conductance of the bilayer, given by the slope of the *I–V* plot. As an example, in the case of 0.5 M NaCl, $\sigma_{b,NaCl}$ is 44.6 $mS \cdot cm^{-1}$. Using equation (4.2), $G_{c,NaCl}$ is calculated to be 1.72 nS. As noted earlier, $N = \frac{G_m}{G_c}$. Using $G_m = 1.59$ nS, gives $N = 0.92$. The calculated values for channel conductance (from the derived equations) are in good agreement with the values obtained by means of current jump under the same solution conditions.

4.4.5 Calculating the Calibration Coefficient for the KCl and NaCl Buffer

The slopes ratios from experiments with KCl and NaCl were calculated under the same number of connector insertions. With 0.5 M, 1.0 M, 1.5 M and 2.0 M KCl and NaCl solutions, the slope ratios of KCl to NaCl were 1.50, 1.71, 1.67 and 1.70, respectively. The average value of 1.65 can then be used as a calibration coefficient for translating the conductance calculated from buffers with K^+ to Na^+ as follows:

$$G_{m,\ KCl} = F \times N \times G_{c,\ NaCl} \tag{4.4}$$

where $G_{m,\ KCl}$ and $G_{c,\ NaCl}$ are the single channel conductance in the same concentration of KCl and NaCl respectively; *F* is the calibration coefficient for the KCl and NaCl salt solutions. The same equation can be applied to approximate the conductance induced by multiple connector channels under the same buffer conditions with different conducting ions, such as K^+, as long as the single channel conductance in NaCl solution is known.

4.4.6 Comparison of Phi29 Connector Channel Conductance with α-Hemolysin

The diameter of the narrow end of the connector channel is 3.6 nm, while the channel formed by α-HL has a diameter of only 1.5 nm [49]. The ratio of the cross-sectional area of the connector and α-HL channels is calculated to be 5.7. The ratio of measured conductance of the connector to α-HL is 5.1 (Table 4.2). Since the conductance of a channel is proportional to its cross-sectional area, it can therefore be concluded that the cross-sectional area of the connector in the buffer solutions was approximately 5.1-fold greater than that of α-HL, which compares well with the ratio of cross-sectional areas from the crystal data of both proteins.

Table 4.2 Comparison of single channel conductance from the GP-10 connector and α-hemolysin

Proteins	Pore diameter (nm)	Cross-section area (nm^2)	Conductance (nS/pore)[a] at 0.5 M NaCl	at 1 M KCl
Connector	3.6	10.2	1.57 ± 0.16	4.84 ± 0.15
α-HL	1.5 [49]	1.8	0.31 ± 0.05[b]	0.94 ± 0.01[c]
Ratio (Connector/α-HL)	2.4	5.7	5.1	5.1

[a]The data for connector conductance at 0.5 M NaCl, and 1 M KCl were obtained from a total of 38 and 36 insertions, respectively. The data for α-HL conductance at both 0.5 M NaCl and 1 M KCl were from a total of four insertions, respectively

[b]Conductance of α-HL at 1 M NaCl has been reported to be 0.68 nS/pore [59]

[c]Conductance of α-HL at 1 M KCl has been reported to be 0.80 nS/pore [60] or 1.0 nS/pore [29]

4.5 Translocation of Double-Stranded DNA

4.5.1 Translocation of Double-Stranded DNA Through Phi29 Connector Channels

The principle of nucleic acid translocation through nanopores was demonstrated almost 15 years ago using α-hemolysin channels. The concept is based on the classic 'coulter-counter' whereby the passage of non-electrolytes through a pore against an electrochemical potential gradient physically blocks the ion flow, which can be detected. In presence of linear dsDNA of varying lengths (12–5,000 bp), numerous transient current blockade events, (each representing the translocation of a single DNA molecule) were observed (Fig. 4.7b). When circular plasmid DNA (Cx43) was added, no such current blockade events were observed, presumably due to the supercoiled nature of plasmid DNA. However, upon digestion of circular plasmid DNA with DNaseI, a burst of DNA translocation events were observed, demonstrating that the phi29 channel is only capable of translocating linear dsDNA (Fig. 4.7c, d).

Occasionally, in presence of multiple pores, blockade events with plateaus at one or more discrete current levels were observed, indicating the simultaneous translocation through multiple pores. The rate of current blockade events was observed to increase with increasing concentration of DNA as well as by increasing the applied voltage, as expected.

4.5.2 Characterization of DNA Current Blockade Events

One of the parameters in individual pore blockade events is the depth of current blocks and is expressed as current blockade percentage, calculated as follows: the size of current blockade resulting from the DNA translocation through

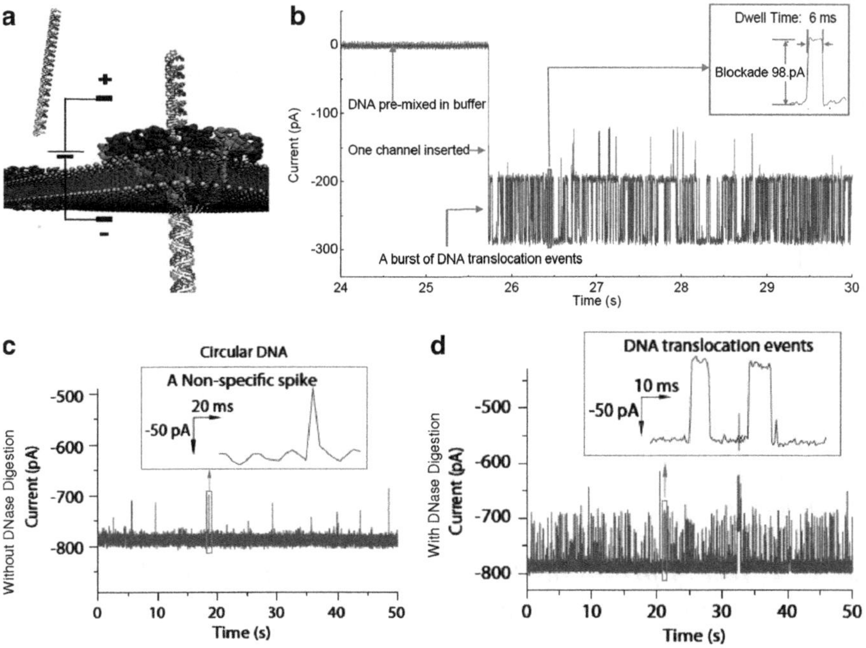

Fig. 4.7 (**a**) Illustration of dsDNA translocation through the BLM connector pore. (**b**) Typical current blockade events from dsDNA tranloscation through the nanopore. Insert: A magnified current blockade event showing the current depth and the dwell time of translocation. Representative blockades caused by 45 pM double-stranded circular plasmid DNA without (**c**) and with (**d**) DNase digestion. Figures reproduced with permissions from: (**c–d**) Ref. [42], © Nature Publishing Group

one channel divided by the step size of the current for one connector insertion. In the presence of dsDNA, the channel blockade percentage was centered at ~32% (Fig. 4.8a), which is consistent with the dimensions of the channel (3.6 nm diameter at its narrowest end) and dsDNA (~2 nm in diameter). That is, blockade $\% = [\pi \cdot (2 \div 2)^2 / \pi \cdot (3.6 \div 2)^2] \times 100$. Furthermore, each transient current blockade event was observed to be nearly identical as demonstrated by a sharp Gaussian distribution (Fig. 4.8a).

Another parameter is the dwell time (τ_p) for DNA translocation events, which represents the time taken by the DNA to translocate through the channel (Fig. 4.8b). The dwell time distribution follows an exponential decay function. The peak dwell time represents the most probable event duration. For instance, under −75 mV and −40 mV, the peak dwell time for 5 kbp dsDNA was 9.2 ms and 22.1 ms, respectively, demonstrating that the dwell time of DNA translocation was affected by applied voltage. In addition, shorter dwell times were observed for shorter lengths of DNA, as shown in Fig 4.8b comparing the dwell time for 35 bp and 5 kbp dsDNA.

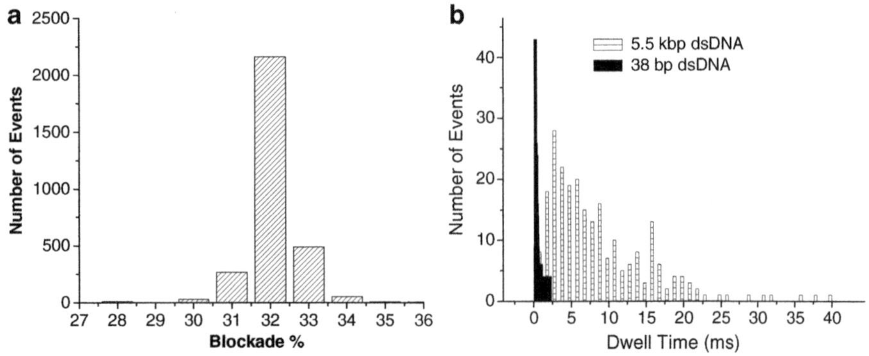

Fig. 4.8 (**a**) A histogram of current blockade percentage induced by linear dsDNA (2 kbp) with a total 3,264 events in the presence of 1 M NaCl, pH 7.8 under −75 mV holding potential. (**b**) Comparison of dwell times for translocation of 38 bp and 5.5 kbp ds-DNA. Figures reproduced with permissions from: (**a**) Ref. [43], © The Royal Society of Chemistry; (**b**) Ref. [42], © Nature Publishing Group

4.5.3 *Effect of pH on the Stability and Translocation Behavior of Connector Channels*

The stability of the connector channel was investigated under extreme pH conditions in the presence of a 20 bp DNA in both chambers. Current traces were recorded under a constant voltage at each of the pH conditions investigated in presence of 1 M NaCl (Fig 4.9a–c). The respective conductance values are 2.50 ± 0.11 nS ($N = 6$) at pH 2(5 mM phosphoric acid buffer), 2.73 ± 0.07 nS ($N = 6$) at pH 7 (5 mM Na_2HPO_4 buffer) and 2.78 ± 0.12 nS ($N = 6$) at pH 12 (5 mM $Na_2HPO_4{\cdot}7H_2O$ buffer). The channel under the extreme pH environments remained open with well-behaved uniform channel conductance. Furthermore, DNA translocation events are observed. At pH 2, formation of apurinic acid occurred leading to short current blockade events (Fig. 4.9a). Although the structure of DNA is affected by extreme pH conditions, the phi29 connector retains its stable channel properties under strong acidic or alkaline conditions.

4.5.4 *Possible Artifacts in Current Signals*

Artifacts can arise from random noise. The connector channel by itself is exceptionally stable; the conductance is uniform; and, does not display voltage gating properties under the reported conditions of −150 mV to +150 mV. Occasionally, unspecific blockade events with detectable time very close to the limit of sampling frequency were observed. These were attributed to interactions of connector reconstituted liposomes with the planar membrane since these events occurred in absence of DNA and their frequency was observed to increase after the addition

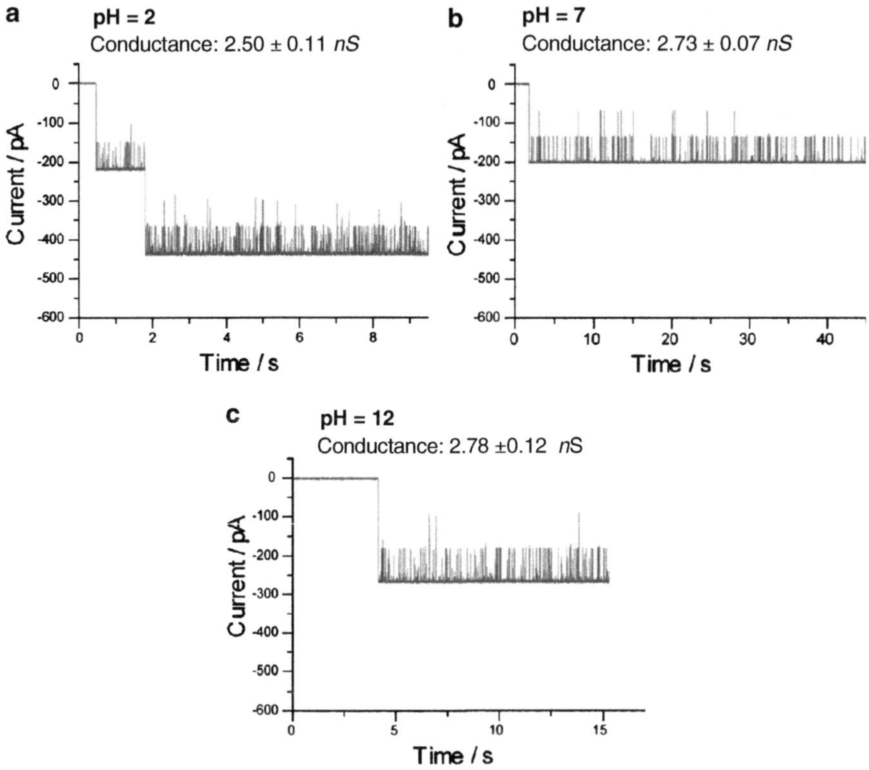

Fig. 4.9 Current traces showing connector insertion accompanied by DNA translocation (20 bp dsDNA) through the channel under (**a**) pH 2; (**b**) pH 7; and (**c**) pH 12 in presence of 1 M NaCl at a constant voltage of −75 mV. Figures reproduced with permissions from: Ref. [43], © The Royal Society of Chemistry

of more liposomes to the chamber. The occurrence of such events can be minimized simply by diluting the connector reconstituted liposomes.

DNA can also collide with the pore without actually traversing through it giving rise to non-specific blockade events. This is especially prevalent in longer length DNA, since an optimum conformation is necessary before the translocation can start. However, once the translocation starts, the likelihood of DNA to diffuse back against the potential gradient is extremely unlikely.

4.5.5 Quantitative PCR Verifying the Translocation of dsDNA

To verify that the current blockade events were indeed the translocation of dsDNA through the connector channels, quantitative PCR (Q-PCR) using the SYBR Green kit (Bio-Rad) was used to quantify the translocation of DNA under a constant voltage (Fig. 4.10a, top). 141-bp DNA was added to the *trans-* side and samples were taken from the *cis-* side for quantification at 30 min intervals. The number of DNA molecules in the *cis*-chamber increased over time ($N = 9$ experiments) and furthermore, DNA translocation rate was affected by the number of inserted connectors. For comparison, control experiments were performed in the absence of connectors (Fig. 4.10a, bottom) and in this case, the number of DNA molecules

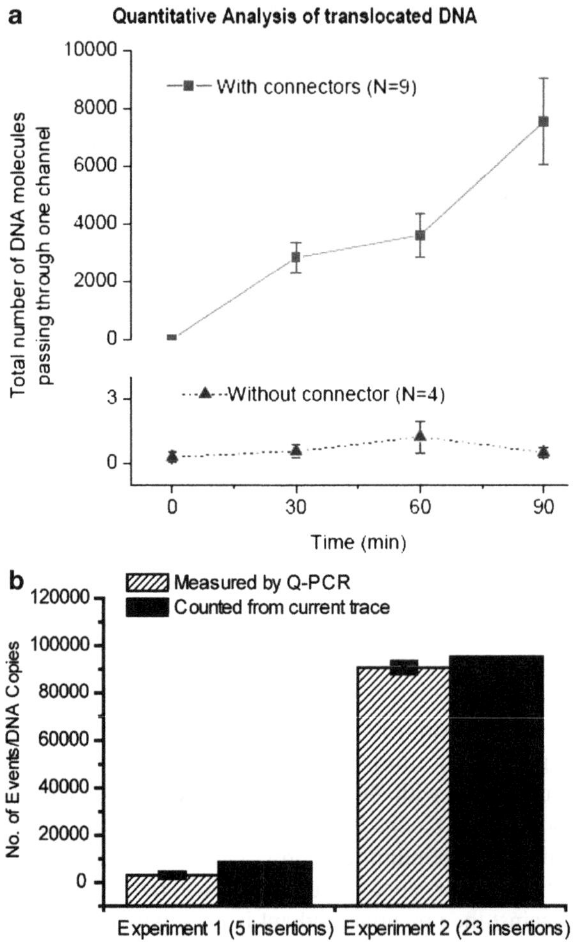

Fig. 4.10 Quantitative PCR analysis of DNA translocation events. (**a**) Quantitative analysis of the total number of DNA passing through one of the connectors in the lipid membrane from the *trans*-chamber to the *cis*-chamber (*top*). Negative controls (*bottom*) were carried out under the same condition but without connectors. The error bars represent standard deviations of the mean from nine independent experiments and four negative control experiments. (**b**) A comparison of copy number of translocated dsDNA measured by Q-PCR with the relevant blockade events counted from current trace recorded in two independent experiments. The error bars represent the standard deviation of three independent Q-PCR measurements. Figures reproduced with permissions from: (**a**) Ref. [42], © Nature Publishing Group; (**b**) Ref. [43], © The Royal Society of Chemistry

in the *cis*-chamber remained undetectable over the 90 min time course ($N = 4$ experiments). It was further observed that the blockade events from the current trace closely correlated with the number of DNA molecules passing through the pores quantified by Q-PCR (Fig. 4.10b).

To rule out the possibility that the increase in the DNA copy number in the *cis*-chamber was due to membrane leakage, three additional experiments were carried out under known conditions of BLM or partition leakage. When leaking occurred, the copy number of DNA per microliter of solution in the *cis*-chamber was approximately 10^4 to 10^5-fold higher than those experiments without leakage.

4.6 Phi29 Motor Channel Exercises a One-Way DNA Traffic Mechanism

The connector channel exercises a one-way traffic property for dsDNA translocation from N-terminal entrance (narrower-end) to C-terminal exit (wider-end), as demonstrated by voltage ramping, electrode polarity switching, and sedimentation force assessment. This is the first instance of a native wild type protein channel exhibiting a rectifying behavior with regard to DNA translocation. The connector channel, however retains a two-way traffic property for ions with equal conductance under both positive and negative *trans*-membrane potentials [42–44].

4.6.1 One-Way Traffic of dsDNA Probed by Ramping Potential to a Membrane with a Single Channel

As noted earlier, the connector channel is inserted into the BLM via vesicle fusion of the liposome/connector complex with the planar lipid membrane. The orientation of the connector channel in the BLM is therefore random, with either the C-terminus facing the *cis*-chamber or the N-terminus oriented towards the *cis*-chamber.

The application of a ramping potential (2.2 mV s^{-1}) from −100 mV to +100 mV to the BLM containing a single connector channel revealed a unidirectional translocation of dsDNA (Fig. 4.11). Since the phosphate backbone of DNA is negatively charged, DNA tends to migrate from the negative to the positive potential. Figure 4.11a shows a control experiment in the absence of DNA. When dsDNA was pre-mixed in both the *cis* and *trans*-chambers at identical concentrations, DNA translocation was observed at either negative (Fig. 4.11b) or positive (Fig. 4.11c) *trans*-membrane voltage, depending upon the orientation of the connector in the BLM. The same phenomena were observed with dsDNA of varying lengths (from 12 bp to 5,500 bp) as well as for ssDNA.

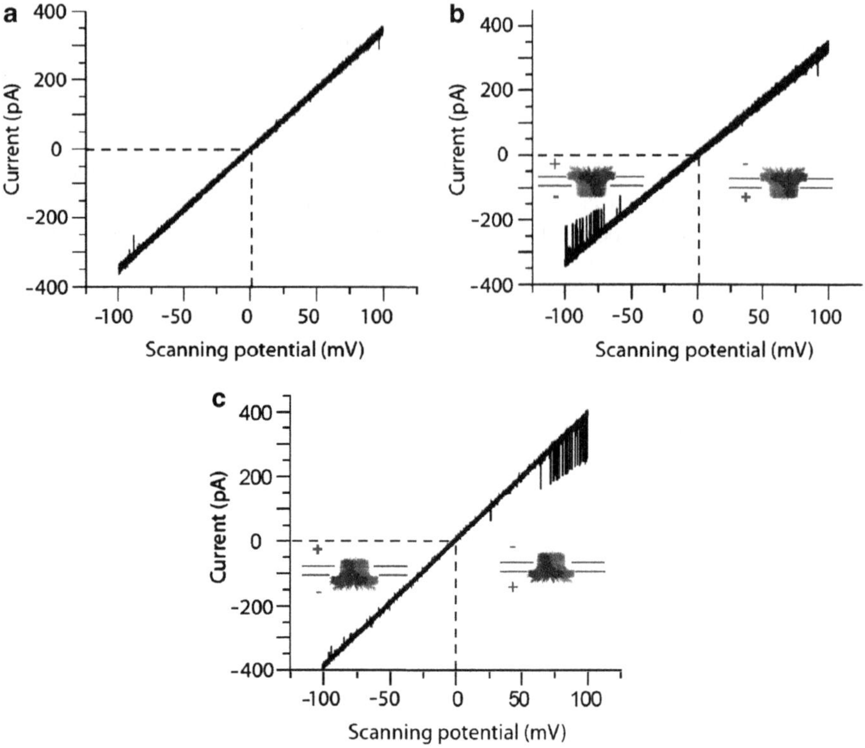

Fig. 4.11 One way-traffic in DNA translocation through a single connector channel in a lipid bilayer under a ramping potential from −100 mV to +100 mV (2.2 mV/s): (**a**) Without DNA; (**b**, **c**) DNA added in both chambers with a single connector. Figures reproduced with permissions from: Ref. [44], © American Chemical Society

4.6.2 *One-Way Traffic of dsDNA Probed by Applying a Constant Holding Potential and by Switching the Voltage Polarity to a Membrane with a Single Channel*

Under a constant potential, switching of voltage polarity revealed that the connector channel allowed only a unidirectional translocation of dsDNA (Fig. 4.12). DNA was pre-mixed in both chambers at equal concentrations. The polarity at which DNA translocation was observed was dependent upon the orientation of the connector in the BLM. Figure 4.12a shows the control experiment in absence of DNA. For example, when there was no DNA translocation under positive potential, switching the voltage to negative potential resulted in dsDNA translocation (Fig. 4.12b). Conversely, when there was no DNA translocation under negative potential, switching the voltage to positive potential resulted in dsDNA translocation (Fig. 4.12c). Similarly,

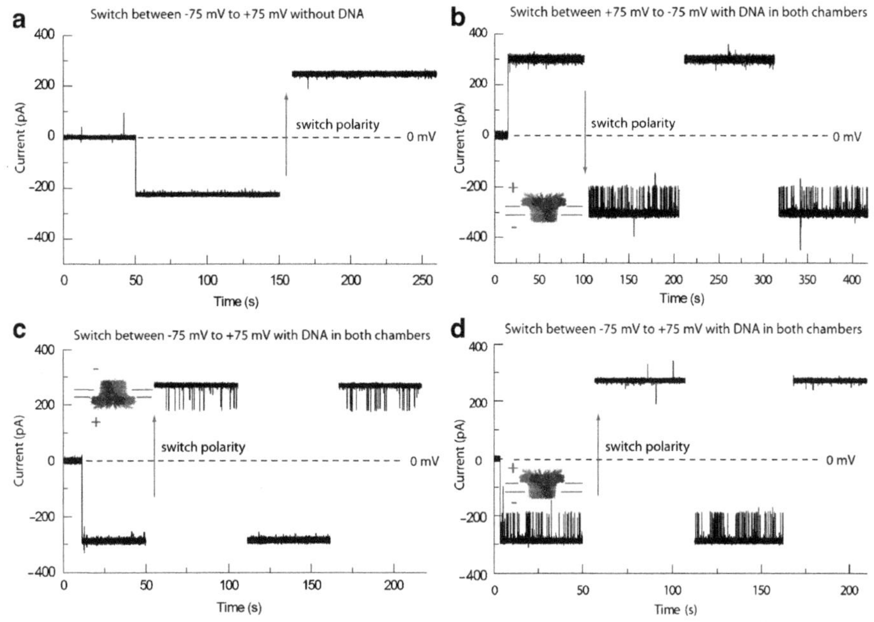

Fig. 4.12 One way-traffic in DNA translocation through a connector channel in a lipid bilayer verified by switching the voltage polarity: (**a**) No DNA in either chamber; (**b–d**) The presence of DNA in both chambers where a single connector was inserted. Figures reproduced with permissions from: Ref. [44], © American Chemical Society

when dsDNA translocation was observed under negative potential, switching the voltage to positive potential resulted in no dsDNA translocation (Fig. 4.12d). The same results were also confirmed under different voltages.

4.6.3 One-Way Traffic of dsDNA Was Probed by Quantification of DNA Translocation Frequency in the Presence of Multiple Channels

Since dsDNA only traverses through connector channels from one direction, the DNA translocation rate would therefore be affected by different arrangements of connector in the BLM in presence of multiple channels, regardless of the DNA concentration. Accordingly, the experiments were carried out under a constant holding potential of −75 mV with varying concentrations of dsDNA (Fig. 4.13). The current traces with multiple connector insertions reflect the change in the frequency of DNA translocation events (represented in the histograms in Fig. 4.13). When one additional connector was inserted with an appropriate orientation to allow DNA to pass, the frequency of DNA translocation increased

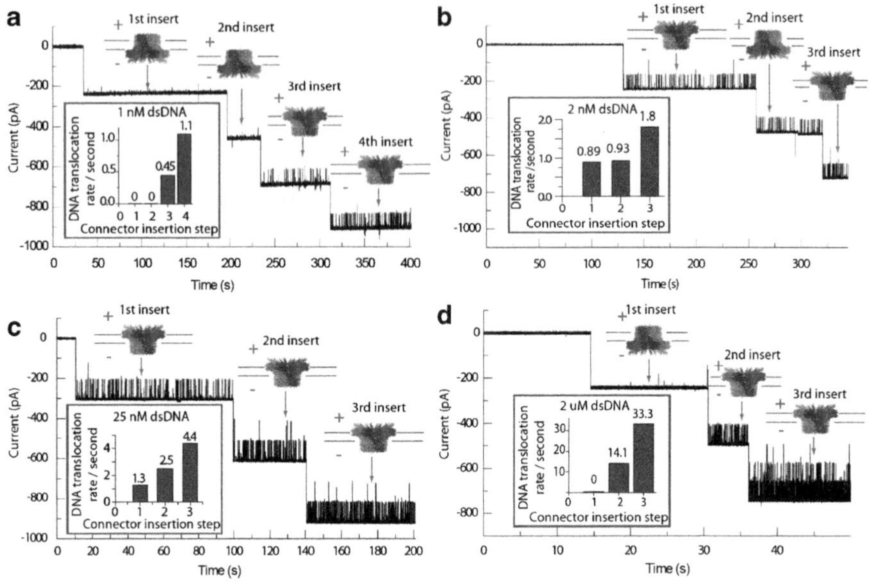

Fig. 4.13 One way-traffic in DNA translocation through a connector channel reflected in the change of DNA translocation frequency related to the orientation of connectors in bilayers with multiple connector insertions. Various concentrations of DNA were premixed in both chambers before applying the voltage. Insert: For the individual experiments (**a**–**d**), the DNA translocation frequencies after the insertion of each connector channel in the BLM are shown. Figures reproduced with permissions from: Ref. [44], © American Chemical Society

(Fig. 4.13a, c, d). When one additional connector was inserted with an opposite orientation that did not allow DNA to pass, the frequency of DNA translocation remained unchanged (Fig. 4.13a, b).

4.6.4 The Orientation of the Connector Was Probed Using Ni-NTA Nanogold Targeted at C-Terminal His-Tag

To elucidate the orientation of the connector in the BLM, reengineered connectors containing a His-tag incorporated at the wider C-terminal end of each connector subunit were used. The incorporation of the terminal His-tag did not affect the folding or DNA packaging activity of the connector channels [42]. The conduction experiments were carried out under asymmetric ionic conditions (Fig 4.14) [50]. A 1.8 nm Ni-NTA nanogold particle (150 pM) was used to bind to the His-tag at the C-terminus. The experiment was designed such that the nanogold was premixed with the buffer on one side of the chamber only. For instance, if the nanogold is present in the *trans-* side under single or multiple connector channel insertion conditions (Fig. 4.14), it will only bind to the connector with its His-tagged

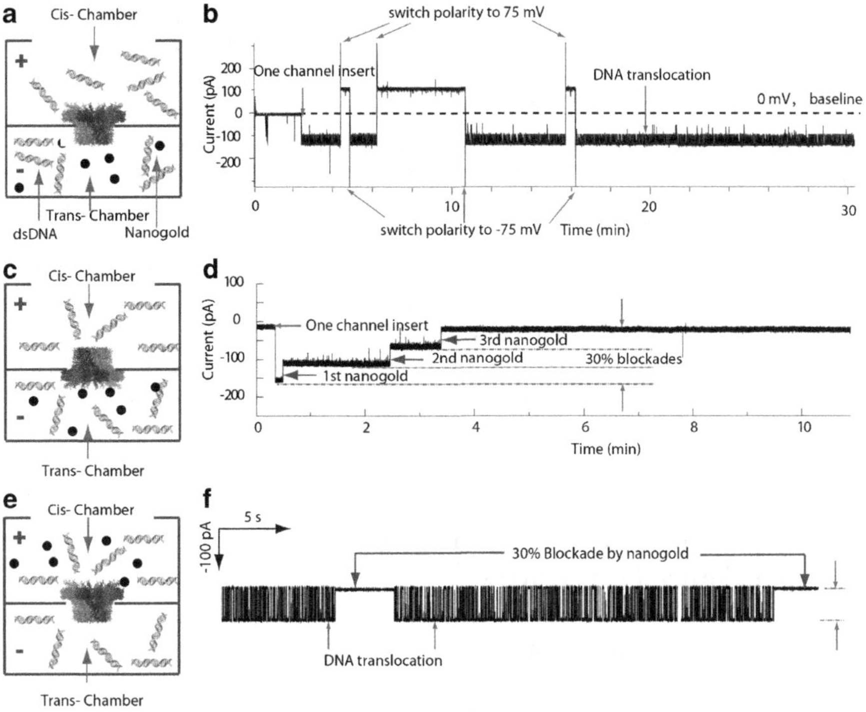

Fig. 4.14 Probing the orientation of the connector using Ni-NTA Nanogold targeted at C-terminal His-tag. (**a**, **c**, and **e**) Illustration of strategy for orientation probing with equal concentration of dsDNA added to both *cis*- and *trans*-chamber and same amount of Ni-NTA Nangold for each experiment. (**b**) A 30-min current trace showing the translocation of dsDNA without blocking by Ni-NTA Nanogold added to the *trans*-chamber; (**d**) A current trace showing three binding steps of Ni-NTA Nangold to C-His connector after addition of Ni-NTA to the bottom *trans*-chamber. (**f**) A current trace showing the translocation of dsDNA and blocking of DNA translocation by Ni-NTA Nanogold added to the top *cis*-chamber. For all these experiments, a *trans*-membrane potential of −75 mV was applied. Figures reproduced with permissions from: Ref. [44], © American Chemical Society

C-terminal end oriented towards the *trans*-chamber (Fig. 4.14c), and will not bind to connectors with the N-terminus facing the *trans*- side (Fig. 4.14a). Discrete stepwise blockage of the channel with a corresponding decrease in conduction was observed when a single connector with appropriate orientation for nanogold binding was present (Fig. 4.14d). The binding of each nanogold resulted in uniform ~30% blockage in current (Fig. 4.14d). The same results were obtained in the case of membranes with multiple channels (data not shown).

The same concentration of dsDNA was added in both sides of the chamber to aid the appraisal of the DNA translocation direction, since dsDNA only migrates from the negative towards the positive potential due to its negatively-charged phosphate backbone. Based on the placement of the nanogold in either the *cis*- or *trans*- side of the chamber (experimental design shown in Fig. 4.14a, c, e) under a constant negative *trans*-membrane potential, the orientation of the connector embedded in the membrane could be determined. It was found that dsDNA translocated one way from the N-terminal narrow end toward the C-terminal wide end. For example, when nanogold was placed in the bottom *trans*-chamber, nanogold binding was not observed while dsDNA translocation occurred, indicating that in this case, the C-terminus containing the His-tag was facing the upper *cis*-chamber and the dsDNA translocated from the N-terminus at the bottom *trans*-chamber that was negative (Fig. 4.14b). Under the same conditions, in the other case where the connector is oriented such that the C-terminus is facing the *trans*-chamber, stepwise binding of nanogold was observed without any DNA translocation (Fig. 4.14d). In another scenario, when the nanogold was added to the upper *cis*-chamber, dsDNA translocation occurred initially and was subsequently blocked, indicating that in this case the His-tagged C-terminus was facing the upper *cis*-chamber, where nanogold binding occurs, and dsDNA translocated from the N-terminus facing the bottom *trans*-chamber (Fig. 4.14f).

4.7 Applications of Phi29 Motor Channel and Future Perspectives

The ingenious, intricate and elegant design of the connector and its channel motivated its application in nanotechnology. A major concern in using protein nanopores relates to their soft nature, fluctuations in folding, dynamics in structure, and the reproducibility of the signal generated. We demonstrated that the phi29 connector channels overcome these aforementioned limitations. The channels are robust and stable under a wide range of experimental conditions. The conductance is uniform, demonstrating a perfect linear relationship with respect to the applied voltages and does not display voltage gating properties under the reported conditions. The dsDNA channel blockade events are nearly identical. By virtue of its channel size and well-behaved properties, the phi29 connector is ideally suited for reengineering a system that can operate outside its natural environment and has tremendous potential to impact biology, engineering, medicine, and other nanotechnological fields.

4.7.1 A Novel System for Studying the Mechanism of Viral DNA Packaging

The work reported here represents a new system for understanding the mechanism underlying DNA translocation through the motor channel in viral DNA packaging. While the one-way traffic phenomenon in the lipid membrane embedded system is a passive transport process mediated by electric potential, it would be worthwhile exploring whether the active motor also exercises a one-way traffic mechanism [44].

During the dsDNA packaging process, the active motor was turned off by adding a non-hydrolysable ATP analog, γ-S-ATP [51]. Both the completely and partially packaged dsDNA remained in the procapsid and did not show a reverse exit in the gradient under all centrifugation force tested (e.g., 149,000 g) [51, 52]. The finding agrees with the phenomena observed previously that dsDNA did not eject from the channel and remained within the channel of the DNA-packaging intermediates [19, 21, 51]. DNA-packaging by the motor is an entropically unfavorable process and chemical energy from ATP hydrolysis is required for DNA translocation [19, 53]. The pressure within viral procapsids gradually increases when longer and longer dsDNA are packaged as the space within the cavity of the procapsid becomes smaller and smaller [19, 21, 51, 54].

To argue for the one-way traffic property of phi29 DNA-packaging channel in a biologically active motor and to rule out the possibility that dsDNA remaining inside the channel is due to the contributions of pRNA, gp16, and/or γ-S-ATP, sedimentation assays were performed. The fully packaged DNA and the DNA packaging intermediates were treated individually and in combination with DNase I and/or RNase A (Fig. 4.15). DNase treatment excludes the possibility of any contact of motor components that could restrain the dsDNA from exiting, while RNase A digestion specifically excludes the possibility of pRNA and gp16 from contacting and holding the DNA. Since pRNA serves as a bridge between the motor and gp16 [20], RNase digestion will eliminate the fulcrum to prevent the exercise of force, if any, from pRNA and gp16 to dsDNA. When the

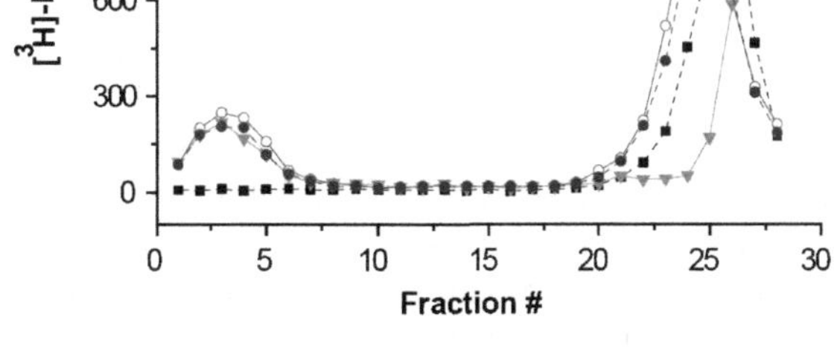

Fig. 4.15 Sedimentation assays for DNA packaging and retention within the procapsid with [3H]-phi29 DNA via 5–20% sucrose gradient. DNA was packaged into the procapsid through the channel of phi29 DNA packaging motor with the in vitro phi29 DNA packaging system [57, 58]). After DNA packaging, the samples were divided and digested with either DNase I or RNase A, respectively. All samples were loaded onto the top of the gradient and sedimented from right to left by ultracentrifugation followed by scintillation counting. Figures reproduced with permissions from: Ref. [44], © American Chemical Society

DNA-packaging intermediates, (with partial dsDNA within the channel and partial dsDNA outside the channel) were treated with DNase I or RNase A, or both, the partially packaged dsDNA stayed inside the procapsid and did not show a reverse exit in the gradient under highest centrifugation force (Fig. 4.15). The data strongly support the conclusion that the motor channel in an active motor also exercises the one-way traffic property.

The studies reported here provide direct evidence to prove the pushing or valving model proposed recently by Guo and Lee [9]. In this model, the connector remains static; DNA translocation is induced by a DNA packaging enzyme or terminase, which pushes a certain length of DNA into the procapsid, and then shifts to bind to a far distal region of the DNA and inserts an additional section. This model does not exclude the *socket wrench* rotating motion by the enzyme. The one-way traffic makes the DNA enter the procapsid and does not allow it to come out, similar to the pumping of blood into the heart and the use of valves to control the flow of blood.

4.7.2 Nanopore Based Stochastic Sensing

Nanopore-based stochastic sensing is an emerging analytical technique that enables measurements of analytes at the single molecule level. When a molecule of interest passes through the channel, the ionic current through the pores would reflect the amplitude, the duration and the rate of the resultant blockade events. By statistical analysis, information on concentration and identity of molecular species can be derived. The DNA translocation experiments with the robust phi29 connectors have revealed its potential for nanopore analysis.

The connector has a larger diameter (~3.6 nm) at its narrowest constriction as compared to some well-established channels, such as the 1.4-nm α-haemolysin channel which only allows the passage of ssDNA. More importantly, for the development of the phi29 connector based target-selective stochastic sensors, the larger channel of phi29 connector would provide more flexibility in the choice of bulky ligands covalently bound inside the channel. Since its crystal structure is known, a systematic and engineering-based approach can potentially be applied to develop a system with enhanced analytical capabilities.

There are many bacterial outer membrane porins with larger pores that can be potentially used as stochastic sensors. However, the gating of these channel proteins induced by voltage or solution conditions, such as protons, anions or cations, may cause transient current blockades in single channel recording that would interfere with detection of analytes. For example, OmpG porins show spontaneous gating under ± 40 mV [55]. In contrast, the phi29 connector channel showed stable channel properties in the voltage range of -150 mV to +150 mV, even at extreme pH conditions. Therefore, the phi29 connector channel would be an excellent candidate for biological pore based stochastic sensors.

4.7.3 Nanopore Based Sequencing of DNA

Nanopore-based DNA sequencing applications holds considerable promise and has the potential to operate orders of magnitude faster and cheaper than traditional methods. In addition, this is a label-free, single molecule recognition approach and requires no sample preparation or amplification. To date, there have been a number of reports demonstrating successful applications of nanopore-based nucleic acid analysis using either protein pores or solid state analogs [23–26, 28–31, 56]. However, the limiting factor for the single pore DNA sequencing technology is that the ultra-fast DNA passage is beyond the temporal resolution of currently available optical and electrical technologies for detecting individual nucleotides with high sensitivity and confidence. The key challenge therefore is to be able to slow the passage of DNA in a controlled fashion while maintaining high signal-to-noise ratio for accurately discriminating the bases. The large channel of the phi29 connector provides room for modifications, conjugations and placement of barriers for better signal detection.

4.8 Conclusions

Herein, an engineered form of the phi29 connector has been incorporated into a lipid bilayer, forming a highly conductive nanopore. In addition, the connector is a biological pore that is extremely reproducible and easily engineered, making it ideal and suitable for future biomedical applications. The robust properties of the phi29 connector with a large uniform size support the expectation that viral components are a new generation of nanomaterials or nanoscale building blocks.

Acknowledgments We thank Feng Xiao and Ying Cai for constructing the recombinant connectors; Rong Zhang for Q-PCR analysis, Jing Peng and Jia Geng for assistance in BLM experiments, and Dan Shu for the in vitro packaging assays. The research was supported by NIH grants GM059944, EB003730, and NIH Nanomedicine Development Center: Phi29 DNA Packaging Motor for Nanomedicine, through the NIH Roadmap for Medical Research (PN2 EY 018230) to P.G., who is also a co-founder of Kylin Therapeutics, Inc.

References

1. Black, L. W. DNA Packaging in dsDNA bacteriophages. *Ann Rev Microbiol* **43**, 267–292 (1989).
2. Guo, P. Introduction: Principles, perspectives, and potential applications in viral assembly. *Seminars in Virology (Editor's Introduction)* **5(1)**, 1–3 (1994).
3. Zhang, W. & Imperiale, M. J. Interaction of the adenovirus IVa2 protein with viral packaging sequences. *J. Virol.* **74**, 2687–2693 (2000).
4. Moss, B. *Virology*. Fields, B. N. & et al. (eds.), pp. 685–703 (Raven Press, New York,1985).

5. Scheffczik, H., Savva, C. G., Holzenburg, A., Kolesnikova, L. & Bogner, E. The terminase subunits pUL56 and pUL89 of human cytomegalovirus are DNA-metabolizing proteins with toroidal structure. *Nucleic Acids Res.* **30**, 1695–1703 (2002).
6. Salmon, B. & Baines, J. D. Herpes simplex virus DNA cleavage and packaging: association of multiple forms of U(L)15-encoded proteins with B capsids requires at least the U(L)6, U(L) 17, and U(L)28 genes. *J. Virol.* **72**, 3045–3050 (1998).
7. Olkkonen, V. M. *et al.* In vitro assembly of infectious nucleocapsids of bacteriophage phi 6: formation of a recombinant double-stranded RNA virus. *Proc Natl Acad Sci USA* **87**, 9173–9177 (1990).
8. Earnshaw, W. C. & Casjens, S. R. DNA packaging by the double-stranded DNA bacteriophages. *Cell* **21**, 319–331 (1980).
9. Guo, P. X. & Lee, T. J. Viral nanomotors for packaging of dsDNA and dsRNA. *Mol. Microbiol.* **64**, 886–903 (2007).
10. Rao, V. B. & Feiss, M. The Bacteriophage DNA Packaging Motor. *Annu. Rev. Genet.* **42**, 647–681 (2008).
11. Guo, P., Erickson, S. & Anderson, D. A small viral RNA is required for *in vitro* packaging of bacteriophage phi29 DNA. *Science* **236**, 690–694 (1987).
12. Guo, P., Grimes, S. & Anderson, D. A defined system for *in vitro* packaging of DNA-gp3 of the *Bacillus subtilis* bacteriophage phi29. *Proc. Natl. Acad. Sci. USA* **83**, 3505–3509 (1986).
13. Garcia, J. A., Mendez, E. & Salas, M. Cloning, Nucleotide-Sequence and High-Level Expression of the Gene Coding for the Connector Protein of Bacillus-Subtilis Phage-Phi-29. *Gene* **30**, 87–98 (1984).
14. Meifer, W. J. J., Horcajadas, J. A. & Salas, M. Phi29 family of phages. *Microbiol. Mol. Biol Rev.* **65(2)**, 261–287 (2001).
15. Simpson, A. A. *et al.* Structure determination of the head-tail connector of bacteriophage phi29. *Acta Cryst* **D57**, 1260–1269 (2001).
16. Guasch, A. *et al.* Detailed architecture of a DNA translocating machine: the high-resolution structure of the bacteriophage phi29 connector particle. *J. Mol. Biol.* **315**, 663–676 (2002).
17. Guo, P., Zhang, C., Chen, C., Trottier, M. & Garver, K. Inter-RNA interaction of phage phi29 pRNA to form a hexameric complex for viral DNA transportation. *Mol. Cell.* **2**, 149–155 (1998).
18. Zhang, F. *et al.* Function of hexameric RNA in packaging of bacteriophage phi29 DNA in vitro. *Mol. Cell.* **2**, 141–147 (1998).
19. Guo, P., Peterson, C. & Anderson, D. Prohead and DNA-gp3-dependent ATPase activity of the DNA packaging protein gp16 of bacteriophage ϕ29. *J Mol Biol* **197**, 229–236 (1987).
20. Lee, T. J. & Guo, P. Interaction of gp16 with pRNA and DNA for genome packaging by the motor of bacterial virus phi29. *J. Mol Biol.* **356**, 589–599 (2006).
21. Smith, D. E. *et al.* The bacteriophage phi29 portal motor can package DNA against a large internal force. *Nature* **413**, 748–752 (2001).
22. Bazinet, C. & King, J. The DNA translocation vertex of dsDNA bacteriophages. *Ann. Rev. Microbiol.* **39**, 109–129 (1985).
23. Thieffry, M., Chich, J. F., Goldschmidt, D. & Henry, J. P. Incorporation in Lipid Bilayers of A Large Conductance Cationic Channel from Mitochondrial-Membranes. *EMBO J.* **7**, 1449–1454 (1988).
24. Hinnah, S. C., Wagner, R., Sveshnikova, N., Harrer, R. & Soll, J. The chloroplast protein import channel Toc75: Pore properties and interaction with transit peptides. *Biophys J* **83**, 899–911 (2002).
25. Alcayaga, C., Venegas, R., Carrasco, A. & Wolff, D. Ion Channels from the Bacillus-Subtilis Plasma-Membrane Incorporated Into Planar Lipid Bilayers. *FEBS LETTERS* **311**, 246–250 (1992).
26. Benz, R., Schmid, A., Nakae, T. & Vosscherperkeuter, G. H. Pore Formation by LamB of *Escherichia coli* in Lipid Bilayer-Membranes. *J Bact* **165**, 978–986 (1986).

27. Helenius, A. & Simons, K. Solubilization of membranes by detergents. *Biochimica et Biophysica Acta (BBA) – Reviews on Biomembranes* **415**, 29–79 (1975).
28. Kasianowicz, J. J., Brandin, E., Branton, D. & Deamer, D. W. Characterization of individual polynucleotide molecules using a membrane channel. *Proc. Natl. Acad. Sci. USA* **93**, 13770–13773 (1996).
29. Vercoutere, W. *et al.* Rapid discrimination among individual DNA hairpin molecules at single-nucleotide resolution using an ion channel. *Nat Biotech* **19**, 248–252 (2001).
30. Howorka, S., Cheley, S. & Bayley, H. Sequence-specific detection of individual DNA strands using engineered nanopores. *Nat. Biotechnol.* **19**, 636–639 (2001).
31. Vercoutere, W. A. *et al.* Discrimination among individual Watson-Crick base pairs at the termini of single DNA hairpin molecules. *Nucleic Acids Res.* **31**, 1311–1318 (2003).
32. Bezrukov, S. M. Ion channels as molecular Coulter counters to probe metabolite transport. *Journal of Membrane Biology* **174**, 1–13 (2000).
33. Butler, T. Z., Pavlenok, M., Derrington, I. M., Niederweis, M. & Gundlach, J. H. Single-molecule DNA detection with an engineered MspA protein nanopore. *Proceedings of the National Academy of Sciences* **105**, 20647–20652 (2008).
34. Szabo, I., Tombola, F., Martinucci, S. & Zoratti, M. DNA interacts with Bacillus subtilis mechano-sensitive channels in native membrane patches. *Cell Physiol Biochem.* **12**, 127–134 (2002).
35. Mobasheri, H. & Lea, E. J. Biophysics of gating phenomena in voltage-dependent OmpC mutant porin channels (R74C and R37C) of *Escherichia coli* outer membranes. *Eur. Biophys. J.* **31**, 389–399 (2002).
36. Carneiro, C. M. *et al.* Probing the volume changes during voltage gating of Porin 31BM channel with nonelectrolyte polymers. *Biochim. Biophys. Acta.* **1612**, 144–153 (2003).
37. Smeets, R. M. *et al.* Salt dependence of ion transport and DNA translocation through solid-state nanopores. *Nano Lett.* **6**, 89–95 (2006).
38. Wang, H. & Branton, D. Nanopores with a spark for single-molecule detection. *Nature Biotechnology* **19**, 622–623 (2001).
39. Iqbal, S. M., Akin, D. & Bashir, R. Solid-state nanopore channels with DNA selectivity. *Nat Nano* **2**, 243–248 (2007).
40. Cai, Y., Xiao, F. & Guo, P. N- or C-terminal alterations of motor protein gp10 of bacterial virus phi29 on procapsid assembly, pRNA binding and DNA packaging. *Nanomedicine* **4**, 8–18 (2008).
41. Guo, Y., Blocker, F., Xiao, F. & Guo, P. Construction and 3-D computer modeling of connector arrays with tetragonal to decagonal transition induced by pRNA of phi29 DNA-packaging motor. *J. Nanosci. Nanotechnol.* **5**, 856–863 (2005).
42. Wendell, D. *et al.* Translocation of double-stranded DNA through membrane-adapted phi29 motor protein nanopores. *Nature Nanotechnology* **4**, 765–772 (2009).
43. Jing, P., Haque, F., Vonderheide, A., Montemagno, C. & Guo, P. Robust Properties of Membrane-Embedded Connector Channel of Bacterial Virus Phi29 DNA Packaging Motor. *Molecular BioSystems* **6**, 1844–1852 (2010).
44. Jing, P., Haque, F., Shu, D., Montemagno, C. & Guo, P. One-Way Traffic of a Viral Motor Channel for Double-Stranded DNA Translocation. *Nano Lett.* **10 (9)**, 3620–3627 (2010).
45. Sun, J., Cai, Y., Moll, W. D. & Guo, P. Controlling bacteriophage phi29 DNA-packaging motor by addition or discharge of a peptide at N-terminus of connector protein that interacts with pRNA. *Nucleic Acids Res.* **34(19)**, 5482–5490 (2006).
46. Robinson, M. A. *et al.* Affinity of molecular interactions in the bacteriophage phi29 DNA packaging motor. *Nucleic Acids Res.* **34**, 2698–2709 (2006).
47. Danilo D. Lasic *Liposomes in gene delivery*. CRC Press LLC., (1997).
48. Bertil Hille *Ion Channels of Excitable Membranes*. Sinauer Associates, Inc., (2001).
49. Song, L. *et al.* Structure of Staphylococcal alpha-Hemolysin, a Heptameric Transmembrane Pore. *Science* **274**, 1859–1865 (1996).

50. Wanunu, M., Morrison, W., Rabin, Y., Grosberg, A. Y. & Meller, A. Electrostatic focusing of unlabelled DNA into nanoscale pores using a salt gradient. *Nature Nanotechnology* **5**, 160–165 (2010).
51. Shu, D. & Guo, P. Only one pRNA hexamer but multiple copies of the DNA-packaging protein gp16 are needed for the motor to package bacterial virus phi29 genomic DNA. *Virology* **309(1)**, 108–113 (2003).
52. Guo, P., Peterson, C. & Anderson, D. Initiation events in *in vitro* packaging of bacteriophage ϕ29 DNA-gp3. *J Mol Biol* **197**, 219–228 (1987).
53. Moffitt, J. R. *et al.* Intersubunit coordination in a homomeric ring ATPase. *Nature* **457**, 446–450 (2009).
54. Tsay, J. M., Sippy, J., Feiss, M. & Smith, D. E. The Q motif of a viral packaging motor governs its force generation and communicates ATP recognition to DNA interaction. *Proc. Natl. Acad. Sci. USA* **106**, 14355–14360 (2009).
55. Conlan, S., Zhang, Y., Cheley, S. & Bayley, H. Biochemical and biophysical characterization of OmpG: A monomeric porin. *Biochemistry* **39**, 11845–11854 (2000).
56. Movileanu, L., Howorka, S., Braha, O. & Bayley, H. Detecting protein analytes that modulate transmembrane movement of a polymer chain within a single protein pore. *Nature Biotechnology* **18**, 1091–1095 (2000).
57. Lee, C. S. & Guo, P. Sequential interactions of structural proteins in phage phi29 procapsid assembly. *J. Virol.* **69**, 5024–5032 (1995).
58. Lee, C. S. & Guo, P. *In vitro* assembly of infectious virions of ds-DNA phage ϕ29 from cloned gene products and synthetic nucleic acids. *J. Virol.* **69**, 5018–5023 (1995).
59. Braha, O. *et al.* Designed protein pores as components for biosensors. *Chem. & Biol.* **4**, 497–505 (1997).
60. Wong, D., Jeon, T. J. & Schmidt, J. Single molecule measurements of channel proteins incorporated into biomimetic polymer membranes. *Nanotechnology* **17**, 3710–3717 (2006).

Chapter 5
Solid State Nanopores for Selective Sensing of DNA

Waseem Asghar, Joseph A. Billo, and Samir M. Iqbal

Abstract This chapter focuses on the functionalized solid state nanopores for the purpose of rapidly and accurately sensing specific sequence of DNA. Fabrication processes are described, consisting of standard photolithography followed by using either a transmission electron microscope or a plasma polymer film to create and shrink the nanopore. The molecular dynamics of DNA-nanopore interactions are also discussed. Smaller pore diameter results in slower translocation of DNA through the nanopore, but increases van der waals force on the DNA and decreases the ionic current. Increase in applied voltage decreases the van der Waals force while increasing the ionic current and translocation velocity. Chemical functionalization of nanopores is then discussed. This allows a nanopore to be selective with translocating specific DNA sequence. This is done by modifying the surface in an attempt to control its surface charges and hydrophobicity. Probe DNA is used to functionalize the pore and achieve selectivity. In terms of sensing, perfect complementary DNA translocates faster than single-base-mismatch DNA. The flux can be measured from the current pulses when the translocating DNA blocks the nanopore under applied voltage.

Keywords Biosensors • Selective sensing • Functionalization • Pulsed plasma polyerization • Surface composition • Molecular dynamics • Van der waals force • Translocation velocity • Self-assembled monolayer • Single base mismatch • Nucleotide polymorphism • Hairpin loop DNA • Electrochemical signals

---

S.M. Iqbal (✉)
Department of Electrical Engineering, Nanotechnology Research and Teaching Facility, Joint Graduate Studies Committee of the Bioengineering Program, University of Texas at Arlington and University of Texas Southwestern Medical Center at Dallas, TX 76019, USA
e-mail: smiqbal@uta.edu

S.M. Iqbal and R. Bashir (eds.), *Nanopores: Sensing and Fundamental Biological Interactions*, DOI 10.1007/978-1-4419-8252-0_5,

5.1 Introduction

Solid stare nanopores show great promise for application as biosensors. Though nanopores can be created by way of a bio-protein channel, solid state nanopores have the advantage of being easier to mass produce on a wide and inexpensive scale. Solid state nanopores could be used to rapidly sequence DNA, thus greatly reducing the time it takes to do so currently and reducing the workload of super-computers tasked with this. They also have the potential to detect anomalies in a DNA strand, giving vital forewarning to doctors about a patient's susceptibility to diseases like cancer.

Methods of fabricating and functionalizing solid state nanopores for various purposes have already been developed. They can be fabricated on wafers using photolithography and many special techniques. Though they can then be functionalized with a variety of ways, this chapter discusses functionalization of solid state nanopores for the purpose of acting as selective DNA sensors through the use of DNA probe hairpin loops. It has already been shown that this functionalization of nanopores has the ability to discriminate between perfect complementary and single-base-mismatch DNA sequences [1]. This illustrates that there is indeed the possibility of creating a reliable DNA sensor.

The goal of this chapter is to explore and explain the current techniques and methods which are currently being or have been used in order to create and functionalize solid state nanopores for selective sensing of DNA. A synopsis of fabrication is given at start followed by descriptions of ways to shrink the nanopores. After this is a discussion of DNA-nanopore molecular dynamics, which is necessary knowledge in order to understand DNA translocation through nanopores. Next a review is provided on DNA translocation and what has been done to functionalize solid state nanopores to accomplish this. The chapter then closes with an overview of biological applications of the functionalized nanopores.

5.2 Fabrication Processes

In the last decade or so, researchers have used different processes to fabricate solid state nanopores (SSNs). In one report, Li et al. at Harvard University made use of an Ar^+ ion beam to drill a SSN in Si_3N_4 membranes [2, 3]. They found that ion interactions with the thin membrane simultaneously involved two physical processes. One is surface erosion, which results in the opening of the pore. The other is surface diffusion, which involves matter transportation resulting in the shrinking of the pore. Both of the processes depend on temperature. Since then a number of techniques have emerged such as electron beam sculpting [4], laser heating [5], feedback controlled chemical etching [6], and etching of silicon-on-insulator (SOI) wafers [7].

Similar efforts have been made by Iqbal and co-workers. Their process involved a series of recursive and continuous sub-processes as follows:

- Dry oxidation of Silicon (Si) wafers
- Conventional photolithography
- Wet etching of SiO_2 using buffered hydrofluoric acid (BHF)
- Tetramethylammonium hydroxide (TMAH) anisotropic wet etching of Silicon
- Reactive ion etching (RIE)
- Focused ion beam (FIB) Drilling
- Scanning electron microscope (SEM) or Transmission electron microscope (TEM) pore shrinking and characterization.

The fabrication process started with the oxidation of Si wafers. The whole process is explained with the help of the process flow diagram in Fig. 5.1. The oxidation process depends on the furnace temperature and the diffusion rate coefficient. After oxidation, the oxidized wafers were coated with a thin layer of positive photoresist (PR) by using a spinner. After soft baking, the wafer was ready for photolithography, where a contact mask aligner was used for exposure. After development of the PR, small squares hundreds of micrometers on a side were opened on one side of the wafer.

Wet etching was then performed using buffered hydrofluoric acid at room temperature to etch the silicon dioxide (SiO_2) while the other side was protected against buffered hydrofluoric acid (BHF) etching by application of PR. At this point, the wafer was ready to go into a tetramethylammoniumhydroxide (TMAH) etch solution at 90 °C. TMAH is an anisotropic etchant; it etched the Si while the

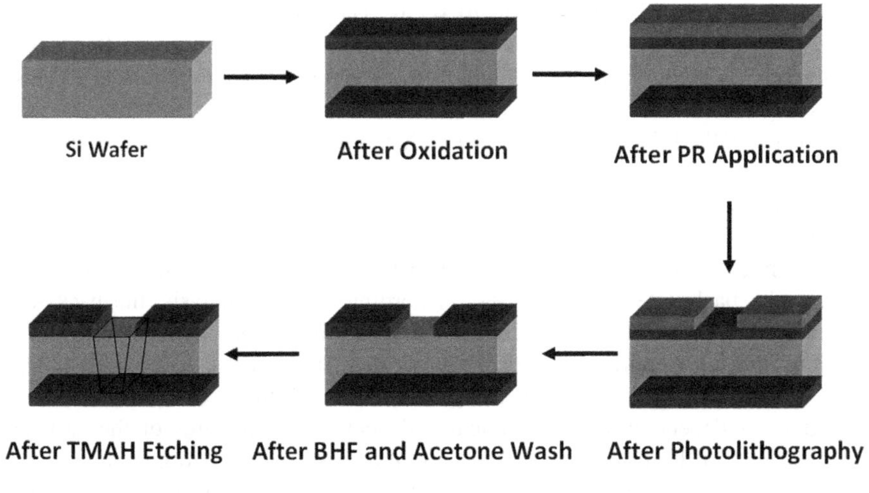

Fig. 5.1 The process flow for fabrication of a membrane which is used for FIB SSN drilling. The process starts with a bare silicon wafer which is oxidized at a temperature > 900 °C. PR is then applied at one side of the wafer. Photolithography, BHF etching, Acetone Wash, and TMAH etching (90 °C and 20% concentration) completes the process flow

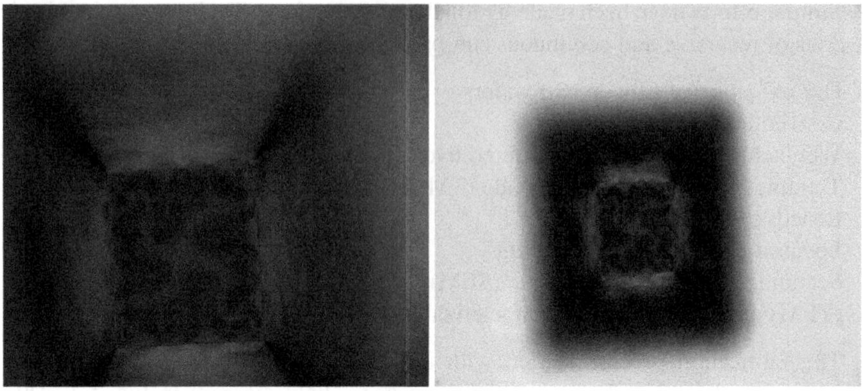

Fig. 5.2 Figure shows the confocal microscope images of the thin SiO_2 membrane. These images were taken after 8 h etching of Si in TMAH

SiO_2 served the purpose of a hard mask. The etching rate of TMAH for Si is 1 μm per minute at 90 °C while the etching rate for SiO_2 is just one Angstrom per minute [8]. The roughness of the etched surface was improved by adding isopropyl alcohol in TMAH solution [9]. The wafer was etched through all of its thickness until a thin membrane of SiO_2 was left at other side. The optical microscope image of membranes are shown in Fig. 5.2.

After fabricating the membranes, the thickness of the SiO_2 was reduced by using reactive ion etching (RIE). In previous reports, fluorine and chlorine have been used as feed gases. Iqbal and coworkers had used CF_4 gas in an RIE chamber at 100 W with a gas flow rate of 20 sccm. They had the option to use high power, but the membranes were already very thin; high power could break the membranes. In addition to this, a higher power yields a higher etching rate. This makes it difficult to control the final membrane thickness.

By using RIE, Iqbal and co-workers reduced the membrane thickness to tens of nanometers. A thin membrane was desirable, because it was essential to drill smaller nanopores with the help of FIB. Also, the thickness of the membrane would finally be transformed into the length of nano channel. Normally a smaller nano channel is desirable. They have also taken into account the RIE etching of from the back side the membrane. According to their analysis, the backside etch rate is roughly 15% as compared to that of the front side.

Gierak et al. reported the drilling of SiC membranes using FIB [10]. They fabricated 15 nm pores by using FIB. They explicate that the limitation of FIB is its dependability on physical and geometrical features of the surface [11]. However, FIB allows direct patterning of the membranes and good reproducibility. The acceleration voltage used in the FIB system was in the range of 30–50 KV. When accelerated ions struck the surface of the membrane, they sputtered the target. Energy of incident ions was transferred to the target atoms and electrons. The colliding ions were deflected from their paths, and surface

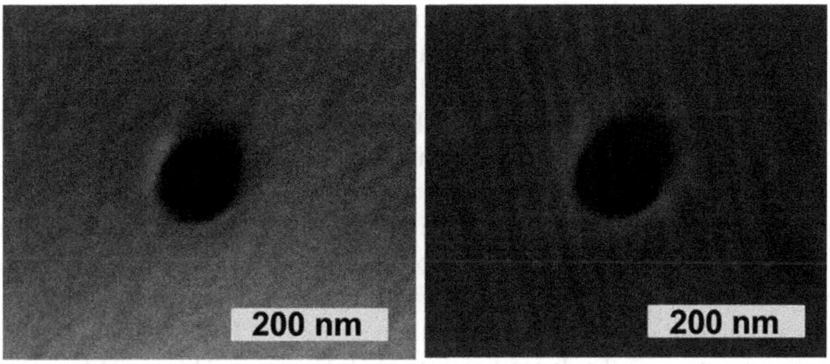

Fig. 5.3 Figure shows the SEM images of the SSNs drilled with the help of FIB in SiO_2 membrane. The *left* and *right images* of the SSNs show the pores with diameters of 100 and 160 nm respectively

atoms were also displaced from their positions [12]. The penetration depth R_p at depth D has the form as given below

$$\frac{Exp\left[-\left(D-R_p\right)^2 * \left(\Delta R_p\right)^2\right]}{2} \tag{5.1}$$

where ΔR_p is range straggle. The ejection of atoms and the number of atoms ejected from the surface depend on the energy of the striking ions as well as the target material of the membrane. The typical yield is 1–10 atoms/ion. The yield also depends on the incident angle between the ion beam and normal to the surface. Yield increases with $1/cos\theta$, where θ is the incident angle.

The initial diameter of the nanopore strongly depends on the membrane thickness and ion dose (ions/cm^3). Typically, a thicker membrane results in a bigger initial pore. It was very challenging to reduce the thickness of the SiO_2 membrane below 20 nm, as the membrane becomes very fragile and can break down during drilling and further processing. Iqbal and co-workers had also used FIB for their nanopore fabrication. They used ZEISS 1540XB FIB equipment, which had Gallium (Ga^+) ions accelerating at a voltage of 30 KV. Typically, the milling current used for SSN fabrication was 5–10 pA. They used the FIB in manual mode, and they fabricated the SSNs with a diameter range of 100–200 nm in SiO_2 membranes. The SEM images are shown in Fig. 5.3.

5.2.1 Deposition Processes for Size Reduction

In order to have a SSN which can mimic the function of a transmembrane protein channel in a lipid bi-layer, researchers have struggled to reduce the pore diameter to sub-ten nanometers. In previous reports, researchers have used different shrinking and deposition processes to reduce the diameter of SSNs. Danelon et al. had drilled a nanopore of 50–80 nm in a thin Si_3N_4 membrane, and they then reduced it to less than 10 nm by applying electron-beam-assisted silicon dioxide deposition to the selected pore area [13]. The gas precursor, tetraethyl orthosilicate (TEOS) decomposed by electron beam and deposited all over and inside the pore surface, resulted in reduction of the pore diameter. The typical values used for the electron beam voltage and current were 10 KV and 130 pA respectively. The TEOS reservoir was maintained at a constant temperature in order to have stable gas pressure inside the system. This had ensured a constant flow rate of gas and a controlled deposition.

Nilsson et al. had used the same procedure as used by Danelon, with H_2O vapor introduction to increase the deposition rate [14]. However, Nilsson et al. used ion-beam-assisted deposition of SiO_2 instead of using electron-beam. They successfully reduced the pore diameter to 25 nm. Harrell et al. used a single, conically-shaped gold tube embedded inside a polymer membrane [15]. They reduced the diameter of the pore via thiol-modified linker. Similar attempts had been made by Chen et al. [16]. They used atomic layer deposition to fine tune the surface properties and reduce the diameter of the pore. Alumina layers were deposited one by one and each cycle required only 12 s; it was proven to be a very controlled process. Alumina coating also had a few additional advantages: it covered the surface defects made by FIB, it eliminated the surface charge, and it caused a reduction of $1/f$ noise during IV measurements.

5.2.1.1 TEM Shrinking

As reported by Heng et al. [17] TEM was used to fabricate a single nanopore in thin membranes of tens-of-nanometers thickness. The same TEM was used to shrink the SSN diameter. When the electron beam is focused on a small membrane area, it melts the surface and drills a hole in the membrane. When the same beam is focused on a larger area, it shrinks the pore due to controlled surface melting.

Dekker and co-workers at Delft university also used TEM to reduce the pore diameter and then in other reports to drill pores in Si_3N_4 to single-nanometer precision [3]. The advantages of their process are in situ visualization of the pore diameter and the fact that the composition of the material remains the same after the process. After TEM melting, when the desired morphology was obtained, the control to lower the electron beam intensity was established; the soft material solidified immediately upon doing this.

They reported that the pore would shrink only when the pore radius satisfied the condition of $r < h/2$, where "r" is the radius of the pore and "h" is the thickness of the membrane. The soft membrane material tried to solidify to a morphology where it satisfied the minimum surface free energy, and the final pore diameter depended on the ratio of "r" and "h" of the membrane. Kim et al. reported similar results, but they updated the relation as $r < h/3$ for pore shrinking to occur under the electron beam [18, 19].

In another report, Chang et al. demonstrated similar results by using a conventional field emission scanning electron microscope (FESEM) [7]. They made the membrane using SOI wafers and made initial pores of 50–200 nm in diameter. They reported shrinking the pore diameter to 10 nm using FESEM. Chang et al. proposed a different mechanism for pore shrinking, which they called as radiolysis. When a high energy beam of electrons struck the surface of the porous Si or polycrystalline Si, it caused defects and dehydrogenation due to disruption of Si/Si and Si/H bonds. This resulted in the shrinking of the pore under FESEM. As a result, they claimed that previous assumptions of pore shrinking due to TEM irradiation by Storm et al. [4] could no longer applied on polycrystalline silicon.

5.2.1.2 Plasma Polymer Film Deposition

Plasma deposited polymers have an additional advantage over other deposition processes in the area of SSN diameter reduction: we can achieve different surface cross linker chemistries, improve surface wetability, and control surface densities of functional groups. In previous reports, polymer films have been used to reduce the diameter of nanopores fabricated in polycarbonated track-etched (PCTE) membranes [20]. PCTE filters are available commercially with 50–100 nm pores. Low pressure plasma discharge was produced using radio frequency (RF) electric fields to accelerate the electrons for collisions with the polymer monomer and creation of radical species. When the radical monomer species touched the surface of the target, they got deposited there and made a thin film of the polymer [21].

Timmons and co-workers at University of Texas at Arlington have been working on deposition of polymer films using 1-Vinyl-2-Pyrrolidone monomer [22]. They have discovered that the density of surface groups and cross linker chemistry could be controlled by using pulsed plasma instead of continuous plasma. They have used variable duty cycle pulsed plasma conditions and have found that the composition of the resultant films show increased retention of the functional groups in the starting monomer. The crosslinked densities of the polymer films were also reduced by decreasing the duty cycle. They applied an RF power of 13.56 MHz for plasma formation. Similar results had also been reported by Zhang et al. but they deposited pyrrole using plasma polymerization process [23].

In another report, Rinsch et al. had used a low pressure pulsed plasma polymerization process for deposition of allyl alcohol as monomer gas [24]. They noticed that when the RF duty cycle is lowered, the –OH functionality of the monomer is increasingly retained in the resulting organic polymer films. They reported increased surface adhesion with the help of controlled density and gradient layering technique.

Chapman et al. had reduced the nanopore diameter from 50 nm to 20 nm in PCTE membrane using RF pulsed plasma deposition [20]. They used vinyl acetic acid ($CH_2 = CHCH_2$ COOH) and perfluorohexane (nC_6 F_{14}) monomers for their process. The vinyl acetic acid (VAA) and perfluorohexane are hydrophylic and hydrophobic respectively. Various pulse condition were applied in order to have variable polymer composition at the surface. The monomer pressure ranged from 40 mTorr to 160 mTorr and peak powers ranged between 100–150 W. They controlled the permeability of O_2 and CO_2 through these membranes and found the permeation rate of the gases as a function of pore diameter and polymer chemistry.

5.3 DNA–Nanopore Interactions by Molecular Dynamics

To achieve optimal design of SSNs for Deoxyribonucleic acid (DNA) analysis, we need to understand DNA translocation dynamics under different applied voltages and electrolytic solution conditions. During translocation, it is essential to understand the effect of different forces which influence the DNA movement through the SSN. Due to the smaller lengths of DNA strands and SSN channels, DNA dynamics during translocation are still not precisely understood.

Cross et al. demonstrated the effect of SSN diameter on DNA mobility [25]. They concluded that in order to understand the molecular dynamics, the SSN diameter should be comparable to the size of the translocating species. The dependence of DNA kinetics on temperature has also been reported [26]. Iqbal and co-workers have developed a molecular dynamics model where they considered the effect of ion accumulation at the entrance of the SSNs [27]. In their simulations, they showed that higher voltages could compensate for the effects of ion accumulation. They used nano-scale molecular dynamics NAND for their simulations [28].

A CHARMM force field was used for DNA-SSNs interaction [29]. Si_3N_4 nanopores of different diameters ranging from 1.5 nm to 4 nm were used for simulations at room temperature as shown in Fig. 5.4. The supercomputer cluster POPEL was used for all the simulations. They simulated the effect of SSN diameter and applied voltage on DNA translocation kinetics.

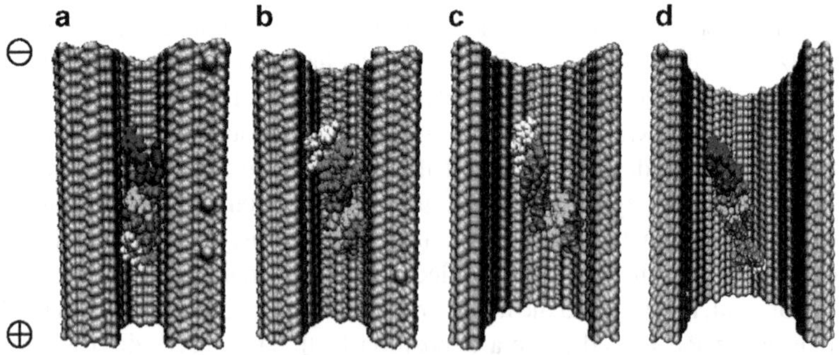

Fig. 5.4 Figure shows DNA translocation through the nanopore with different diameter: (**a**) 1.5 nm, (**b**) 2 nm, (**c**) 3 nm and (**d**) 4 nm [27]. Reprinted by permission from New World Publishing International: American Journal of Biomedical Sciences, copyright (2009)

5.3.1 Nanopore Diameter Dependant Interactions

In the presence of a frictionless environment, the electrophoretic translocation of DNA is independent of the SSN diameter. Here it is also assumed that there are no charges confined inside the nanopore, and length of the DNA strand does not play any role in the translocation. In one of the previous reports, Cross et al. proved that translocation of DNA did depend on the length of the DNA and nanopore diameter [25]. They suggested that the notion of frictionless and free electrophoresis did not exist in nanopores with diameters comparable to the size of the DNA.

Iqbal and co-workers reported size dependant DNA translocations [27]. They simulated the effects of DNA diameter on translocation velocity, van der Waals forces, and ionic current at higher applied voltage; these effects are shown in Fig. 5.5. The van der Waals forces decrease with increasing pore diameter due to

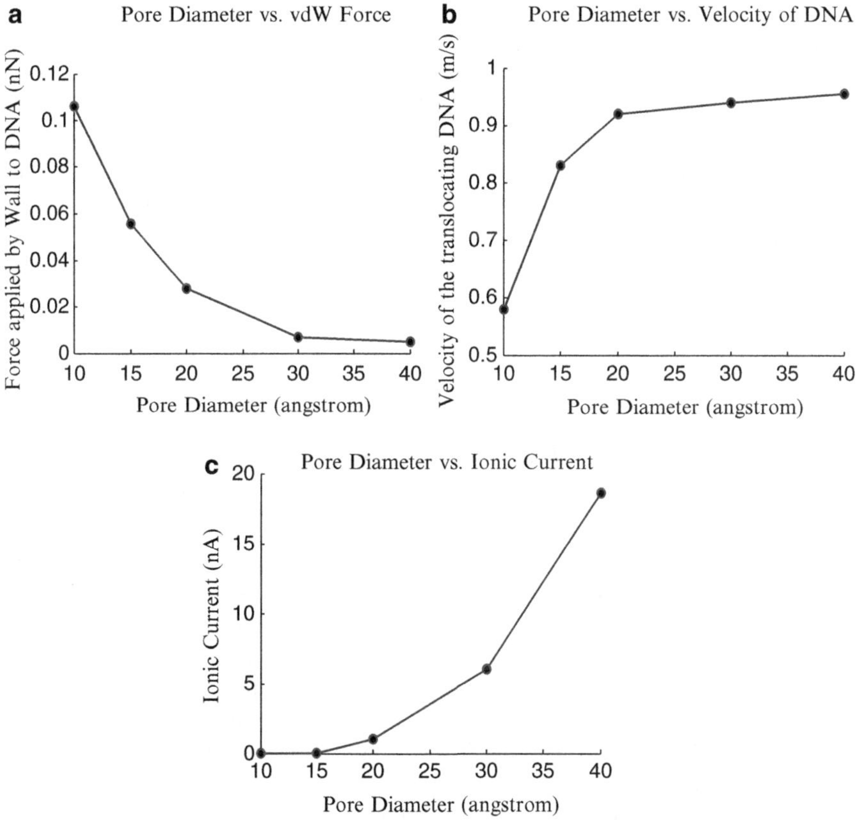

Fig. 5.5 Figure shows the effects of SSN diameter on (**a**) van der Waals force (**b**) velocity of translocation and (**c**) ionic current [27]. Reprinted by permission from New World Publishing International: American Journal of Biomedical Sciences, copyright (2009)

the confinement effect. Accordingly, electrophoretic translocation velocity of DNA decreases with reducing pore diameter. Both of the parameters have an exponential relation with the SSN diameter. At lower diameter, they showed linear behavior before finally trying to stabilize at higher diameter values. This analysis is very helpful in determining the optimal pore diameter for a better signal yield. Ionic current showed a parabolic relation with increasing diameter as shown in Fig. 5.5c.

They found that the translocation time is the same for different DNA sequences at certain applied conditions, which is contrary to previous experimental reports [30, 31]. The discrepancy could be from the smaller lengths of 8 bases of single-stranded DNA (ss-DNA) considered.

5.3.2 Applied Voltage Dependant Interactions

The ionic current is linearly dependant on applied voltage. Iqbal and co-workers found that ionic current showed a parabolic relation with the applied voltage as shown in Fig. 5.6c. They explained that this might be due the presence of DNA inside the SSN during simulations. The van der Waals forces also showed a parabolic dependence on the voltage (Fig. 5.6a). At higher applied voltages the translocation velocity of DNA is very high due to its linear behavior and translocation time is observed to be less than 100 ns. Such a small value of translocation time can be limiting factor on analyzing the DNA interactions. They suggested that a coarse grain model would be required to simulate the behavior of DNA at lower applied voltage.

5.3.3 Applied Voltage Dependant Effective Nanopore Diameter

The effective diameter of a functionalized nanopore depends on the applied voltage. In a bare nanopore, the ions of the electrolyte form a double layer which reduces the DNA translocation speed through the pore. But in the cases of ss-DNA attachment or polymer deposition on the inner pore surface, the SSN shows different kinetics. Liu et al. considered 8-base long ss-DNA coating on the inner walls of the nanopores [27]. The tethered DNA has a negative charge due to the phosphate backbone. The separation between the two attached DNA strands can be assumed to be approximately equal to the radius of gyration. They showed that the tethered DNA began to straighten in the direction of the oppositely-charged electrode due to negative charge on its backbone. It increased the effective diameter of the nanopore.

Besides applied voltage, the effective pore diameter is also influenced by ion condensation and ion flow. The state of the pore after ss-DNA attachment, as well as the effect of applied voltage on the effective pore diameter, is shown in

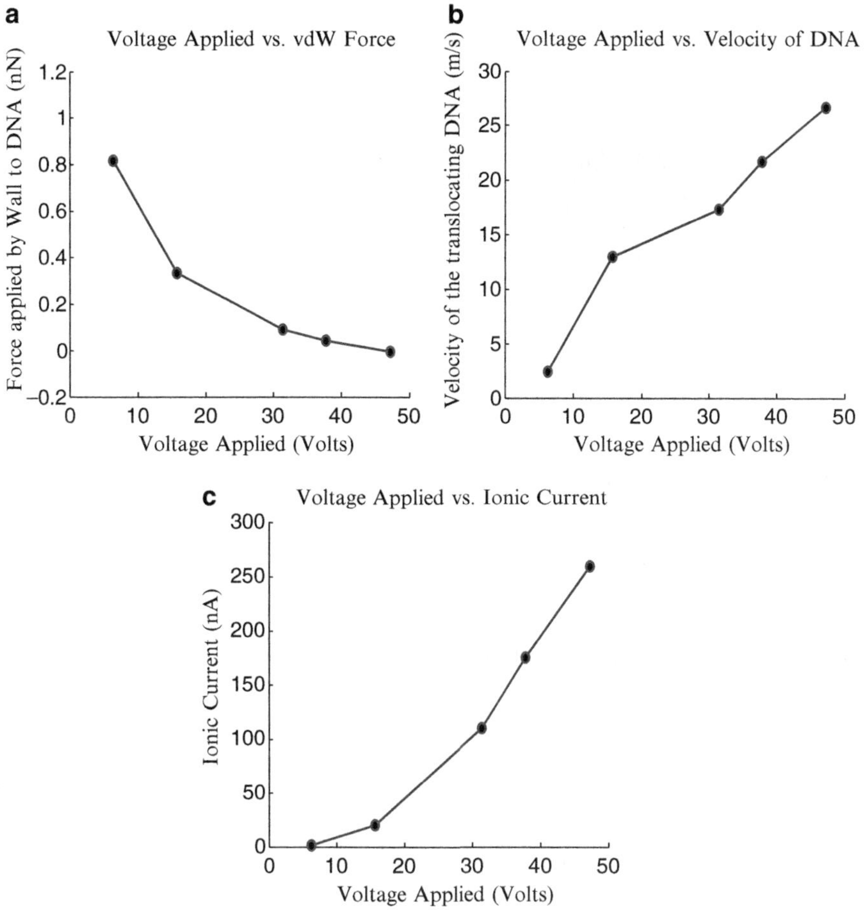

Fig. 5.6 Figure shows the effect of applied voltage on (**a**) van der Waals force (**b**) translocation velocity and (**c**) ionic current. Van der Waals force and ionic current display parabolic behavior, and translocation velocity shows nearly linear dependence on the applied voltage. Reprinted by permission from New World Publishing International: American Journal of Biomedical Sciences, copyright (2009)

Fig. 5.7. The reorientation of the tethered DNA plays very important role in estimating the effective diameter under varying applied potential. In Fig. 5.7b, it can be seen that applied potential plays a very vital role in estimating the pore diameter. Pore diameter increases almost linearly with the increasing applied potential. The effect of potential is more considerable when in the sub-nanometer range. In the case of polymer coating, a new model will be required to estimate the pore diameter at a certain voltage. The model of ref [27] is limited due to having a translocation time of less than 100 ns. For translocation times in milliseconds, they suggested a new model for low-voltage DNA kinetics.

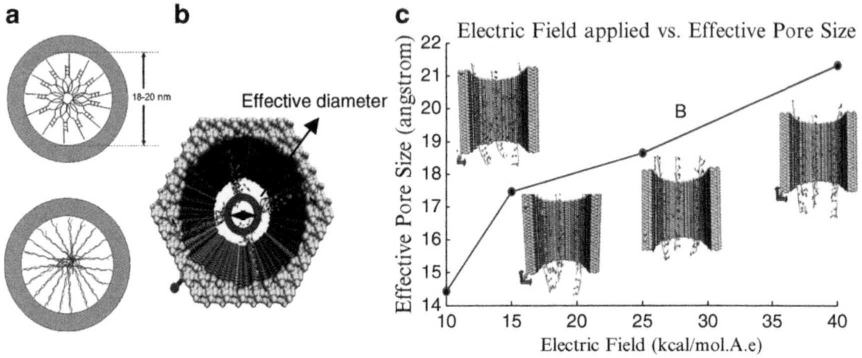

Fig. 5.7 The ss-DNAs coated on nanopore surface. (**a**) Radii of gyration for same length ss-DNA in hairpin and linear conformation, coated inside a pore; (**b**) Effective pore diameter; (**c**) DNA re-orients under applied electric field of various strengths and effective pore diameter as a function of electric field strength. The plot in (**c**) shows the linear effect of electric field on the pore diameter. This type of estimation is very important in determining the optimal condition of DNA analysis. Reprinted by permission from New World Publishing International: American Journal of Biomedical Sciences, copyright (2009)

5.4 Selective DNA Translocation

Functionalizing the inner nanopore walls is essential for selective DNA translocation. The nanopores can be chemically modified for bio-sensing [32–36]. For selective nanopore sensing, we have to not only provide selective probes but also have to control the surface charges and hydrophobicity of the nanopores. Surface charges play very important role in determining the ionic current and DNA interactions during translocation [37]. The surface modification improves the physicochemical properties and is the basic approach for selective sensing. Selectivity of nanopore transport channels can be controlled with a tethered ss-DNA attachment for selective molecular recognition [38]. Based on the material in which the SSN is fabricated, different strategies are employed depending on the surface functionalities of nanopore. The second goal to functionalize the nanopore is to make it more biologically friendly.

5.4.1 Functionalization, Sensing and Selectivity

Polymer deposition as described in Sect. 2.1.2 forms the basis for functionalization of the solid state nanopore. Depending on the substrate material and target molecules, different approaches are used for surface modification. Siwy et al. explored the thiol chemisorption on the gold surface of nanotubes [39]. They suggested an abiotic mimic with a single conical gold nanotube in a synthetic membrane.

They proposed a new idea of protein biosensors for channel stochastic sensing. They attached biotin at the entrance of the gold nanotube. This resulted in the complete blockage of ionic current, rather than transient charge, upon the attachment of the target molecule to the biochemical molecular-recognition agent [40]. They investigated biotin-streptavidin, protein-G to immunoglobulin, and antibody to protein ricin interactions. In another report, the same group used thiol-modified single strand DNA for attachment on the gold nanotube. They related the rectification of the ionic current through artificial ion channels to mimic the functionality of biological channels [15].

Saline chemistry was used by Nillson et al. for local functionalization of SiO_2 nanopores with DNA probes [14]. The localized deposition for nanopore diameter reduction was achieved with Tetraethylorthosilicate (TEOS) gas in an FIB system. Alkanethiol linkers were used for selective derivatization of nanopore entrance. The nanopores were then functionalized with acrylamide-terminated ss-DNA probes. Each funcationalization step was performed by immersing each chip in the appropriate solution. Contact angle measurement and X-ray photoelectron spectroscopy were used for confirming the DNA attachment. They explained that the ss-DNA attachment on the pore entrance had reduced the nanopore diameter, which was confirmed by IV measurements. Iqbal et al. used a similar method and implemented a bilayer scheme [1]. They used 3-aminopropyltrimethoxysilane, forming a silane layer, and the nanopore channel was further functionalized with a homo-bifunctional agent [41]. An amine-modified DNA probe was attached to the pore surface. In order to make a hair pin loop out of the probe DNA, the probe was engineered to have complementary sequences at both ends of the molecule. The hair-pin loop orientation was required for analyzing single-mismatch base pairs of DNA during translocation. When the target DNA passed through the functionalized nanopore, it perturbed the ionic current due to blockade and DNA-nanopore interactions.

The structure of a single translocation molecule or DNA segment can't be determined until the current signal magnitude is high and translocation velocity is low. At higher translocation velocity, it's very difficult to get useful information about a single base pair of DNA. The typical translocation velocity in a 10 nm nanopore at 120 mV is 27 nucleotides/μs [42]. This is not desirable due to the limitations of the electronic detection systems. DNA translocation speed can be decreased by varying the electrolyte temperature, applied potential, and electrolyte viscosity [31, 43, 44], but these variations also decrease the signal-to-noise ratio resulting from reduced ion mobility.

At this point the idea of sizing, quantifying and analyzing short DNA seems unrealistic. Kim et al. explored the idea of reducing the translocation time by functionalizing the pore with positively charged material [45]. First they reduced the pore diameter by atomic layer deposition (ALD) from 100 nm to 30 nm in a 250 nm free standing membrane. They used alumina (Al_2O_3) for the ALD process. The alumina provided the conformal coating on all sides of the nanopore walls [16]. The membrane thickness was increased to 320 nm after Al_2O_3 deposition. After cleaning with piranha the chip was immersed at room temperature for 1 h in ethanolic solution containing 1% (v/v) γ-aminopropyltriethoxysilane (APTES) [46]. Due to APTES functionalization, the nanopore walls had positive charge [47, 48], which would slow down the speed of negatively charged DNA translocating through it. Kim et al. reported 56.7 ± 2.6 μs and 104.6 ± 28.6 μs for 539 bp and 910 bp ds-DNA without sacrificing the current blockade magnitudes. They reported that cation selectivity of the nanopore would be decreased after modification with APTES due to excess positive charge inside the nanopore.

Similar types of silanization have been reported by other groups [49, 50]. Umehara et al. reported ion current rectification with a quartz nanopipette coated with positively charged poly-L-lysine (PLL). Adsorption of PLL molecules on the negatively-charged surface was followed by baking to achieve a stable coating layer. Wanunu et al. reported that the modification of the surface of nanopore below 5 nm tended to clog upon simple immersion of the chip in saline solution [51]. They suggested conductive cell salination for this problem. In conductive cell salination, organic electrolyte is mixed with saline chemistries and certain voltage is applied. The flow of ions disturbs any clogging of unbound saline molecules. They verified their results by ionic current measurements, which were in agreement with the actual nanopore diameter measurements.

There have been other chemistries proposed for nanopore functionalization. For example, the amide terminal can be functionalized with homo-bifunctional agent whose one active side can be attached to the NH_2 terminal of the chip. The other side can be attached to an amine-terminated DNA probe. The DNA attachment can be verified by hybridization with complementary florescent tagged ss-DNA followed by imaging. Ellipsometry or contact angle measurements can be used for verifying the functionalization of the chip with DNA probes. The SiO_2 surface without functionalization is more hydrophilic as compared to the chip functionalized with DNA. A lesser number of OH^- groups will be present on functionalized chips, which results in reduced hydrophilicity. In one recent report, Jang et al. showed that the ratio of N-(3-dimethylaminopropyl)-N-ethylcarbodiimide hydrochloride (EDC) and N-hydroxysuccinimide (NHS) for functionalization is very critical [52]. They suggested that for protein immobilization a ratio of 3:1 worked best for EDC:NHS. They showed that EDC diluted with ethanol instead of DI water and NHS diluted with PBS gave the best fluorescent intensity after protein attachment. One other important factor was the time duration of the reaction. They got the best results when the chips were immersed for 4 h in functionalizing mixture.

When a nanopore device is functionalized, a SAM layer is formed all over the chip surface and inside the nanopore. After DNA probe attachment, some of the probes would attach to the outer walls of the nanopores, which would also bind to specific target molecules but do not participate in signal yield. This reduces the sensitivity of the selective nanopore. In one report, it has been suggested that a soft lithography technique using polydimethylsiloxane (PDMS) can be used for selective functionalization of the outer nanopre membrane with hexaethyleneglycol thiol derivative [53]. Selective functionalization would thus block the non-specific attachment of the DNA molecules on the outer membrane while only the inner walls of the nanopore are functionalized. During detection the target molecules would not bind to the outer membrane and this would help in pushing the sensing of the nanopore to the next level.

Another important requirement is to improve the wetability of the nanopore by functionalization with hydrophilic molecules. Hydrophobic nanopores, even at hundreds of nanometers diameter, form a high potential barrier to the penetration of ions and charged molecules. SiO_2 and Si_3N_4 nanopores can be made hydrophilic by treating these with UV ozone plasma for 1 min. Immersion of a SiO_2 nanopore in piranha solution also makes the pore surface hydrophilic by the formation of OH^- functional groups.

Nanopore devices having membranes with more than one channel have also been used for selective sensing of biological molecules, but single-nanopore-based solid state devices are preferred because of real time stochastic sensing of individual hybridization and binding events. If localized nanopore functionalization is performed, a lesser number of probe molecules will be required for sensing. During sensing, the time interval between adding the target molecules to the solution and its binding with probe molecules depends on the concentration of target molecules. Depending on the time interval, we can analyze the concentration of the target molecules by first making our system intelligent by known concentrations. Ideally SSN can detect if we have just one target molecule in the solution, but the probability of binding events will be much less.

Single nucleotide polymorphism (SNP) occurs when one gene has two different alleles. With the addition of restriction enzymes, solid state nanopores can be used for identifying SNPs. Restriction enzymes are used for cleaving double strand DNA with specific sequences. Enzyme-DNA complex was used to detect SNPs in one report by Zhao et al. [54]. When DNA passed through silicon nitride nanopore, some threshold voltage is required to apply shear force. They found a different threshold voltage during the permeation of DNA and enzyme-DNA complex. For enzyme-DNA complex, more threshold voltage is required to dissociate the bond between enzyme and DNA which allows DNA to pass through the pore. Restriction enzyme was reported to show different binding forces depending on the sequence of DNA. Based on required threshold voltage differences, SNPs can be detected. Their approach can be used for genotyping instead of analyzing the complete DNA sequence.

5.4.2 *Perfect Complementary and Mismatch DNA Discrimination*

One of the motives behind solid state nanopores was to have a device which can be used to analyze DNA precisely without the loss of any useful information about its structure and dynamics. Many reports have aimed at improving the sensitivity and selectivity of nanopores. The inner walls of nanopores have been modified with different chemistries. In one report, Kohli et al. used synthetic membranes with gold nanotubes [55]. They functionalized the membrane pores with specific DNA hair-pin probes and used it for selective DNA translocation. They referred to their hair-pin loop DNA as "transporter". They found different fluxes for perfect complementary (PC) DNA and single-base-mismatch (SBM) DNA target molecules.

Similar results have been reported by others as well [56]. Peptide nucleic acids (PNA) were used as probes by one group [57]. PNA molecules are uncharged, synthetic DNA analogs. They gave better ionic strength dependant hybridization and biochemical stability. When PNA molecules were attached as probe to the nanopore walls they give more flux modulation on translocation of complementary DNA sequence. The increased flux modulation was the result of the charge repulsion mechanism when complementary DNA (−ve charge) hybridized to PNA (uncharged). At higher electrolyte concentrations, decreases in flux were not distinguishable. At much lower concentrations, the pore gave one fold modulation of flux upon DNA-PNA binding.

Using biological ion channel α-Hemolysin (α-HL), discrimination between complementary DNA and single base mismatch targets has been reported before [31, 58]. A longer translocation time was reported when a PC target passed through the α-HL channel as compared to a single base mismatched DNA target. Iqbal et al. used functionalized Si nanopores and reported the change in translocation time based on target DNA sequence, but interestingly they found longer translocation pulses for SBM ss-DNA sequences [1]. The nanopore surface was functionalized with hair-pin loop DNA as a selective probe. The fabricated nanopore and functionalized nanopore schematic is shown in Fig. 5.8. PC and

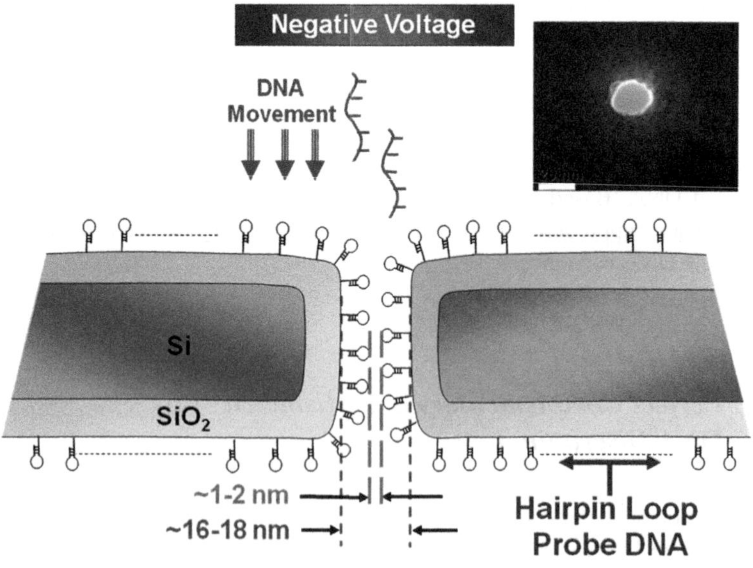

Fig. 5.8 The nanopore diameter functionalized with hairpin-loop probe DNA. The nanopore diameter was 16–18 nm which reduced to 2 nm after hairpin-loop immobilization inside the pore. Reprinted by permission from Macmillan Publishers Ltd: Nature Nanotechnology, copyright (2007)

SBM DNA targets were then introduced inside the negative compartment of the measuring setup. They found the translocation time was less for the PC DNA target when compared with the SBM DNA target. This result was exactly opposite to the previous reports using α-HL ion channels [58].

They also reported that the flux of the target DNA was lesser in the case of SBM as compared with perfect complementary DNA probes. The complementary DNA probe showed higher blockade current, faster translocation, and higher flux values in contrast to SBM targets. The data in Fig. 5.9 shows translocation behavior of different molecules.

Iqbal et al. reported the similar results while using nanopores with reduced diameters. With the reduced diameter of the nanopore, the channel gave better discrimination between SBM and PC DNA. They noticed no significant changes in the cases of 1-base-mismatch, 2-base-mismatch, or 3-base-mismatch. They also mixed both of the targets (PC and SBM) in a ratio of 1:1 before introducing into the negative compartment of the measuring setup. They reported that during first 10 min, SBM behavior was prominent. After 40 min, the nanopore lost its selectivity and PC behavior was prominent. The PC DNA target hybridized with the hairpin loop DNA and transversed at a faster rate and time while

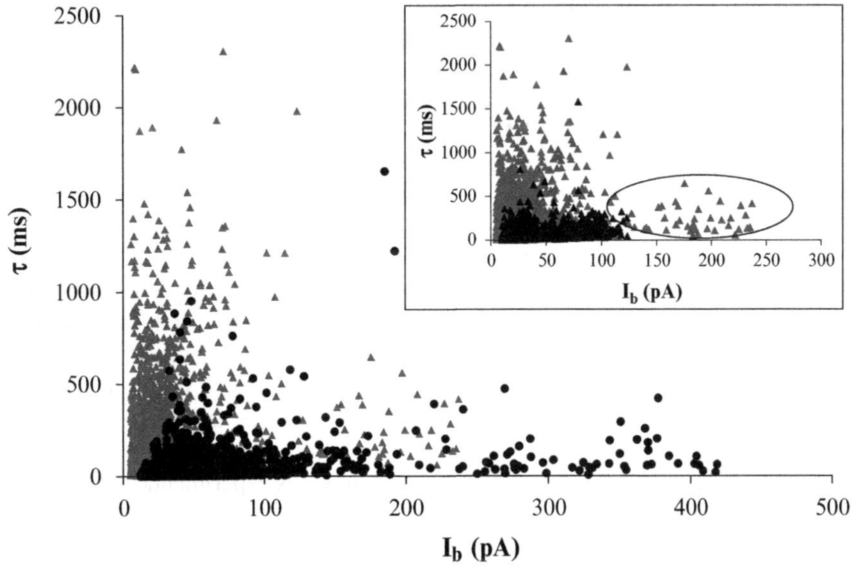

Fig. 5.9 Figure shows the average translocation time τ (ms) vs I_b (pA) blockage current. *Gray triangles* show the SBM and *black circles* show PC DNA targets. It clearly depicts the higher translocation times and lower flux in the case of SBM ss-DNA targets [1]. The *circle* in inset shows the subsequent addition of SBM target again without flushing the older sample. It means that as all DNA probes are hybridized with PC DNA before introducing SBM DNA, the channel has lost its selectivity and gives longer translocation time and higher blockade current in case of SBM DNA also. Reprinted by permission from Macmillan Publishers Ltd: Nature Nanotechnology, copyright (2007)

SBM DNA target faced repulsive potential due to same-charge. DNA flux and average translocation time was explained with the help of diffusion molecular transport theory and nearest neighbor thermodynamics calculations [59–62]. It was proven that the nanopore channels provided repulsive potential to the SBM DNA target, which caused its translocation time to increase and gave lesser flux compared with PC DNA target. The magnitude of repulsive potential for the SBM target was larger than attractive potential offered to PC DNA targets. Due to single-base-mismatch, the value of ΔG_{SBM} was greater in comparison to that for PC DNA target.

Energy-based mesoscale DNA translocation model was also developed by Liu and Iqbal [63] for optimizing the experimental parameters. Hairpin loop DNAs were modeled as independent potential wells inside the nanopore. When PC DNA passes through the functionalized nanopore, the hybridization lowers the potential of the well which facilitates the translocation of PC DNA through the nanopore. But the SBM DNA faces repulsive force from the constant potential well. Translocation kinetics of DNA were shown to largely depend on the applied potential. At lower applied electric field, the hybridization dominates the translocation process of PC DNA, while at higher applied potential, the electric field dominates the translocation process. In the case of SBM DNA, at lower applied potential the DNA can not pass through the nanopore due to repulsion from the potential wells, while at higher potential, the external force dominates the repulsive force and facilitates the DNA translocation process. So at higher applied potential (>200 mV), the single base selectivity of DNA diminishes and both types of DNA (PC and SBM DNAs) translocate with almost same speed. The model describes nonlinear behavior between the translocation speed and applied potential in the case of PC DNA. The model predicts blocking behavior at lower potential and direct pass behavior at higher potential during SBM DNA translocation. The model describes that at lower applied electric field, although the translocation velocity of PC DNA is significantly more than the SBM DNA but the output signal strength would be small. At higher applied potential, there would not be significant velocity difference. The model was optimized for electric field strength to obtain clear and distinctive signal for PC and SBM DNA during translocation. The model suggested that the applied potential in the range of 2.5–5 mV/nm would be ideal for this type of nanopore sensing setup. The output of the model agreed with the previous experimental results.

5.5 Biological Applications

At present we don't have ultimate knowledge about the self assembled monolayer (SAM) density, orientation and degree of order. In that respect protein functionalized nanopores are preferable because the single target attachment at desired location is achievable through mutagenesis. In a number of reports, nanocapillary array membranes (NCAMs) having arrays of nanopores ranging from 100 nm to 500 nm have also been used [64, 65]. An antibody-based recognition element was incorporated inside the nanopores for capturing the target proteins. The electroless plating of gold was performed on nanopores, and antibodies were immobilized by using thiol-gold linkage chemistry. Human Insulin was used as a target molecule. Mass spectroscopy was used to analyze the selective capturing of the insulin. A 20-fold increase in mass spectroscopy intensities were reported. Brogan et al. reported on the immobilization of rabbit anti-calf alkaline phosphates (RACAP) Immunoglobulin G (IgG) on a surface modified with gold and made use of thiol-gold interaction [64]. They showed that immunosurfaces with higher antigen binding ratios can be made with specifically immobilized fragments as compared with random immobilized fragments.

Transducing ligand binding mechanisms into electrochemical signals using nanoporous polycarbonate membranes have also been reported [66, 67]. The nanopores were first coated with gold and the diameter was reduced to 30 nm. To improve wetting, the sample was immersed into 10 mM Tris solution for 2 h. Solution enriched with galactose/glucose receptor (GGR) protein was added on the porous membrane. The chip was left for a few hours for the diffusion of protein inside the nanopore. After washing, new solution with D-glucose was added to one compartment of the conductive cell. Upon ligand binding, hinge-twist motion was observed which reduced the nanopore diameter and subsequently decreased the ionic current. They also reported a reduction in sensitivity with increasing nanopore diameter. They called their biosensor a "reagent-less electrochemical biosensor".

SSNs also have a number of applications to be used as a biosensor for detecting relatively large targets of size 10–300 nm. Uram et al. used relatively larger nanopores (diameter of 650 nm) for selectively detecting icosahedral chlorella virus [68]. Fabrication of a larger nanopore is a less demanding process. Antibodies were immobilized on the walls of nanopores that helped in selective detection of viruses based on resistive pulse sensing. Increases in the amplitude of the resistive pulse were reported on the selective binding of antibodies and virus particles. They reported that a virus concentration of 5×10^7 virus particle/mL could be successfully analyzed by their device.

5.6 Summary and Targets

It has been shown that nanopores can be created using standard photolithography followed by drilling with FIB. They can then be shrunk down to smaller diameters through the use of either TEM shrinking or plasma polymer film (PPF) deposition. Smaller nanopores are better because of the decreased translocation time, thus allowing for more accurate sensing. Selective sensing is achieved by the application of DNA probes in the nanopores, discriminating down to single-base mismatch in the target molecules.

A target to meet in the future is to increase the detection resolution for the functionalized nanopores. As it stands, DNA translocates very fast through a nanopore. The speed at which it translocates means that many, many trials must be performed in order to see a concrete difference in translocation time such as in Fig. 5.9. Finding ways to emphasize and increase this difference would make the detection resolution much clearer and require fewer trials. The push is on to develop a nanopore sensor which can detect each base of any DNA sequence efficiently, quickly and cheaply.

References

1. Iqbal, S.M., D. Akin, and R. Bashir, *Solid-state nanopore channels with DNA selectivity.* nature nanotechnology, 2007. **2**(4): p. 243–248.
2. Li, J., et al., *Ion-beam sculpting at nanometre length scales.* Nature, 2001. **412**(6843): p. 166–169.
3. Stein, D., J. Li, and J.A. Golovchenko, *Ion-beam sculpting time scales.* Physical review letters, 2002. **89**(27): p. 276106.
4. Storm, A.J., et al., *Fabrication of solid-state nanopores with single-nanometre precision.* Nature materials, 2003. **2**(8): p. 537–540.
5. Wu, S., S.R. Park, and X.S. Ling, *Lithography-free formation of nanopores in plastic membranes using laser heating.* Nano Lett, 2006. **6**(11): p. 2571–2576.
6. Park, S.R., H. Peng, and X.S. Ling, *Fabrication of nanopores in silicon chips using feedback chemical etching.* Small, 2007. **3**(1): p. 116.
7. Chang, H., et al., *Fabrication and characterization of solid-state nanopores using a field emission scanning electron microscope.* Applied Physics Letters, 2006. **88**: p. 103109.
8. Sato, K., et al., *Anisotropic etching rates of single-crystal silicon for TMAH water solution as a function of crystallographic orientation.* Sensors & Actuators: A. Physical, 1999. **73**(1–2): p. 131–137.
9. Sundaram, K.B., A. Vijayakumar, and G. Subramanian, *Smooth etching of silicon using TMAH and isopropyl alcohol for MEMS applications.* Microelectronic Engineering, 2005. **77**(3–4): p. 230–241.
10. Biance, A.L., et al., *Focused ion beam sculpted membranes for nanoscience tooling.* Microelectronic Engineering, 2006. **83**(4–9): p. 1474–1477.
11. Gierak, J., et al., *Sub-5 nm FIB direct patterning of nanodevices.* Microelectronic Engineering, 2007. **84**(5–8): p. 779–783.
12. Gadgil, V.J., et al., *Fabrication of nano structures in thin membranes with focused ion beam technology.* Surface & Coatings Technology, 2009. **203**(17–18): p. 2436–2441.
13. Danelon, C., et al., *Fabrication and functionalization of nanochannels by electron-beam-induced silicon oxide deposition.* Nano Lett, 2005. **5**: p. 403–407.
14. Nilsson, J., et al., *Localized functionalization of single nanopores.* Advanced Materials, 2006. **18**(4): p. 427–431.
15. Harrell, C.C., et al., *DNA-nanotube artificial ion channels.* Journal of the American Chemical Society, 2004. **126**(48): p. 15646.
16. Chen, P., et al., *Atomic layer deposition to fine-tune the surface properties and diameters of fabricated nanopores.* Nano Letters, 2004. **4**(7): p. 1333–1337.
17. Heng, J.B., et al., *Stretching DNA using the electric field in a synthetic nanopore.* Nano letters, 2005. **5**(10): p. 1883.
18. Kim, M.J., et al., *Rapid fabrication of uniformly sized nanopores and nanopore arrays for parallel DNA analysis.* Adv. Mater, 2006. **18**(23): p. 3149–3153.
19. Kim, M.J., et al., *Characteristics of solid-state nanometre pores fabricated using a transmission electron microscope.* Nanotechnology, 2007. **18**: p. 205302.

20. Chapman, C.L., et al., *Plasma polymer thin film depositions to regulate gas permeability through nanoporous track etched membranes*. Journal of Membrane Science, 2008. **318**(1–2): p. 137–144.
21. Timmons, R.B. and A.J. Griggs, *Pulsed plasma polymerizations*. Plasma Polymer Films: p. 217–245.
22. Han, L.M. and R.B. Timmons, *Pulsed-plasma polymerization of 1-vinyl-2-pyrrolidone: Synthesis of a linear polymer*. Journal of Polymer Science Part A Polymer Chemistry, 1998. **36**: p. 3121–3129.
23. Zhang, J., et al., *Investigation of the plasma polymer deposited from pyrrole*. Thin solid films, 1997. **307**(1–2): p. 14–20.
24. Rinsch, C.L., et al., *Pulsed radio frequency plasma polymerization of allyl alcohol: Controlled deposition of surface hydroxyl groups*. Langmuir, 1996. **12**(12): p. 2995–3002.
25. Cross, J.D., E.A. Strychalski, and H.G. Craighead, *Size-dependent DNA mobility in nanochannels*. Journal of Applied Physics, 2007. **102**: p. 024701.
26. Fyta, M.G., et al., *Multiscale coupling of molecular dynamics and hydrodynamics: application to DNA translocation through a nanopore*. Arxiv preprint physics/0701029, 2007.
27. Ramachandran, A., et al., *Characterization of DNA-Nanopore Interactions by Molecular Dynamics*. American Journal of Biomedical Sciences, 2009. **1**(4): p. 344–351.
28. Kalé, L., et al., *NAMD2: Greater Scalability for Parallel Molecular Dynamics* 1*. Journal of Computational Physics, 1999. **151**(1): p. 283–312.
29. MacKerell Jr, A.D., et al., *All-atom empirical potential for molecular modeling and dynamics studies of proteins*. Journal of Physical Chemistry B-Condensed Phase, 1998. **102**(18): p. 3586–3616.
30. Aksimentiev, A., et al., *Microscopic kinetics of DNA translocation through synthetic nanopores*. Biophysical journal, 2004. **87**(3): p. 2086–2097.
31. Meller, A., et al., *Rapid nanopore discrimination between single polynucleotide molecules*. Proceedings of the National Academy of Sciences of the United States of America, 2000. **97**(3): p. 1079.
32. Xiao, K.P., et al., *A chloride ion-selective solvent polymeric membrane electrode based on a hydrogen bond forming ionophore*. Anal. Chem, 1997. **69**(6): p. 1038–1044.
33. Minami, H., et al., *An evaluation of signal amplification by the ion channel sebsir based on a glutamate receptor ion channel protein*. Analytical Sciences, 1991. **7**(Supple): p. 1675–1676.
34. Kuramitz, H., et al., *Electrochemical immunoassay at a 17-estradiol self-assembled monolayer electrode using a redox marker*. The Analyst, 2003. **128**(2): p. 182–186.
35. Aoki, H. and Y. Umezawa, *Trace analysis of an oligonucleotide with a specific sequence using PNA-based ion-channel sensors*. The Analyst, 2003. **128**(6): p. 681–685.
36. Gadzekpo, V.P.Y., et al., *Development of an ion-channel sensor for heparin detection*. Analytica Chimica Acta, 2000. **411**(1–2): p. 163–173.
37. Ali, M., et al., *Chemical modification of track-etched single conical nanopores inducing inversed inner wall polarity*. GSI Annu. Rep. 2006, 2007. **1**: p. 323.
38. Lee, S.B., et al., *Antibody-based bio-nanotube membranes for enantiomeric drug separations*. Science, 2002. **296**(5576): p. 2198.
39. Siwy, Z., et al., *Conical-nanotube ion-current rectifiers: the role of surface charge*. Journal of the American Chemical Society, 2004. **126**(35): p. 10850.
40. Siwy, Z., et al., *Protein biosensors based on biofunctionalized conical gold nanotubes*. Journal of the American Chemical Society, 2005. **127**(14): p. 5000.
41. Manning, M., et al., *A versatile multi-platform biochip surface attachment chemistry*. Materials Science & Engineering C, 2003. **23**(3): p. 347–351.
42. Chen, P., et al., *Probing single DNA molecule transport using fabricated nanopores*. Nano Letters, 2004. **4**(11): p. 2293–2298.
43. Meller, A., L. Nivon, and D. Branton, *Voltage-driven DNA translocations through a nanopore*. Physical Review Letters, 2001. **86**(15): p. 3435–3438.

44. Fologea, D., et al., *Slowing DNA translocation in a solid-state nanopore*. Nano Lett., 2005. **5**: p. 1734–1737.
45. Kim, Y.R., et al., *Nanopore sensor for fast label-free detection of short double-stranded DNAs*. Biosensors and Bioelectronics, 2007. **22**(12): p. 2926–2931.
46. Guo, Z., et al., *Direct fluorescence analysis of genetic polymorphisms by hybridization with oligonucleotide arrays on glass supports*. Nucleic Acids Research, 1994. **22**(24): p. 5456.
47. Balladur, V., A. Theretz, and B. Mandrand, *Determination of the main forces driving DNA oligonucleotide adsorption onto aminated silica wafers*. Journal of colloid and interface science, 1997. **194**(2): p. 408–418.
48. Fang, Y. and J.H. Hoh, *Early intermediates in spermidine-induced DNA condensation on the surface of mica*. J. Am. Chem. Soc, 1998. **120**(35): p. 8903–8909.
49. Umehara, S., et al., *Current rectification with poly-L-lysine-coated quartz nanopipettes*. Nano Lett, 2006. **6**(11): p. 2486–2492.
50. Wang, G., et al., *Electrostatic-gated transport in chemically modified glass nanopore electrodes*. J. Am. Chem. Soc, 2006. **128**(23): p. 7679–7686.
51. Wanunu, M. and A. Meller, *Chemically modified solid-state nanopores*. Nano Letters, 2007. **7**(6): p. 1580–1585.
52. Jang, L.S. and H.K. Keng, *Modified fabrication process of protein chips using a short-chain self-assembled monolayer*. Biomedical Microdevices, 2008. **10**(2): p. 203–211.
53. Gyurcsányi, R.E., T. Vigassy, and E. Pretsch, *Biorecognition-modulated ion fluxes through functionalized gold nanotubules as a novel label-free biosensing approach*. Chemical Communications, 2003. **2003**(20): p. 2560–2561.
54. Zhao, Q., et al., *Detecting SNPs using a synthetic nanopore*. Nano letters, 2007. **7**(6): p. 1680.
55. Kohli, P., et al., *DNA-functionalized nanotube membranes with single-base mismatch selectivity*. Science, 2004. **305**(5686): p. 984.
56. Vlassiouk, I., P. Takmakov, and S. Smirnov, *Sensing DNA hybridization via ionic conductance through a nanoporous electrode*. Langmuir, 2005. **21**(11): p. 4776–4778.
57. Pretsch, E., *The new wave of ion-selective electrodes*. Trends in Analytical Chemistry, 2007. **26**(1): p. 46–51.
58. Howorka, S., S. Cheley, and H. Bayley, *Sequence-specific detection of individual DNA strands using engineered nanopores*. Nature biotechnology, 2001. **19**(7): p. 636–639.
59. Berezhkovskii, A.M. and S.M. Bezrukov, *Optimizing transport of metabolites through large channels: molecular sieves with and without binding*. Biophysical journal, 2005. **88**(3): p. 17–19.
60. Bauer, W.R. and W. Nadler, *Molecular transport through channels and pores: Effects of in-channel interactions and blocking*. Proceedings of the National Academy of Sciences, 2006. **103**(31): p. 11446.
61. Chaara, M. and R.D. Noble, *Effect of convective flow across a film on facilitated transport*. Separation Science and Technology, 1989. **24**(11): p. 893–903.
62. Noble, R.D., *Generalized microscopic mechanism of facilitated transport in fixed site carrier membranes*. Journal of membrane science, 1992. **75**(1–2): p. 121–129.
63. Liu, Y. and S.M. Iqbal, *A mesoscale model of DNA interaction with functionalized nanopore*. Applied Physics Letters, 2009. **95**: p. 223701.
64. Brogan, K.L., et al., *Direct oriented immobilization of F (ab) antibody fragments on gold*. Analytica Chimica Acta, 2003. **496**(1–2): p. 73–80.
65. Kim, B.Y., et al., *Direct Immobilization of Fabin Nanocapillaries for Manipulating Mass-Limited Samples*. J. Am. Chem. Soc, 2007. **129**(24): p. 7620–7626.
66. Benson, D.E., et al., *Design of bioelectronic interfaces by exploiting hinge-bending motions in proteins*. Science, 2001. **293**(5535): p. 1641.
67. Tripathi, A., et al., *Nanobiosensor Design Utilizing a Periplasmic E. coli Receptor Protein Immobilized within Au/Polycarbonate Nanopores*. Biosens. Bioelectron, 2003. **19**: p. 249–259.
68. Uram, J.D., et al., *Submicrometer pore-based characterization and quantification of antibody-virus interactions*. Small (Weinheim an der Bergstrasse, Germany), 2006. **2**(8–9): p. 967.

Chapter 6
Sensing Single Protein Molecules with Solid-State Nanopores

Bradley Ledden, Daniel Fologea, David S. Talaga, and Jiali Li

Abstract This chapter is focused on the development of experiments and theory of using solid-state nanopores for sensing single protein molecules in their native and unfolded states. Proteins serve diverse roles such as transport carriers, catalysts, molecular motors, cellular structural support, and others that make life possible. Because of these widely differing roles, proteins have an enormously diverse set of shapes, sizes, and charge structures as compared to polynucleic acids. Solid-state nanopores are particularly suitable for characterizing single protein molecules because they can be fabricated with adjustable dimensions and are stable under conditions that denature proteins. This chapter describes the nanopore experimental setup, signal recording, data analysis, and basic principles related to the experiments and the theory connecting the electrical signal with the properties of proteins. Examples of experimental results illustrate the ability of solid-state nanopores to differentiate proteins in their folded and unfolded states. Native-state protein nanopore translocation follows biased one-dimensional diffusion of charged particles that is sensitive to size and electrical charge. Due to the heterogeneous charge sequence of polypeptides, unfolded proteins obey a coupled electrophoretic and thermally activated process that is sequence specific. The chapter concludes with a discussion of future directions and open challenges for single protein characterization using solid-state nanopores.

Keywords Protein capture • Protein shape during translocation • Protein's Charge variation with pH • Parameters affecting translocation

---

J. Li (✉)
Department of Physics, University of Arkansas, 226 Physics Bldg., Fayetteville, AR 72701, USA
e-mail: jialili@uark.edu

S.M. Iqbal and R. Bashir (eds.), *Nanopores: Sensing and Fundamental Biological Interactions*, DOI 10.1007/978-1-4419-8252-0_6,

6.1 Introduction

The combination of variable polypeptide sequence and nearly arbitrary chain length results in an astronomical number of possible proteins. The resulting variety of structures and functions of proteins are so complicated that in some ways the diversity of life on Earth can be viewed as a consequence thereof. Eukaryotic life forms must transport functional biopolymers across membranes to survive. This fact motivates study of protein translocation through nanometer-scale pores. Such understanding is of fundamental importance not only for basic science but also in biotechnological applications that seek to mimic the selectivity and sensitivity of biological translocation [1, 2].

The physiochemical properties of the two most important classes of biopolymers, polynucleic acids and polypeptides, are substantially different. Therefore the cellular transport machinery is different depending on the nature of the biopolymer being transported. Proteins in particular present special challenges for cells, as they must be transported in a way that is compatible with attainment of a correctly folded three-dimensional native structure. The experimental and theoretical approaches to both natural and artificial ion channel translocation must accommodate these differences.

In this chapter we summarize the recent development of single nanopore measurements and theory to enable measurement of the physical properties of proteins at the single molecule level. These physical properties include: the protein size or volume, electrical charge, and conformational states.

6.1.1 *Polypeptides Measured by Protein Pores*

Protein channels or protein pores such as α-hemolysin have well defined structure and dimensions. However, due to their small fixed diameter, only polypeptides or denatured proteins are able to pass through the pores. Several research groups have studied polypeptide and protein pore interactions and the results have been presented in many publications. Starting in the 1990s, experiments began revealing that some peptide chains could reside inside in the lumen of protein pores or channels [3]. Later, studies in Lee's lab demonstrated that resistive pulse signals from a single α-hemolysin pore could differentiate between single, double, and collagen-like triple helices and pulses from α-hemolysin or aerolysin pores could reveal differences between wild type and mutant Histidine containing protein [4, 5]. The Auvray and Pelta research groups have studied the interaction of dextran sulfate and maltose binding protein with α-hemolysin pores [6, 7]. The Movileanu and Bayley laboratories have used wild type and mutated α-hemolysin pores to examine the effect of electrostatics on the interaction between peptide sequences and the α-hemolysin pore [8, 9].

6.1.2 Proteins Measured by Solid-State Nanopores

Solid-state nanopores are capable of measuring proteins of any conformation or size due to their tunable dimensions. Using 30–55 nm diameter, 20 nm thick nanopores formed by *e*-beam lithography in a free standing silicon nitride membrane, Han et al. measured bovine serum albumin (BSA), ovalbumin, avidin, streptavidin, human chorionic gonadotropin β (β-HCG), and monoclonal anti-β-HCG proteins [10, 11]. Using ~15 nm diameter silicon nitride nanopores, Fologea et al. compared the current blockage signal of BSA with fibrinogen, and measured the pH dependence of the BSA current blockage signal. In this work we confirmed that BSA indeed translocated through a nanopore using a chemiluminescent method [12]. Talaga and Li have studied unfolding of bovine β-lactoglobulin variant a (βLGa) and Histidine-containing phosphocarrier protein (HPr) [13]. Recently, Firnkes et al. reported on translocation of avidin [14], and Niedzwiecki et al. have reported on the adsorption of BSA in silicon nitride nanopores [15].

6.1.3 Parameters to Be Measured in a Nanopore Experiment

As illustrated in Fig. 6.1a, the main component of a nanopore sensing system is a single nanopore in a silicon nitride membrane separating two chambers connected electrically only by the electrolyte solution inside the nanopore. When a voltage is applied across the membrane, negatively (or positively) charged protein molecules added to the *cis* chamber near the nanopore are captured by the electric field, and driven through the nanopore to the positively (or negatively) biased *trans* chamber.

The translocation process of a protein molecule in a nanopore can be quantitatively described in terms of the nanopore and protein geometries. The geometric parameters required are illustrated in Fig. 6.1a and are: the average diameter (d_m) and the length (l_m) of a protein molecule, the mean diameter (D_p) or area (A_p) and the effective thickness (H_{eff}) of a nanopore.

At the macroscopic level, a nanopore's electrical behavior in an electrolyte solution of conductivity σ obeys Ohm's Law for an electrolytic resistor to good approximation. For an approximately cylindrical nanopore under applied voltage Ψ, the open pore current measured when the nanopore is not occupied by a protein molecule is $I_o = \Psi / R_0 = \Psi \sigma A_p / H_{eff}$. When a protein molecule is in the nanopore, it partially blocks the flow of ions [Fig. 6.1a, b (right)] producing a transient decrease in the open pore current. The current trace in Fig. 6.1c was recorded when a laminin (L6274, Sigma-Aldrich) protein sample was added to the *cis* chamber. The pH 7 electrolyte solution contained 1 M KCl and 3 M guanidine. The laminin protein was partially denatured in 3 M guanidine. The *trans* chamber was negatively biased for the recording; laminin is positively charged at pH 7.

Each current blockage event in Fig. 6.1c represents a laminin protein molecule interacting with or translocating through the nanopore. The transient decrease in

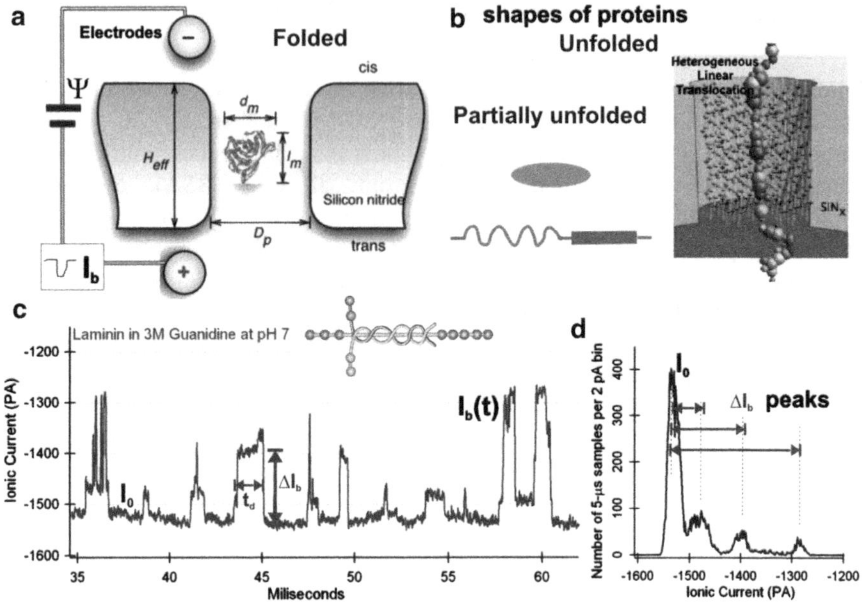

Fig. 6.1 (**a**) Schematic diagram of protein nanopore translocation experiment. (**b**) Illustration the shapes of proteins and an unfolded protein translocation in a nanopore. (**c**) Several recorded Laminin (L6274, Sigma-Aldrich) current blockage events in partially denatured condition (in 3 M Guanidine HCl at pH 7 and 1 M KCl). The insert is an illustrated structure of Laminin at its native state (taken from http://www.sigmaaldrich.com). (**d**) The Ipoint histogram or the instantaneous time distribution of blockade current ΔI_b over all events

ionic current caused by the translocating protein is a time dependent quantity, $\Delta I_b(t)$. Separating the translocation-relative variability in $\Delta I_b(t)$ from the noise-related variability in $\Delta I_b(t)$ is an ongoing challenge. In this chapter, we consider the mean current blockage amplitude ΔI_b and the dwell time t_d while a protein molecule stays in a nanopore.

6.1.4 Event Classification

Analysis of protein translocation current blockage events has been a challenging task due to their complexity. For initial analysis of a data set, we usually first generate the instantaneous time distribution of blockade current ΔI_b over all events (Fig. 6.1d), called I_{point} histogram [16]. The I_{point} histogram can help us to identify possible peak values of ΔI_b that can be used to set a trigger line to classify events. To classify events, a MATLAB based program has been developed. Using a trigger line at approximately half the value of an interested peak, the MATLAB program can select a group of events, calculate the mean blockage current ΔI_b and the time duration t_d of each event. When the current, $I(t)$, crosses the trigger line the event is considered to start, and the event stops when the current crosses the trigger line again. The time difference between the stop and start times is calculated as the event duration t_d. The average current of all points between the start and stop times is calculated as ΔI_b. The events with t_d that are very short (beyond the time resolution of the measuring system) and too long are discarded.

6.2 Principles of Protein Translocation

The translocation of proteins through nanopores differs from that of the more commonly studied polynucleic acids. The physical properties of proteins differ from those of polynucleic acids in several important ways that directly impact their nanopore capture, insertion, and translocation. These differences result in changes in data analysis and interpretation for protein translocation in nanopores.

6.2.1 Protein Capture by Nanopores

A charged molecule near the entrance of a properly biased nanopore will be captured. Entrance into the pore depends on the molecule's ability to attain a sterically compatible geometry for translocation during its encounter with the nanopore opening. For a polynucleotide this typically requires threading one end of the chain or forming a bend in the chain. Since a polynucleotide is uniformly charged, bend insertion can occur essentially anywhere in the sequence. By contrast, proteins have both pre-formed loops and charges of both sign distributed along their length suggesting that particular locations of bend insertion will be preferred. Protein C and N termini can have opposite polarity and therefore may also exhibit selectivity during the insertion event.

6.2.2 Protein Shape or Geometry During Translocation

In contrast to polynucleotides, proteins most often have a single well-defined three-dimensional native-state structure. This structure can be partly or completely disrupted by denaturants, temperature, or the application of electric fields. Tertiary contacts in proteins can be stabilized by both covalent and non-covalent interactions. Thus, the structure of the protein during translocation can be globular (Fig. 6.1a), looped, or a completely unfolded linear chain and (Fig. 6.1b) depending on the conditions during the measurement.

When the protein translocates as a loop or linear chain, only a segment of the amino acid chain will typically be inside the nanopore (Fig. 6.1b) and be exposed to the influence of the electric field therein. The molecular volume of amino acids varies much more than that of nucleotides. As a result, the magnitude of the current drop can vary more for proteins and is potentially more sensitive to particular features in the sequence. Moreover the distribution of charges along the polypeptide chain is sequence dependent; the net charge inside the pore can fluctuate as a function of translocation position. The net electrical driving force can change direction when the net charge of the local segment changes sign, which could drive the protein back and forth near the electrically neutral charged region. The polypeptide chain could be transiently stalled or trapped in an electrical potential well due to protein heterogeneous charge sequence.

The variability of protein physical properties influences the behavior of the main translocation observables derived from nanopore resistive pulse measurements. The complexity in the translocation physics of proteins is reflected in both the mean current drop and current blockage duration (translocation time).

6.2.3 The Mean Current Drop Amplitude of an Event

The intrusion of a protein or a polypeptide segment into the nanopore reduces its current flow capacity. To relate the transient current blockage amplitude ΔI_b to the physical properties of protein molecules, Ohm's law can be exploited based on the volume displacement of the electrolyte solution from the pore [17–20]. A translocating molecule that is much smaller than an idealized cylindrical nanopore will cause a transient drop in current that can be written as

$$\Delta I_b(t) = -I_0 \frac{\Lambda(t)}{H_{eff} A_p} [1 + f(d_m/D_p, l_m/H_{eff})] \tag{6.1}$$

when the nanopore accounts for nearly all the resistance in the circuit. Here $A_p H_{eff} = V_p$, the volume of the pore and $f(d_m/D_p, l_m/H_{eff})$ is a correction factor that depends on the shape of a protein molecule and the relative values associated with the dimensions of a molecule and a nanopore. (Note, this equation neglects nanopore surface charge effects, which might be significant at low salt concentrations [21].) The correction, f, contains higher order terms in the ratios of the molecule to pore diameter (d_m/D_p) and molecule to effective pore length (l_m/H_{eff}) [18]. For example, $f(d_m/D_p, l_m/H_{eff}) = (4/5)(d_m/D_p)^2$ for a spherical shaped particle that is smaller than the pore but not at the small particle limit [18]. For short molecules that fit entirely inside the length of the pore like small folded proteins, these ratios are less than one and contribute little to the current drop [18, 22–24]. For molecules such as polynucleotides and unfolded proteins that are much longer than the pore, a different derivation based on Ohm's law produces a relation that contains only the first term for the absolute, $\Delta I_b \approx I_0(\Lambda/V_p)$, or relative current drop, $\Delta I_b/I_0 \approx \Lambda/V_p$. These equations relate the measured current drop amplitude to the volume of the molecule transiently blocking the pore.

Equation (6.1) shows that the instantaneous excluded volume of a molecule, $\Lambda(t)$, can be estimated by measuring $\Delta I_b(t)$, however the correction term will vary depending on the conformation of the protein during the translocation which can vary as shown in Fig. 6.1b. A folded globular protein will require a correction appropriate for a spherical or ellipsoidal particle (Fig. 6.1a) and the expected current drop is larger than for a linear particle with the same volume inside the nanopore. The partially unfolded, and completely unfolded protein translocations will be closer to the linear case where the correction, $f(d_m/D_p, l_m/H_{eff}) \sim 0$. However, even in the linear case, the variability of the amino acid side chain volume suggests that there will be instantaneous excluded volume $\Lambda(t)$ changes as illustrated in Fig. 6.1b.

6.2.4 Protein Translocation Times

The time it takes for a charged protein molecule to pass a voltage biased nanopore or the dwell time, t_d, involves many phenomena. To simplify the problem, here we first assume protein molecules are rigid particles with a total charge Q and once the molecules enter the pore, they move along the center line of the pore of length H_{eff} under the electrical field strength $E = \Psi/H_{eff}$, and we further ignore complex issues such as protein-pore interactions and electro-osmotic flow. Under these assumptions, the total force exerted on a protein molecule is the electric driving force opposed by a viscous drag plus a term of random force caused by collision with molecules in solution, we can approximate the translocation time with a 1-D Langevin equation,

$$m\frac{dv}{dt} = F_e(x) - F_{drag} + kW(t) \tag{6.2}$$

where v is the velocity of the molecule, $F_e = Q_{in}\Psi/H_{eff}$ is the driving force due to the electric field, $F_{drag} = \alpha v$ where α is the drag coefficient related to the diffusion coefficient by $\alpha = k_bT/D$, k is defined by the fluctuation-dissipation theorem, and $W(t)$ is a 'noise term' or Wiener process that represents the random thermal forces on the molecule. The variable x is the position of the first part of the molecule that enters the pore.

If we assume a protein molecule translocate a nanopore with a terminal or an average speed, $dv/dt = 0$, and the average dwell time (mean first passage time) t_d is long, the mean value of the fluctuating force is zero, then $F_e(x) = F_{drag}$. Using this approximation, we can derive the translocation time for a uniformly charged long chain polymer like a DNA molecule and a charged globular shaped protein molecule.

Globular protein translocation. If the passing protein molecule is much smaller ($d_m << D_p$, $l_m << H_{eff}$) than the pore, and if we assume the interaction between a protein molecule and the pore can be neglected (i.e. free translocation), and further we assume the protein translocation process is driven by an electrophoretic force, $F_e = Q\Psi/H_{eff}$, opposed by a viscous drag, $F_{drag} = \alpha v = \eta C_f v$, with a terminal speed $v = H_{eff}/t_d$, the t_d can be written as

$$t_d = C_f\frac{\eta H_{eff}^2}{Q\Psi} \tag{6.3}$$

Here $\alpha = \eta C_f$ is the friction coefficient, η is the solution viscosity, C_f is a constant for a protein in a specific shape, and Q is the total net effective charge of a protein.

Both ΔI_b and t_d (Fig. 6.1c) also depend on the geometry and electrical properties of a nanopore, on the properties of solution, and the bias potential.

The distribution function for the translocation time can be derived from the Fokker-Planck equation equivalent to Eq. (6.2) [13].

$$\mathrm{P}_{fpt}(t) = \frac{e^{-\frac{(d-tv)^2}{4tD}}(d+tv)}{t\sqrt{4\pi tD}} \tag{6.4}$$

Here d is the distance to be translocated. For a long polymer with $l_m >> H_{\mathrm{eff}}$, $d = l_{\mathrm{m}} + H_{\mathrm{eff}}$; for a small spherical particle $d = H_{\mathrm{eff}}$. In this formulation, the width of the distribution arises due to thermal fluctuations. Equations (6.3) and (6.4) are appropriate for uniformly charged polymers and particles that behave as a point charge. The prediction under these assumptions is that the translocation time should decrease when electrostatic bias across the nanopore is increased.

When a protein molecule is unfolded and it passes a nanopore as a linear amino acid chain, very different translocation kinetics is expected due to the inhomogeneous charge distribution. A protein molecule that has neutral regions bracketed by a positively charged and a negatively charged region can experience a net zero electrical force when the net charge of the local segment chain in the pore is zero. At this stall point the unfolded protein molecule is electrostatically trapped; increasing the voltage only serves to increase the electrostatic trap barrier height. The molecule can escape the trap by thermal fluctuations, thus an inhomogeneous charge polymer translocation could be thermally activated if it has zero net charged regions. Since larger applied bias voltages, Ψ, would result in deeper traps, the dwell time, t_d, is predicted to increase with Ψ, the opposite prediction from the uniformly charged translocation model of Eqs. (6.3) and (6.4). The consequences of unfolded protein translocation are discussed below, in context with the experimental evidence.

6.3 Experimental Setup and Sample Preparation

6.3.1 Experimental Setup

The results discussed in this chapter were measured with a solid-state nanopore sensing system as illustrated in Fig. 6.2a. The main components of this system include a nanopore chip, two PDMS chambers (*cis* and *trans*), a pair of Ag/AgCl electrodes, and an Axopatch (200B) single channel recording system. The nanopore chip has a dimension of 3 mm by 3 mm and is sandwiched between two PDMS chambers. A freestanding silicon nitride membrane widow supported by a silicon substrate contains a single nanometer size pore at the center of the chip. The thickness of the freestanding membrane is ~275 nm as illustrated in the expanded view of the region around the pore in Fig. 6.2b. The nanopore is the

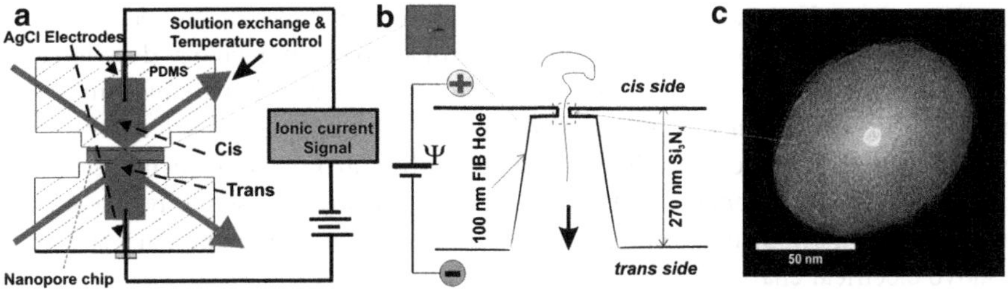

Fig. 6.2 (**a**) The Schematic of a solid-state nanopore detection system for protein translocation experiments. (**b**) The nanopore cross-section profile (across the center) illustrated. (**c**) An ion beam sculpted ~8 nm diameter pore imaged by TEM

only connection between two electrolyte filled chambers. The bandwidth of the Axopatch 200B was set at 10 or 100 kHz depending on the time duration of the protein signal and noise level of the nanopore.

6.3.2 *Solid-State Nanopore Fabrication*

Solid-state nanopores have been fabricated by low energy ion beams and by electron beams [25, 26] with several insulating materials such as silicon nitride [27, 28], silicon dioxide [29, 30], and aluminum oxides [31, 32]. The data discussed in this chapter were measured with silicon-nitride nanopores made by low energy noble gas ion beams. The typical size of nanopores used for protein measurements are between 4 and 30 nm selected according to the size and conformational states of proteins to be measured. Practically, nanopores with diameters 2–3 times larger than the protein are better choices for sensing protein molecules. The 275 nm low stress silicon nitride membrane was deposited by LPCVD at CNF (Cornel Nanofabrication Facility). The freestanding membrane window was created by photolithography, reactive ion etching, and anisotropic wet KOH etching. A single ~100 nm hole or a FIB hole (Fig. 6.2b) is milled through the freestanding membrane by a 50 keV Ga ion beam from a focused ion beam machine. The nanopores are made by shrinking the ~100 nm FIB hole to a desired diameter using low energy noble gas ion beam sculpting [27, 29, 33, 34]. The nanopores made by this method have a length or thickness between 10 and 20 nm depending on ion beam parameters selected [34]. A TEM image of a nanopore made by this method is shown in Fig. 6.2c.

6.4 Measuring Native State Proteins

The volume (Λ) a protein molecule occupies in space or solution is approximately proportional to the protein's molecular weight (M) or size, $\Lambda \propto M$. When a protein molecule is at its native or folded state and if the protein length is smaller than the nanopore, $l_m < H_{eff}$, the measured current blockage amplitude ΔI_b is expected to be directly proportional to the excluded volume Λ of a protein molecule as described in Eq. (6.1). In addition, when a native state protein molecule is passing through a nanopore driven by an electric field, the time duration t_d is expected inversely proportional to its electrical charge Q as described in Eq. (6.3). In this section, based on our experimental results and data analysis, we discuss the resolution of using ion beam sculpted silicon nitride nanopores on measuring a protein's size and relative electrical charge.

6.4.1 Sizing Protein Molecules

Figure 6.3 demonstrates the use of a solid-state nanopore to measure and discriminate the size of native state proteins. Using a 22 nm nominal diameter pore, three different proteins were measured sequentially: BSA (66.4 kDa, 607 aa, −18 *e*), Fibrinogen (340 kDa, ~1,500 aa, −16 *e*), and Laminin M (850 kDa, ~3,110 aa, +34 *e*). Event density plots in Fig. 6.3a, b, c show that the current blockage amplitude was correlated to protein size (Fig. 6.3d): ΔI_b (Laminin)=84 pA > ΔI_b (Fibrinogen)= 74 pA > ΔI_b (BSA)=50 pA. Figure 6.3e shows that the time duration followed the same trend: t_d(Lam)=154 μs > t_d(Fib)=101 μs > t_d(BSA)=64 μs. Laminin has twice the electrical charge than BSA and Fib but its t_d is the longest, which suggests that the time t_d depends on both the electrical charge and molecular size, $t_d \sim \Lambda/Q$,

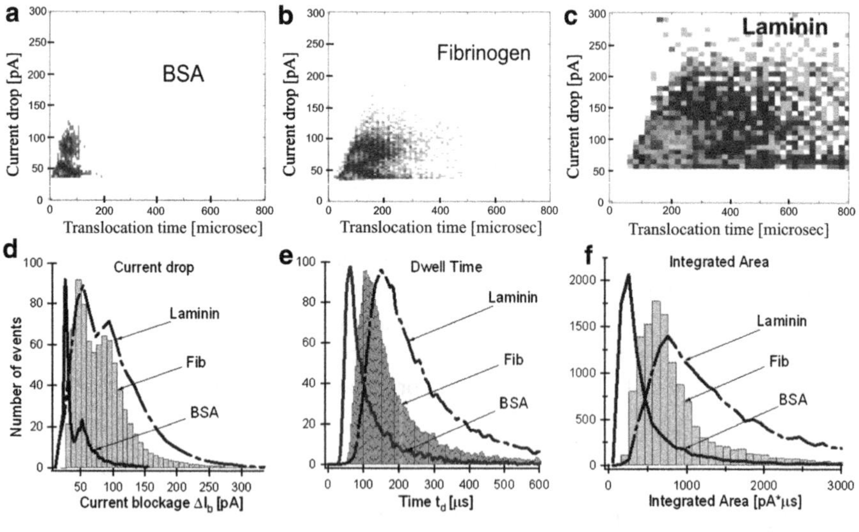

Fig. 6.3 Event Number density plots for BSA (**a**), Fibrinogen (**b**), and Laminin (**c**) in 1 M KCl, 40% Glycerol, pH=7. The current drop (**d**), time duration (**e**), and the integrated area of events for all three proteins (**f**). The pore used was 22 ± 2 nm made by Ar at 3 kV. The low pass filter was set at 100 kHz for this set of data

possibly due to an increased drag force when the molecular volume is larger. The integrated area histogram (Fig. 6.3f) is a measure of total charge of ions in solution displaced by the protein, $A_{ecd} = \int_{event} \Delta I_b(t)dt$, and clearly shows that the nanopore measurement differentiated these three proteins. Even though BSA and Fibrinogen had similar total charges (-16 e vs -18 e at pH=7.0), the difference in their molecular weights (and therefore volumes) allowed the current drop signals of fibrinogen to be clearly differentiated from those of BSA [12].

Advantages of using solid-state nanopores to characterize native state proteins include that the measurement is performed in aqueous salt solution and no marker or modifications of proteins are required. However, the data also show resolution in sizing different native state proteins is relatively low. One possible reason for this is that the distribution of the peaks including both ΔI_b and t_d are broad, limiting the precision of the excluded volume measurement. Another possible limitation in application to native state proteins is that most proteins have positive and negative charged residues. In the strong electric field in a nanopore, positive and negative charges will be driven to opposite directions possibly inducing structural changes in native state proteins. Both transient and long-lived structural changes could occur and the contribution of these phenomena to the nanopore events has yet to be fully characterized. When present, these nanopore-induced conformational changes could lead to systematic bias in the estimation of the volume Λ when using Eq. (6.1).

6.4.2 Measuring a Protein's Relative Charge at Different pH

Native state proteins, depending on their structure and charge distribution, can be partially or fully unfolded by the strong electrical field strength (~10^5V/cm) in a nanopore [35]. However, proteins with many disulfide bonds are the most likely to maintain their native state form during electrophoretic translocation. Native state BSA has 17 disulfide bonds, which should make it relatively stable in a nanopore; it is expected to behave like a simple charged particle during translocation and not experience large long-lived conformational changes [36]. Therefore BSA is a good model system for measuring the electrical charge change as a function of pH.

BSA (Fig. 6.4a) has an isoelectric point (pI) ranging from pH 5.1–5.5 [37]. The protein has an overall negative charge (-18 e) at pH 7. The charge of BSA can be altered by varying the pH of the solution. Using a ~16 nm diameter pore in a solution of 0.5 M KCl at pH 7.0, with ψ=120 mV, I_0 ~ 7.4 nA was measured. After addition of BSA to the negatively biased *cis* chamber, downward blockage events occurred (Fig. 6.7a) indicating that BSA molecules were negatively charged. When the *cis* chamber was positively biased, no blockage events were observed at the beginning of the experiment. The cumulative results are presented in an event number distribution plot (Fig. 6.7d).

When the solution pH of the chambers was lowered to acidic conditions (pH < 5) below the pI of BSA, current blockages disappeared if the *trans* chamber remained

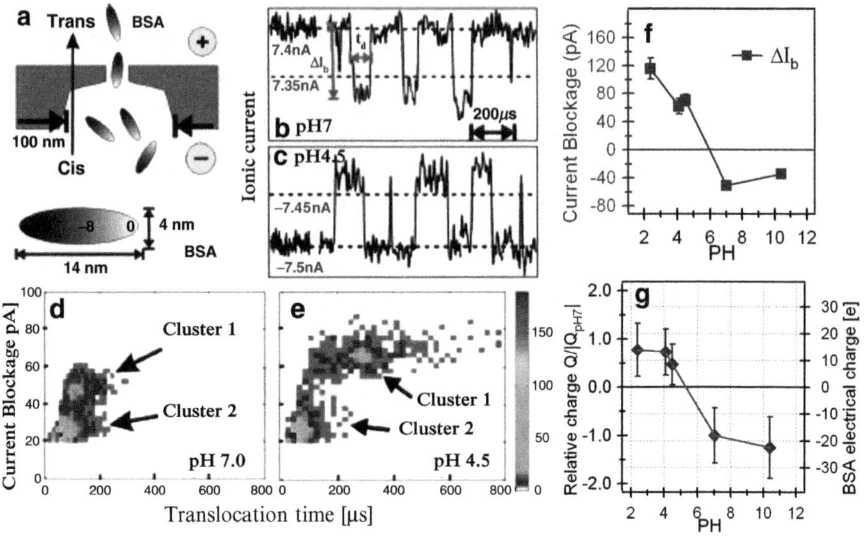

Fig. 6.4 BSA protein in 0.5 M KCl solution measured as a function of pH at ψ=120 mV. Panel (**a**) shows a sketch of BSA translocation through a solid-state nanopore with BSA dimensions shown. Panel (**b**) shows current drops due to BSA translocation at pH 7. Panel (**c**) shows the current drops from BSA translocation at pH 4.5. Note that BSA is positively charged and the bias across the nanopore has been reversed. Panels (**d**) and (**e**) are density plots of BSA translocation through a solid-state nanopore at these two different pH values. (**f**) The most probable values of ΔI_b, (**g**) the relative BSA electrical charge to pH 7. The open pore current was I_0~7.4 nA at ψ=120 mV for all pHs measured

positively biased. However, when the *trans* chamber was switched to negative bias, current blockages appeared again as shown in Fig. 6.7c, indicating the net charge of BSA protein had changed to positive at pH < 5. This measurement is consistent with the fact that BSA is positively charged when the pH is lower than its pI [38]. We studied the translocation of BSA through the same nanopore at three different acidic pH values (4.5, 4.1 and 2.4). The BSA molecules proved to be positively charged at all these pH values. When the pH was 4.5, near the pI of BSA, the most probable values are ΔI_b~71 pA and t_d ~269 μs as shown in Fig. 6.7e. The same measurement was performed at a higher pH value of 10.4. The most probable values of ΔI_b (Cluster 2 events were used) varied as solution pH changed (Fig. 6.4f) indicating the shape of the BSA molecule was not the same. The open pore current I_0 was approximately a constant at all the pH values measured.

A decrease in ΔI_b (Fig. 6.4f) suggests that the local excluded volume Λ of BSA molecules decreased at higher pH indicating a change in conformation or dimensions of BSA molecules took place. These results are consistent with reports of BSA denaturation at high pH, and formation of dimers at low pH [39] or BSA volume expansion at acidic pH [37]. A change in the charge of BSA (Q) can be estimated by the change in t_d. The relative charges Q/Q_{pH7} estimated from Eq. (6.3) are shown in Fig. 6.4g. A pI $\approx$ 5.3 was estimated for the BSA by this nanopore experiment.

This result demonstrated that a nanopore measurement is sensitive to the electrical charge of proteins and can be used to probe the pI of an unknown native state protein in an aqueous solution close to their native conditions.

6.5 Linear Amino Acid Chain Translocation

The electrophoretic force on an unfolded protein will tend to stretch it out during translocation. If the chain length l_m is longer than the pore length H_{eff}, only a segment of the chain is occupying the pore as illustrated in Figs. 6.1b and 6.5a. The electrophoretic translocation kinetics of a driven linear amino acid chain is expected to depend on the details of the primary sequence of charged amino acids rather than the total net charge as is the case for a globular proteins (Fig. 6.1a) or polynucleic acids. Using bovine β-lactoglobulin variant a (βLGa) and a histidine-containing phosphocarrier protein (HPr) as model proteins [35], we now discuss the consequences of unfolded βLGa and HPr translocation in silicon nitride nanopores. In addition, we discuss and compare the difference between folded and unfolded protein translocation kinetics.

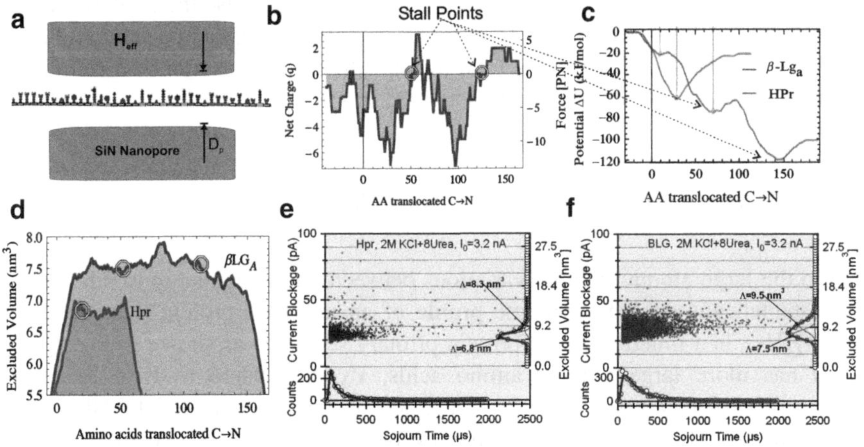

Fig. 6.5 (**a**) Schematic of linear translocation for βGLa. (**b**) The net charge of βGLa as predicted by treating the ionization of the individual residues as independent at pH 7.0 as a function of AA number. (**c**) The electrostatic contribution to the potential energy of βGLa and Hpr as a function of AA through the nanopore. This potential is for the C terminus entering first. (**d**) The calculated excluded partial volume Λ profile for ββGLa and Hpr proteins as a function of number of amino acids (AA) translocated. The slopes on the *left* and *right sides* show the insertion and exit of the linear chain. The *circles* mark the location of stall points (charge zero) during the translocation. Event Scatter Plot and marginal distributions of Sojourn Times and Calibrated Excluded Volume for βGLa and (**e**) HPr (**f**) in a small-diameter nanopore (D_p~4 nm) under denaturing conditions (8 M urea). The open pore current was the same (I_o=3.2 nA) during these measurements. βGLa (PDB file 2AKQ, 73,549.60 kDa) has 162 amino acids and it has a charge of ~8*e* at pH 7. HPr (PDB file 1POH, 9121.54 Da) has 85 aa and it has a charge of −2e at pH 7. The pore effective thickness H_{eff}=20 nm was used for all calculations

6.5.1 Stall Point Potential Well Model

Due to the heterogeneous distribution of charged amino acids, the calculated net charge of a segment of a βLGa polypeptide chain, $Q_{AA}=\sum q_i \, in\, the\, pore$, can be positive, negative and zero as a function of the number of residues translocating in a pore as shown in Fig. 6.5b. Locations where the net charge or force is zero, we call stall points. The positive and negative net charge positions in the curve imply that at these positions during the linear translocation the electrophoretic force, $Fe = Q_{AA}\Psi/H_{eff}$, can either oppose or drive the translocation. This can be seen more clearly in the translocation potential plot in Fig. 6.5c. The potential profile is calculated by

$$\Delta U(x) = \int_0^x \frac{\Psi}{H_{eff}} q(x)dx. \qquad (6.5)$$

The potential profile plot shows that βLGa has two stall points and Hpr has one. Near the stall points, the peptide chain is likely thermally fluctuating in the potential well that is formed due to the polypeptide charge sequence. If the potential well is deep enough, the unfolded protein will be metastably trapped in the nanopore; the translocation process would still go forward eventually because the total electrical driving force is downhill. While the unfolded protein is trapped at a stall point, its escape is thermally activated. The total translocation will therefore be a combination of electrophoretic and thermally activated processes.

6.5.2 The Excluded Volume at Stall Points

Due to the large atomic volume variations between amino acids (up to 3.8 times) the instantaneous excluded volume profile of an unfolded protein translocation, $\Lambda_{AA}=\sum V_{Ai}$, is a function of the protein primary sequence as shown in Fig. 6.5d. βLGa has more large volume amino acids, V_{Ai}, compared to HPr, thus the calculated excluded volume (Λ_{AA}) for βLGa is larger compared to HPr. The measured excluded volume is a time average of the local segment volume. Since the polypeptide chain dwells in the pore longer at the stall points, the corresponding excluded volume at the stall points (red circles in Fig. 6.5d) should have more weight. Since the purely electrophoretic part of the translocation should contribute ~1–2 μs to the translocation time of these small single-domain proteins, the net translocation time is likely dominated by the sojourns at the stall points. Thus the experimentally determined excluded volume should correspond to the stall point volumes.

The experimental data measured in the same ~4 nm diameter pore for βLGa (Fig. 6.5e) and HPr (Fig. 6.5f) under denatured conditions (8 M urea) show that the peak value of the mean ΔI_b is indeed larger for βLGa than HPr supporting this analysis. The volumes measured from the peak values of ΔI_b data match quantitatively with the volume calculated at the stall points for the βGLa and HPr in Fig. 6.5d (right axis). The broader distribution of ΔI_b for βLGa is likely caused by contributions from the translocation of looped configurations due to the presence of disulfide bonds in βLGa [35].

6.5.3 Parameters Affecting the Linear Translocation Potential

The net charge profile Q_{AA} is not only a function of protein primary charge sequence, it also depends on the nanopore effective length H_{eff} and solution pH as illustrated in Fig. 6.6. Changing the thickness of the pore changes the length of the segment of the unfolded protein that is exposed to the nanopore environment. The translocation potential changes significantly with pore thickness because of the changes in coarse grain sampling of the local structure. In this case, changing the pore thickness will change the presence or absence of stall points. In the limit of thick pores, $l_m < H_{eff}$, there will be no stall points.

Varying solution pH systematically changes the charge sequence of the same protein to vary the electrostatic translocation potential. Changing the pH alters the number, depth, and location of stall points along the polypeptide chain as shown in Fig. 6.6b for bacterial L-lactate dehydrogenase.

In addition, the depth of the potential well $\Delta U^{\ddagger}$ is deeper or the barrier height is larger at a higher voltage. This analysis implies the linear translocation potential profile depends on the protein charge sequence, nanopore thickness, solution pH, and applied voltage.

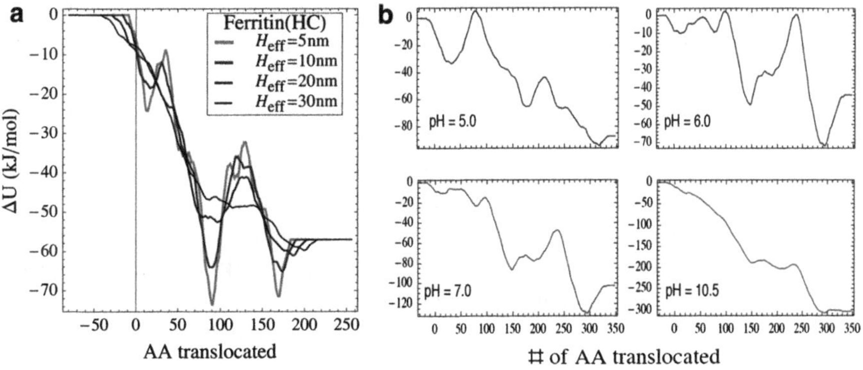

Fig. 6.6 Nanopore length H_{eff} and solution pH effects on translocation potential profile: (**a**) translocation potential profile at four different pore thicknesses (H_{eff}=5, 10, 20, 30 nm) for ferritin heavy chain (PDB file: *2Z6M, 176 aa*). (**b**) The predicted potentials for linear translocation of bacterial L-lactate dehydrogenase (PDB file: *1LLD, 319 aa*) through a 10 nm pore at pH=5.0, 6.0, 7.0, and 10.5 as labeled in the figure

6.6 Comparison of Time Histograms

For a linear amino acid chain translocation as shown in Fig. 6.5a, if there are neutral regions or stall points, Eq. (6.3) is no longer valid. If the potential wells are deep enough, Kramers reaction rate theory [40] predicts that the distribution of sojourn times should be multi-exponential according to the number of barriers present. When there is only one barrier present, using a simple one dimensional transition state theory, the predicted escape time from one electrostatic trap with barrier height ($\Delta U^{\ddagger}$) will be

$$t^{-1} \propto \exp(-\Delta U^{\ddagger}/k_B T) \quad (6.6)$$

This neglects protein conformational changes that could provide alternate lower-barrier pathways.

As shown in Fig. 6.5c, there is one stall point for Hpr and two for βLGa. The time histograms measured for these two proteins under denaturing conditions show that the t_d histogram of Hpr (Fig. 6.7a) fits well with a single exponential, and the t_d histogram of βLGa (Fig. 6.7b) fits better with two exponentials. This is consistent with the presence of no more than two stall points during translocation of these two proteins.

It is enlightening to compare the time histograms of heterogeneously charged polypeptides with homogeneously charged dsDNA. Without any stall points, due to their uniform charge density, the time histograms shown for a 2.7 kb (Fig. 6.7c) and a 7 kb (Fig. 6.7d) DNA fit well to the analytical model predicted in Eq. (6.4).

The DNA molecules translocate through a nanopore moving on average at the molecule's terminal velocity with variance increasing linearly with t_d according to $<x^2>=2\mathrm{D}t_\mathrm{d}=2(\mathrm{k_b T}/)\mathrm{t_d}$. In this case, the most probable time will be the length divided by its average velocity, $t_d=l_m/v$ ($l_m>>H_{eff}$). Thus a longer t_d is expected and observed for the 7 kb DNA (Fig. 6.8d) compare to the 2.7 kb.

The time distribution described in Eq. (6.4) derived for a charged particle is appropriate to fit the time distribution of BSA. In this case, l_m~14 nm$<H_{eff}$=20 nm, $d=H_{eff}$=20 nm. Indeed the Eq. (6.4) fits well for the time histograms of BSA at pH 7 (Fig. 6.7e) and pH 4.5 (Fig. 6.7f). These time histograms of the native state BSA at pH 7 (Fig. 6.7e) and at pH 4.5 (Fig. 6.7f) are from the cluster one events in Fig. 6.4d, e, respectively. These fits suggest that native state BSA translocation can be treated under the simple charged particle model.

The above analyses have made many simplifying assumptions and neglected many complex issues including neglect of: conformational changes during the translocation process, protein interactions with the nanopore wall, dynamics of long, floppy segments of the molecule outside of the pore, the surface charge of a nanopore and electroosmosis. Nevertheless, the simple models have thus far been able to quantitatively explain the protein translocation data.

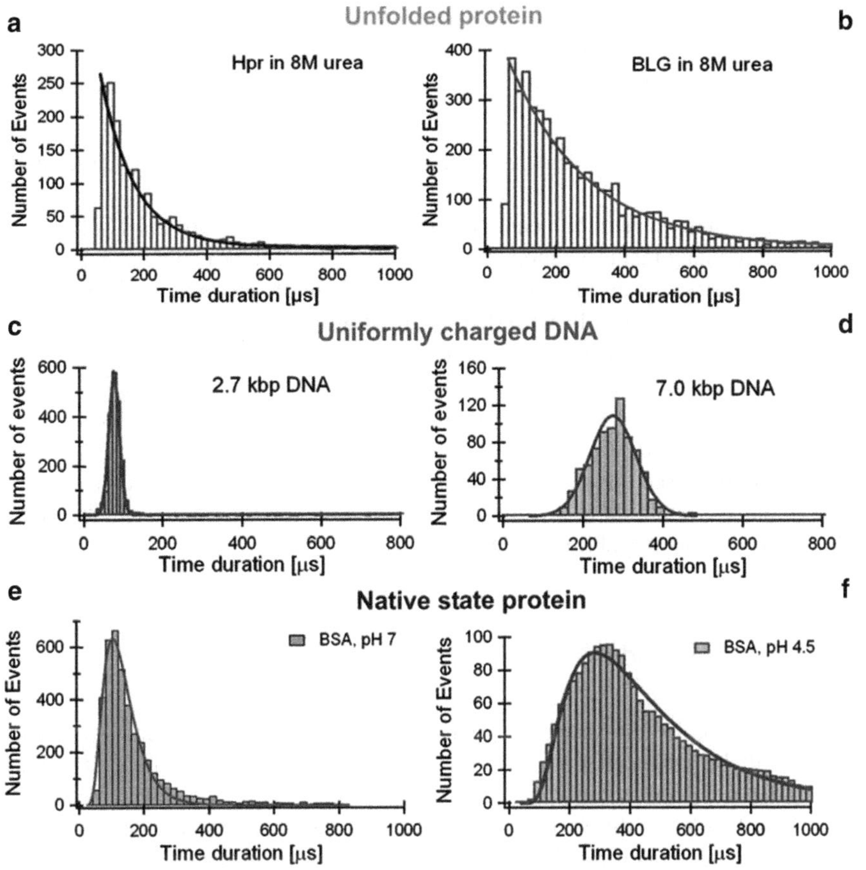

Fig. 6.7 *Time duration histograms of* unfolded proteins with Hpr that has one stalling point (**a**) and βGLa that has two stalling points (**b**); uniformly charged dsDNA with no stall point: 2.7 kb (**c**) and 7.0 kb (**d**); native state BSA as a charged particle at pH 7 (**e**) and pH 4.5 (**f**). The *solid curves* are fittings to exponentials for unfolded proteins (**a**) and (**b**). The rest of the panels are fittings to Eq. (6.6) and the pore effective thickness $H_{\text{eff}}=20$ nm was used for the fittings

6.7 Summary

Nanopore translocation of native state protein molecules can be treated as the biased 1D diffusion of simple charged particles if the proteins are stable with respect to electrostatically induced denaturation. Uniformly charged polymer translocation also follows 1D biased diffusion. The kinetic behavior of unfolded protein translocation is highly dependent on the sequence of charged residues. These observations are in sharp contrast to DNA translocation. When calculating the driving force for DNA translocation, the thickness of the pore cancels out because of the uniform charge density; there is essentially no DNA sequence effect on the translocation driving force. The distribution of translocation times depends strongly on protein sequence, applied voltage ψ, as well as H_{eff}.

In summary, our studies of single protein translocation in solid-state nanopores have shown that:

1. A nanopore experiment can measure a protein's size, electrical charge, and conformation state.
2. If a native state protein has enough disulfide bonds (e.g. BSA has 17) to keep the protein intact inside a nanopore in spite of the high electric field strength, the protein translocates through the nanopore like a simple charged particle.
3. If a protein is completely unfolded as an amino acid chain, the translocation kinetics is highly sequence dependent.
4. It is possible to distinguish proteins based on their nanopore translocation signal profile in their native and unfolded state as functions of pH (charge state) and applied voltage (driving force).
5. Furthermore, based on our analysis, an advantage of the nanopore experiment is that it has the potential to distinguish proteins with single or multi-site mutants. We describe this possibility in details below.

6.8 Future Trends

The sensitivity of the ionic current signal $\Delta I_b(t)$ to the charge and volume of specific segments of the polypeptide chain present inside the nanopore at the stall points suggests that unfolded protein translocation could ultimately provide enough contrast to routinely distinguish different proteins in complicated mixtures. The compatibility of nanopores with microfluidics and their ability to obtain data from zeptomole samples suggest that these approaches could be used to screen single cell or subcellular samples. To illustrate these ideas we report some proof-of-concept calculations.

6.8.1 Nanopore Protein Mixture Screening

Using staphylococcal nuclease (SNase) as a model protein, we predict the nanopore translocation profiles of several SNase mutants and envision the results from a set of nanopore measurement as illustrated in Fig. 6.8.

Staphylococcal nuclease (SNase, Fig. 6.8a) Consists of a polypeptide chain of 149 amino acid residues without disulfide binds [41, 42]. SNase has been used as a model protein to study protein folding and unfolding. More than 500 mutants have been made and characterized in order to study the sequence dependence of its structure and function.

Varying Protein charge Sequence allows control of the magnitude of specific barriers in the translocation profile and can provide insight into the translocation time distribution. Fig. 6.8b illustrates how protein charge sequence changes alter

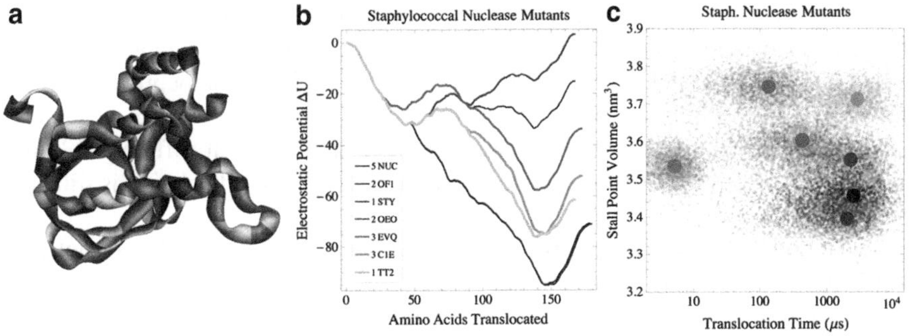

Fig. 6.8 (**a**) Structure of wild type SNase with basic (positive at pH 7.8) residues labeled *dark* and acidic (negative at pH 7.8) residues labeled *light grey*. (**b**) Electrostatic contribution to translocation potential for several mutants of SNase at pH 7.8 illustrating how sequence can change the location, number, and depth of electrostatic barriers. (**c**) Future nanopore measurement with improved signal to noise ratio would allow of distinguishing single site mutant proteins

the barriers present during translocation. The linear translocation of seven selected SNase mutants in a nanopore at pH. 7.8 (Fig. 6.8b) shows their electrical potential profiles vary significantly. The potential well depth or barrier height could vary by a factor of 5 between the SNase mutants.

Varying the electrostatic potential allows changing the electrical potential barrier height $\Delta U^{\ddagger}$ values systematically. Systematically changing the electrical potential barrier height $\Delta U^{\ddagger}$ values will change the protein dwell time t_d in a pore as described in Eq. (6.5). Our data analysis tools have enabled us to predict a 2D map of stall point volumes (Λ_{AA}) vs. time durations (t_d) shown in Fig. 6.8c from a future nanopore measurement of these SNase mutant proteins.

A 2D map illustrated in Fig. 6.8c, in principle, would be able to detect single site or multi-site mutants. Furthermore, a 2D map of excluded volume (Λ or ΔI_b) and time duration (t_d) measured under different pH and voltage could allow identification and resolution of proteins.

6.8.2 Open Challenges

Several challenges must be overcome to make the 2D map in Fig. 6.8c become a reality. One of such challenges is to find the correlation between protein primary charge sequence and the dwell time or the theoretical connection between the measured distributions of translocation times with the sequence and structural properties of the proteins.

At the present time, a quantitative theory for phenomenon of protein dwell time in a nanopore is not available and systematic studies of the sequence and condition effects on protein translocation are almost non-existent. As a result, a full explanation of the dwell time distribution remains an unsolved problem.

The current blockage signal from a nanopore measurement contains the information on the protein's shape or folding state and on the sequence of the amino-acid volume and charge of the local segment of the polypeptide chain in the pore if the protein molecule is unfolded. The research on using solid-state nanopores to measure the shape of single protein molecules at different folding states and to probe its charge sequence when it is unfolded in salt solution is still at a very early stage.

Equations (6.1), (6.3) and (6.4) were developed using several assumptions including: (1) a uniform distribution of charge on the protein, (2). the protein is a small hard sphere, (3) the protein does not interact with the pore walls, and (4) there is only one ionization state of the protein. These assumptions limit the general applicability of these equations. Future work must evaluate the need for corrections for these assumptions to enable a better description of protein translocation.

Complicated assemblies of proteins such as laminin produce complex signals that cannot be uniquely interpreted. For example, the recorded current trace (Fig. 6.1c) for Laminin protein shows the complexity of the signal measured for a partially denatured protein. At the present time, we are still developing analysis routines to process these more complicated data sets. In this chapter we only attempted to discuss the current blockage signal produced by proteins in two simplest conformations: in their native state and in their completely unfolded state.

Future developments of nanopore techniques for sensing proteins should be focused on improvement of signal-to-noise, understanding of protein translocation signal and protein structure, dynamic adjustment of DC bias potential, control of nanopore surface chemistry, in situ characterization of nanopore geometry and electrical response, analysis improvements to provide richer translocation data, and incorporation of other single molecule methods into the nanopore experiments.

Acknowledgments We thank Professor J. Golovchenko and Harvard nanopore group for nanopore fabrication, Ryan Rollings, Edward W. Graef Jr., Denis F. Tita, and Errol Porter for nanopore fabrication and characterization. We acknowledge the funding support provided by NHGRI/NIH R21HG003290, NHGRI/NIH R21HG00477, NSF/MRSEC 080054, ABI-111/710, and NIH R01GM071684 to DST.

References

1. Rapoport, T.A., *Protein translocation across the eukaryotic endoplasmic reticulum and bacterial plasma membranes.* Nature, 2007. 450(29): p. 663–669.
2. Wickner, W. and R. Schekman, *Protein Translocation Across Biological Membranes.* Science, 2005. 310(5753): p. 1452–1456.
3. Simon, S.M. and G. Blobel, *A protein-conducting channel in the endoplasmic reticulum.* 1991. 65(3): p. 371–380.
4. Sutherland, T.C., Y.-T. Long, R.-I. Stefureac, I. Bediako-Amoa, H.-B. Kraatz and J.S. Lee, *Structure of Peptides Investigated by Nanopore Analysis.* Nano Lett, 2004. 4(7): p. 1273–1277.

5. Stefureac, R., L. Waldner, P. Howard and J.S. Lee, *Nanopore Analysis of a Small 86-Residue Protein*. Small, 2008. 4(1): p. 59–63
6. Oukhaled, G., J. Mathe, A.L. Biance, L. Bacri, J.M. Betton, D. Lairez, J. Pelta and L. Auvray, *Unfolding of Proteins and Long Transient Conformations Detected by Single Nanopore Recording*. Physical Review Letters, 2007. 98(15): p. 158101–4
7. Pastoriza-Gallego, G.G. M., B. Thiebot, J.-M. Betton and J. Pelta, *Polyelectrolyte and unfolded protein pore entrance depends on the pore geometry*. Biochimica et Biophysica Acta - Biomembranes 2009. 1788: p. 1377–1386.
8. Mohammad, S. Prakash, A. Matouschek and L. Movileanu, *Controlling a Single Protein in a Nanopore through Electrostatic Traps*. Journal of the American Chemical Society, 2008. 130(12): p. 4081–4088.
9. Movileanu, L., S. Howorka, O. Braha and H. Bayley, *Detecting protein analytes that modulate transmembrane movement of a polymer chain within a single protein pore*. Nat Biotech, 2000. 18(10): p. 1091–1095.
10. Han, A., G. Schurman, G. Mondin, R.A. Bitterli, N.G. Hegelbach, N.F. de Rooij and U. Staufer, *Sensing protein molecules using nanofabricated pores*. APPLIED PHYSICS LETTER, 2006. 88: p. 093901–3.
11. Han, A., M. Creus, G. Schurmann, V. Linder, T.R. Ward, N.F. de Rooij and U. Staufer, *Label-Free Detection of Single Protein Molecules and Protein–Protein Interactions Using Synthetic Nanopores*. Analytical Chemistry, 2008. 80(12): p. 4651–4658%U http://dx.doi.org/10.1021/ac7025207.
12. Fologea, D., B. Ledden, D.S. McNabb and J. Li, *Electrical Characterization of Protein Molecules by a Solid-State Nanopore*. APPLIED PHYSICS LETTERS, 2007. 91.
13. Talaga, D.S. and J. Li, *Single-Molecule Protein Unfolding in Solid State Nanopores*. Journal of American Chemical Society, 2009. 131(26): p. 9287–9297.
14. Firnkes, M., D. Pedone, J. Knezevic, M. Döblinger and U. Rant, *Electrically Facilitated Translocations of Proteins through Silicon Nitride Nanopores: Conjoint and Competitive Action of Diffusion, Electrophoresis, and Electroosmosis*. Nano Letters, 2010. 10(6): p. 2162–2167.
15. Niedzwiecki, D.J., J. Grazul and L. Movileanu, *Single-Molecule Observation of Protein Adsorption onto an Inorganic Surface*. Journal of the American Chemical Society, 2010. 132(31): p. 10816–10822.
16. Li, J., M. Gershow, D. Stein, E. Brandin and J.A. Golovchenko, *DNA Molecules and Configurations in a Solid-state Nanopore Microscope*. Nat. Mater., 2003. 2: p. 611–615.
17. Bezrukov, S.M., *Ion Channels as Molecular Coulter Counters to Probe Metabolite Transport*. Journal of Membrane Biology, 2000. 174(1): p. 1–13.
18. DeBlois, R.W. and C.P. Bean, *Counting and Sizing of Submicron Particles by the Resistive Pulse Technique*. Review of Scientific Instruments, 1970. 41(7): p. 909.
19. Gregg, E.C. and k.D. Steidley, *Electrical Counting and Sizing of Mammalian Cells in Suspension*. Biophysical Journal, 1965. 5(4): p. 393–405.
20. Henriquez, R.R., T. Ito, L. Sun and R.M. Crooks, *The resurgence of Coulter counting for analyzing nanoscale objects*. The Analyst, 2004. 2004(129): p. 478–482.
21. Smeets, R.M., U.F. Keyser, D. Krapf, M.-Y. Wu, D. Nynke H and C. Dekker, *Salt Dependence of Ion Transport and DNA Translocation through Solid-state nanopores*. Nano Lett., 2006. 6 (1): p. 89–95.
22. King, G.M. and J.A. Golovchenko, *Probing Nanotube-Nanopore Interactions*. Physical Review Letters, 2005. 95(21): p. 216103.
23. Levadny, V., V.M. Aguilella and M. Belaya, *Access resistance of a single conducting membrane channel*. Biochimica et Biophysica Acta (BBA) - Biomembranes, 1998. 1368(2): p. 338–342.
24. Vodyanoy, I. and S.M. Bezrukov, *Sizing of an ion pore by access resistance measurements*. Biophysical Journal, 1992. 62(1): p. 10–11.
25. Dekker, C., *Solid-state nanopores*. Nature Nanotechnology, 2007. 2: p. 209–215.

26. Healy, K., B. Schiedt and A.P. Morrison, *Solid-state nanopore technologies for nanopore-based DNA analysis.* Nanomedicine, 2007. 2(6): p. 875–897.
27. Li, J., D. Stein, C. McMullan, D. Branton, M.J. Aziz and J.A. Golovchenko, *Ion-beam sculpting at nanometre length scales.* Nature, 2001. 412(12 July): p. 166–169.
28. Gierhart, B.C., D.G. Howitt, S.J. Chen, Z. Zhu, D.E. Kotecki, R.L. Smith and S.D. Collins, *Nanopore with Transverse Nanoelectrodes for Electrical Characterization and Sequencing of DNA*, in *The 14th International Conference on Solid-State Sensors, Actuators and Micro-systems.* 2007, Transducers & Eurosensors: Lyon, France.
29. Stein, D., J. Li and J.A. Golovchenko, *Ion-Beam Sculpting Time Scales.* Physical Review Letters, 2002. 89(27).
30. Storm, A.J., J.H. Chen, X.S. Ling, H.W. Zandbergen and C. Dekker, *Fabrication of solid-state nanopores with single-nanometre precision.* Nature Materials, 2003. 2: p. 537–540.
31. Venkatesan, B.M., B. Dorvel, S. Yemenicioglu, N. Watkins, I. Petrov and R. Bashir, *Highly Sensitive, Mechanically Stable Nanopore Sensors for DNA Analysis.* Adv. Mater., 2009. 21: p. 1–6.
32. Venkatesan, B.M., A.B. Shah, J.-M. Zuo and R. Bashir, *DNA Sensing Using Nanocrystalline Surface-Enhanced Al_2O_3 Nanopore Sensors.* Adv. Funct. Mater., 2010. 20: p. 1266–1275.
33. Stein, D.M., C.J. McMullan, J. Li and J.A. Golovchenko, *Feedback-controlled ion beam sculpting apparatus.* Review of Scientific Instruments, 2004. 75(4): p. 900–905.
34. Cai, Ledden, Krueger, Golovchenko and Li, *Nanopore sculpting with noble gas ions.* Journal of Applied Physics, 2006. 100.
35. Talaga, D. and J. Li, *Single-molecule protein unfolding in solid state nanopores.* J. Am. Chem. Soc., 2009. 131: p. 9287–9297.
36. Fologea, D., B. Ledden, D.S. McNabb and J. Li, *Electrical Characterization of Protein Molecules in a Solid-State Nanopore.* Appl. Phys. Lett., 2007. 91.
37. Peters, T., Jr., *Serum Albumin.* Adv. Protein Chem., 1985. 37: p. 161–245.
38. Collins, B.E., K.-P.S. Dancil, G. Abbi and M.J. Sailor, *Determining Protein Size Using an Electrochemically Machined Pore Gradient in Silicon.* Advanced Functional Material, 2002. 12 (3): p. 187–191.
39. Bloomfield, V., *The Structure of Bovine Serum Albumin at Low pH.* Biochemistry, 1966. 5(2): p. 684–689.
40. Kramers, H.A., *Brownian motion in a field of force and the diffusion model of chemical reactions.* Physica (Utrecht), 1940. 7: p. 284–304.
41. Cotton, F.A., J. Edward E. Hazen and M.J. Legg, *Staphylococcal nuclease: Proposed mechanism of action based on structure of enzyme—thymidine 3′,5′-bisphosphate—calcium ion complex at 1.5-Å resolution* Proc. Nati. Acad. Sci. USA, 1979. 76(6): p. 2551–2555.
42. Tucker, P.W., E.E. Hazen and F.A. Cotton, *Staphylococcal nuclease reviewed: A prototypic study in contemporary enzymology* Molecular and Cellular Biochemistry, 1979. 23(3).

Chapter 7
Simulation of Electronic Sensing of Biomolecules in Translocation Through a Nanopore in a Semiconductor Membrane

Maria E. Gracheva, Amandine Leroux, Jacques Destiné, and Jean-Pierre Leburton

Abstract A two-level computational model for simulation of the electric signal detected on the electrodes of a Semiconductor-Oxide-Semiconductor (SOS) capacitor forming a nanoscale artificial membrane, and containing a nanopore with translocating DNA are presented. At the device level, a three-dimensional self-consistent scheme involving snapshots of the DNA charge distribution, as well as the electrolytic charge and the charge in the semiconductor membrane compute the electrostatic potential over the whole solid-liquid system. With this numerical approach we investigate the possibility of resolving individual nucleotides as well as their types in the absence of conformational disorder. At the system level, we develop a circuit-element model for the SOS semiconductor membrane where the membrane is discretized into interconnected elementary circuit elements to assess the response of the DNA away from the pore. The model is tested on the translocation of 11 base single-stranded C_3AC_7 DNA molecule, for which the electric signal shows good qualitative agreement with the multi-scale device approach of Gracheva et al. also described in the first part of this chapter (Gracheva et al., Nanotech. 17, 622–633, 2006), while quantifying the low-pass filtering in the membrane.[1]

Keywords DNA sequencing • Solid-state nanopore • Poisson equation • Multi-scale approach • Circuit modeling • SPICE simulation

[1] The material used in this chapter was in part published in the Nanotechnology Journal (Gracheva et al., Nanotech. 17, 622–633, 3160–3165, 2006). This material is reproduced with permission from the Publisher (Institute of Physics Publishing Limited).

J.-P. Leburton (✉)
Department of Electrical and Computer Engineering and Department of Physics,
Beckman Institute, University of Illinois at Urbana-Champaign, Urbana, IL 61801, USA
e-mail: jleburto@uiuc.edu

S.M. Iqbal and R. Bashir (eds.), *Nanopores: Sensing and Fundamental Biological Interactions*, DOI 10.1007/978-1-4419-8252-0_7,

7.1 Introduction

In the middle of the 1990s, single molecule translocation through a proteinaceous nanopore was proposed as a new tool for DNA characterization [1]. Single stranded DNA molecules have a diameter of 1–2 nm [2] which is comparable to the pore size of biological ion channels [3]. It was suggested that by utilizing the natural nano-scale features of a proteinaceous pore, one can obtain direct information on the DNA molecular structure. Indeed, when an electric field drives a DNA molecule through the proteinaceous nanopore in a lipid bilayer placed in an electrolyte solution, the DNA molecule partially or totally blocks the pore reducing the ionic current through it. Hence, the passage of each DNA molecule can be detected as a transient decrease of ionic current whose duration is proportional to the DNA length. Many experiments have been done in this context [1, 4–10] demonstrating the correlation between the length of the DNA chain and the duration of the ionic current blockade. Later works revealed additional factors such as temperature [5], driving voltage (applied electric field) [9], and DNA pairing [4, 6, 7, 10] affecting the DNA translocation rate. Moreover, in the case of mono-nucleotide chains, this technique has been able to discriminate between polymers of different molecular compositions [4, 6, 10].

Although proteinaceous nanopores are convenient DNA characterization tools, they have shortcomings as the pore is of a fixed size, and its stability and low noise characteristics are restricted to well defined chemical, mechanical, electrical and thermal conditions [11]. In recent years, artificial solid-state nanopores made of silicon and silicon compounds began to replace the original bio-nanopores, and promise to overcome limitations of the latter [11–16]. As with proteinaceous nanopores, the correlation between DNA translocation and ionic current blockage has been established by both simulations and experiments [13, 17, 18]. By measuring the duration and magnitude of the blocking current transient, the polymer length of single-stranded DNA can be determined [13]. However, for both bio-nanopore and solid-state nanopore systems, complete characterization of DNA molecules is still at its early stage. Nevertheless, quoting our colleagues: "with further improvements, the method could in principle provide direct, high-speed detection of the sequence of bases in single molecules of DNA or RNA" [1].

In addition to ionic current blockage, translocation of DNA molecules through nanopores in Semiconductor-Oxide-Semiconductor (SOS) membrane induces a change of the electrostatic potential on the semiconductor layers of the membrane. Hence, as DNA permeates the pore, its electronic signature can be recorded in the form of a voltage trace.

Sequencing DNA strands with a nanopore device has not been achieved yet. Although it has been demonstrated that voltage signals resulting from DNA translocation through a capacitor membrane can be recorded [18, 19], the temporal resolution of such measurements precludes from recording signals that could be associated with translocation of individual nucleotides. While new experimental

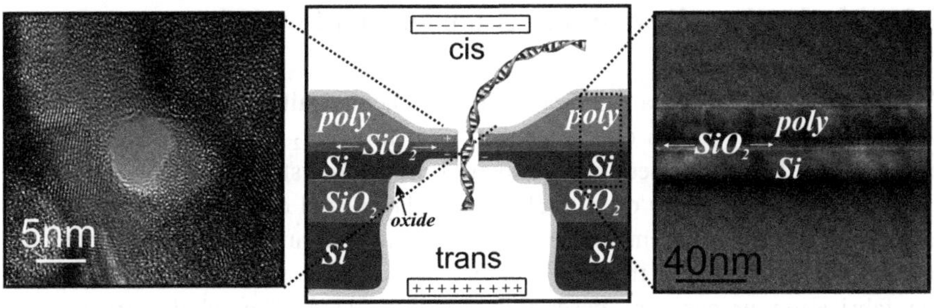

Fig. 7.1 (*Center panel*) Schematic of a novel biosensor consisting of a nanopore in a capacitor membrane. (*Right panel*) High resolution TEM through the capacitor membrane structure. (*Left panel*) High resolution lattice image of a nanopore sputtered into the capacitor membrane. Reproduced from [19] with permission

methods for measuring fine signals with nanopore sensors are being developed [18, 20–23], the theoretical resolution of such measurements can be estimated through computation.

Following a preliminary experimental demonstration, we have developed a computational model that simulates the electrical response of DNA translocation through a nanopore in a SiO_2 membrane sandwiched between two conducting layers of heavily doped Si [Fig. 7.1 (center panel)] [19]. In this approach, the DNA translocation through the nanopore is simulated by using "snapshots" of the molecular charge distribution obtained from atomistic modeling [19, 24] while the voltage response in the solid-state capacitor is modeled by a 3D self-consistent Poisson Solver (PS). We evaluate the possibility of resolving individual nucleotides as well as their types. We show that the new device has the potential to identify DNA molecules by using the voltage induced on the electrodes without utilizing the signal of ionic current fluctuations.

In addition, to account for the ohmic and capacitive loading in the membrane and to assess the magnitude of the signal on a detector far from the nanopore we developed a simple membrane-nanopore model based on a circuit-element approach to assess the electric response of a DNA translocation along the membrane. In this approach, the electrostatic induction from each of the charges in the DNA strand moving through the pore and felt by the semiconductor membranes is modeled as a current source, while the membrane equivalent circuit is simulated by SPICE to assess the voltage traces collected on the remote electrodes of the nanopore-capacitor.

This chapter is organized as follows: Section 7.2 describes the structure of the membrane. Section 7.3 describes the computational model to simulate the electrical response of the solid-state capacitor to a DNA translocating through the nanopore. The model of the DNA translocation and the elements of the equivalent electric circuit are detailed respectively in Section 7.5. Sections 7.4 and 7.6 present and discuss the results obtained by our models.

7.2 Motivation: Device Structure Description

A synthetic nanopore in a SOS capacitor is schematically illustrated in Fig. 7.1 (center panel). It consists of a very thin (<5 nm) SiO_2 membrane, sandwiched between two heavily doped silicon layers (one polycrystalline silicon; the other crystalline silicon) sitting on a thicker oxide layer, and covered by a thin oxide layer. A nanopore of diameter smaller or equal to 2 nm is etched through the membrane layers in the thinnest (center) part of the structure. The details of the fabrication process are described elsewhere [14]. The whole structure is immersed in a KCl buffer solution containing two external electrodes under electric bias, which can also measure the ionic current flowing through the nanopore, as proposed in other DNA molecule detectors [4, 10, 14]. The device membrane forms a nanoscale electric capacitor, which can record voltage fluctuations induced by the sequence of charges carried by a DNA strand when it translocates through the nanopore.

The use of semiconductor materials allows also direct integration of nanoscale Metal-Oxide-Semiconductor (MOS) amplifiers with high sensitivity on the nanopore layer structure to improve the voltage signal. Owing to the short distance between the two capacitor plates, one can expect the voltage signal to be characteristic of the electronic structure of the DNA molecule, potentially with resolution up to a single base pair when the capacitor dielectric and nanopore diameter are ~1 nm. Because of the solid-state nature of the nanopore, the system is robust. It operates over a wide range of electrolyte concentrations, pH [25], temperature and external voltages [26], unlike proteinaceous pores that operate under stringent chemical and electrical conditions. This last consideration provides wide latitude to optimize the system for better detection.

The membranes are manufactured using conventional semiconductor processing techniques, starting with a 200 mm Semiconductor-Oxide-Insulator (SOI) wafer with a silicon layer 20 nm thick on a 150 nm thick buried oxide. Rapid thermal oxidation of the heavily-doped (*n*-type or *p*-type) crystalline silicon (SOI) layer is used to produce a high-integrity gate dielectric <5 nm thick. Subsequently, heavily-doped polysilicon is deposited on the oxide and chemical-mechanical polishing is used to reduce the thickness of the poly electrode to about 20 nm (see Fig. 7.1 (center panel)). A high resolution Transmission Electron Microscope (TEM) micrograph shows cross-section through the membrane structure in Fig. 7.1 (right panel). A membrane, comprised of the poly-SiO_2-silicon capacitor, is formed by creating a through-wafer via using optical lithography in conjunction with reactive ion and wet chemical etching on sacrificial layers. After electrical contacts to the top and bottom silicon electrodes are formed, a nanopore is produced in the membrane using a tightly focused, high-energy electron beam to sputter atoms from the capacitor. A top-down lattice image of the 6–7 nm diameter pore through the membrane is shown in Fig. 7.1 (left panel). The figure shows a TEM of an $R_p = 3.5 \pm 0.1$ nm radius pore produced in a poly-SiO_2-silicon membrane 45 nm thick, taken at a tilt angle of 0°. This image represents a two-dimensional (2D) projection through the membrane.

The shot noise observed in the area identified as the pore is indicative of perfect transmission of the electron beam through the pore in the membrane. The three-dimensional (3D) structure can be inferred from 2D projections of the pore taken at various tilt angles. Although it is not unique, one simple model for the structure consists of two intersecting cones each with a cone angle of ~30° in this case. To ensure the integrity of the capacitor, we measured the tunneling current between the electrodes prior to and after the pore had been sputtered [14, 18]. Further details regarding the device description can be found in [19].

7.3 Computational Model and Nanopore Device Modeling

In order to achieve an understanding of the electrostatics at play between the different materials, i.e. the electrolytic solution and the DNA molecules interacting with the semiconductor/oxide structure, we have developed a computational model of the device that reproduces realistically the bio-electronic system shown in Fig. 7.1. In our model the top poly-Si and bottom Si capacitor films have been replaced by two heavily doped n^+ – Si layers. The nanopore is placed in the center of the membrane and assumed to have cylindrical symmetry. The n^+ – Si layers have conical shape above and below the nanopore as a result of the electron beam sputtering.

A schematic of the idealized device geometry is shown in Fig. 7.2. We chose the coordinate system in which the *y*-axis is parallel to the nanopore axis and the *xz* plane is parallel to the device structure layers including the SiO_2 layer.

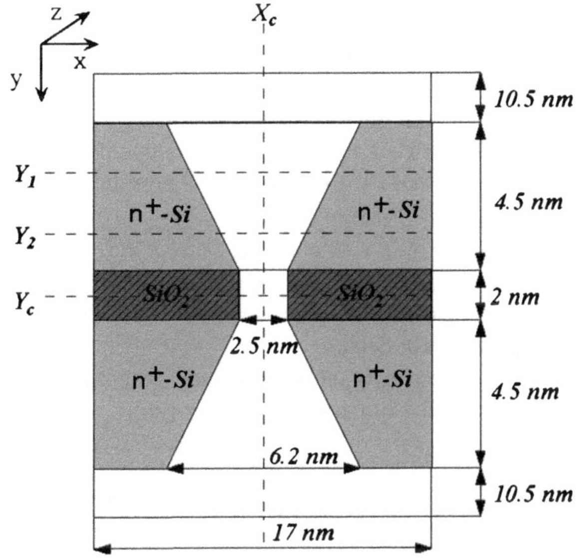

Fig. 7.2 Schematic of the device geometry: *xy* cross-section through the center of the nanopore (*yz* cross-section is similar) provides a side view of the modeled device and cross-sections for which the electron, negative and positive ion concentrations are represented on the following figures. The drawing is not to scale

7.3.1 Computational Approach

We assume a quasi-static behavior of electrical response, and consider a step-by-step motion of DNA through the nanopore. An atomic-level description of the DNA strand with its charge distribution has been computed by using the NAMD code as described in [24]. In this method, a molecular system is approximated by an ensemble of atoms interacting with each other according to a molecular force field that has been developed and calibrated to reproduce quantitatively physical properties of the simulated system [17, 27]. For each DNA position in the nanopore we compute the voltage response on the capacitor electrodes self-consistently by a 3D Poisson Solver (PS) incorporating the charge variation in the electrolytic solution as well as in the semiconductor materials and the oxide [19].

7.3.2 3D Self-Consistent Modeling of the Capacitor Response

By tracking the motion of every atom in the DNA strand during its translocation, molecular dynamics provides the temporal charge distribution of the polymer in the nanopore. In this work we use the molecular charge distribution for a "snapshot" of the DNA conformations in a continuum charge model of the electrolyte and the solid state materials to compute the electrostatic potential in the whole region, and specifically in the Si layers to obtain the voltage response due to DNA translocation. Here, we assume the snapshot of the initial DNA charge distribution moves rigidly through the pore. In the future we can analyze the behavior of a succession of such "snapshots".

7.3.2.1 Charge Model of the Buffer Solution

At room temperature, we assume that all KCl molecules in the electrolyte solution are fully dissociated. Hence, in the absence of external potential, there is an equal number of K^+ and Cl^- ions when the electrolyte solution is at equilibrium:

$$[K^+]_0 = [Cl^-]_0 = c \tag{7.1}$$

where c is the concentration of the buffer solution, which is assumed to be constant. See Appendix for details.

In the presence of an electrostatic potential $\phi(\mathbf{r})$, the ion concentrations in solution obey the Boltzmann statistics [28]:

$$[K^+](\mathrm{r}) = [K^+]_0 \exp\left(\frac{q\phi(\mathrm{r})}{kT}\right), \tag{7.2}$$

$$[Cl^-](\mathbf{r}) = [Cl^-]_0 \exp\left(-\frac{q\phi(\mathbf{r})}{kT}\right), \tag{7.3}$$

where T is temperature and k is Boltzmann constant. These virtual solid-state parameters for the solution permit us to formulate a whole semiconductor model for the charge and electric potential.

7.3.2.2 Semiconductor Charge Model

In the solid-state material regions, the electron and hole concentrations are given by the basic semiconductor physics theory [28]. In the Si-layers we assume the donor doping concentration $N_d^+ = 2 \times 10^{20} cm^{-3}$, so the carriers are degenerate, and their distribution follows the Fermi-Dirac distribution. The electron concentration $n(\mathbf{r})$ and the hole concentration $p(\mathbf{r})$ are given by [28]

$$n(\mathbf{r}) = N_c \frac{2}{\sqrt{\pi}} F_{1/2}(\eta_c(\mathbf{r})), \tag{7.4}$$

$$p(\mathbf{r}) = N_v \frac{2}{\sqrt{\pi}} F_{1/2}(\eta_v(\mathbf{r})), \tag{7.5}$$

where N_c and N_v are the density of states in the conduction and valence bands of the solid-state material; $F_{1/2}$ is the 1/2 order Fermi-Dirac function, the parameters $\eta_c(\mathbf{r})$ and $\eta_v(\mathbf{r})$ are related to the local potential $\phi(\mathbf{r})$ by

$$\eta_c(\mathbf{r}) = \frac{E_f - E_c(\mathbf{r})}{kT}, \text{ with } E_c(\mathbf{r}) = -q\phi(\mathbf{r}) - E_g, \tag{7.6}$$

$$\eta_v(\mathbf{r}) = \frac{E_v(\mathbf{r}) - E_f}{kT}, \text{ with } E_v(\mathbf{r}) = -q\phi(\mathbf{r}), \tag{7.7}$$

where E_c $(\mathbf{r})$ is the solid-state material conduction band edge profile and E_v $(\mathbf{r})$ is the corresponding valence band edge profile; E_g is the band gap of the material; E_f is the Fermi level, which, because of the nature of the system, is assumed to be constant and is taken as our reference energy level.

To model the electrolyte/semiconductor interface we also use the conduction band offset between materials with respect to the Si:

$$\Delta E_c^{SiO_2} = 3.2\, eV, \quad \Delta E_c^{solution} = -0.3\, eV. \tag{7.8}$$

This essential feature of the solid-state materials accounts for Fermi level pinning and band bending in the presence of the positive space charge inside the semiconductor and close to the interface. $\Delta E_c^{SiO_2}$ is a measured value, whereas $\Delta E_c^{solution}$ is a model parameter. The choice of its value does not affect the induced voltage dramatically.

We also include in our model a thin (2 Å) layer of negative charge at the surface of the semiconductor structure with the density of $N_{surface} = 10^{21}$ cm^{-3}, resulting from the etching process [14]. While the density of this surface charge is not known presicely, the amplitude of the calculated voltage signal due to DNA translocation varies by 5% if the magnitude of the charge present on the surface of the nanopore is reduced to $N_{surface} = 10^{10}$ cm^{-3}. The surface charge density value is selected to accommodate the experimental data [14]. In our treatment we do not consider the tunneling current between the electrodes through the DNA.

7.3.2.3 Self-Consistent Scheme

The system under investigation consists of several material regions which are the Si and SiO_2 layers, and the buffer solution containing the DNA. The charge in different regions of the device originates from different sources. In the Si layers the charge comes from the doping ions, electrons and holes. The charge density in the solid-state regions is given by:

$$\rho_{solid-state}(\mathrm{r}) = q\{N_d^+(\mathrm{r}) - N_a^-(\mathrm{r}) + p(\mathrm{r}) - n(\mathrm{r})\}, \tag{7.9}$$

where N_d^+ is the donor density, assumed to be fully ionized, and $N_a^-(\mathrm{r})$ is the acceptor density which is zero in our model.

In the buffer solution, contributions to the charge include the K^+ and Cl^- ions, along with the charge distribution on the DNA strand, $\rho_{DNA}(\mathbf{r})$. The charge density of the solution is then

$$\rho_{solution}(\mathrm{r}) = q\{[K^+](\mathrm{r}) - [Cl^-](\mathrm{r})\} + \rho_{DNA}(\mathrm{r}). \tag{7.10}$$

Each material is characterized by its relative permittivity, i.e., $\varepsilon_{Si} = 11.7$, $\varepsilon_{SiO_2} = 3.9$. For the buffer solution we chose the dielectric constant to be that of water, i.e. $\varepsilon_{solution} = 78$. We also assume the same value inside the nanopore ($\varepsilon_{pore} = 78$), although due to the size of the pore, the dielectric constant may exhibit significant local variation [29]: the relative permittivity inside the pore may vary from 78 to 1 depending on whether or not the water is completely excluded from the pore during the DNA translocation. The exclusion of water may have a dramatic effect on the signal and therefore we are motivated to study translocations in narrow (~1 nm) diameter pores.

Poisson's equation

$$\nabla \cdot (\varepsilon(\mathrm{r})\nabla\phi(\mathrm{r})) = -\rho(\mathrm{r}) \tag{7.11}$$

is solved self-consistently by a multigrid method [30] on the whole volume of simulated structure with the following boundary conditions: Dirichlet boundary conditions at the top and the bottom bias gate regions ($\phi_{top-gate} = \phi_{bottom-gate} = 0$). Neumann boundary conditions were applied for other regions.

The self-consistent three-dimensional Poisson Solver is set up as following: upon entering the initial guess charge distribution, i.e., DNA charge plus electrolyte ionic charge, the numerical solution starts from solving the Poisson equation to obtain the potential distribution. The local potential is then substituted into the materials constructive equations to obtain a corresponding local charge concentration (ionic concentration in the buffer solution region, electrons in other regions). The charge concentration is then inserted back into the Poisson equation and the new potential distribution is calculated. The procedure is repeated for every input charge distribution until convergence is obtained. For the potential, the convergence tolerance was 1 μV. Due to the DNA strand translocation through the nanopore, the DNA charge distribution is position-dependent.

In order to calculate the voltage induced on the Si electrodes due to a charge translocation we average the potential over the circular perimeter of the pore at the intersection of the SiO_2 and Si layers. Thus, we record the potential at the nearest possible distance between the electrode and the DNA. In this way we obtain two potential values, one for the upper (*cis*) electrode and one for the lower (*trans*) electrode. We subtract the corresponding values of the potential for the empty nanopore (pore without translocating charge) to obtain the potential difference induced on the electrodes by the translocating charge.

7.4 Results and Discussion for the 3D Self-Consistent Modeling of the Capacitor Response

We first characterized the modeled device, as well as study the translocation of the negative test charge and two linear sequences of charges, the results were published previously in [19]. We further follow these tests with simulation of the electrical response of the layered membrane to a ssDNA translocating through the pore.

We can estimate the number of ions in the volume of the nanopore as $N_{pore} = CV_{pore}$, where $C = 1$ M is the average electrolyte solution concentration in the nanopore as a cylinder of height $h = 2$ nm with radius $R_{pore} = (2.5/2)$ nm. Thus, $N_{pore} = 7.53$. This value compares very well with the average number of potassium and chlorine ions in the pore during the course of molecular dynamics simulation, where there are 5–10 potassium and chlorine ions on average present in the pore [17].

The charge on the surface of the pore was calculated as $[N_{surface}\ x\ 2\text{Å}][h\ x\ 2\pi R_{pore}]e^{-} \approx 3.14e^{-}$, which is comparable to the value inferred from the measured electrolytic conductivity [14]. This number is consistent with the number of ions in the pore, so that there is an overall neutrality of the region. For comparison, a DNA carries an excessive charge of $1e^{-}$ per base and several bases may be trapped between the electrodes at any particular time during a DNA translocation.

7.4.1 Translocation of a Single-Stranded DNA in a Helical Conformation

The translocation of a 20 nucleotide long poly(dC)$_{20}$ single-stranded DNA molecule in a helical conformation [2] through the nanopore was simulated first [19], followed by computations of the voltage signal induced on the Si electrodes. As described above, we extract the DNA charge distribution from molecular dynamic simulation by taking into account every atom's contribution of the DNA strand (the DNA data used here and below were obtained from Dr. Aleksei Aksimentiev) and map the charge distribution into the Poisson Solver. For this purpose, every atom's charge on the DNA strand is distributed over a sphere of 3 Å in diameter, but solution charge was excluded from this volume as it was also excluded in the two previous cases. (Two DNA snapshots were used: one in the helical conformation 20 nucleotide long poly(dC)$_{20}$ (this subsection), and one in the stretch conformation 11 nucleotide long poly(dC)$_{11}$ with mutated base #4 (next subsection)).

Figure 7.3 shows the potential contours in the middle of the pore for four different DNA positions as it proceeds along the pore axis. The DNA translocates parallel to the axis on the nanopore and is slightly off the nanopore center, a situation which is likely to happen in a real experimental set-up. Figure 7.3 shows the merging of equipotential caused by the DNA molecule with the solid state nanopore which induces a potential variation in the Si layers.

Figure 7.4a depicts the increased concentration of positive ions around the DNA, which is dominantly negatively charged, while Fig. 7.4b shows the repulsion of negative ions from the molecule. We note that K^+ ions are mainly localized around the negatively charged backbone of the molecule. There is a mix of K^+ and Cl^- ions in the surrounding DNA bases solution. Of interest is also the change in ion concentrations caused by the DNA when compared to the concentrations in the empty pore.

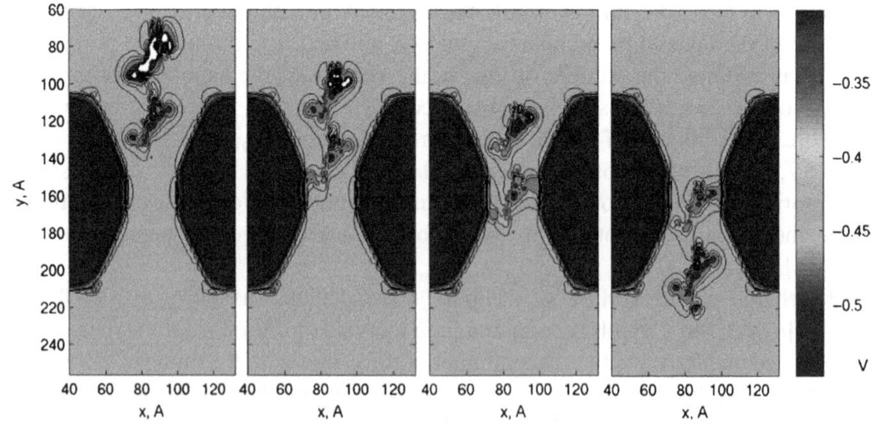

Fig. 7.3 Potential contours taken in the middle of the nanopore for the single-stranded DNA translocating through the nanopore

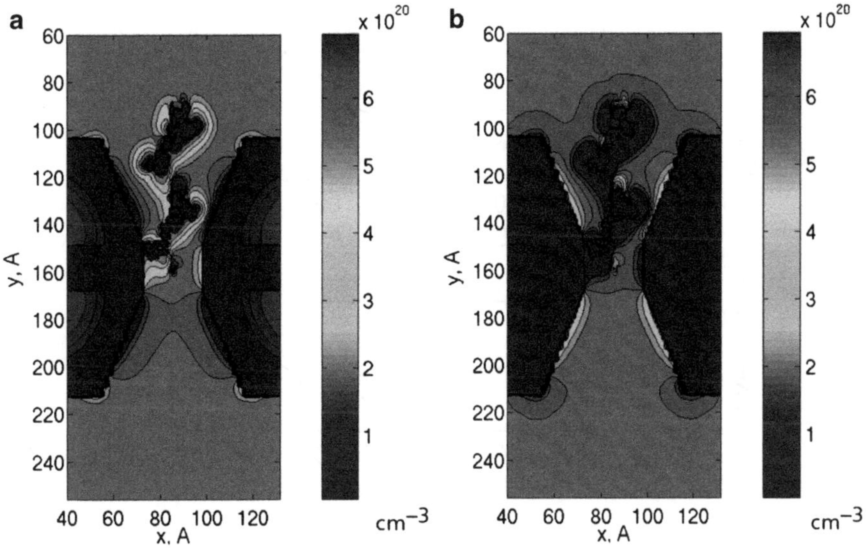

Fig. 7.4 (**a**) Cl^- ion and (**b**) K^+ ion distribution in the surrounding salt solution in the presence of the DNA

Figure 7.5 shows the electrode voltage signal obtained from the rigid translocation of the DNA molecule through the nanopore. One notices the magnitude of the signal on the order of 10 mV, which is much larger than that in the case of the linear chain due to the more densely packed charge on the DNA chain. Also, owing to the rigid translocation, the signal at the lower electrode is shifted by 2 nm relative to the upper electrode signal. From the non-zero voltage trace one can conclude that the translocated DNA is about 75 Å long. In addition, the recorded signal reflects the DNA strand structure and conformation with distinguishable features that are about 4 Å apart (compared to 3.4 Å separation between the bases on a DNA strand). This implies that it is possible to identify individual bases as they pass by the electrodes. However, the off-center translocation of modeled DNA helix does not allow one to identify all of the bases in the strand, but only those that come close to the electrodes. Indeed, as the translocating DNA consists of 20 identical cytosine bases, the variation in the recorded signal between 10 and 15 mV can be looked at as the variation due to *positional noise* of a single cytosine base in the pore, which we estimate to be ~2–3 mV.

Narrower nanopores would constrict the translocating molecules to smaller volume and, thereby, reduce the distance between nucleotides and Si-electrodes, resulting in larger and higher resolution signals. Also, it is well known that in electrolytic solutions, each DNA base is characterized by a specific dipole moment, which is electrically identifiable. Here again, the constriction of the DNA to very narrow pore could be used to reduce the stochastic orientation of dipole moment resulting from molecular dynamics, thus considerably lowering the conformational noise of the bio-molecule.

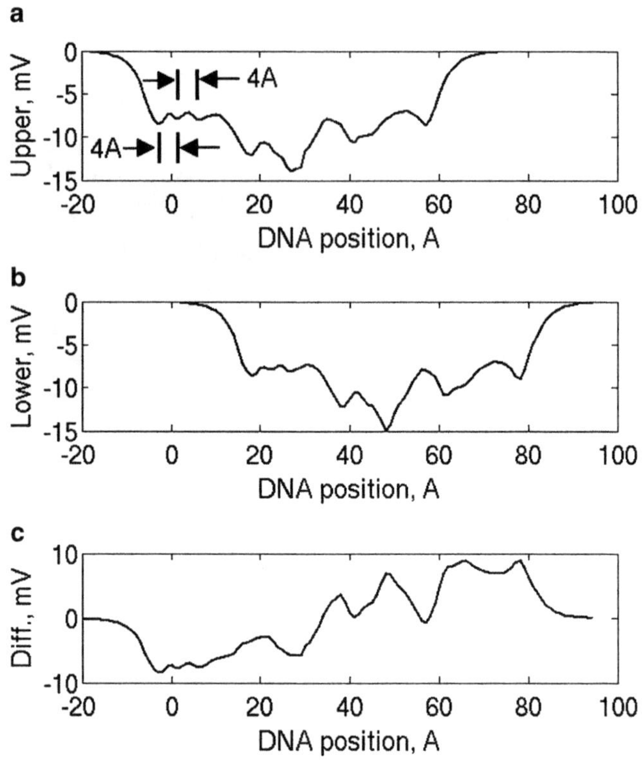

Fig. 7.5 Voltage signals on the two electrodes for a single-stranded DNA poly(dC)$_{20}$ passing through the nanopore (in angstroms): (**a**) upper electrode voltage signal, (**b**) lower electrode voltage signal, and (**c**) voltage difference between (**a**) and (**b**). The horizontal axis shows the position of the DNA front in the pore

7.4.2 Translocation of a One Mutated Base Single-Stranded DNA in a Stretch Conformation

The modeled capacitor membrane consists of a 2 nm thick SiO_2 layer sandwiched between two heavily doped n^+ – Si layers of 4.5 nm thickness each. The nanopore in the capacitor membrane has a double-conical shape with the diameter of the narrowest part of 1 nm (see details in [19, 31]). Taking the distribution of the DNA's charge from atomistic simulations, we solve self-consistently the Poisson equation, assuming the ions in the electrolyte obey the Boltzmann distribution, whereas electrons and holes in the semiconductor are ruled by the Fermi-Dirac statistics. This model was described elsewhere in details [19, 31]. The average ion concentrations that we obtain in the pore correlate with the data obtained by molecular dynamics simulations. Since we are interested in the average induced signal on the electrodes we believe that our solution model accurately describes the

Fig. 7.6 (**a**) DNA strand enveloped by a cylindrical surface, a phantom pore, initially wide enough to accommodate the strand entirely. (**b**) The conformation of a DNA strand is 1 nm diameter pore. The DNA backbone is stretched, DNA bases are tilted towards the 5′ end of the strand. (**c**) Contour plots of negative charge concentration in the capacitor membrane and electrolyte solution. (**d**) Same for the positive charge concentration. Reproduced from [31] with permission

solution dynamics. The concentration of the KCl electrolyte is 1 M and the doping density in silicon is $N_d^+ = 2 \times 10^{20} cm^{-3}$. We also include in our model a thin (2 Å) layer of negative charge at the surface of the semiconductor structure with the density of 10^{21} cm^{-3}, resulting from the etching process [14].

The calculated contour plots of negative and positive charge concentration in the capacitor membrane and electrolyte solution are shown in Fig. 7.6c and d, respectively, for one of the DNA position in the pore. The cross-section was taken through the middle of Si — SiO_2 — Si layers and the center of the nanopore. It can be seen that the central SiO_2 layer is void of charge carriers. There is a distribution of electrons in the Si layers with a depletion region near the surface due to the fixed surface charge (Fig. 7.6c). The accumulation of the positive charge is present around the DNA as it is negatively charged (Fig. 7.6d).

While DNA is moved in the fixed conformation along the axis of the nanopore in 1 Å increments, the electrostatic potential induced by the DNA charge is recorded on the lower and upper Si electrode as a function of the DNA position in the nanopore (see [19] for full description). The DNA is at "zero" position when the front tip of the molecule is at the nanopore center. As the signals obtained from the upper and lower electrodes do not differ significantly, we focus on the voltage traces calculated on the upper Si electrode, similar to a different single electrode approach [20].

In Fig. 7.7a the recorded voltage trace for 11 base DNA with the sequence 5′-CCCACCCCCCC-3′ is shown in dotted line. To simplify the notation, we abbreviate this DNA sequence as C_3AC_7. The other studied DNA sequences are abbreviated in a similar way. The voltage trace of the DNA's backbone which was obtained by removing the bases is shown in solid line on the same figure. The difference between these two signals is displayed in Fig. 7.7b. From the two signals

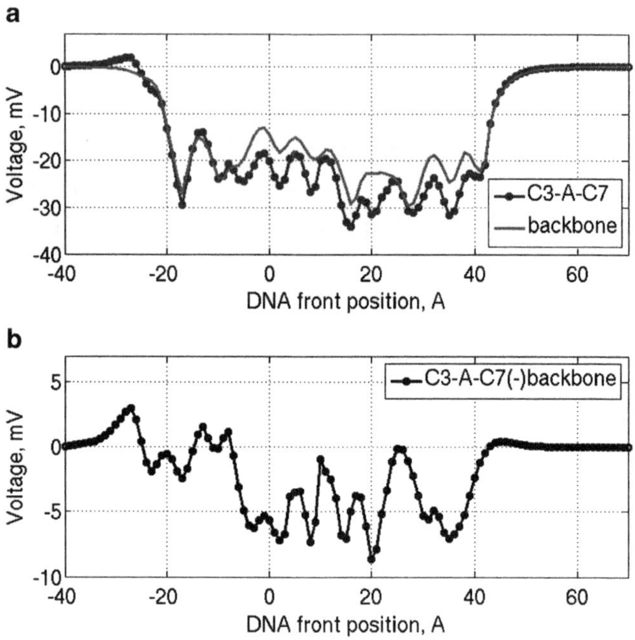

Fig. 7.7 (**a**) Calculated voltage trace from C_3AC_7 strand (*dotted line*) and the backbone signal of the same strand (*solid line*). (**b**) The difference between the whole DNA signal and the backbone signal

shown in Fig. 7.7a we conclude, on one hand, that the maximum recorded voltage caused by the DNA translocation is around 35 mV. This signal is appreciable and is due to the DNA confinement in the constriction where screening by the solution is minimized. In addition, due to the narrowness of the pore the DNA translocates in close proximity to the electrodes. On the other hand, the maximum voltage induced by the DNA's backbone is about 30 mV, resulting in a maximum signal from a base of approximately 8 mV (Fig. 7.7b). We point out that these numbers are obtained in the absence of the conformational disorder. Also, measured from the non-zero voltage trace, the DNA length is ~70 Å, which is in agreement with the actual stretched DNA length. There are several distinct fluctuations on the voltage trace, which, as we show later, correspond to the 11 segments of the backbone with a DNA base in every segment.

In order to better understand the observed variations in the voltage trace we separate the C_3AC_7 DNA into 11 fragments, each containing a sugar ring, a phosphate group and a base (with the exception of the first fragment which does not contain the phosphate group). Each fragment was driven through the pore separately and the corresponding voltage signal was calculated at the electrode. The composite plot of traces of the 11 fragments and the trace of the whole C_3AC_7 sequence is shown in Fig. 7.8. We observe that 10 of 11 fragments of the DNA produce similar traces in the form of a negative dip. The remaining one

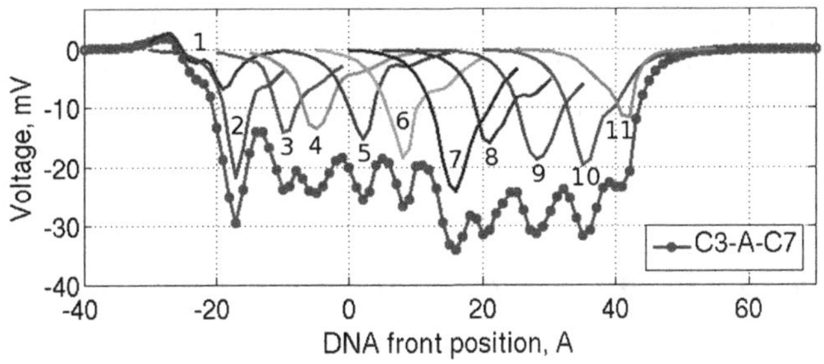

Fig. 7.8 Voltage traces of the 11 single-nucleotide fragments and the signal of the whole C_3AC_7 DNA sequence. The contribution from individual nucleotides can be seen in the total trace, and, thus one can count the number of nucleotides in the strand

(trace #1 on Fig. 7.8) corresponds to the first fragment of the DNA with no phosphate group in it. The absence of the phosphate group, which is negatively charged, allows for a positive rise in this trace caused by the positive charge of the sugar ring and the base in this segment. We also observe that the traces of the DNA fragments overlap significantly, as the average width of each dip extends over 8–12 Å. This means that in the whole DNA trace the signals from as many as three nearby fragments (and therefore bases) are mixed, which is to be expected as the Coulomb interaction is a long range interaction. In addition, the trace #4, which contains the "A" base does not have any obvious distinct features from other segments containing the "C" bases, except maybe a slight widening of the signal compared to the nearest fragments, which may be due to the different distribution of charge on the base as well as to the different orientation of the base # 4 on the backbone, or, most likely, both.

To study the effect of single nucleotide substitution on the signal induced on the capacitor structure, we replace one base (# 4) in the C_3AC_7 strand by C, G, and T nucleotides respectively, as described in [27, 31]. We keep the conformations of the three resulting strands identical to that of the original strand, with the exception of the replaced nucleotides. The resulting DNA strands have sequences C_3CC_7, C_3GC_7 and C_3TC_7, correspondingly. The orientation of their aromatic rings as well as the backbone structure are identical. The calculated voltage traces for the four strands obtained by a single base mutation at the fourth place differ in the vicinity of the fourth base over a position range between −15 and 5 Å (results are not shown). As expected, the traces are identical away from the fourth base as the DNA conformations there remained the same.

7.5 Circuit Element Modeling of the Membrane

In this section we assess the influence of the membrane capacitances and resistances on the electrical response of the DNA translocation by using an equivalent electric circuit based on the membrane geometry and material properties. In this approach, two external oxide layers have been added to the top and bottom of the membrane layers to decrease the parasitic capacitance resulting from the electrolyte double layer at the inner interface between the solution and the semiconductor layers. As the upper and lower electrodes are identical, we only describe the modeling of the upper layer as detailed below.

7.5.1 Modeling of the n-Doped Silicon Electrodes

The combination of the electrolyte double layer [32–34] and the depletion zone in the semiconductor layer at the inner interface of the nanopore behaves as a cylindrical capacitor. In this work, the depletion thickness W is assumed to be constant and equal to 1 nm as estimated from multi-scale simulation for a $N_D = 2 \cdot 10^{20}\ \text{cm}^{-3}$ Si doping in the Si layer and a surface potential $\Phi_S \cong 0.16$ V [19]. As the double layer capacitor connected in series with the depletion capacitor presents a quite large value, its effect is negligible with respect to the lower capacitance of the depletion layer. In order to account for the 30° slanted angle, the n-doped Si part is divided into five mini-layers, each defining a cylindrical capacitor characterized by a radius r_i (see Fig. 7.9). The wall slope inside each mini-layer is assumed to be negligible. The depletion capacitance in each mini-layer is computed as:

$$C_{dep,i} = \varepsilon_0 \varepsilon_{Si} \frac{2\pi h r_i}{W}. \tag{7.12}$$

The silicon electrode is modeled by a network of resistors which is obtained by dividing each mini-layer into five elementary cells so that the whole electrode is comprised of 25 electric cells. One elementary cell is therefore a ring characterized by its height h, its external radius $r_{i,ext}$, and its internal radius $r_{i,int}$. It is made of two radial resistors, $R_r/2$, and two so called "thickness" resistors, $R_{th}/2$, all interconnected as shown on Fig. 7.9. The corresponding resistances are given by:

$$R_r = \frac{\rho_{Si}}{2\pi h} \ln \frac{r_{i,ext}}{r_{i,int}}, \tag{7.13}$$

$$R_{th} = \frac{\rho_{Si} h}{\pi (r_{i,ext}^2 - r_{i,int}^2)}, \tag{7.14}$$

where p_{Si} is the resistivity of the doped silicon layer.

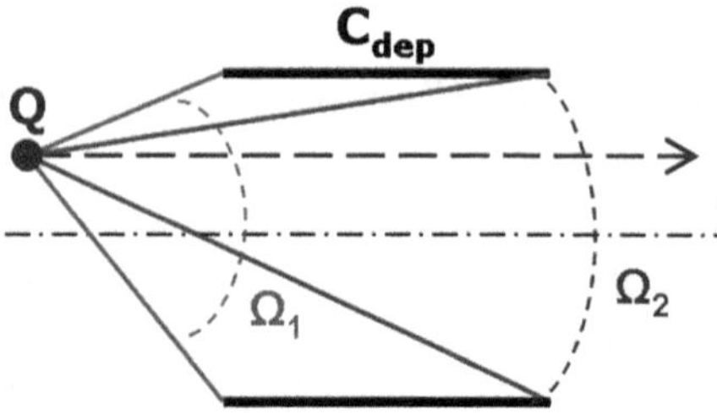

Fig. 7.9 Schematic of the solid angles defined by a point charge and sustained by the edge surfaces of the cylindrical capacitor C_{dep}

7.5.2 Modeling of the SiO_2 Layers

Both upper and lower external SiO_2 layers are assumed to be 2 nm thick. They separate the upper and lower electrodes from the electrolyte solution and are modeled by five capacitors, each of them connected to the resistor terminals of the upper (lower) mini-layer of the upper (lower) Si electrode. For the middle SiO_2 layer, the same design is used: 5 parallel capacitors connecting the upper electrode to the lower electrode. As each SiO_2 capacitor figures a ring shape, again defined by its height h, its external radius r_{ext} and its internal radius r_{int}, it is expressed as:

$$C_{SiO_2} = \frac{\varepsilon_0 \varepsilon_{SiO_2}}{h} \pi (r_{ext}^2 - r_{int}^2). \tag{7.15}$$

7.5.3 Screening

The presence of the KCl solution screens the electrostatic induction of the DNA charge on the semiconductor layers in two ways: first, by the polarization of the water molecules that is accounted for by the electrical permittivity of the liquid. This factor is presently considered as a parameter of the model, and plays an important role in the amplitude of the signal. Indeed, as the DNA strand fills the nanopore, it expells water molecules which reduces the dielectric screening. As a consequence, the factor is likely to vary locally, from the electrical permittivity of water to a much lower value, if water is absent. Second, K^+ and Cl^- ions behave as counterions over the charges of the ssDNA, thereby screening their electrostatic induction on the electrodes. This screening impacts the electrodes mainly for the four upper mini-layers where the electrostatic induction is reduced by a damping factor arbitrarily fixed to 0.1. Because the last layer, close to the intermediate SiO_2 layer has a 1 nm radius, the screening effect is assumed to be negligible for the corresponding inner capacitor. Indeed single-stranded DNA features a 1 nm diameter, which matches the 1 nm diameter of the pore, thus, as the DNA molecule occupies the whole space, it minimizes the presence of counterions, as shown by the current blockade. Moreover, since the DNA molecule expells the electrolyte from the nanopore, the electrolyte-oxide interface [35] disappears, and is consequently not simulated in this model.

7.5.4 Circuit Element Modeling of the DNA Translocation

As mentioned in the Introduction, the coupling between the translocating DNA strand and the device relies on the electrostatic induction felt by the sensing capacitors through the depletion zone at the inner interface of the nanopore. In the context, we model the electrostatic induction of the DNA molecule on the semiconductor layers as current sources $I(t)$ given by [36]

$$I(t) = \sum_i I_i(x_i) = \sum_i q_i \cdot F'(x_i) \cdot v, \tag{7.16}$$

where

$$x_i = x_{i_0} + vt \tag{7.17}$$

and x_{i0} is the original position of the charge i for time $t = 0$.

Here,

$$F'(x) = \frac{dF}{dx}, \tag{7.18}$$

where the geometrical factors F are connected to the inner capacitors of the nanopore by the solid angles Ω_1 and Ω_2 sustained by the edges of the cylinder, as shown in Fig. 7.9. F is associated to the difference between these solid angles:

$$F(x) = \frac{\Omega_i(x) - \Omega_2(x)}{4\pi}, \tag{7.19}$$

and consequently accounts for the magnitude of the electrostatic induction of the charge on the capacitor plate and varies with its position in the capacitor.

Figure 7.10 illustrates the equivalent circuit of the detection process (the complete membrane model). One can see each couple current source – inner capacitor corresponding to each mini-layer connected with the 25-cell resistors network. In order to produce a clear scheme, we represent the five capacitors related

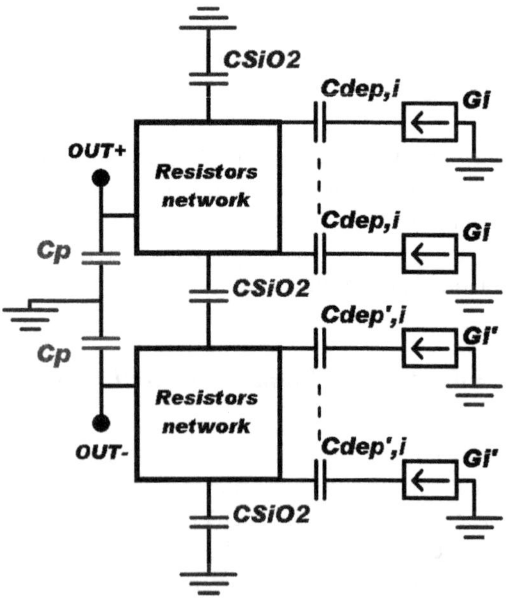

Fig. 7.10 Equivalent electric circuit of the overall nanopore-capacitor: current sources G_i accounting for the physical modeling of the DNA strand translocation are connected to the depletion capacitors $C_{dep,i}$. SiO_2 layers and Sj electrodes are modeled by a resistor-capacitor network

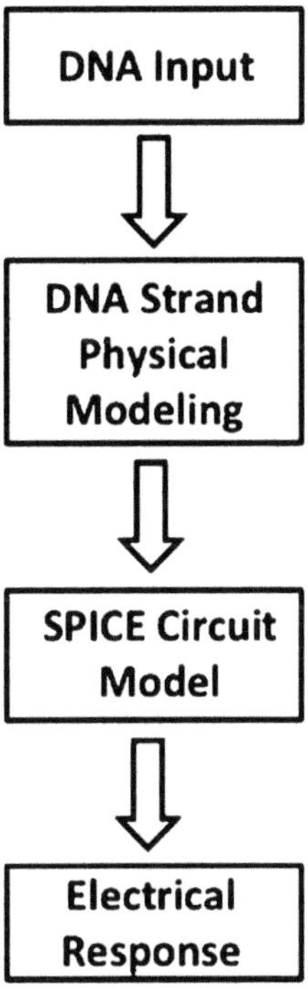

Fig. 7.11 Chart diagram of the proposed electric circuit approach to compute the response of a DNA strand translocation through a nanopore

to each SiO_2 layer by a symbolic single capacitor, connected to both resistor networks in the case of the intermediate layer and to one resistor network and the electrolyte solution in the case of the external layers. Finally we introduce two parasitic capacitors connected in parallel with the output of both electrodes; they account for the capacitors that will inevitably appear with the connection to any measurement device such as instrumentation amplifiers, and for which we assumed a fixed value of 0.01 pF.

The whole approach is depicted in Fig. 7.11. The atomistic charge distribution of the DNA strand obtained from the NAMD [24] is mapped on our grid in order to calculate the electrostatic induction of the strand. Then we use our SPICE circuit model and simulate the translocation of the ssDNA molecule to obtain the electrical signature of the strand.

7.6 Results and Discussion of the SPICE Model

7.6.1 DNA Strand Translocation

In Fig. 7.12a we show the calculated electrical response (solid line) on the upper electrode of a 11 bases ssDNA sequence, $5' - CCCACCCCCCC - 3'$ (C_3AC_7).

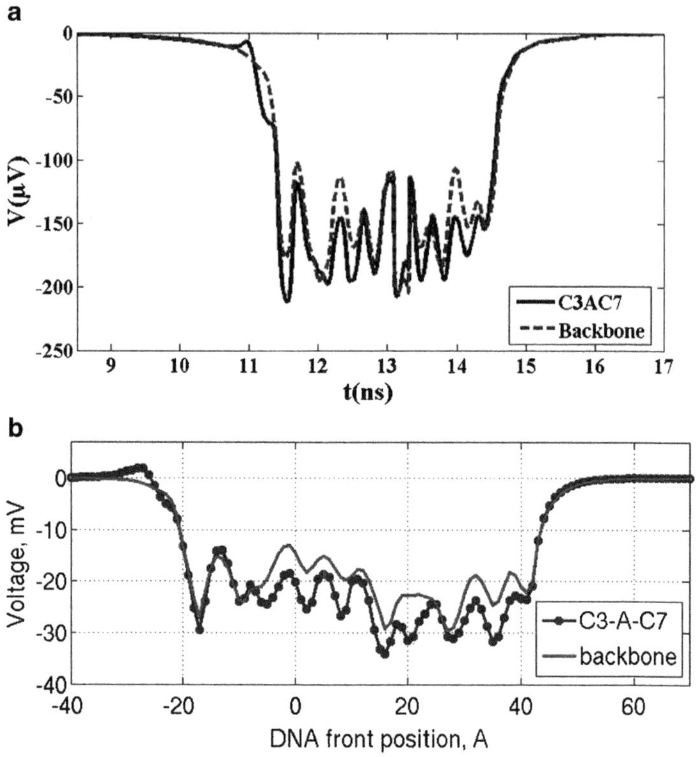

Fig. 7.12 (**a**) Electric response of the C_3AC_7 sequence (*solid line*) and its backbone (*dashed line*) computed on the upper electrode by the SPICE circuit model. (**b**) Electric response of the C_3AC_7 sequence and its backbone computed on the upper electrode by the multi-scale approach [19, 31]

On this figure we also display the calculated electrical response (dashed line) of the molecule backbone. The molecule axis coincides with the axis of the pore while the strand moves through the pore at a constant velocity of 2 m/s in a fixed conformation.

The simulation starts at position $x = -25$ nm and ends at $x = 25$ nm with the same convention as above. At first sight, the trace in Fig. 7.12a is comprised of ten dips including the first one characterized by a smaller magnitude and a shoulder. Actually, the third dip is a superposition of the two dips corresponding to the third and fourth bases of the C_3AC_7 molecule: It is clearly seen in Fig. 7.13 that displays the simulation of each fragment of the strand translocating independently through the pore. Each base of the C_3AC_7 molecule is associated with a sugar ring and a phosphate group except the first one which does not include a phosphate group. As a consequence, the first base signature exhibits a short positive shoulder that arises from the missing, negatively charged, phosphate group that usually offset the positive charge of the sugar ring and the base thereby resulting in a positively charged fragment. The maximum of the recorded fluctuations reaches 212 μV for

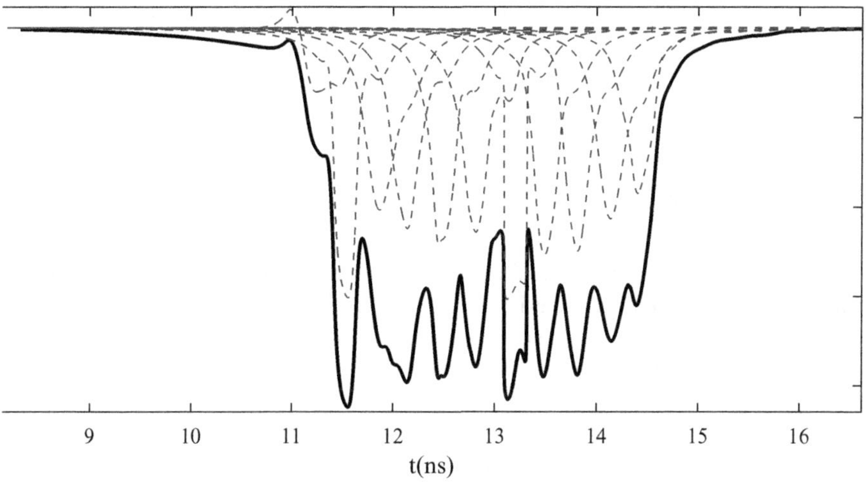

Fig. 7.13 Electric responses of the C_3AC_7 strand (*black solid line*) and its 11 fragments (*grey dashed line*), each translocating independently through the nanopore, computed on the upper electrode

the whole C_3AC_7 strand and 205 μV for the backbone, while the maximum voltage for a single base is about 35 μV. For the sake of comparison, we show on Fig. 7.12b the voltage traces computed by the multi-scale approach of Gracheva et al. [19, 31]. Note that we display voltage trace versus time as Gracheva et al. plot their trace versus position, which are equivalent given the constant velocity $v = 2$ m/s used in both models. Our results feature a good qualitative agreement regarding the shape of the curves. There is however a major discrepancy in the magnitude of the signal which is due to the fact that Gracheva et al. compute the voltage signal over a ring located at the inner constriction of the nanopore by 3D self-consistent Poisson Solver, while our SPICE model, which computes the voltage signal on the overall electrode (Fig. 7.2), accounts for the circuit losses in the membrane leading to a significant reduction of the signal amplitude read by an external observer. Moreover, the response greatly depends on the parasitic capacitors connected in parallel on the output as they act as a capacitor divider. As a consequence, these parasitic capacitors should be minimized as much as possible.

As mentioned earlier, we show in Fig. 7.13, the simulation of the 11 molecular fragments of the single-stranded C_3AC_7 molecule, each translocating independently through the nanopore-capacitor. Each fragment is composed of a sugar ring, a phosphate group and a base (except the first one which does not include the phosphate group) and results in a voltage dip. These voltage dips look quite similar and overlap significantly. Moreover the fourth trace related to the fragment containing the *A* base is not significantly different from the *C* base signatures. As discussed above, the first fragment signature exhibits a short positive shoulder. Our qualitative observations are consistent with the analysis of Gracheva et al. [31] as their corresponding simulation features the same characteristics such as the 11 quite similar voltage dips and the shoulder. However, one of the advantages of the Spice model over multi-physics simulations [19, 31] is the ability to analyze the signal response as a function of parameters external to the nanopore.

7.7 Conclusion

We performed a three-dimensional multi-scale/multi-material numerical simulation of the quasi-electrostatic characteristics of a novel solid state nano-biophysical device which can be used as a DNA detector. The simulation demonstrates the possibility of a nanopore device to function as a DNA detector. In particular, we show that the DNA translocation causes electrodes' voltage to exhibit changes in excess of 10 mV that can be recorded experimentally. From the recorded voltage trace we also show the possibility to measure DNA length and observe fine features that correspond to single nucleotides.

These experiments and simulations are only the first step in our exploration of the nanopore-capacitor mechanism of DNA single molecule electrical sequencing. Only one DNA snapshot was analyzed with respect to the electric response of the nanopore and was presented in this work. Several other DNA snapshots that were also analyzed (but not presented) produced similar results. The analysis of more realistic stochastic DNA translocation is a subject of our current research.

We also studied the electrical signals produced on a capacitor membrane containing a narrower 1 nm nanopore by single stranded DNAs with a single base mutation. The calculated maximum voltage due to the different bases varies from 2 to 9 mV, which is experimentally detectable. Signals from individual nucleotides can be identified in the recorded voltage traces, suggesting a 1 nm diameter pore in a capacitor can be used to accurately count the number of nucleotides in a DNA strand. Our data show the possibility of distinguishing DNAs with different sequences in the ideal situation when the strands have the same conformation. However, the influence of stochastic DNA translocation on the induced signal is an important issue for realistic modeling of the nanopore – DNA system. The stochastic "wiggling" of the DNA, although restricted by the nanopore, may reduce the resolution of single nucleotides or even make nucleotides indistinguishable. This issue is currently under investigation and will be the subject of a subsequent publication. Nevertheless, our results are an important milestone in DNA detection because techniques could be used to reduce or annihilate DNA stochastic movement in a nanopore. One of such recently proposed techniques involves wrapping DNA on a carbon nanotube [20], which is then lowered into a pore to detect bases. Another approach for reducing the effect of the DNA stochastic motion on the recorded voltage, is the method of "DNA flossing" through a nanopore to sample segments of the molecule and collect the average reading [37].

In order to assess the influence of the membrane capacitances and resistances on the signal level we developed a SPICE circuit model and compared our results with the multi-scale model described in [19, 31] to point out a good qualitative agreement between both methods, thus leading to the model validation. On the quantitative side, the magnitude of the strand signatures differs by a factor of the order of 10^2, mainly due to losses induced by the electrical circuit and the parasitic capacitors included in the model. In this context, our results still show the resolution of the 11 bases even in the worse conditions.

Hence, our model enables the optimization of the device by exploring its physical features (geometry, electrical material properties) or for instance, by adding MOS amplifiers to the structure to improve the recorded signals. We can already point out the importance of minimizing the parasitic capacitor value as well as the external radius of the membrane, which are key parameters in terms of signal magnitude. Moreover, the results provided by the Spice model can be used, in the electric circuit domain, as input data for the analysis of analog amplification, numerical conversion and processing of the electrical signal in the vicinity of the nanopore.

Although these results are encouraging, they are quite coarse assessments as assumptions such as the fixed conformation of the molecule during the entire translocation, the constant velocity or the electrical permittivity of the solution arbitrarily set to water electrical permittivity, have been made. In this respect the noise induced by ionic stochasticity and the DNA conformation dynamics needs to be assessed, but this topic is beyond the scope of the chapter.

Acknowledgments This work was funded by NIRT-NSF grant #NSFCCR02-10843, DARPA grant #392FA9550-04-1-0214, NIH grants ROI-HG003713-01 and P41-PR05969. The authors gratefully acknowledge the use of the supercomputer time at the National Center for Supercomputer Applications provided through Large Resource Allocation Committee grant MCA05S028. We are grateful to Dr. G. Timp for useful discussion and to Dr. A. Aksimentiev for supplying NAMD data.

Appendix

Ionic Concentrations

The ionic concentrations in KCl electrolyte solution are similar to electron and hole concentrations in an intrinsic semiconductor. Because of this similarity, we can consider the $K^+ Cl^-$ electrolytic solution as an intrinsic semiconductor and introduce virtual semiconductor parameters, i.e. a virtual energy band gap E_{geff}, virtual density states of K^+ ions and Cl^- ions, N_{K^+} and N_{Cl^-}, and virtual effective masses, $m^*_{K^+}$ and $m^*_{Cl^-}$ for potassium and chlorine ions, respectively [38]. With these virtual parameters, we can calculate the ion concentrations of the electrolytic solution as follows:

$$[K^+]_0 = [Cl^-]_0 = \sqrt{N_{K^+} N_{Cl^-}} \exp\left(-\frac{E_{geff}}{2kT}\right), \tag{7.20}$$

where $[K^{+}]_0$ and $[Cl^{-}]_0$ are the bulk ion concentration, and the virtual density of states N_{K^+} and N_{Cl^-} are given by

$$N_{K^+} = 2\left(\frac{m^*_{K^+}kT}{2\pi\hbar^2}\right), \; N_{Cl^-} = 2\left(\frac{m^*_{Cl^-}kT}{2\pi\hbar^2}\right). \tag{7.21}$$

References

1. Kasianowicz, J.J., Brandin, E., Branton, D., Deamer, D.W., Characterization of individual polynucleotide molecules using a membrane channel, PNAS **93**, 13770–13773 (1996).
2. Saenger, W. 1984. *Principles of Nucleic Acid Structure*. Springer Verlag, New York.
3. Song, L., Hobaugh, M.R., Shustak, C., Cheley, S., Bayley, H., Gouaux, J.E., Structure of staphylococcal alpha-hemolysin, a heptameric transmembrane pore, Science **274**, 1859–1865 (1996).
4. Akenson, M., Branton, D., Kasianowicz, J.J., Brandin, E., Deamer, D.W., Microsecond timescale discrimination among polycytidylic acid, polyadenylic acid, and polyuridylic acid as homopolymers or as segments within single RNA molecules, Biophys. J. **77**, 3227–3233 (1999).
5. Meller, A., Nivon, L., Brandin, E., Golovchenko, J., Branton, D., Rapid nanopore discrimination between single polynucleotide molecules, PNAS **97**, 1079–1084 (2000).
6. Deamer, D.W., Akeson, M., Nanopores and nucleic acids: prospects for ultrarapid sequencing, Trends in Biotech. **18**, 147–151 (2000).
7. Vercoutere, W., Winters-Hilt, S., Olsen, H., Deamer, D., Haussler, D., Akeson, M., Rapid discrimination among individual DNA hairpin molecules at single-nucleotide resolution using an ion channel, Nature Biotech. **19**, 248–252 (2001).
8. Meller, A., Nivon, L., Branton, D., Voltage-driven DNA translocations through a nanopore, Phys. Rev. Lett. **86**, 3435–3438 (2001).
9. Meller, A., Branton, D., Single molecule measurements of DNA transport through a nanopore, Electrophoresis **23**, 2583–2591 (2002).
10. Deamer, D.W., Branton, D., Characterization of nucleic acids by nanopore analysis, Acc. Chem. Res. **35**, 817–825 (2002).
11. Li, J., Gershow, M., Stein, D., Brandin, E., Golovchenko, J.A., DNA molecules and configurations in a solid-state nanopore microscope, Nature materials **2**, 611–615 (2003).
12. Heng, J.B., Dimitrov, V., Grinkova, Y.V., Ho, C., Kim, T., Muller, D., Sligar, S., Sorsch, T., Twesten, R., Timp, R., Timp, G., The detection of DNA using a silicon nanopore, IEDM Tech. Digest **8**, 767–770 (2003).
13. Heng, J.B., Ho, C., Kim, T., Timp, R., Aksimentiev, A., Grinkova, Y.V., Sligar, S., Schulten, K., Timp, G., Sizing DNA using an artificial nanopore, Biophys. J. **87**, 2905–2911 (2004).
14. Ho, C., Qiao, R., Heng, J.B., Chatterjee, A., Timp, R.J., Aluru, N.R., Timp, G., Electrolytic transport through a synthetic nanometer-diameter pore, PNAS **102**, 10445–10450 (2005).
15. Storm, A.J., Chen, J.H., Ling, X.S., Zandbergen, H.W., Dekker, C., Fabrication of solid-state nanopores with single-nanometer precision, Nature Materials **2**, 537–540 (2003).
16. Chang, H., Kosari, F., Andreadakis, G., Alam, M.A., Vasmatzis, G., Bashir, R., DNA-mediated fluctuations in ionic current through silicon oxide nanopore channels, Nano Lett. **4**, 1551–1556 (2004).
17. Aksimentiev, A., Heng, J.B., Timp, G., Schulten, K., Microscopic kinetics of DNA translocation through synthetic nanopores, Biophys. J. **87**, 2086–2097 (2004).
18. Heng, J.B., Aksimentiev, A., Ho, C., Dimitrov, V., Sorsch, T., Miner, J., Mansfield, W., Schulten, K., Timp, G., Beyond the gene chip, Bell Labs Tech. J. **10**, 5–22 (2005).

19. Gracheva, M. E., Xiong, A., Aksimentiev, A., Schulten, K., Timp, G., Leburton, J.-P., Simulation of the electric response of DNA translocation through a semiconductor nanopore-capacitor, Nanotech. **17**, 622–633 (2006).
20. King, G.M., Golovchenko, J.A., Probing nanotube-nanopore interactions, Phys. Rev. Lett. **95**, 216103–216107 (2005).
21. Fan, R., Karnik, R., Yue, M., Li, D., Majumdar, A., Yang, P., DNA translocation in inorganic nanotubes, Nano Lett. **5**, 1633–1637 (2005).
22. Fologea, D., Uplinger, J., Thomas, B., McNabb, D.S., Li, J., Slowing DNA translocation in a solid-state nanopore, Nano Lett. **5**, 1734–1737 (2005).
23. Zwolak, M., Ventra, M.D., Electronic Signature of DNA Nucleotides via Transverse Transport, Nano Lett. **5**, 421–424 (2005); Lagerqvist, J., Zwolak, M., Di Ventra, M., arXiv:cond-mat/0601394 (2006).
24. Aksimentiev, A., Heng, J. B., Timp, G., Schulten, K., Microscopic kinetics of DNA translocation through synthetic nanopores, Biophysical Journal **87**, 2086–2097 (2004).
25. Heng, J.B., Aksimentiev, A., Ho, C., Marks, P., Grinkova, Y.V., Sligar, S., Schulten, K., Timp, G., The electromechanics of DNA in a synthetic nanopore, Biophys. J. **90**, 1098–1106 (2006).
26. Heng, J., Aksimentiev, A., Ho, C., Marks, P., Grinkova, Y., Sligar, S., Schulten, K., Timp, G., Stretching DNA using the electric field in a synthetic nanopore, Nano Lett. **5**, 1883–1888 (2005).
27. Phillips, J.C., Braun, R., Wang, W., Gumbart, J., Tajkhorshid, E., Villa, E., Chipote, C., Skeel, R.D., Kale, L., Schulten, K., Scalable molecular dynamics with NAMD, J. of Computational Chemistry **26**, 1781–1802 (2005).
28. Sze, S.M. (1981). *Physics of semiconductor devices*. Wiley-Interscience publication.
29. Sansom, M.S., Smith, G.R., Adcock, C., Biggin, P.C., The dielectric properties of water within model transbilayer pores, Biophys. J. **73**, 2404–2415 (1997).
30. Press, W.H., Teukolsky, S.A., Vetterling, W.T., Flannery, B.P. (2001). *Numerical Recipes in Fortran 77*. Cambridge University Press.
31. Gracheva, M.E., Aksimentiev, A., Leburton, J.-P., Electrical signatures of single-stranded DNA with single base mutations in a nanopore capacitor, Nanotechnology **17**, 3160–3165 (2006).
32. Vidal, J. , Gracheva, M.E., Leburton, J.-P., Electrically tunable solid-state silicon nanopore ion filter, Nanoscale Res. Lett. **2**, 61–68 (2007).
33. Geddes, L.A., Baker, L.E., *Principles of Applied Biomedical Instrumentation*, 3rd ed. New York, USA, Willey - Interscience Publication (1989).
34. Bard, A.J., Faulkner, L.R., Electrochem. Meth., New York, USA, John Wiley & sons - Inter-science Publication (1980).
35. Martinoia, S., Massobrio, G., *A behaviorial macromodel ofthe ISFET in Spice*, Sensors and Actuators **62**, 182–189 (2000).
36. Leroux, A., Destine, J., Vanderheyden, B., Gracheva, M.E., Leburton, J.-P., SPICE-circuit simulation of the electrical response of semiconductor membrane to a single-stranded DNA translocating through a nanopore, to be published in the IEEE Transactions on Nanotechnology (2010).
37. Kasianowicz, J.J., Nanopores: flossing with DNA, Nature Materials **3**, 355–356 (2004).
38. Muller, R.S., Kamins, T.I., Chan, M. (2003). *Device electronics for integrated circuits.* John Wiley and sons Inc.

Chapter 8
Solid-State Nanopores: Methods of Fabrication and Integration, and Feasibility Issues in DNA Sequencing

Xinsheng Sean Ling

Abstract In this chapter, we review several strategies for the fabrication of solid-state nanopores in the context of the ultimate goal of a DNA sequencing and biomolecular analysis technology using these devices. The feasibility of a proposed DNA sequencing platform using solid-state nanopores and hybridization probes is discussed.

Keywords Nanopore DNA sequencing • Solid-state nanopores • TEM-drilling • DNA hybridization • Hybridization-assisted sequencing • HANS • Mean first-passage time • Fokker-Planck equation • 1D driven diffusion

8.1 Introduction

In his renowned lectures given in the 1960s, theoretical physicist Richard Feynman once famously declared that "everything that living things do can be understood in terms of jiggling and wiggling of atoms." Nearly half century later, many solid-state physicists are venturing into the realm of biological systems with the goal of developing solid-state techniques to probe the "jiggling and wiggling" atoms and to uncover new insights into molecular biology. The emergence of nanobioscience in recent years is a natural evolution for solid-state physicists for whom making small devices is a way of life. In the push for smaller and faster electronic devices, many micro and nanofabrication techniques have become widely accessible. It is now possible to make devices that are of the length scales of biomolecules. Solid-state nanopores are one of these latest examples.

X.S. Ling
Department of Physics, Brown University, 182 Hope Street, Providence, RI 02912, USA
e-mail: xsling@brown.edu

S.M. Iqbal and R. Bashir (eds.), *Nanopores: Sensing and Fundamental Biological Interactions*, DOI 10.1007/978-1-4419-8252-0_8,

The field of solid-state nanopores has to thank mother nature for the inspiration. In biology, on cell membranes, there are plenty of tiny nanometer-sized holes, called pores. Cells need them to regulate ion contents, DNA and RNA molecules move through them, all as parts of the intricate machinery of life. It is now possible to make pores of similar dimension in solid-state materials. There have been intense interests in recent years in using these solid-state nanopores to explore biology, to read out DNA sequences, and to perhaps build artificial cells mimicking life.

Perhaps the single most important applications scientists had in mind for nanopores is for the development of a new method for DNA sequencing. The concept of direct DNA sequencing using nanopores was first introduced by Kasianowicz et al. [1] in a seminal paper on DNA translocation through a biological pore α-hemolysin, a toxin secreted by *Staphylococcus aureus*. Their basic idea, patented by Church et al. [2], is simple and appealing: since the ionic current through an ion channel is a measure of the opening of the pore, it seems possible that if a macromolecule like DNA enters the pore, the ionic current will be suppressed by an amount which is a measure of the size of the molecule. Since the four nucleotides comprising a single-stranded DNA (ssDNA) or RNA have slightly different physical sizes, they may give rise to measurable fluctuations in the ionic current through the pore if the molecule "translocates" in a linear fashion under the drive force of an applied electric field across the ion channel. In this method, the DNA sequence is read out by the temporal variation of ionic current through the pore.

The subsequent experiments by a number of groups [3, 4] using α-hemolysin failed to detect any sign of individual nucleotides. There are a number of reasons that the proposed direct sequencing approach may be difficult. Among these is the fact that the spacing between nucleotides is about 0.4 nm, but the α-hemolysin channel length is about an order of magnitude longer, hence the current blockade effect is an averaged signal from several nucleotides.

Motivated by the desire for a more robust and more tunable system for studying the DNA translocation phenomena, the first solid-state nanopore device was developed by Li et al. [5] using ion-beam techniques. Since then, several alternative methods have been reported (Storm et al. [6]; Chang et al. [7]; Wu et al. [8]; Park et al. [9]).

8.2 Fabrication Techniques for Single Solid-State Nanopores

With the standard electron-beam lithography techniques [10], one can routinely produce features at 20–100 nm. To produce pores at 1–20 nm range, one has to resort to non-conventional methods. The ion-beam technique developed by Li et al. [5] was the first such example. Here, the latter methods are discussed including the TEM-drilling technique [6] which is more popular at present with researchers. It should be emphasized that the solid-state nanopore technology is an evolving field, there is no "best" method of fabricating nanopores. It is hoped that the readers will be inspired to develop even "better" techniques.

8.2.1 Controlled Shrinking by e-Beam or Laser Heating

First a brief recount of the history will be discussed on the methods developed at Delft using a transmission electron microscope (TEM). It was first discovered by Jianghua Chen and Henny Zandbergen in mid 2002 in the SiO_2/Si_3N_4 membranes prepared by Arnold Storm in Cees Dekker's group at Delft, that under the electron beam (or e-beam) of a TEM, a pore often spontaneously shrinks all the way to complete closure. It was quickly realized that this is a typical surface-tension driven mass flow effect: the electron beam fluidizes the amorphous materials which flows under the force of surface tension. As shown in Fig. 8.1, the pore shrinks as a function of time during the TEM imaging process. The TEM images in Fig. 8.1 are time-lapsed (from left to right). The average diameter of the pore decreases almost linearly with time. The TEM used was a commercial transmission electron microscope (TEM), Philips CM-30UT, operated at an accelerating voltage of 300 kV. A good level of vacuum in the microscope was maintained, at ~ 10^{-7} mbar, such that one can rule out the pore closing as being an effect of carbon contamination.

Several features of the observed effect can be understood using the Laplace's law of surface tension. As shown in a simplified model in Fig. 8.2, once the glassy silicon oxide material is fluidized by high energy electrons in a TEM, the material

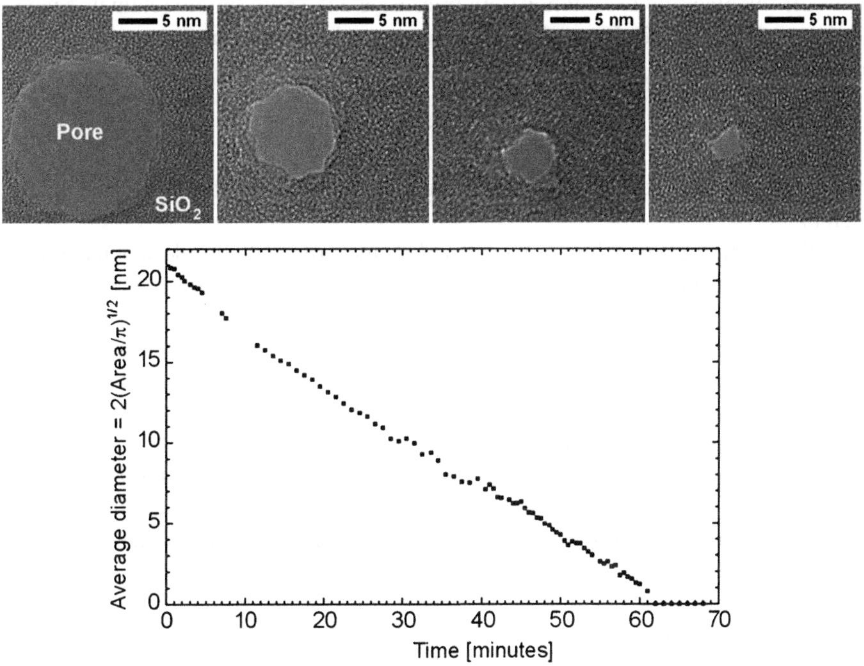

Fig. 8.1 (*Top*) Time-lapsed TEM images of a SiO_2 nanopore; (*Lower*) The average diameter of the pore vs. time with the pore region being exposed to the electron beam under a TEM [6]. Used with permission. Copyright Nature 2004

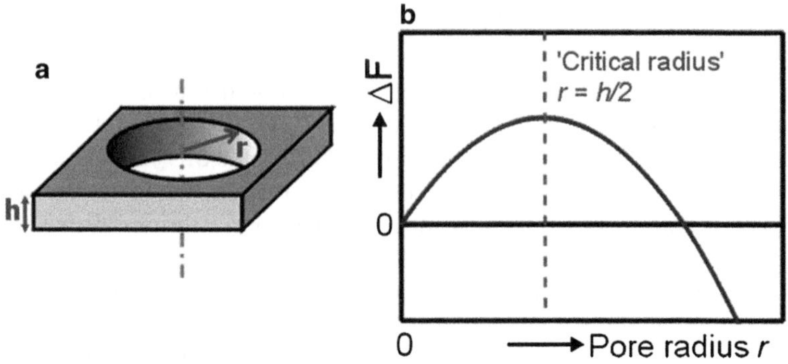

Fig. 8.2 A model for pore shrinking and expansion based on Laplace's surface tension mechanism: (**a**) A simplified sketch of a pore inside a SiO_2 membrane. (**b**) Free energy of the pore [6]. Used with permission. Copyright Nature 2004

behaves like a fluid. As such, if a hole was formed inside the SiO_2 membrane prior to fluidization, it will shrink or expand responding to the unbalanced Laplace pressures due to the finite radius of the hole and the finite thickness of the membrane. A small hole below the critical radius will shrink while a large one will expand without bound. When the electron beam is switched off, the material cools off (by black-body radiation) and freezes, and the pore should retain its shape. In the model shown in Fig. 8.2, the free energy change of forming a pore inside a membrane is,

$$\delta F = \gamma(2\pi rh - 2\pi r^2), \tag{8.1}$$

where γ is surface tension, r is the radius of the pore, h the thickness of the membrane.

If the starting pore radius is less than $h/2$, the pore will shrink spontaneously under the surface tension; on the other hand, if the starting pore diameter is greater than $h/2$, the pore expands without bound, causing the membrane to rupture. The "critical" diameter is the thickness of the sheet. This scaling argument is valid at any scale. The key prediction that a pore with a starting diameter larger than the thickness of the membrane will expand upon heating was verified in the subsequent experiments of Storm et al. [6].

We must point out that the exact mechanism of the pore shrinking induced by an e-beam is likely to be more complicated than the simple picture above. For example, the inner rim of the pore during shrinking is not circular, as would be expected from a simple surface-tension mechanism, see Fig. 8.1 (top). It is quite possible that the SiO_2 material is no longer uniform during the exposure to the e-beam. Later studies indeed revealed significant loss of oxygen especially after prolonged exposure to the high energy e-beam (see Kim et al. [11]).

The surface-tension model in Fig. 8.2 predicts that the physics of a critical diameter for pore shrinking or expanding is scale independent. Indeed this was confirmed by Wu et al. [8] in a laser-heating experiment on plastic pores (holes).

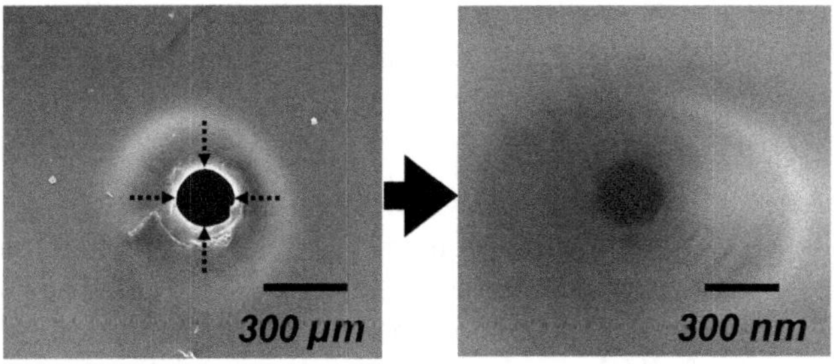

Fig. 8.3 SEM images. (*Left*) A mechanically punched hole, ~200 μm in diameter, in a membrane made of Apiezon-W wax. The *dashed lines* here indicate the direction of Laplace pressure when the material is fluidized. (*Right*) After exposing to a laser beam (514 nm in wavelength), the hole shrinks to a pore of 200 nm in diameter while the hole (pore) size being monitored optically. Below 200 nm, a feedback mechanism using ionic current is needed [11]. Used with permission. Copyright Nano Letters 2006

Figure 8.3 shows SEM images of a 200 μm plastic hole being shrunk to a 200 nm pore. The nearly perfect circular geometry at the inner rim of the 200 nm pore is a clear indication the driving mechanism here is surface tension. In contrast to SiO_2 in TEM, the plastic material is expected to be uniform under laser heating.

It was noticed that the linear time-dependence of the pore diameter (or radius) shown in Fig. 8.1 was also seen by Wu et al. [8] in their laser-induced shrinking of plastic pores, suggesting that the linear time dependence is a universal behavior of pore closing. Wu et al. [8] extended the surface-tension model to account for the time dependence.

As shown in Fig. 8.3, at any given point on the innermost surface of the pore, the Laplace pressure is given by the Laplace–Young equation

$$P = \gamma\left(-\frac{1}{r} + \frac{1}{r_d}\right) \tag{8.2}$$

where γ is the surface tension, r the radius of the pore and r_d the radius of the curvature along the pore channel direction. For clarity, the signs of the radii of opposing curvatures are given explicitly. During pore shrinking, the thickness of the membrane is large compared with the pore radius, so the first term in Eq. (8.2) dominates.

To derive an equation of motion to describe pore shrinking, Wu et al. [8] showed that, assuming the membrane is made of an incompressible fluid and it thins uniformly as the pore closes, by balancing the viscous drag with surface tension, one arrives at:

$$\dot{r} = \frac{\gamma}{4\eta}\left(-1 + \frac{r}{r_d}\right). \tag{8.3}$$

where η is the viscosity of the fluidized material.

This equation can be used to interpret all behaviors described earlier (Storm et al. [6], Wu et al. [8]). Since $r << r_d$ during the entire shrinking process, Eq. (8.3) can be reduced to

$$\dot{r} = -\gamma/4\eta. \tag{8.4}$$

According to Eq. (8.4), the radius (or diameter) of a pore in a fluidized membrane will decrease linearly with time if the starting diameter is less than the membrane thickness. The rate at which it shrinks is determined by the ratio of surface tension to viscosity, both of which are dependent on the material and temperature (and other physical conditions). However, the dynamics is universal.

8.2.2 TEM "Drilling"

It started out as a curiosity while considering the simplified surface-tension model in Fig. 8.2. According to Fig. 8.2b, the energy barrier to thermally nucleating a pore in an uniform membrane of thickness h is ~h^3. Thus in very thin membranes, it should be possible to nucleate a pore by locally heating up the membrane. This hypothesis was quickly put to the test, and it was quickly found (Storm et al. [6]) that indeed it is possible to form a nanopore in a thin uniform SiO_2 membrane.

The exact mechanism of the pore forming under a local focused e-beam is still uncertain. The careful element analyses (Wu et al. [11] and Kim et al. [12]) of the material near the pore showed the loss of oxygen. Thus it is common in the solid-state nanopore field to consider the pore forming mechanism under a focused e-beam to be equivalent to "drilling", or removal of materials locally (Fig. 8.4).

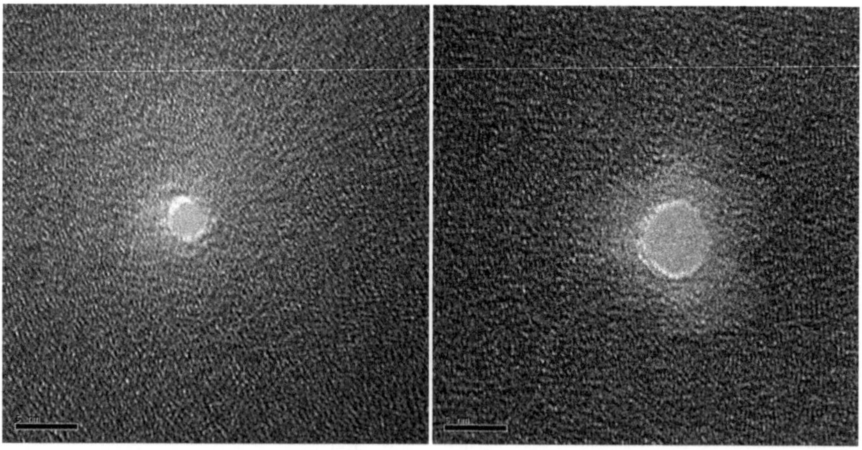

Fig. 8.4 TEM images of two nanopores formed by TEM "drilling." The membrane is Si_3N_4, with thickness ~20 nm. The TEM used is a JEOL 2010 F system with a field-emission source operating at 200 keV. The scale bars = 5 nm. (Images taken by V. Balagurusamy, unpublished, 2010)

It is quite possible that the e-beam removal (or "sputtering") of the materials is only to thin down the membrane locally to the point where the probability of spontaneously nucleating a pore become unity. The fact that the diameter of the TEM "drilled" is critically sensitive to the electron beam shape (Venkat Balagurusamy and Paul Weinger, private communications, 2010) lends support to this view.

8.2.3 Feedback Chemical Etching

Here a low-cost method in fabricating nanopores in silicon chips is discussed. The basic concept behind this method (Park et al. [9]) is similar to that in the work of Apel et al. [13], but the method of Park et al. [9] is more precise and is unique to silicon processing.

The starting material is a <100> silicon wafer. The first step is to use standard silicon processes to fabricate an inverted pyramid with a sharp tip on one side of the silicon wafer, using the well-known techniques of anisotropic etching of Si in alkaline solutions such as KOH, or $(CH_3)_4NOH$ (tetramethylammonium hydroxide or TMAH). After the first step, the silicon chip is mounted as a partition in an electrochemical setup separating two chambers filled with KCl and KOH solutions, with KCl on the front side with the pyramid, and KOH on the back, as shown in Fig. 8.5 (left, inset). While KOH is etching away silicon from the back side, the electrical current across the silicon chip is monitored using two Pt electrodes, one in KCl and the other in KOH. Under proper bias voltages, it is possible to detect the

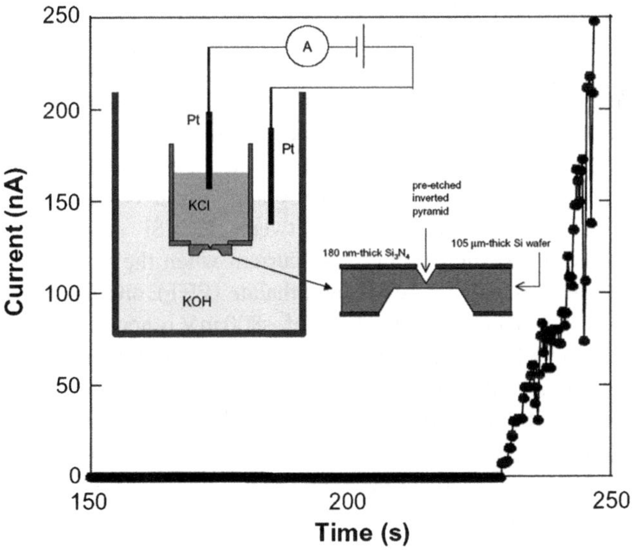

Fig. 8.5 Current vs. time during the feedback etching process [9]. Used with permission. Copyright Small 2007

opening of a nanometre-scale pore in the silicon chip at the tips of the inverted pyramid. Thus one can stop the etching process promptly and prevent over-etching.

For this study, a 105 μm thick <100> undoped silicon wafer (Virginia Semiconductor, Inc.) was used. First, a 180 nm layer of Si_3N_4 was grown by standard Plasma Enhanced Chemical Vapor Deposition (PECVD) or Low Pressure Chemical Vapor Deposition (LPCVD) on both sides of the wafer and was used as an etching mask. For inverted pyramid fabrication, the front side of the Si wafer had an exposed area of 30 μm × 30 μm square (pattern generated by photolithography) with Si_3N_4 layer removed by CF_4 plasma etching. On the back side, a larger exposed area (1 mm × 1 mm) was patterned by the same method. The front and back patterns were aligned such that the two square patterns were approximately concentric.

The exposed bare silicon in the squares, both front and back, were etched into invert pyramid shapes in 45% (w/w) KOH at 85 °C. The etching was continued until the inverted pyramid on the front surface attained a sharp tip. At this point, the tip of the front inverted pyramid had not reached the back surface (the top of the pyramid on the back).

After extensive cleaning with de-ionized (DI) water, the silicon-chip sample was attached using a chemical resistant wax (Apiezon W) on an opening of a Teflon tube such that the front side with the sharp invert pyramid was in contact with an ionic solution (1 M KCl), while the back side with large area of bare silicon was exposed to KOH etchant. As shown in the inset of Fig. 8.5, one Pt electrode was inserted in each solution, respectively, and they were connected to a power supply and an ammeter in series.

The first key step in fabricating a nanopore is to control the opening process. Here, one monitors the opening processes by measuring the time dependence of the electrical current flowing through the silicon chip. However, there is a large zero-bias electric current spontaneously flowing through the silicon chip as soon as the silicon sample is brought into contact with two electrolytes (KCl and KOH). This is due to electrons generated at the interface of Si and KOH solution during the etching process, $Si + 4OH^- \rightarrow Si\ (OH)_4 + 4\ e^-$. Fortunately it was found that the zero-bias current can be minimized by applying a voltage to cancel out the electrochemical potential difference between KCl/Si and KOH/Si surfaces. In contrast, there is no such background current when the materials being etched are insulating such as polyethylene terephthalate (PET), and Kapton.

In the study of Park et al. [9], a voltage of ~800 mV is applied with cathode being the Pt electrode in KCl, and anode in KOH. The etching rate of KOH in <100> at room temperature is approximately 18 nm/min, which is slow enough such that the etching process can be stopped manually (or by robotic arms). While the back side of the silicon chip is being etched by KOH, the electric current across the chip is monitored and recorded in real time. When a pore is etched through at the tip of the inverted pyramid, the electric current increases sharply, as shown in Fig. 8.5. Depending on the desired pore size, at certain current level, one can terminate the etching process, and remove the silicon-chip sample from the electrochemical cell and clean it with DI water.

Besides the abrupt increase in current, it was found that one can measure the simultaneous cyclic voltammogram at a certain rate to check the existence of a pore. The current–voltage (*I–V*) curves show different characteristics before and after the opening of a pore. In the absence of a pore, the *I–V* curve is non-ohmic over a wide range of bias voltage, in agreement with prior studies of electrochemical KOH etching of silicon. On the contrary, in the presence of a pore, the *I–V* curve shows ohmic characteristic.

It was found that the post-etch pore ionic resistance gives a good measure of the pore size, correlating strongly with the stoppage current, the electric current level at which the etching process is terminated. To properly determine the ionic resistance (or conductance) of the pore, the *I–V* measurements are performed with both sides of the silicon chip filled with 1 M KCl solutions and using Ag/AgCl wires as electrodes. Figure 8.6 shows TEM pictures of a sample obtained at a stoppage current of 248 nA (at 800 mV) and after 18-min oxidation. It was found that the pores can be shrunk with additional oxidation time. In fact a similar surface-tension driven shrinking process was invoked [9].

Park et al. [9] have demonstrated that this simple benchtop method can be used for fabricating nanopores in silicon chips in the range of 3–50 nm. This technique only relies on standard silicon processing equipment, with no requirement of specialized ion-beam and electron-beam facilities, making solid-state nanopore technology affordable for a wide range of users.

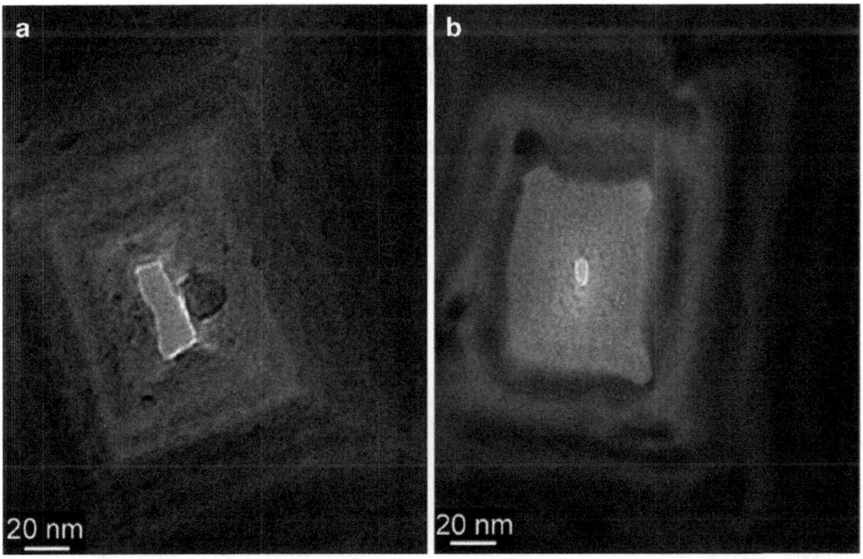

Fig. 8.6 TEM images showing the morphology of (**a**) an as-etched and (**b**) an oxidized pore. The as-etched silicon sample has a resistance of 6.7 MΩ (1 M KCl, Ag/AgCl electrodes). (**a**) *Rectangular-shaped pore* is attained with dimensions of 15 nm × 38 nm. (**b**) After oxidation at 1,100 °C for 1 min, the pore becomes smooth and round (10 nm long axis and 5 nm short axis) [9]. Used with permission. Copyright Small 2007

8.3 An Integration Strategy for Multiple Nanopores: Electrically Addressable Pores

The field of solid-state nanopores has experienced significant growth during the past few years. However, for many biotechnology applications, single nanopore devices are inadequate. For example, in the proposed DNA sequencing platform, hybridization-assisted nanopore sequencing or HANS (Ling et al. [14]), the sequencing is achieved by combining DNA hybridization and solid-state nanopores. In HANS, one needs to analyze large number of DNA probes. For example, for 8-mer DNA probes, there are $4^8 = 65{,}536$ possible combinations. Thus developing integrated nanopore devices will be crucial to many applications. Here, the effort in developing addressable nanopores is discussed.

It was proposed (Ling [15]) that one can realize an electrically addressable nanopore array (EANA) device by utilizing the unique property of anistropic etching of silicon crystals. As shown in Fig. 8.7, long V-shaped grooves will result due to the anisotropy of the etching rates of a <100>-oriented silicon wafer. An example is shown in Fig. 8.7 (mid). Thus if a silicon wafer is etched from both sides to form V-shaped grooves at the right angle, when the tips of V-grooves are etched just open (this can be done to within 20 nm accuracy by controlling the etching rate and etching time), a linear array of nanopores will be formed.

After etching, the whole device can be oxidized to form SiO_2 to insulate all silicon surfaces. The next step is to seal both the front and the back sides of the silicon chip with glass slides and curable polymer (e.g. Dow CYCLOTENE) with proper microfluidic tubings (e.g. Upchurch Scientific) and electrodes (Ag/AgCl) connected. The resulting device is sketched in Fig. 8.7 (right).

The unique feature of the EANA device is that one can use an electric field to bias a pore while at the same time use pressure to control the ionic solutions on both sides of the pore. By placing the electrodes in different V-grooves on the topside of the wafer, one can measure the electrical current through each pore independently. This feature is essential for high throughput applications.

Figure 8.8 shows a sample with over-etched pores for visualization purposes. The device shows $3 \times 5 = 15$ over-etched pores of size about 5 μm. The process

Fig. 8.7 (*Left*) A linear array of electrically-addressable nanopores formed by etching a silicon wafer. (*Mid*) A SEM micrograph of a V-groove on the silicon <100> surface after wet etching with KOH. The *scale bar* is 500 nm. (*Right*) A sketch of a possible EANA nanopore array system. For clarity, only one V-groove is drawn for the upper side of the device

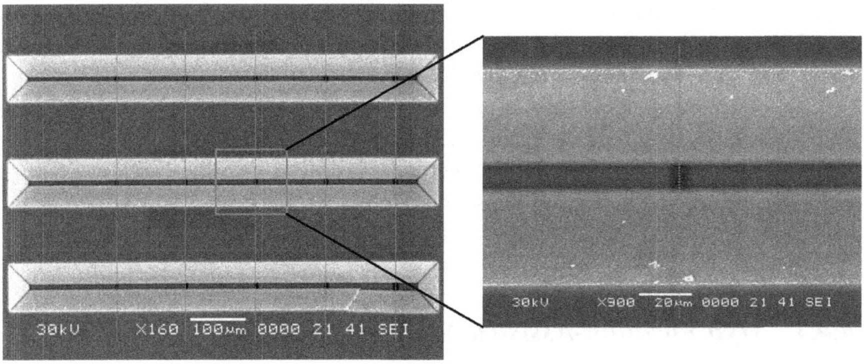

Fig. 8.8 A silicon chip etched from both sides into V-grooves. The *dashed lines* indicate the V-grooves from the backside. The pore sizes can be controlled down to 20 nm by optimizing the etching rate and etching time. (Images taken by S.R. Park, unpublished, 2004)

begins with a 100 µm thick Si wafer (<100> oriented). First, a silicon nitride Si_3N_4 protective layer is deposited, about 200 nm in thickness, on both sides of the wafer using PECVD (Plasma Enhanced Chemical Vapor Deposition). One then spin-coats PMMA, bake, and make a two sets of parallel lines using e-beam lithography on both sides. The two sets of line are perpendicular to each other. The patterns are developed in PMMA developer (MIBK:IPA 1:1). The PMMA pattern is transferred onto the silicon nitride layer using RIE (Reactive Ion Etcher). Repeat the same RIE procedure on the other side of the wafer. Now one has a silicon wafer sandwiched by silicon nitride films having patterns of vertical lines on one side and horizontal lines on the other side. KOH anisotropic etch is carried out at 85 °C with 45% (KOH) solution using silicon nitride as a mask. This gives about 0.8 µm/min etching rate. Approximately one hour later, the pores are opened.

It was found by Adam Politzer [16] that the inhomogeneous wafer thickness can pose a serious problem in fabricating large number of nanopores in a single chip. A careful analysis of the pore openings showed that the pore sizes are not directly correlated with the lithography feature sizes [16]. To overcome this problem, it has been suggested by Robert Riehn [17] that instead of using a regular silicon wafer, a thin crystalline silicon membrane on a SOI wafer should be used as the starting material. Since the thickness of the silicon layer on a SOI wafter can be precisely controlled, the issue of inhomogeneities may be avoided.

8.4 Considerations on Using Solid-State Nanopores for DNA Sequencing

The challenge of DNA sequencing is due to the simple fact that the basic units of the nucleic acids, the four nucleotides of adenine, cytosine, guanine, and thymine, are spaced at a very short physical distance, about 4 Å (0.4 nm) apart, on the sugar-phosphate backbone of a DNA. It is truly an amazing physical phenomenon that DNA polymerases, a complex molecular machinery, can resolve individual bases with high accuracy during the DNA synthesis of a cell cycle.

The DNA sequencing techniques developed by Sanger [18], Gilbert [19], and others take advantage of this unique capability of DNA polymerases. However, the very process that makes Sanger method possible is also the limiting factor in reducing the cost while raising the speed of DNA sequencing.

8.4.1 Hybridization-Assisted Nanopore Sequencing (HANS)

It was proposed by Ling et al. [14] that the solid-state nanopore technology developed so far may be adequate to implement a low-cost DNA sequencing strategy, namely, the hybridization assisted nanopore sequencing (HANS). The HANS approach borrows the idea of a known technique called sequencing-by-hybridization (SBH) derived from the ideas in "Southern blot" [20], but HANS differs from SBH in terms of the way DNA sequences are extracted.

In SBH, small pieces of short oligonucleotides with known sequences are used as "probes" to hybridize an unknown DNA. From the overlapping parts of hybridizing oligonucleotides, one can in principle reconstruct the sequence of the unknown DNA using a computer if there are no repetitions of DNA sequences. However, the genomic DNA is known to have large amount of repeats, preventing SBH from becoming a viable sequencing technology.

As shown in Fig. 8.9, a single-stranded ssDNA with unknown sequence is hybridized with two identical short strands of oligonucleotides and is driven through a solid-state nanopore. The ionic current through the nanopore is suppressed when the DNA enters the pore, as usual. The parts of the DNA with matching oligonucleotides strand should suppress the ionic current by a larger magnitude. It is expected

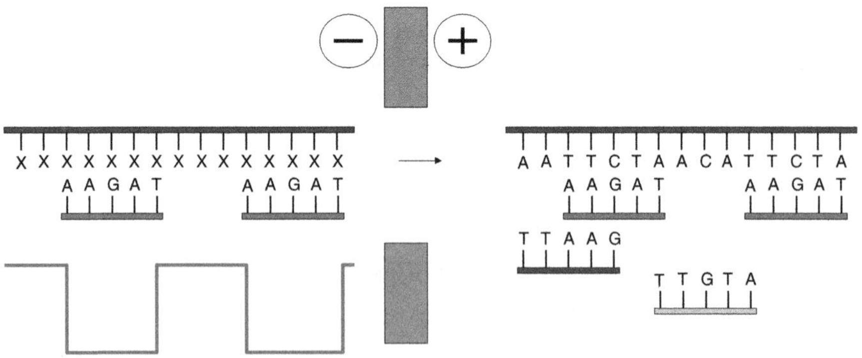

Fig. 8.9 The basic concept of HANS in obtaining the DNA sequence of the ssDNA based on the positions (*top panel*) of the hybridizing oligonucleotides measured (*lower panel*) by the ionic current traces for each oligonucleotide [14]

that the ionic current versus time trace should reveal two features corresponding to the two matching oligonucleotides. The sequencing is achieved by aligning the current blockades signals and their corresponding probe sequences similar to the concept of sequencing by hybridization (SBH). The key difference between HANS and SBH is that the main framework of sequence information in HANS is formed by the locations of the DNA probes, thus free from the problems of repeat sequences in a DNA.

The major difference between the HANS approach and that of the direct nanopore sequencing of Kasianowicz et al. [1] is that the HANS approach bypasses the harsh requirement of single-base spatial resolution (0.4 nm) which is smaller than the pore length for both solid-state and biological nanopores.

8.4.2 *Error Analysis of Positional Measurement Using DNA Translocation Through a Nanopore*

Even with the relaxed demand for positional accuracy in the HANS approach, it is still worthwhile to consider the major obstacles. The first concern is that DNA molecules are entropic springs (Marko and Siggia [21]), as such their contour confirmation fluctuates violently at ambient temperatures. Fluctuations between possible polymer configurations change the effective distance between the probes on the ssDNA seen by the nanopore. The simplest solution will be to hold the DNA under tension: the electric field inside the pore pulls the DNA forward, while the magnetic force or optical force pulling on the bead backward, as shown in Fig. 8.10.

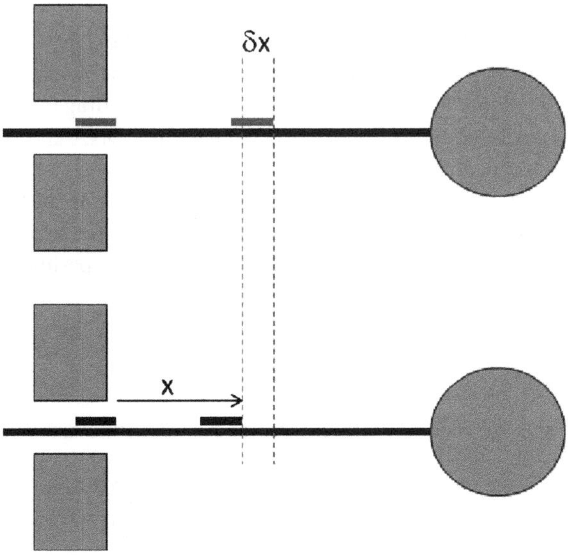

Fig. 8.10 A simple model for controlled DNA translocation. The long black lines are target ssDNA molecules with unknown sequences. The shorter bars indicate two sets of oligonucleotides with known sequences from a n-mer hybridization library. The basic issue is whether the nanopore technique is ultimately capable of resolving the difference in their positions

Using the effective spring constant published in literature, one can give a quantitative estimate on the magnitude of error from this source. According to the measurement of Bustamante et al. [22], for a λ-dsDNA molecule with a contour length of 16.4 μm, the force at which the Watson-Crick pairs start to break is 65 pN. To prevent the overstretching transition such that DNA hybridization still takes place, one expects not to exceed this value. For λ-dsDNA, the effective spring constant k just before this overstretching transition takes place is about 53 pN/μm. At this effective spring constant, the root-mean-square displacement of the bead can be estimated, using equipartition principle, to be about $(x_B{}^2)^{1/2} = 8.8$ nm. However, one expects the actual thermal displacement of the DNA probes $(x_p{}^2)^{1/2}$ relative to the pore is significantly smaller since the effective spring constant between the probe and the pore is enhance by a factor of L/x, where L is the contour length and x the distance between the probe and the pore. Here it is assumed that the DNA behaves like a linear spring that its effective spring constant scales as $1/L$.

Secondly, in the limit that the entropic spring effects can be ignored, that the DNA translocation can be viewed effectively as a biased random walk by the pore on the DNA, thermal fluctuation effects will still appear as dictated by the fluctuation-dissipation theorem. One expects that a spread in the first-arrival time (or first-passage time) will occur. If this thermal smearing is too large (see below for what is "too large"), the HANS sequencing method (as well as other nanopore-based sequencing methods) will be fundamentally prohibited.

Following the analysis by Lubensky and Nelson [23], one can write down a Fokker–Planck equation for the probability $P(x, t)$ of finding the pore at distance x on the DNA as:

$$\frac{\partial P(x,t)}{\partial t} = D\frac{\partial^2 P(x,t)}{\partial x^2} - v\frac{\partial P(x,t)}{\partial x}, \tag{8.5}$$

where D is the diffusion constant of the DNA, v is the translocation velocity (proportional to applied voltage across the nanopore). According to the Einstein relation, $D = k_B T/\gamma$, where γ is the Stokes' drag coefficient of the DNA (which may be altered by the interaction between the DNA and the wall of the nanopore during translocation).

If the two adjacent probes are spaced at x apart, the Fokker–Planck equation can predict [24] the mean-first-passage time of the second probe, in large v limit which is satisfied for typical DNA translocation experiments in solid-state nanopores,

$$<t> \approx \frac{x}{v} - \frac{D}{v^2}, \tag{8.6}$$

and the second moment of first-passage time as:

$$<t^2> \approx \frac{x^2}{v^2} - \frac{4D^2}{v^4}. \tag{8.7}$$

From Eqs. (8.6) and (8.7), one can give a quantitative constraint on the feasibility of DNA sequencing using HANS approach.

Imagine two ssDNAs, as shown in Fig. 8.10, each has two probes hybridized to it, with their 5′ ends attached to beads. Assume both molecules have a probe (oligo) at the same distance away from the free 3′ ends. However, the oligos near the 5′ ends differ by a distance of δx. The distance between the two probes in lower DNA is x.

In order to resolve δx in the temporal signals from the translocation data, the change $\delta <t>$ in the mean-first-passage time $<t>$ must be greater than the variance in the mean-first-passage time, i.e.

$$\delta <t> \geq \sqrt{<t^2> - <t>^2}. \tag{8.8}$$

Condition (8.8) can be re-written as,

$$\frac{\delta x^2}{2D} \geq \frac{x}{v}. \tag{8.9}$$

This says that, in order to resolve the location of a probe to precision δx, the time it takes for thermal diffusion to occur over distance δx has to be greater than the time it takes to translocate across distance x. This condition makes physical sense: thermal diffusion is always there, but one can still resolve the location of the probe accurately as long as the translocation is fast enough. A similar conclusion was reached by Jene Golovchenko [25] using a different argument.

Another way to look at condition (9) is that the diffusion length over time x/v has to be less than δx, or

$$\delta x \geq \sqrt{2D\frac{x}{v}}. \tag{8.10}$$

In fact, Eq. (8.10) is the same principle as that in gel electrophoresis: the separation between two bands (due to two different lengths of DNA fragments) grows linearly with time, due to the electric-field driven drift, but the spread of each band due to thermal diffusion grows as $t^{1/2}$. Thus after long enough time, the travel distance between the bands will exceed the width of each band.

Next, a simple back-of-envelope calculation will show that it is not trivial to satisfy condition (8.10) in standard DNA translocation experiment. For $x = 40$ nm (100 bases apart between two adjacent probes), $\delta x = 0.4$ nm (single-base resolution), $D \sim 10^8$ nm^2/s (using that for single-base nucleotides in water), the translocation velocity, $v > 2Dx/\delta x^2$, would have to be greater than 50 m/s, or 200 kb/μs. This is 10,000 times faster than that in typical nanopore experiments (Storm et al. [26]) using Si_3N_4 pores. Besides, at this high translocation speed, the existing patch-clamp electronics will not be able to detect the current signals. A simple solution to this problem is to use the "reverse translocation" concept (Peng and Ling, [27]). If the DNA is held under tension, the DNA moves with the bead, the effective diffusion constant of the DNA is now set by the bead, which is a lot smaller. For a 1 μm polystyrene bead in water, $D \sim 10^5$ nm^2/s, 1,000-fold less. In this case, the required translocation velocity v is about 200 bp/μs for single-base resolution. If one relaxes the resolution to 3 bases, the desired translocation speed is ~10 bp/μs, which is about half what is typical for solid-state nanopores (~20–25 bp/μs at 120 mV).

The above estimates are encouraging for HANS sequencing using solid-state nanopores: if 6-mer probes are used, one should be able to resolve the probe location to within half its size. Combined with the known sequence information of each probes, reconstruction of the sequence for the test DNA should be entirely feasible. The absolutely necessary requirement here is the 1,000-fold reduction in diffusion constant when the DNA is held under tension in the "reverse translocation" procedure. It should be pointed out that the direction of translocation can still be forward, in the direction of electric force, as long as the DNA is held under tension.

8.5 Recent Experiments on Constrained DNA Translocation and Hybridization Detection

A key issue in the field of nanopore technology is how to control the DNA translocation processes. Here the discussions will be on the DNA-on-bead approach as the basic concept in controlling DNA translocation through a nanopore. One can control the bead motion using laser trap (Keyser et al. [28]) or magnetic field (Peng and Ling, [27]). The optical tweezers approach is more convenient, but the absorption of laser light by the buffer causes the ionic current through the pore to fluctuate dramatically (Keyser et al. [28]). Thus it is difficult to simultaneously measure ionic current while the DNA is being dragged in a dynamic fashion.

8.5.1 Reverse DNA Translocation Through a Solid-State Nanopore by Magnetic Tweezers

The basic concept of the experiment by Peng and Ling [27] is shown in Fig. 8.11. DNA molecules are attached to magnetic beads via the standard streptavidin–biotin bonds. The free end of the DNA can be captured into the nanopore by the applied electric field. Subsequently, one can apply a precisely controlled magnetic force on the magnetic bead to balance the electrical force on the trapped DNA, i.e., the DNA is in a tug-of-war between the magnetic bead and the nanopore. By increasing the magnetic force further, or reducing the bias voltage, until the magnetic force exceeds the electrical force, the DNA can be pulled out of the nanopore from the *cis* side of the nanopore. Since one can construct a magnetic field gradient over a large space, this technique is inherently applicable to large number of addressable nanopores.

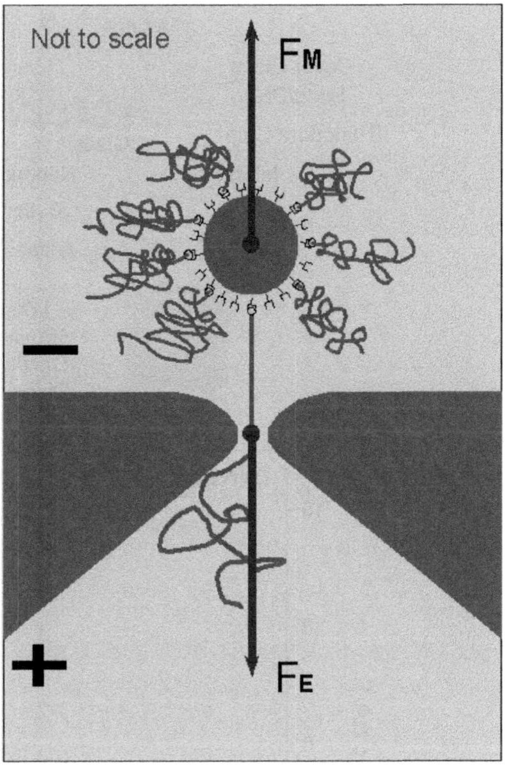

Fig. 8.11 The concept of reverse translocation [27]. Used with permission. Copyright Nanotechnology 2009

By ramping the magnetic field slowly, the DNAs in all the nanopores can be pulled out slowly during one ramping step.

In the experiment of Peng and Ling [27], as shown in Fig. 8.12, the ionic current signals were used to detect DNA being captured into or being pulled out of the nanopore. To make sure that the ionic current data is not due to bead blocking the pore or leaving the pore, 0.1 M KCl is used instead of the more standard 1.0 M KCl buffer. It is known (Chang et al. [29] and Smeets et al. [30]) that the presence of a DNA in a nanopore has two competing effects for the nanopore conductance: the physical volume of the DNA leads to a reduction in total ion population in the nanopore, thereby reducing nanopore conductance; the negatively charged DNA brings in extra counterions, leading to a conductance enhancement. The net effect of a translocating DNA on the nanopore conductance depends on the ionic strength of the buffer solution. At 1.0 M KCl, the DNA entry is signaled by a drop in ionic current, while at 0.1 M KCl the ionic current signal due to a DNA entering the pore is an increase [29, 30]. At 0.1 M KCl, DNA (attached to a bead) captured into the nanopore and then pulled out of the nanopore by the bead will be indicated by a current increase and then decrease. This "Chang-Bashir effect" [29] provides a convenient way to distinguish a true DNA capture signal from the bead blocking the nanopore.

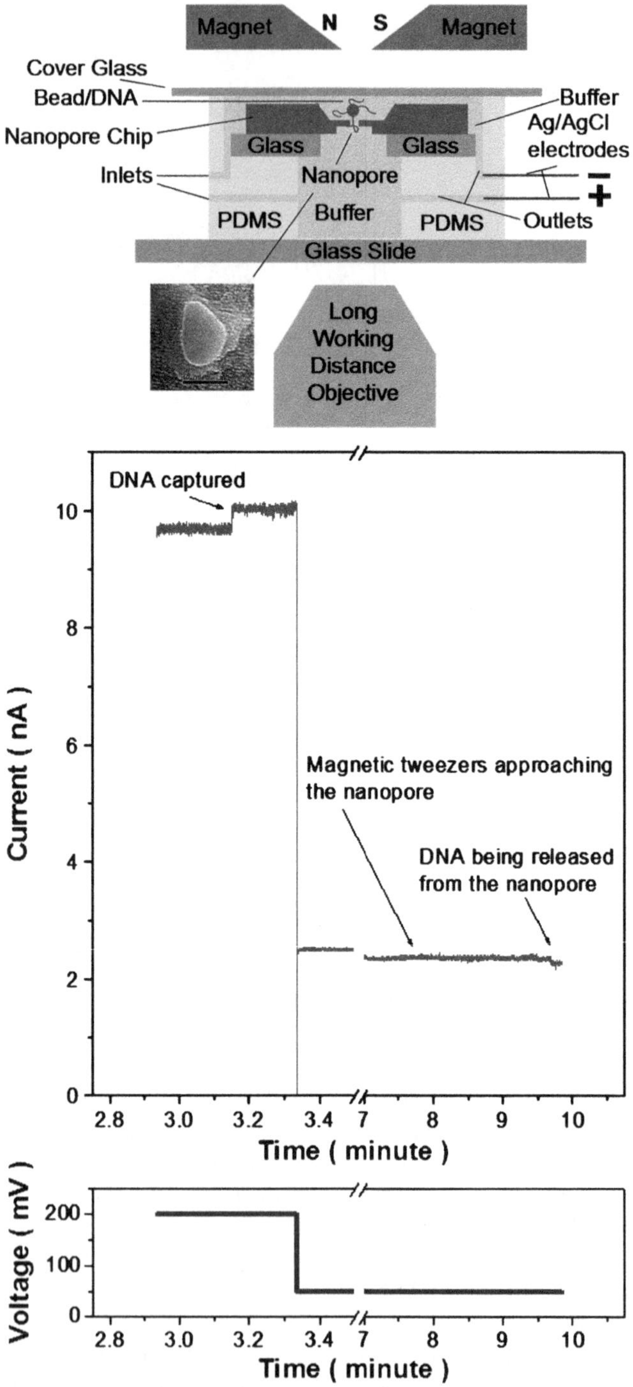

Fig. 8.12 The main panel shows the ionic current vs. time of a DNA capture and release. Inset: experimental configuration and a TEM picture of the nanopore used. *Lower panel* shows the applied voltage as a function of time [27]. Used with permission. Copyright Nanotechnology 2009

It was found (Peng and Ling, [27]) that, after a DNA is captured, it is possible to trap indefinitely a single DNA by reducing the applied voltage quickly to prevent multiple DNAs being captured into the nanopore, as well as the breakage of the streptavidin–biotin bond. At 50 mV bias, we found that a DNA can be trapped inside the pore for a long period of time (hours), as also observed by Keyser et al. [28].

In contrast to standard DNA translocation experiment, the driving force is *localized* at the pore region, extending only over a few nanometers. While in reverse translocation, the driving force is tension which is *extended* over the entire DNA length between the pore and the bead. The entropic spring effects (Marko and Siggia, [21]) of the DNA can be suppressed. In addition, the center-of-mass of the DNA moves with the bead, thus acquiring a smaller diffusion constant. As discussed in Sect. 4.2 above, suppressing the diffusive motion of the DNA inside the nanopore will be crucial to positional measurements on the nanoscale features on a DNA.

To pull the DNA out of the nanopore mechanically, a pair of magnetic tweezers was used. For fine-tuning the magnetic force on the bead, the magnets with calibrated field profile are mounted on a Burleigh Inchworm nano-positional stage. The gradual increase of magnetic force is achieved by moving the magnets slowly towards the nanopore chip. In the distance range over which the DNA is pulled out, the force increase rate is about 0.2 fN per step on the Burleigh stage. (The Burleigh stage can move 20 nm per step, and it can traverse at a speed from 0.01 to 2,062 μm/s.) Once the magnetic force exceeds the electric force (at 50 mV), the magnetic bead moves away from the pore region, as indicated from the real-time video images. Coincidentally, the nanopore ionic conductance drops.

This technique is compatible with simultaneous ionic-current measurements and is suitable for multiple nanopores, paving the way for large scale applications. Since the DNA is held under tension during the translocation, the center-of-mass diffusion constant of the DNA is controlled by the bead, and is thus suppressed.

The current increase when a DNA is captured by the nanopore was due to the low ionic strength of the buffer used. This effect was first discovered by Chang et al. [29]. Here it is used as a reliable method for identifying DNA capture.

8.5.2 Nanopore DNA Translocation Experiment for Hybridization Detection

To investigate the feasibility of electrical detection of hybridization probes on a DNA using nanopores, a 12 mer hybridization experiment was carried out (Balagurusamy et al. [31]). The choice of 12 mer hybridization is based on the stability consideration. Shorter probes are less stable at room temperature. In future experiments on shorter probes, a cooling device for the buffer will have to be used. For HANS (hybridization-assisted nanopore sequencing) studies, the solid-state nanopores are preferred over α-hemolysin pores because the latter do not allow the passage of double-stranded DNA.

Using the Zucker mfold software (http://mfold.bioinfo.rpi.edu/), available at the Rensselaer bioinformatics web server, a model DNA molecule was designed for the experiment. The details of the molecule are shown in Fig. 8.13. For this experiment,

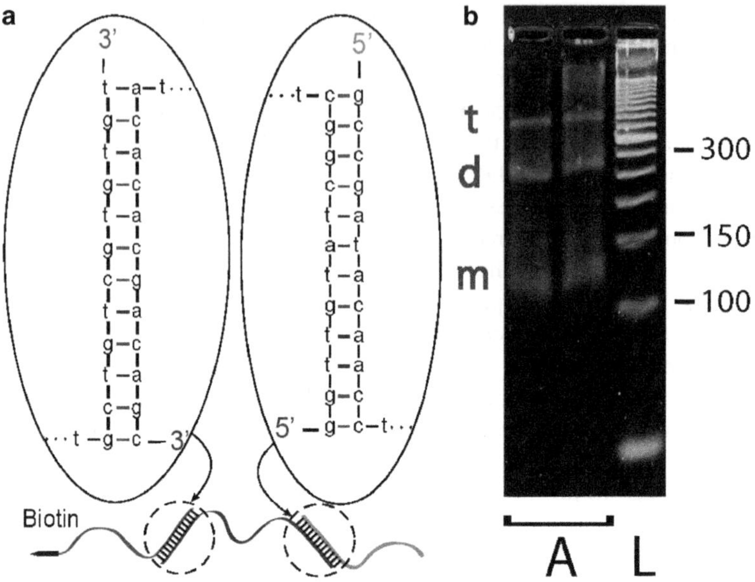

Fig. 8.13 (**a**) The structure of the designed model DNA molecule formed by hybridization between three oligonucleotides. The middle single-stranded segment of poly-T (thymine) has different 12 base long sequences on each end that are complementary to the same number of bases on one end of other two strands. This facilitates the formation of two double strand regions by Watson-Crick base-pairing scheme when these single strands are mixed under suitable conditions. The sequence of the bases in the double-strand regions is shown in the inset. One of the single strands is connected to a biotin molecule at the free end that is used to attach the complex to a streptavidin-coated polystyrene bead through biotin–streptavidin binding. (**b**) The gel electrophoresis pattern obtained when running the solutions of mixtures containing one, two and three designed single-stranded DNA molecules under an applied voltage of 25 V (25 min) and then 75 V (200 min) with a 50 bp DNA ladder reference. Lanes L and A refer to the 50 bp increment DNA ladder and the ssDNA complex samples. Bands labelled *t*, *d*, and *m* are formed by the monomers, dimers and trimers of the ssDNA [31]. Used with permission. Copyright Nanotechnology 2010

a model DNA system was constructed that consisted of three single-stranded poly-T (thymine) oligonucleotides: 3′-TGTGTGCTGTCG-$(T\text{-}\,\text{-}\,\text{-}T)_{130}$-(Biotin)-5′; 3′-CGACAGCACACA-$(T\text{-}\;\text{-}\;\text{-}T)_{120}$-CGGCTATGTTGG-5′; 3′-$(T\text{-}\;\text{-}\;\text{-}T)_{130}$- - CCAACATAGCCG-5′. The oliogmers were synthesized and PAGE-purified by a commercial source (Midland Certified Reagent Co, Midland, TX, USA).

The biotin linker at the 5′ end of the first oliogmer is for attaching a bead, see Fig. 8.14. The three single-stranded molecules join by Watson-Crick pairing at the sticky ends to form a trimer structure with ss–ds–ss–ds–ss segments (see Fig. 8.13a). In order for these single-strands to form a trimer complex, the single strands that form the ends of this complex have only one sticky end whereas the middle strand has two sticky ends. Each end of the middle molecule is complementary to the sticky end of only one of the other two strands. In this design, before base-pairings form, two of the oligonucleotides with one sticky end are 142 bases long, and the third with two sticky ends is 144 bases long.

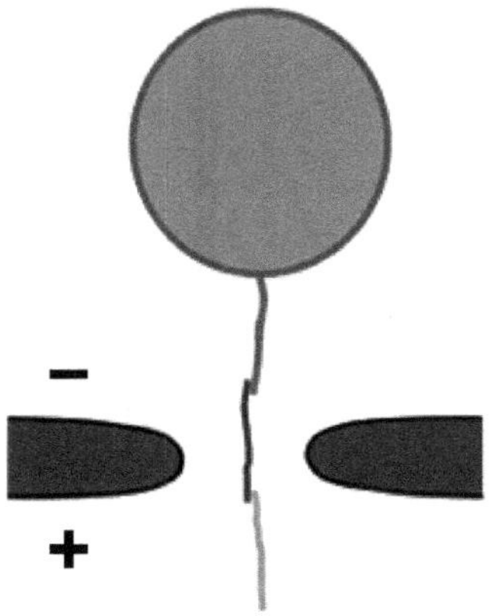

Fig. 8.14 The schematic of the ssDNA trimer complex translocates through the nanopore, driven by an applied voltage. For clarity, the length of the single-strands, the bead diameter and the thickness of the pore have been drawn to scale to illustrate that when the first double-strand region exits the nanopore, the second region is still far away from the nanopore. The bead is not drawn to scale, however [31]. Used with permission. Copyright Nanotechnology 2010

The base-pairings are designed to be stable at room temperature. As such, in equilibrium, the buffer sample should contain monomers of 142 bases, and dimers of 274 bases and the trimers 404 bases in length respectively. The formation and stability of the trimers have been confirmed using gel electrophoresis. Figure 8.13b shows the presence of three different bands, one between 100 and 150, another between 250 and 300, and another one between the 400 and 450 bp bands of a 50-base increment DNA ladder reference. Also, the UV absorption spectrum measurements at wavelength 260 nm from the single-stranded DNA complex mixture in 1 M SPSC buffer from 25 to 70 °C show the melting transitions of the *ss1–ss2* and *ss2–ss3* DNA segments at temperatures close to 51 and 57 °C, respectively. The melting transitions of individual dimers were confirmed from measurements on buffer solutions containing only one type of dimers. The measured melting temperature is closer for *ss2–ss3* double-stranded segment with the computationally obtained value of 60.8 °C while it is somewhat lower for *ss1–ss2* segment, 64.2 °C. The calculated contour length of these individual polymer chains is ~49 nm that is nearly twice the optimum thickness of the silicon nitride that bears the nanopore. This relatively large contour length compared to the membrane thickness would allow the location of the second double-stranded region to be well away from the nanopore when the first double-stranded segment just exits the nanopore during their electric-field driven translocation.

The nanopore device is fabricated using the TEM "drilling" approach [6]. For the pore used for collecting the data below, the pore dimensions are 10 nm × 7.8 nm obtained from TEM. The measured nanopore resistance is close to the calculated value using these pore parameters and the optimum membrane thickness. The open-pore current shows linear *I–V* characteristics typical of a properly wetted solid-state nanopore. Two typical translocation events for DNA trimers are shown in Fig. 8.15, along with two computer-simulated translocation events.

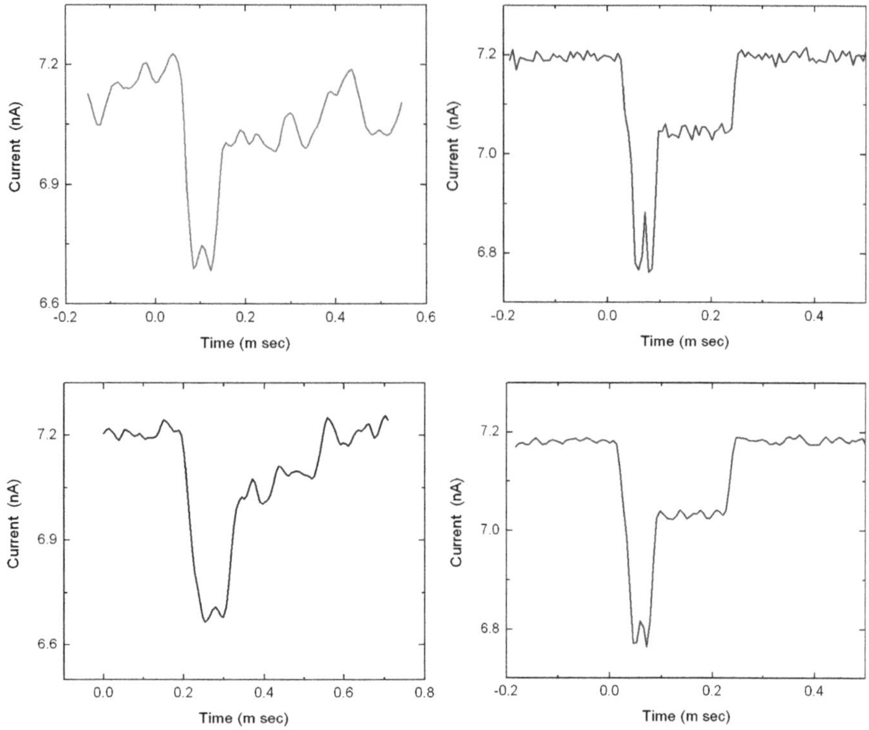

Fig. 8.15 Two nearly identical trimer events are shown in left. On the right are two simulated translocation events. The simulation was done using a hard rod model consisting of alternating single and double-stranded DNA regions (*ss–ds–ss–ds–ss*) corresponding to the trimer complex used. An effective circular nanopore is assumed corresponding to the size of the nanopore used in the experiment with an effective membrane thickness of 10.68 nm and in 1 M KCl buffer. A white Gaussian noise was added to the simulated current and then filtered with a Bessel filter with 10 kHz cut-off frequency. The simulation was done in MATLAB [31]. Used with permission. Copyright Nanotechnology 2010

The DNA-bead complex translocation events were collected with a nanopore setup immersed in an ionic buffer at a pH of 8.0 with 1 M KCl, 10 mM Tris, 1 mM EDTA (TE) concentration using an Axopatch-200 current amplifier in voltage-clamp mode. After filling the DNA-bead complex sample in the *cis* side of the nanopore, the current was recorded at a voltage of 100 mV until DNA translocation events were observed. The repeated runs of translocation experiments have yielded large amount of interesting data. In some time intervals, the events are so dense in their arrival that it is difficult to identify individual events. Then there are also relatively isolated, "clean," events that, with the assistance of computer simulation, can be reasonably interpreted. Here, two such events are shown in Fig. 8.15 above. The most striking features of these events are the two double dips at the bottom of the curves. Using the parameters of the nanopore used here, and the lengths of the different segments of the DNA, as well as the noise level of the background ionic current, by assuming a constant translocation velocity during transit, one can calculate the expected current traces, as shown in the right side panels. The only free parameter is the speed of translocation. The agreement is remarkable. (The eventual recovery of current is understood to be the breakage of biotin–streptavidin bond, see Balagurusamy et al. [31]).

8.6 Summary

In this chapter, the recent advances in the fabrication of solid-state nanopores have been discussed. It is clear that for individual nanopores, there are several adequate methods of making them. There are interesting materials physics issues to be discovered in this effort. Solid-state physicists are well equipped for this task. The progress in making addressable nanopore devices is beginning and the key issues have been identified. The concept of nanopore DNA sequencing by combining nanopore techniques with sequencing-by-hybridization is discussed. The theoretical limit in achieving the HANS sequencing platform is discussed. Latest nanopore experiments on detecting hybridization probes are discussed.

Acknowledgments The author got involved in the nanopore field during a sabbatical visit (2002–2003 academic year) to the Molecular Biophysics Group of Professor Cees Dekker in the Delft University of Technology of the Netherlands. He enjoyed a stimulating collaboration with Arnold Storm, Jianghua Chen, Hanny Zandburgen and Cees Dekker, as well as discussions with Ulrich Keyser, Liviu Movileanu, Derek Stein and Nynke Dekker. For his sabbatical, the author acknowledges gratefully the financial support from the Guggenheim Foundation and the Dutch Science Foundation (FOM). Since the author's return to Providence in 2003, he owes a great deal of gratitude to his postdoctoral associates Dr. Sang Ryul Park and Dr. Venkat Balagurusamy, graduate students Hongbo Peng, Shanshan Wu, Paul Weinger, Sungcheol Kim, and undergraduate student Adam Politzer. Outside Brown, the author owes a great deal of gratitude to Professor Robert Riehn of NC State University on the recipes of nanofabrication, and especially to Dr. Alexandros Pertsinidis of Stanford University for the stimulating discussions that led to the concept of HANS which sustained the author's interest in this field. Grants from the National Science Foundation under the National Nanotechnology Initiative (NNN) helped launch the nanopore program at Brown, a R21 grant from the NIH-NHGRI allowed the author to continue this pursuit. The author also gratefully acknowledges encouraging discussions with Professors Daniel Branton, Jene Golovchenko, and especially Jiali Li and Amit Meller during a visit to their labs at Harvard in the summer of 2001.

References

1. Kasianowicz, J.J., Brandin, E., Branton, D. & Deamer, D. W. (1996), Characterization of individual polynucleotide molecules using a membrane channel. *Proc. Nat. Acad. Sci. USA 93*, 13770–13773.

2. Church; G., Deamer; D.W., Branton; D., Baldarelli; R., Kasianowicz; J., Characterization of individual polymer molecules based on monomer-interface interactions, US Patent No. 5,795,782 (issued on 8-18-1998).
3. Akeson, M., Branton, D., Kasianowicz, J. J., Brandin, D. & Deamer, D. W. (1999) Microsecond time-scale discrimination among polycytidylic acid, polyadenylic acid, and polyuridylic acid as homopolymers or as segments within single RNA molecules. *Biophys. J. 77*, 3227–3233.
4. Meller, A., L. Nivon and D. Branton (2001) Voltage-Driven DNA Translocations through a Nanopore. *Phys. Rev. Lett. 86*, 3435–3438.
5. Li, J., Stein, D., McMullan, C., Branton, D., Aziz, M.J., & Golovchenko, J.A. (2001) Ion-beam sculpting at nanometre length scale. *Nature 412*, 166–169.
6. Storm, A. J., Chen, J. H., Ling, X. S., Zandbergen, H. W. & Dekker, C. (2003) Fabrication of solid-state nanopores with single-nanometre precision. *Nature Mater. 2*, 537–540.
7. Chang, H., Kosari, F., Andreadakis, G., Alam, M.A., Vasmatzis, G., Bashir, R., *Nano Lett.* (2004), *4*, 1551.
8. Wu, S.., Park, S.R., and Ling, X.S. (2006) Lithography-Free Formation of Nanopores in Plastic Membranes using Laser Heating, *Nano Letters 6*, 2571–2576.
9. Park, S.R., Peng, H. and Ling, X.S. (2007) Fabrication of Nanopores in Silicon Chips Using Feedback Chemical Etching, *Small 3*, 116–119.
10. McCord, M. A.; M. J. Rooks (2000). *SPIE Handbook of Microlithography, Micromachining and Microfabrication.*
11. Wu, M.Y., Krapf, D., Zandbergen, M., Zandbergen, H. and Baton, P.E. (2005) Formation of nanopores in a SiN/SiO_2 membrane with an electron beam. *Appl. Phys. Lett. 87*, 113106-1-3.
12. Kim, M-J., B. McNally, K. Murata and A. Meller (2007) Characteristics of solid-state nanometer pores fabricated using transmission electron microscope. *Nanotechnology, 18*, 205302 (5pp).
13. Apel, P.Y., Korchev, Y.E., Siwy, Z., Spohr, R., Yoshida, M. Nucl. Instrum. Methods Phys. Res. Sect. B (2001), *184*, 337–346.
14. Ling, X.S., Bready, B., and Pertsinidis, A., "Hybridization-Assisted Nanopore Sequencing", United States Patent Application No. 20070190542.
15. Ling, X.S., "Addressable nanopores and micropores including methods for making and using same", United States Patent 7678562.
16. Politzer, A., Sc.B. thesis (2005), Brown University.
17. Riehn, R., (2008), private communications.
18. Sanger, F, Nicklen, S, Coulson, A.R. (1977), "DNA sequencing with chain-terminating inhibitors". *Proc. Natl. Acad. Sci. U.S.A. 74*, 5463–7.
19. Maxam, A.M., Gilbert, W., (1977), "A new method for sequencing DNA". *Proc. Natl. Acad. Sci. U.S.A. 74* (2): 560–4.
20. Southern, E. M. (1975), "Detection of specific sequences among DNA fragments separated by gel electrophoresis". *Journal of Molecular Biology 98* (3): 503–517.
21. Marko, J.F., and Siggia, E.D., (1995), "Statistical mechanics of supercoiled DNA", *Phys. Rev.* E *52*, 2912–2938.
22. Bustamante, C., Marko, J.F., Siggia, E.D., Smith, S. (1994), "Entropic elasticity of lambda-phage DNA", *Science, 265*, 1599–1601.
23. Lubensky, D. K., Nelson, D. R., (1999). Driven Polymer Translocation Through a Narrow Pore, *Biophys. J. 77*, 1824.
24. Redner, S. (2001). A Guide to First-Passage Processes, Cambridge University Press.
25. Golovchenko, J. (2010), private communications.
26. Storm, A.J., Storm, C., Chen, J.H., Zandbergen, H.W., Joanny, J.F., and Dekker, C., (2005),"Fast DNA translocation through a solid-state nanopore", *Nano Lett. 5*, 1193–1197.
27. Peng, H. and Ling, X.S., (2009), "Reverse DNA translocation through a solid-state nanopore by magnetic tweezers", *Nanotechnology*, *20*, 185101 (8pp).
28. Keyser, U.F., Koeleman, B.N., Krapf, D., Smeets, R.M.M., Lemay, S.G., Dekker, N.H., Dekker, C., (2006), "Direct force measurements on DNA in a solid-state nanopore", *Nature Physics*, *2*, 473.

29. Chang, H., Venkatesan, B.M., Iqbal, S.M., Andreadakis, G., Kosari, F., Vasmatzis, G., Peroulis, D., Bashir, R. (2006) "DNA counterion current and saturation examined by a MEMS-based solid state nanopore sensor", *Biomed Microdevices 8*, 263–269.
30. Smeets, R.M.M., Keyser, U.F., Krapf, D., Wu, M.Y., Dekker, N.H., and Dekker, C., (2006), "Salt-dependence of ion transport and DNA translocation through solid-state nanopores", *Nano Lett. 6*, 89–95.
31. Balagurusamy, V., Weinger, P., Ling, X.S., (2010) Detection of DNA hybridizations using solid-state nanopores", *Nanotechnology 21*, 335102 (5pp).

Chapter 9
Nanopore Recordings to Quantify Activity-Related Properties of Proteins

Erik C. Yusko, Yazan N. Billeh, Jerry Yang, and Michael Mayer

Abstract Electrical current recordings through electrolyte-filled nanopores (so called resistive pulse-sensing experiments) are attracting increasing attention for identifying and characterizing biomolecules. The majority of the work employing this method so far has focused on detection of oligonucleotides, polymers, and viruses. Most recently nanopores have been used to detect single proteins. This chapter reviews the very first attempts to use nanopores for characterizing properties of proteins that relate to their activity. The emphasis lies on those studies that provided quantitative information on activity-related properties of proteins, such as protein conformation, ligand binding, and enzyme activity. Nanopore-based studies have tremendous potential for investigating the function of proteins because the technique is capable of interrogating individual proteins at high-throughput without requiring labeling.

Keywords Protein • Virus • Nucleotide • Polymer • Protein conformation • Protein volume • Protein charge • Electrophoretic mobility • Drift velocity • Isoelectric point • Ligand affinity • Stoichiometry • Association constant • Dissociation constant • Kinetics • On-rate • Off-rate • Binding isotherm • Drug-protein interaction • Bovine serum albumin • Ovalbumin • Avidin • Streptavidin • Antibody • Immunoglobulin G • Immunoglobulin E • Antibody fab fragments • Staphylococcal enterotoxin B • β-Lactoglobulin • E. *coli* maltose binding protein • Lectin • Carbonic anhydrase II • Sulfonamide • RNA aptamer • Fibrinogen • Poly (ethylene glycol) • Biotin • Ricin • Enzyme activity • Protease • Phospholipase • Membrane-active enzyme • Catalytic rate constant • Forward rate constant • Michaelis constant • Phospholipase C • Phospholipase D • Anthrax lethal factor • Trypsin • Alkaline phosphatase • Amyloid-β • Phosphatidylcholine • Phosphatidylinositol • α-hemolysin • Gramicidin • Alamethicin • Resistive pulse • Current blockage • Translocation time • Sojourn time • Biosensing • BLM • Planar lipid bilayer

---

M. Mayer (✉)
Department of Biomedical Engineering, University of Michigan, 1101 Beal Avenue,
Lurie Biomedical Engineering Building, Room 2174, Ann Arbor, MI 48109-2099, USA
e-mail: mimayer@umich.edu

S.M. Iqbal and R. Bashir (eds.), *Nanopores: Sensing and Fundamental Biological Interactions*, DOI 10.1007/978-1-4419-8252-0_9,

9.1 Introduction

This chapter describes the use of nanopores for quantifying activity-related properties of proteins. We review recent advances in characterizing the most basic structural properties of proteins that relate to their activity such as size, charge, and conformation, as well as functional properties such as binding of ligands to proteins and the activity of enzymes.

Sensing proteins in nanopores, in most cases, is based on applying an electric potential difference between the electrolyte solutions separated by a pore-containing membrane and measuring the fluctuations of ionic current through the nanopore. Under the appropriate experimental conditions (i.e. size of the nanopore, ionic strength, and applied electric potential difference), the presence of a single protein in a nanopore displaces a sufficiently large conducting volume of electrolyte to induce a detectable change in the resistance (Fig. 9.1). The resulting decrease in the current through the pore is transient and is recorded with high-gain, low-noise amplifiers, which were originally developed for electrophysiology applications such as patch-clamp recordings of single ion channels in the membranes of living cells [18, 34].

Taking advantage of this simple concept of resistive-pulse sensing, nanopores provide a unique platform for characterizing the function of proteins. For example, the magnitude, frequency, and duration of these current fluctuations can provide information on the volume of the protein, concentration of the protein, and properties such as conformation, surface charge, bound ligands or enzymatic activity of the protein in solution. In addition, nanopore sensing can be accomplished with a simple instrumental setup (Fig. 9.1) compared to other single molecule techniques such as single molecule fluorescence microscopy. Furthermore, nanopore sensing of proteins is attractive because it is rapid (1–20 min), reports on multiple properties from individual proteins in the same recording, and may be applied in a label-free manner. In addition, nanopore sensing can explore the function of proteins in solution without immobilization of the analyte.

Recently, nanopores with ligands immobilized on the nanopore walls have been employed as biosensors to detect the presence and concentration of specific proteins. A recent, excellent review article by Howorka and Siwy outlined these methods [20]. This chapter will focus, however, on methods that detected *single* proteins, quantified the binding of proteins and ligand molecules, or quantified the activity of enzymes with nanopores.

On a fundamental level, the size, charge, and conformation of a protein influences its function, and thus we will begin with a discussion on recent research that employed nanopores to quantify these properties. Next, we will discuss nanopore-based techniques to measure protein function quantitatively. Discussed in order of increasing complexity, these parameters include the equilibrium association and dissociation constants between proteins and ligands, the kinetic on- and off-rates of ligands binding to proteins, and the catalytic rate constant and Michaelis constant, which describe enzyme kinetics. Novel strategies to characterize proteins on a single molecule level rapidly will be useful in a variety of fields ranging from drug

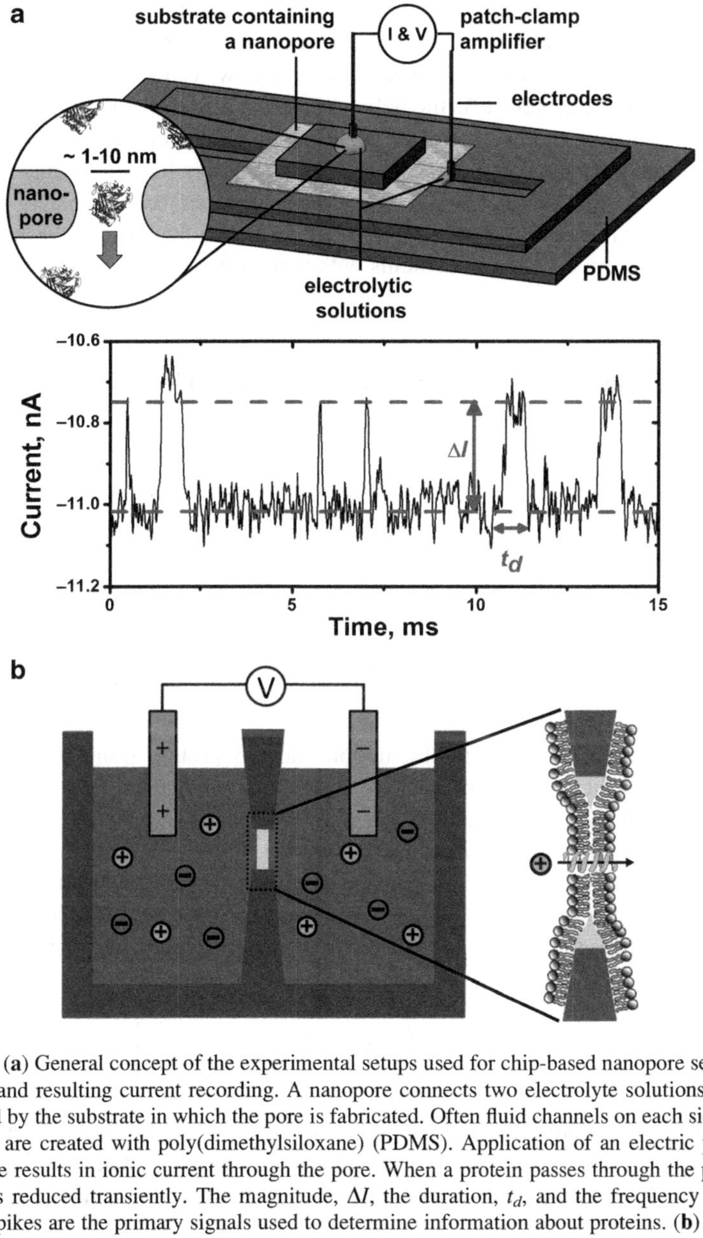

Fig. 9.1 (**a**) General concept of the experimental setups used for chip-based nanopore sensing of proteins and resulting current recording. A nanopore connects two electrolyte solutions that are separated by the substrate in which the pore is fabricated. Often fluid channels on each side of the substrate are created with poly(dimethylsiloxane) (PDMS). Application of an electric potential difference results in ionic current through the pore. When a protein passes through the pore, the current is reduced transiently. The magnitude, ΔI, the duration, t_d, and the frequency of these current spikes are the primary signals used to determine information about proteins. (**b**) Concept for experimental setups used in free-standing lipid bilayer-based nanopore sensing. In this recording platform, a biological nanopore channel connects the two electrolyte solutions separated by a planar lipid bilayer (BLM). Adopted from Macrae et al. with permission [27]

development in the pharmaceutical industry to the fundamental sciences for providing an understanding of protein structure–function relationships.

The field of nanopore science is relatively new, especially in regard to the detection of single proteins. As the field progresses, we envision a shift from simply detecting single proteins to the development of functional protein assays. Furthermore, as many researchers pursue the use of nanopores for sequencing DNA and RNA, it is possible that nanopores will be used in a similar manner to probe the amino acid sequence of proteins. It is in this context, that we review progress in characterizing the function of proteins with nanopores.

9.2 Characterizing the Size, Charge, and Conformation of Proteins from Nanopore Recordings

A series of recent reports have demonstrated the use of nanopores for determining basic characteristics of proteins such as size, charge, and conformation. Here we describe the details of sensing proteins in nanopores and of these reports.

9.2.1 Determining the Size of Proteins with Nanopores

The amplitude of the current fluctuations due to protein translocation through a nanopore is directly proportional to the volume of the protein given that the protein has roughly a spherical shape [11, 13, 16, 40]. The underlying physical model that relates protein size to the amplitude of these current fluctuations was derived by Maxwell [31] and later improved by DeBlois and Bean in 1970 [11]. Based on their findings, the passage of a spherical particle through a *cylindrical* nanopore reduces the current through the pore in a time dependent fasion, $\Delta I(t)$, that is directly proportional to the conducting volume excluded by the protein, $\Lambda(t)$:

$$\Delta I(t) = -\frac{\sigma \cdot E}{l_p^2} \cdot \Lambda(t) \cdot \left[1 + f\left(\frac{d_m}{d_p}, \frac{l_m}{l_p}\right)\right], \tag{9.1}$$

where Λ (m^3) is approximately equal to the volume of the protein, σ ($\Omega^{-1} \cdot m^{-1}$) is the conductivity of the solution, E (V) is the voltage drop across the nanopore, and l_p (m) is the length of the nanopore. A dimensionless correction factor, f, can account for deviations when the diameter of the molecule, d_m, is similar to the diameter of the pore, d_p, or the length of the molecule, l_m, is similar to length of the pore, l_p. This f factor is described in greater detail in DeBlois and Bean; under most experimental conditions it can be neglected [11]. In addition, when l_p is comparable to d_p, the term l_p should be substituted for (l_p + $0.8 \cdot d_p$) to account for the access resistance of the pore (access resistance results from electric field lines converging from the bulk solution to the opening of nanopores) [17]. Finally, it should be noted that machined nanopores are not perfectly cylindrical. Equation (9.1) is, however, suitable for estimating volumes despite possible errors due to the deviation in the shape of nanopores.

Using (9.1), or variations of it, Staufer's group and others estimated the volume of different proteins [13, 15, 16, 38, 46]. One of the first successful experiments to detect single proteins identified unlabeled bovine serum albumin (BSA). In this work, Han et al. used an adaptation of (9.1)[1] to estimate the size of BSA and found it had a diameter of 7–9 nm [16]. In another study, Fologea et al. compared BSA with the similarly charged but larger protein fibrinogen. Here the authors could readily distinguish these proteins because larger peak amplitudes and translocation times were measured with fibrinogen than BSA as expected from (9.1) [13].

A more recent investigation by Han et al. distinguished between four proteins, ovalbumin (OA), avidin (AV), BSA, and streptavidin (SA), using a pore with a diameter of 28 nm based on values of ΔI [15]. Figure 9.2a shows a histogram of the distribution of ΔI for each of these proteins.

Moreover, the linear relationship between $\Delta I(t)$ and $\Lambda(t)$ can be used to identify the size of protein complexes and the binding of proteins to viruses or particles that display ligands. Saleh and Sohn used pores with a diameter of 1 μm to detect

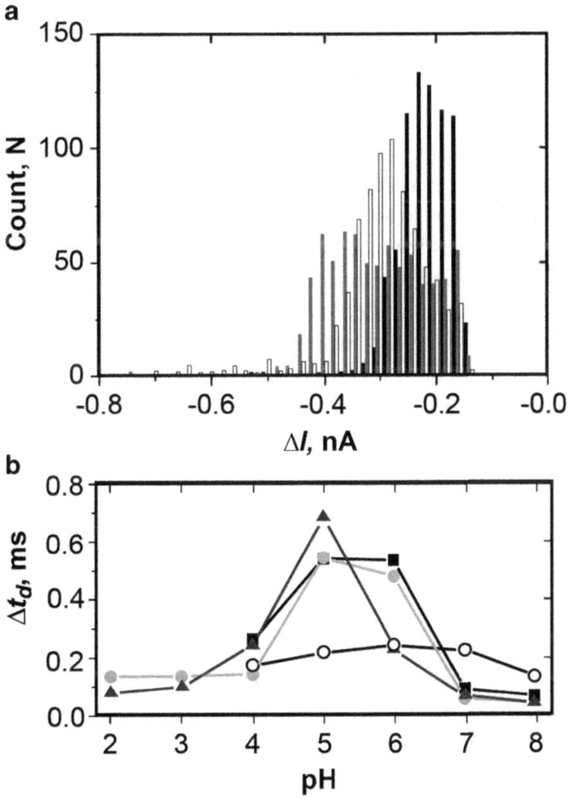

Fig. 9.2 (**a**) Histogram showing ΔI counts for the proteins BSA (white), OA (black), and SA (grey) at pH 6. (**b**) Translocation time of different proteins as a function of pH in the recording electrolyte (*filled square*: OA; *filled circle*: BSA; *filled triangle*: SA; and *open circle*: AV). The range of pH values used was selected such that proteins were soluble and no visible aggregates formed in order to avoid clogging. Adopted from Han et al. with permission [15]

[1]This adapted equation normalized ΔI with respect to the baseline current of translocation events, I. With this equation, the diameter of the molecules, d_m, could be determined: $d_m{}^3 = s\frac{\Delta I}{I}(l_p + 0.8d_p)d_p{}^2$, with $s \approx 1$.

antibodies binding to latex colloids from the resulting increase in ΔI [38]. An investigation with sub-micrometer pores by Uram et al. expanded on this concept to detect immune complexes formed from the potential biowarfare agent staphylococcal enterotoxin B (SEB) with polyclonal antibodies against SEB. Analysis of values for ΔI made it possible to estimate the number of proteins in these immune complexes; this number ranged from 610 to 17,300 proteins [49].

An important point to note about (9.1) is that the net charge of proteins can alter ΔI if the ionic strength in the recording buffer is low (i.e. <300 mM). This effect occurs because ions with an opposite polarity of a charged protein associate with the protein in solution [5, 17, 44]. At low ionic strength, these counterions can affect the conductance of the solution within a pore. In practice, however, low ionic strength solutions are rarely used due to the decreased conductivity of the solution, which reduces the magnitude of ΔI (9.1). The charge of a protein also has a significant effect on the translocation time of proteins as discussed in the subsequent section.

9.2.2 Determining the Charge of Proteins with Nanopores

In addition to volume, the charge of a protein affects the translocation time, t_d, of proteins passing through a nanopore. In general, translocation times are in the range of tens to hundreds of microseconds, though for very large molecules, translocation times have been observed in the range of milliseconds [45, 46]. In particular, the translocation time can be characteristic for certain proteins and thus useful for protein identification. Measurements of translocation time are, however, challenging to predict accurately due to the effect of electroosmotic flow and the possibility of interactions between the pore walls and a protein. Recent research by Talaga and Li, described the distribution of t_d by the distribution function, $P\ (t_d)$:[2]

$$P(t_d) = \frac{\exp\left[-\frac{(d_p - v \cdot t_d)^2}{4 \cdot D \cdot t_d}\right] \cdot (v \cdot t_d + d_p)}{2t_d \cdot \sqrt{4 \cdot D \cdot t_d \cdot \pi}} \text{ with } d_p = \begin{bmatrix} l_p & l_m < l_p \\ l_p + l_m & l_m \geq l_p \end{bmatrix}, \tag{9.2}$$

where v (m s^{-1}) is the electrophoretic drift velocity and D (m$^2 \cdot$ s^{-1}) is the diffusion constant of the protein *within the nanopore* [46]. The electrophoretic drift velocity is given by:

$$v = u_e \cdot \varepsilon, \tag{9.3}$$

[2] The factor of 2 in the denominator of equation (9.2) is not present in the cited work by Talaga and Li. Working with Talaga and Li, we determined that the factor of 2 in the denominator is required for correct normalization such that the area of this probability density function equals 1.

where u_e ($m^2 \cdot v^{-1} \cdot s^{-1}$) is the electrophoretic mobility of a charged protein and ε ($V \cdot m^{-1}$) is the strength of the electric field [14]. This distribution function assumes a biased diffusion, first-passage process based on the one-dimensional translocation of the protein through the length of the pore to an infinite sink. It should also be noted that (9.2) assumes that the charges on the protein are distributed uniformly and that electroosmotic flow does not contribute to transport [46, 53]. Both of these assumptions are not always met in nanopore assays with proteins [46]. Furthermore, early studies have reported values for the drift velocity and the diffusion constant that were one to three orders of magnitude lower than those measured in the bulk solution [40, 46]. These reduced values could be due to interactions of the proteins with the walls of the pore or to a non-homogenous distribution of charges on the proteins [40, 46]. Nevertheless, (9.2) provides a good approximation.

Although accurate prediction of the effect of charge on the translocation time of proteins through nanopores is still not possible, it is clear that the charge on a protein affects t_d. Stroeve and coworkers demonstrated the effects of protein charge on translocation time by studying protein flux through nanopores [6, 7, 25]. Although not detecting single proteins, these investigations showed that the flux of BSA and bovine hemoglobin through nanopores depended on the applied electric potential difference and on the charge of the protein, which was changed by altering the pH of the solution. These results indicated that protein charge influenced protein translocation and that the polarity of the charge and isoelectric point (pI) of proteins can be estimated from translocation experiments. A possible application could be to separate similarly sized molecules of different charges through nanopores for ultrafiltration [6, 7, 25].

Staufer's group provided clear evidence for the dependence of t_d on charge by sensing single proteins with nanopores and solutions of different pH (Fig. 9.2b) [15]. From these experiments, the translocation time of a protein was longest when the pH of the solution was equal to the pI of a protein. This characteristic can be beneficial since slowing the translocation speed of a protein (increasing t_d) improves the accuracy of determining ΔI. In addition, the pH also determined the polarity of the potential difference at which proteins passed through the nanopore. For instance, negatively charged proteins only passed through the pore when the solution on the opposite side of the pore contained the positive anode. Hence, it is necessary to know the pH of a solution and the pI of the protein under investigation.

Recently, Siwy's group took advantage of ion current rectification in nanopores to estimate the pI of streptavidin [1, 2, 41–43, 51].[3] In this work, streptavidin bound to biotin groups which were covalently attached to the nanopore walls. Modifying the pH of the solution changed the charge of the bound streptavidin and hence the charge on the walls of the nanopore. This change in the charge density on the pore walls strongly affected the ion current rectification through the nanopore, which could then be used to estimate the pI of streptavidin.

---

[3]Ion current rectification refers to the condition where the current at one polarity of the electric potential difference is significantly different than the current at the opposite polarity (i.e. the system has non-ohmic behavior). For more information see refs. [40–42].

9.2.3 Determining the Conformation of Proteins with Nanopores

Protein function depends on correct folding of proteins. Nanopore-based investigations have started to use translocation time and current pulse amplitudes to distinguish different states of protein folding. Li and colleagues recently employed nanopores to distinguish folded, partially unfolded, and completely unfolded proteins based on their translocation signatures [46]. Talaga and Li used pores fabricated in silicon nitride membranes with diameters of 4–8 nm to study unfolding of the protein bovine β-Lactoglobulin (βLGa) in solutions containing up to 8 M urea (maintaining the same pH). These authors analyzed the peak amplitudes of the translocation events in combination with (9.1) to calculate the volume excluded by the protein. Smaller excluded volumes and longer t_d for unfolded proteins compared to folded proteins made it possible to distinguish folded, partially folded, and unfolded protein states.

Earlier work by Auvray's group showed that using the biological pore α-HL, it is possible to distinguish between folded and unfolded *E. coli* maltose binding protein (MBP) that was denatured by guanidium chloride [36]. In this work t_d was used to distinguish folded proteins from denatured proteins, as t_d was longer for the folded than the unfolded proteins.

These examples illustrate the potential of nanopores for characterizing proteins with regard to size, charge, and conformation.

9.3 Nanopore Recordings for Characterizing Equilibrium Binding Constants and Stoichiometries of Binding

The function of most proteins involves binding to other proteins or to small molecules. Characterizing these interactions is important, for example, in developing drugs with a high affinity and specificity for a target protein. Table 9.1 lists the protein–ligand interactions that have been investigated quantitatively using nanopores to date.

9.3.1 Using Nanopores to Determine the Stoichiometry of Ligand Binding to Proteins

The stoichiometry of protein–ligand interactions (i.e. the number of ligand molecules that a protein can bind), is one of the most basic properties of protein–ligand binding. Two nanopore-based techniques have been employed so far to determine stoichiometry of binding.

Uram and coworkers recently used current recordings through nanopores to determine the volume of virus particles with and without antibodies bound (Fig. 9.3) [50].

Table 9.1 Equilibrium dissociation constants and rates of binding determined using nanopores

Protein	Ligand	Pore	Parameter measured
Streptavidin-functionalized nanoparticles	Monoclonal anti-streptavidin antibody	Micropore in PDMS	$K_d = 3.5 \pm 0.5 \times 10^{-9}$ M
W120A SA[a]	Biotin	αHL in lipid bilayer	$k_{off} = 0.31 \pm 0.01\ s^{-1}$ $k_{on} = 3.8 \pm 0.3 \times 10^6\ M^{-1}\ s^{-1}$ $K_d = 8.2 \pm 0.2 \times 10^{-8}$ M
Monoclonal anti-biotin IgG	Biotin	αHL in lipid bilayer	$k_{off} = 1.9 \pm 0.03 \times 10^{-2}\ s^{-1}$ $k_{on} = 4.9 \pm 1.0 \times 10^7\ M^{-1}\ s^{-1}$ $K_d = 3.9 \pm 0.6 \times 10^{-9}$ M
Lectin	Gal-β-1,3-GalNAc	αHL in lipid bilayer	$k_{off1} = 160 \pm 20\ s^{-1}$ $k_{off2} = 9.6 \pm 2.4\ s^{-1}$ $k_{on} = 1.0 \pm 0.4 \times 10^{-9}\ M^{-1}\ s^{-1}$ $K_{d1}{}^{b} = 1.7 \times 10^{-3}$ M $K_{d12}{}^{b} = 3.1 \times 10^{-4}$ M
IgE	RNA Aptamer	Nanopore in quartz pipette	$k_{off} \sim 1^{-5}\ s^{-1}$ $k_{on} = 1.9 \times 10^5\ M^{-1}\ s^{-1}$ $K_d \sim 1 \times 10^{-9}$ M
Carbonic anhydrase II	Sulfonamide-alamethicin	Alamethicin in lipid bilayer	$K_d = 1.9 \pm 0.9 \times 10^{-6}$ M

[a]W120A SA is a mutant form of streptavidin. This mutant has a reduced affinity for biotin, which permitted the measurement of k_{off}.
[b]K_{d1} was the dissociation constant calculated for monovalent binding, and K_{d12} was the dissociation constant calculated assuming a monovalent step followed by a divalent step. See ref. [21] for details.

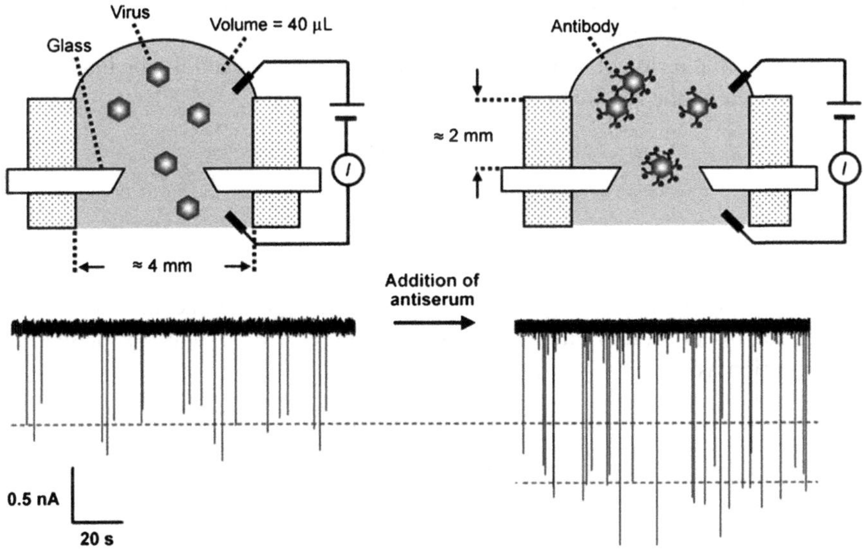

Fig. 9.3 Detecting antibody binding to virus particles. Individual virus particles passing through a conical nanopore (tip diameter of 650 nm) resulted in transient current pulses with a peak amplitude proportional to the volume of the particle. Binding of antibodies to the virus particles resulted in current pulses with an increased amplitude, indicating a larger volume for the antibody–virus complex compared to virus particles in the absence of antibodies. Adopted from Uram et al. with permission. Copyright Wiley-VCH Verlag GmbH & Co. KGaA [50]

Uram et al. calculated the volume of virus particles before and after the addition of antibody from ΔI. Assuming that a typical immunoglobulin G, IgG, antibody has a volume of 347 nm^3 [39], the number of antibodies bound to each virus at equilibrium could be estimated. The results suggested that the majority of the antibodies were bound to the virus via one of their two binding sites. Thus, this method provided information about whether the antibodies were attached by a monovalent or divalent interaction. In addition, this method yielded an estimate for the number of epitopes on a virus for a specific antibody preparation.

Martin's group demonstrated another technique to determine the stoichiometry of protein–ligand interactions [40]. In these experiments, Sexton et al. used the translocation time of the protein–ligand complex to estimate the number of antibody-Fab fragments bound to individual BSA proteins. Upon the addition of increasing concentrations of Fab fragments to a solution containing BSA, the translocation time of BSA increased markedly. Based on these results, Sexton et al. estimated that BSA presents up to three epitopes for the binding of Fab fragments.

9.3.2 Determining Dissociation Constants with Nanopores

Another important characteristic of protein–ligand interactions is the equilibrium dissociation constant. Two methods have been used to determine this constant based on current recordings through nanopores, and both methods employ standard equations for binding equilibria.

For the simple case of a protein, P, binding to a single ligand, L, the binding can be described by the reaction [10, 35]:

$$\mathrm{P} + \mathrm{L} \underset{k_{off}}{\overset{k_{on}}{\rightleftharpoons}} \mathrm{PL}. \tag{9.4}$$

Once binding has reached equilibrium, the strength of the binding can be quantified by an equilibrium association constant, K_a (M^{-1}), which is defined for this reaction as:

$$K_a = \frac{[\mathrm{PL}]}{[\mathrm{P}] \cdot [\mathrm{L}]} = \frac{k_{on}}{k_{off}} = \frac{1}{K_d}, \tag{9.5}$$

where [PL] is the concentration of the protein–ligand complex at equilibrium, [P] is the concentration of unbound protein at equilibrium, [L] is the concentration of unbound ligand at equilibrium, k_{on} ($\mathrm{M}^{-1} \cdot \mathrm{s}^{-1}$) is the association rate constant, k_{off} (s^{-1}) is the dissociation rate constant, and K_d (M) is the dissociation constant.

Since nanopore experiments can make it possible to determine the concentration of protein–ligand complexes, [PL], in solution, (9.6) can be used to obtain K_a and K_d for the interaction between P and L [32, 47].

$$\theta = \frac{[\mathrm{PL}]}{[\mathrm{P_o}]} = \frac{[\mathrm{PL}]}{[\mathrm{P}] + [\mathrm{PL}]} = \frac{[\mathrm{L}]}{[\mathrm{L}] + \frac{1}{K_a}} = \frac{[\mathrm{L}]}{[\mathrm{L}] + K_d}. \tag{9.6}$$

Here θ is the fraction of ligand binding sites that are occupied by ligand, and $[\mathrm{P_o}]$ is the total concentration of the protein. Due to conservation of mass, $[\mathrm{P_o}] = [\mathrm{P}] + [\mathrm{PL}]$ and $[\mathrm{L}] = [\mathrm{L_o}] - [\mathrm{PL}]$ with $[\mathrm{L_o}]$ equal to the initial concentration of the ligand.

Equation 9.6 illustrates that the dissociation constant is equal to the concentration of ligand at which half of the ligand-binding sites are occupied (i.e. $\theta = 0.5$). Thus, a small value of the dissociation constant indicates that a particular ligand has a high affinity for a particular protein. Hence K_d is often used to describe or compare the affinity of a ligand for a protein or in the pharmaceutical industry for characterizing drug–protein interactions and antibody–antigen interactions.

As stated above, in order to apply (9.6) for determining K_d, the nanopore sensing method must be able to determine either the concentration of protein–ligand complexes, the unbound ligand, or the unbound protein at equilibrium. The technique described in Sect. 9.3.1 [50] was able to calculate the number of antibodies bound to each virus particle based on ΔI. In a similar manner, Uram et al. applied this analysis to data obtained by Saleh and Sohn [38, 47]. In these experiments, translocation of colloids functionalized with streptavidin through the pore resulted in transient current fluctuations. As a function of the concentration of monoclonal anti-streptavidin antibodies, the number of antibodies bound to the colloids could be obtained from the peak amplitude data. Knowing the total concentration of antibody, Uram et al. calculated the equilibrium concentration of free antibody and plotted a binding isotherm (Fig. 9.4) [47]. A fit to the data in Fig. 9.4 with a variation of (9.6) determined K_d (Table 9.1).

Whitesides and coworkers demonstrated a second method for determining the equilibrium dissociation constant using a pore-forming peptide, alamethicin [32]. Alamethicin self-assembles in planar lipid bilayers to form ion channels that have multiple, discrete conductance values depending on the number of peptides forming the ion channel at a given time. Here the alamethicin peptides were modified with the ligand sulfonamide, and the binding of carbonic anhydrase II to alamethicin-sulfonamide disrupted the formation of ion channels and reduced the amount of transported charge during the time interval of the recording. Mayer et al. then established a functional relationship between the concentration of unbound alamethicin-sulfonamide peptides and the quantity of transported charge. Thus, the authors could calculate the concentration of unbound alamethicin-sulfonamide peptides in the presence of various concentration of the binding protein, carbonic anhydrase II. Consequently, the authors obtained a binding isotherm and determined the K_a and K_d values for the interaction (Table 9.1).

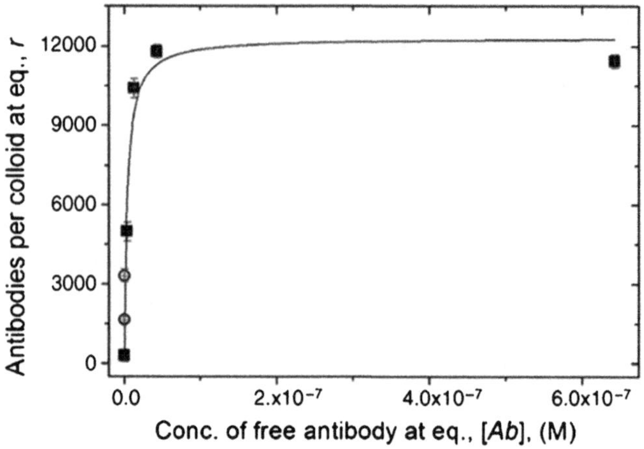

Fig. 9.4 Binding isotherm obtained by current recordings through nanopores of particles decorated with monoclonal anti-streptavidin antibodies that bound to streptavidin on the surface of these particles [38]. Equation 9.6 was used to fit the data in order to calculate K_a and K_d. Adopted from Uram and Mayer with permission [47]

9.4 Determining On- and Off-Rates of Binding with Nanopores

Methods for characterizing the kinetics of protein–ligand binding using nanopores employed the biological pore α-hemolysin and solid-state nanopores fabricated in quartz pipettes [12, 21, 24, 33]. These methods share the capability to detect the association or dissociation of individual protein–ligand complexes. Applying (9.5), the value of K_d can be calculated when the values of k_{off} and k_{on} are determined.

To detect these individual binding events, Bayley and coworkers used a ligand attached to the interior wall of α-hemolysin by a polymer chain [22, 33]. In these experiments, the polymer was long enough to protrude from either side of the pore. Consequently, in the absence of binding, the polymer chain freely moved between the *cis* and *trans* chambers. This flexibility resulted in rapid, transient current decreases when the polymer extended through the pore (Fig. 9.5). Figure 9.5 illustrates that the binding of a protein to the ligand prevented the movement of the polymer chain between the *cis* and *trans* chambers. This signal permitted the detection of association and dissociation of individual protein–ligand complexes. Movileanu et al. used a poly (ethylene glycol), PEG, polymer with a biotin moiety at the free end and a mutant form of streptavidin (W120A), which has a lower affinity for biotin than the wild type [33]. Figure 9.5 illustrates that binding of W120A to the biotin-PEG on the *cis* side of the pore prevented the polymer from entering the lumen of α-hemolysin. This binding resulted in a stable, large current value. Hence, the return to rapid current blockages signified the dissociation of the W120A–biotin complex, and the average duration, t_{off} (s), of the large, steady-state current value was related to k_{off} by:

$$\frac{1}{t_{off}} = k_{off}. \tag{9.7}$$

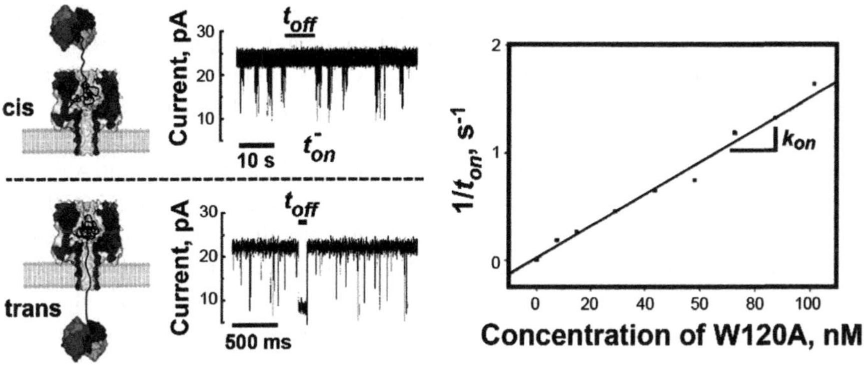

Fig. 9.5 Detecting the binding of a protein to a ligand that was attached to the entrance of a α-hemolysin pore by a polymer chain. Translocation of the polymer resulted in rapid, transient current decreases. Binding of the protein to the end of the polymer chain resulted in a steady current value as long as the protein remained bound, yielding a time to dissociation of t_{off}. The interval between binding events, t_{on}, as a function of the protein concentration was used to calculate k_{on} with (9.8). Adopted from Movileanu et al. with permission [33]

Conversely, the average time interval of rapid current fluctuations that occured between the steady-state current values, t_{on} (s), was related to k_{on} and the concentration of free protein, [P], by [33]:

$$\frac{1}{t_{on}} = [\mathrm{P}] \cdot k_{on}. \tag{9.8}$$

Howorka et al. expanded on this method by covalently linking two ligand-modified polymers within the pore in order to distinguish between monovalent and bivalent binding of a carbohydrate binding protein, lectin, to the disaccharide Gal-β-1,3-GalNAc [21]. Upon addition of lectin, the stochastic response of this assay permitted the determination of two distinct values for k_{off}. The two values for k_{off} resulted from monovalent or bivalent binding of lectin to the disaccharide and permitted the authors to calculate two values of K_d (Table 9.1). Although only demonstrated in genetically engineered α-hemolysin pores, this concept could be applied to solid-state nanopores if a *single* polymer could be immobilized in the pore. Unlike methods that measure protein–ligand complexes under equilibrium conditions, this method has the unique ability to determine values of K_d for monovalent and bivalent binding.

Robertson's group demonstrated another approach to calculate kinetic parameters. In this method ligand-modified polymers were dissolved freely in the bulk solution [24]. Again, in the absence of protein, the translocation of polymers through the pore resulted in rapid current blockages. The frequency of these transient current blockages was related to the concentration of unbound, ligand-modified polymer. Depending on the length of the polymer, binding of a protein to the ligand reduced the frequency of the current blockages or resulted in long current blockages that indicated the capture of a protein–ligand complex (Fig. 9.6). Thus, similar to the approach mentioned above, k_{off} values could be estimated from the duration of the long current blockades by (9.7) [24, 33]. Kasianowicz et al. first demonstrated this technique using wild type α-hemolysin pores and biotinylated, single-stranded DNA in order to detect the binding of the protein NeutrAvidin to biotinylated DNA [24]. The authors used short biotinylated DNA strands (10 mer) to demonstrate that protein binding to the ligand-modified polymers reduced the frequency of the current blockages. In addition, the authors observed that protein binding to long biotinylated DNA (50 mer) strands resulted in permanent current blockages (permanent because NeutrAvidin binding to biotin is essentially irreversible $K_d \sim 10^{-15}$ M) [26].

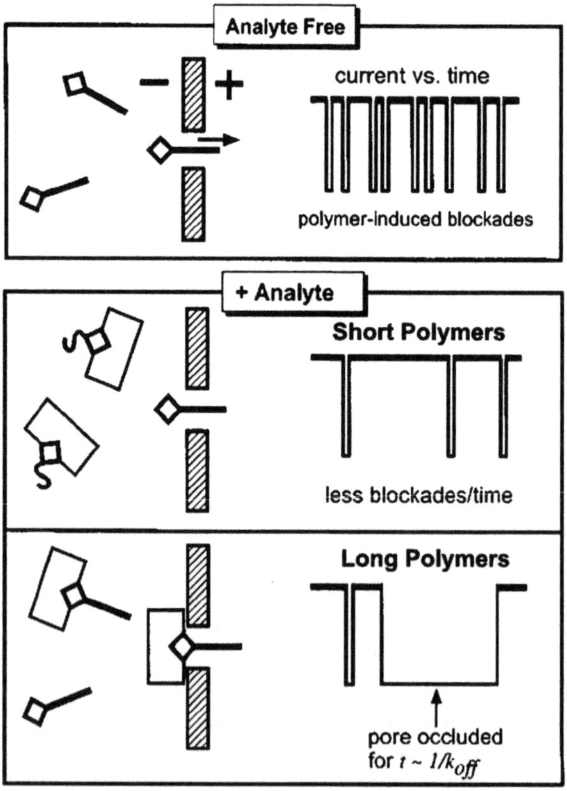

Fig. 9.6 Detecting the binding of proteins to ligand molecules that were attached to polymers dissolved freely in the bulk solution. Unbound polymers passed through the pore and resulted in transient current fluctuations. Depending on the length of the polymer, binding of the protein to the ligand on short polymers (10 mer DNA) reduced the frequency of the current blockades, while binding of protein to the ligand on long polymers (50 mer DNA) resulted in current blockades with a duration proportional to k_{off}. Adopted from Kasianowicz et al. [24]

Due to the potential for large ligand-modified polymers to occlude the pore for the duration of a protein–ligand interaction, this method could be used to determine k_{off} by applying (9.7) [24]. The authors pointed out, however, that due to the high electric field within the nanopore, the dissociation rate of charged proteins and polymers may be affected. In addition, when short polymers are used, this technique could also determine the concentration of unbound ligand, [L], in solution and consequently, the concentration of bound ligand [24]. If the measurement of [L] was taken under equilibrium conditions, the association constant, K_a, could be determined using (9.6). This method requires a nanopore small enough to prevent translocation of the protein and could therefore be implemented with solid-state nanopores if their diameter is smaller than that of the proteins to be detected.

Gu and coworkers reported a third technique that entails the use of ligand immobilized on a nanopore wall [12]. Here the binding of a protein to a ligand molecule in the pore resulted in a stepwise decrease in the current for each protein bound. This information could be used to determine k_{on} and potentially k_{off}. Figure 9.7 illustrates this sensing scheme and a representative current trace. The authors first immobilized aptamers targeted toward immunoglobulin E, IgE, and ricin, in glass nanopipettes.[4] In this method, the time to each binding event (indicated by a stepwise current decrease) was fit with an exponential decay equation with the time constant equal to t_{on} (Fig. 9.7 inset). In combination with (9.8), the authors then determined k_{on} (Table 9.1).

Due to the high affinity of aptamers toward their target ($K_d < 10^{-8}$ M), the authors did not observe dissociation events for the duration of their experiments. Ligands with lower affinity may, however, yield measurement of k_{off} as well [12]. In principle this method could be applicable to protein–ligand interactions if a ligand of interest replaced the aptamer.

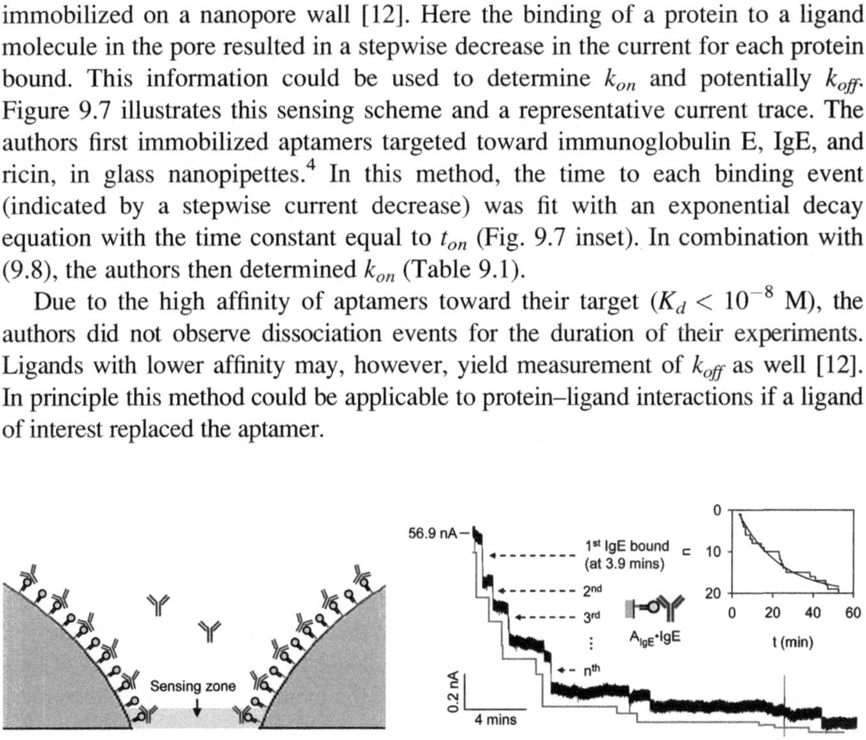

Fig. 9.7 Detecting the binding of individual IgE and Ricin proteins with aptamers immobilized on the surface of a nanopore. Individual binding events resulted in stepwise decrease of the current through the nanopore. (*Inset*) Number of binding events versus time was fit with an exponential decay equation with the time constant equal to t_{on}. Adopted from Ding et al. with permission [12]

[4]Aptamers are short DNA or RNA segments that have been selected from a large pool (>10,000) of molecules and enriched using the SELEX technology. They often have high affinity for their target $K_d \sim 10^{-11}$.

9.5 Quantifying Enzyme Activity with Nanopores

Monitoring the activity of enzymes with nanopores is a very recent development in the field of nanopore sensing. Using α-hemolysin, Cockroft et al. detected the addition of single nucleotides to a DNA template due to the activity of DNA polymerase [9], and Clark et al. detected single nucleotides as a result of exonuclease activity [8]. In this section we will focus on three recent publications that explored strategies of using nanopores to extract *quantitative* information about the activity of enzymes.

In this context, Guan's group described a method that detected protease activity by measuring the translocation of the resulting peptide products though a nanopore [55] (see Chapter 13 also). These authors used a mutant version of α-hemolysin, called $(M113F)_7$, to detect the translocation of a fragment of amyloid-β (Aβ) peptides (residues 10–20) and to report the activity of trypsin, an enzyme that cleaves peptide bonds after arginine or lysine. Upon addition of trypsin, the authors observed translocation of the enzymatic products and the original Aβ fragment. These three molecules could be distinguished based on their different molecular volumes, which resulted in different magnitudes of ΔI (Fig. 9.8). From the fraction of events due to each molecule, the authors could monitor the formation of enzymatic products as a function of time. The authors then calculated the Michaelis constant, K_m, and the catalytic rate constant, k_{cat}, for the reaction between trypsin and Aβ (10–20) (Table 9.2).[5] This method was hence used to study protease kinetics.

Majd et al. described a second assay for enzyme activity in which the conductance through gramicidin, gA, pores that are embedded in a lipid bilayer reported the activity of membrane-active enzymes [28].[6] In this method, the surface charge of the lipid membrane affected the conductance through gA pores [4], and the activity of the enzymes phospholipase D (PLD) and phospholipase C (PLC) altered the net surface charge of the membrane. For example, the substrate for PLD is the electrically neutral lipid phosphotidylcholine (PC), while the enzymatic products are choline and a negatively charged lipid, phosphatidic acid (PA). Figure 9.9 illustrates this reaction. Conversely, the substrate for phospholipase C is the negatively charged lipid phosphotidylinositol, and the enzymatic products are inositol-phosphate and a neutral lipid, diacyl glycerol. Thus, the enzymatic activity of these phospholipases changed the surface charge of the membrane, and in an electrolyte

[5]The catalytic rate constant k_{cat} describes the rate at which the enzyme-substrate complex is converted to the free enzyme and free product. The Michaelis constant, K_m, is the concentration of substrate that results in the half-maximal velocity of the enzymatic reaction.

[6]Gramicidin is a peptide consisting of 15 amino acids that spans one leaflet of a bilayer. If gA peptides are present in both leaflets of a bilayer, they can transiently form a dimer, which conducts monovalent cations through a central pore with diameter of ~4 Å. These transient ion channels result in discrete current values that reflect the number of ions passing through individual gA pores in a planar lipid bilayer at a given instant. Antonenko and coworkers first characterized the K_d for the interaction of two monomers of gA that bind to form dimeric gA ion channels in a lipid bilayer [37].

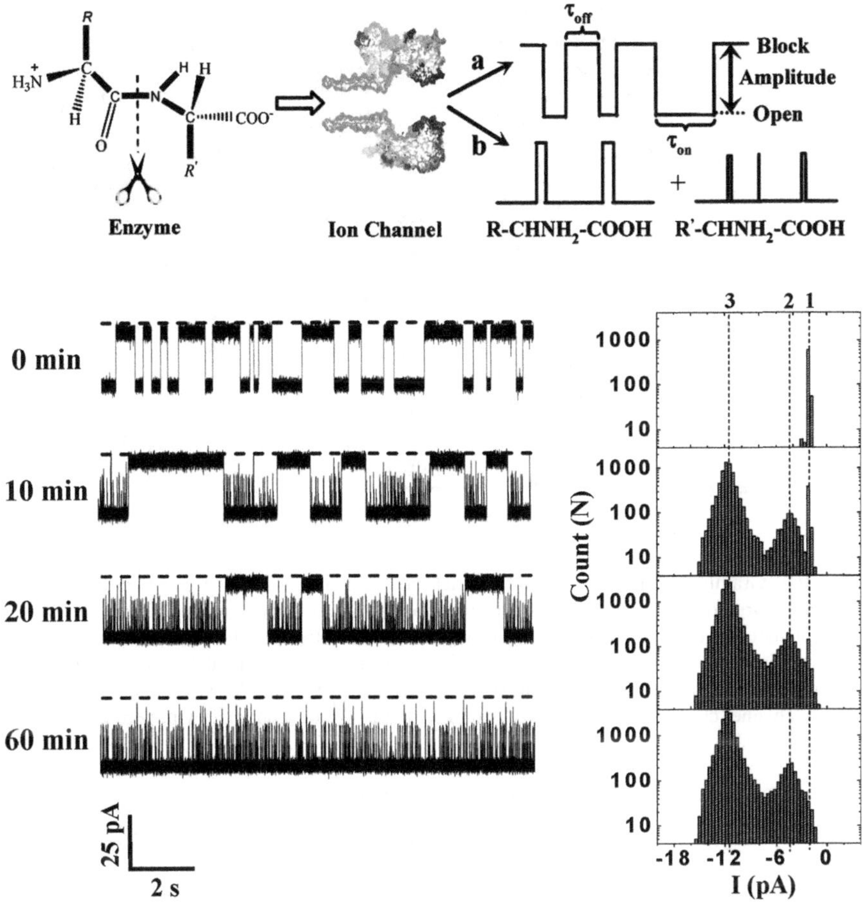

Fig. 9.8 Detecting protease activity with a α-hemolysin pore. Translocation of a ten amino acid fragment of amyloid-β (residues 10–20) through the pore resulted in transient current blockages (shown at time 0). Protease activity cleaved the amyloid-β into two smaller peptides, and translocation of these peptides resulted in current fluctuations with a characteristic peak amplitude. Adopted from Zhao et al. with permission [55]

with low ionic strength, this change in surface charge affected the local concentration of ions near the surface of the membrane and consequently the conductance of gramicidin pores embedded in the lipid bilayer. Hence, the conductance of gramicidin reported the enzyme activity of PLD and PLC. Due to the complex nature of catalysis on a surface, the authors calculated a pseudo-first order rate constant, k_f (s^{-1}) (Table 9.2). This method permitted the detection of the two phospolipases using picomolar to nanomolar amounts of enzyme.

In a third demonstration of quantifying enzyme activity with nanopores, Yang's group and Mayer's group recently attached an enzyme substrate covalently to gA pores and detected enzymatic hydrolysis by a change in the single-channel

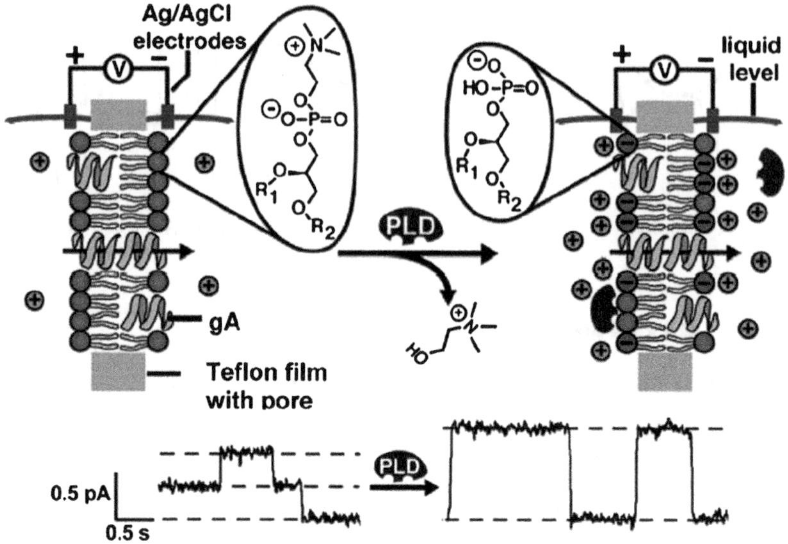

Fig. 9.9 Gramicidin pores can report the activity of membrane-active enzymes that alter the surface charge of membranes. Transient formation of ion pores by gramicidin dimers resulted in the conductance of monovalent cations through the pore. Activity of the enzyme phospholipase D resulted in the conversion of zwitterionic lipids to negatively charged lipids. The subsequent accumulation of cations near the negatively charged membrane was detected by an increase in the single-channel conductance through gA pores. Adopted from Majd et al. with permission [28]

Table 9.2 Quantification of enzyme activity using nanopore recordings

Enzyme	Substrate	Pore	Parameter measured
Trypsin	Aβ 10–20	αHL	$K_m = 59.2 \times 10^{-6}$ M $k_{cat} = 4.43$ s^{-1}
Phospholipase D[a]	Phosphotidylcholine	Gramicidin	Pseudo-$k_f = 1.3 \times 10^{-4}$ s^{-1}
Alkaline phosphatase	gA-Phosphate[b]	Gramicidin	$K_m = 3.0 \pm 0.4 \times 10^{-7}$ M $k_{cat} = 3.1 \pm 0.2 \times 10^{-3}$ s^{-1}

[a]Due to the heterogeneous nature of catalysis on a lipid bilayer surface, the K_m and k_{cat} constants could not be determined. The method provided a pseudo-first order rate constant, k_f, with the rate of formation of product, $dP/dt = k_f$ [E] where [E] is the concentration of enzyme. Additional constants specific to interfacial enzymes were estimated; see ref. [28] for more details

[b]Gramicidamine derivative carrying a glycolic acid-*O*-phosphate [3]

conductance through the pores [3, 27]. Macrae et al. employed this concept to detect the activity of alkaline phosphatase and a protease, anthrax lethal factor. For the detection of alkaline phosphatase, the authors modified the ion channel forming-peptide to contain a phosphate group at the entrance of the pore, gA-phosphate. At low ionic strength, the negatively charged phosphate resulted in increased conductance of monovalent cations through the gA-phosphate pore compared to a gA pore with an electrically neutral entrance. Alkaline phosphatase removed the phosphate from the entrance of gA, which resulted in a gA molecule with an electrically neutral entrance and a decreased single-channel conductance. Consequently, this approach made it possible to distinguish the formation of pores due to the substrate, gA-phosphate, from those due to the uncharged gA (product). This ability enabled the authors to monitor the fraction of events due to the enzymatic substrate and product over time, and consequently, calculate K_m and k_{cat} (Table 9.2).

For the detection of anthrax lethal factor, Macrae et al. linked nine amino acid residues covalently to gA peptides. Anthrax lethal factor is a protease produced by the bioterrorism agent Anthrax, and it cleaves a specific amino acid sequence (Fig. 9.10). The activity of LF removed eight of these residues and resulted in a significant increase in the formation of gA pores and the subsequent transport of charge through the membrane (Fig. 9.10). Thus, this signal could be used to detect and quantify the activity of anthrax lethal factor.

This technique exploited the amplification characteristics of ion channels in that the cleavage of *one* phosphate group on a *single* gA molecule reduced the flux of ions through the pore by *thousands* of ions per second. This method also used only picomolar concentrations of enzyme in volumes that resulted in the use of femtomoles of protein.

All three of these nanopore-based, enzymatic assays open the door for generating miniature platforms to monitor rapidly the activity of enzymes *in situ* and with native substrates.

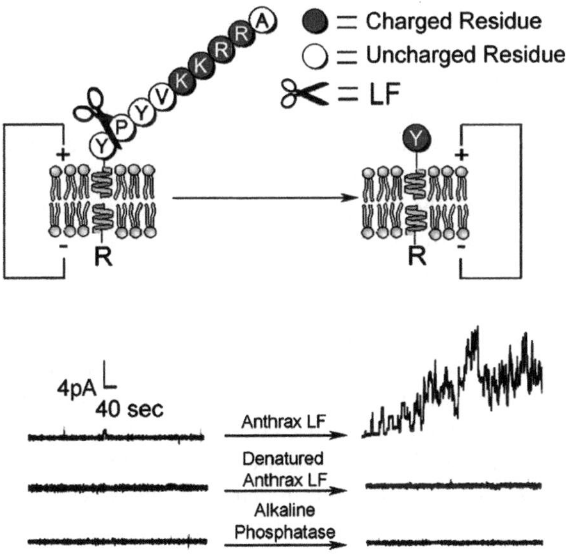

Fig. 9.10 Detecting enzyme activity with gA peptides that are modified covalently with substrate. Anthrax lethal factor (LF) cleaves nine amino acids from the entrance of a modified gA peptide. LF activity resulted in the formation of gA pores and a significant increase in the transport of charge across the membrane over time. Adopted from Macrae et al. with permission [27]

9.6 Future of Protein Sensing with Nanopores

Nanopore-based sensing is a relatively new single molecule technique. Of all the single molecule techniques, it is one of the simplest and most accessible methods. With respect to the characterization of proteins, nanopores have the unique capability to determine the size of proteins in solution rapidly without requiring labels. Furthermore, additional work on characterizing the state of folding of proteins could perhaps lead to quantifying conformational changes of functional proteins. Further developments may even make it possible to sequence amino acids in a similar manner as researchers are trying to use nanopores for sequencing of DNA and RNA based on nanopore recordings.

Nanopore-based techniques for detecting the binding of a protein and ligand do not require a fluorescent label on the protein or ligand, and thus, the techniques have a unique advantage over fluorescence-based methods. Similarly, nanopore-based detection of enzyme activity can quantify the kinetics of enzymes without the use of fluorescent labels or secondary enzymatic reactions. Detection of enzymes from nanopore recordings can also employ nanopores that present a covalently-attached substrate of an enzyme. These assays in particular exploit ion channel amplification, in which the cleavage of one molecule due to enzyme activity may result in a large change in the flow of ions through a pore [17, 27]. This amplified sensitivity, coupled with the small volumes used in nanopore-based assays, makes it possible to work with only femtomoles of enzyme [27, 28]. Together, these characteristic may permit the analysis of filtered but otherwise untreated samples of biomolecules in physiologic solutions.

Although nanopore-based sensing has great potential to explore the function of proteins, several challenges have to be met. For example, reliable quantification and prediction of the translocation times of proteins through nanopores remains difficult. Values calculated to date for the diffusion constant and the electrophoretic drift velocity from translocation times through nanopores are different from the values found in bulk solution. These discrepancies may be due to interactions between the protein and the nanopore walls. Hence minimizing these interactions will be important. Furthermore, the surface charge on many proteins is not uniform, and thus for proteins with a dipole moment, the electrophoretic force acting on the protein may be a function of the position of the protein with respect to the electric field in the nanopore [46]. In addition, proteins are entropically constrained in nanopores with diameters on the order of the protein's dimensions [30]. Further complications include that the electric field is not uniform near the narrowest constriction of the nanopore and that electroosmotic flow (EOF) is typically not considered in calculations based on the translocation time. We showed recently, however, that EOF can be significant in nanopores under certain conditions [53]. In order to realize the full potential of quantitative nanopore recordings these challenges will have to be addressed.

One strategy might be chemical modification of nanopore walls. Currently, the ability to modify the surface of nanopores chemically in an efficient manner is limited. Novel methods that alter the surface chemistry of nanopores with ease and flexibility may accelerate the use of solid-state nanopores for protein–ligand binding assays because the efficiency of immobilizing a variety of ligands is crucial for the acceptance of this technique [54]. In addition, surface chemistries that reduce non-specific interactions of the protein with the surface will minimize clogging of nanopores and solve some of the problems mentioned above.

Developing nanopores that mimic biological pores such as porins and ion channel proteins is an intriguing future perspective [29]. Topics range from replicating the ion current rectifying properties of many ion channels (perhaps the most advanced aspect at the moment) to replicating the natural selectivity of nucleopores and ion channels, and possibly, to gate nanopores for signal amplification in analogy to ligand-gated ion channels in nerve endings. Recent, inspiring work has advanced this area of research in replicating the basic nucleopore complex found in a yeast cell [23]. Furthermore, the ability of ion channel proteins to undergo conformational changes in response to external stimuli is intriguing, since triggers for these responses are not limited to ligand binding but may include electric potentials, mechanical stresses, and chemical reactions such as phosphorylations [19, 52].

Finally, electrical noise remains a challenge for most nanopore sensing experiments; in particular for studying small proteins with very short translocation times. Developing nanofabrication techniques that permit the fabrication of nanopores in materials with low capacitance will help reduce electrical noise and may potentially increase the sensitivity of nanopores for detecting proteins with short translocation times [48].

Acknowledgments The authors acknowledge the following funding sources: National Institutes of Health (M.M., grant no. 1RO1GM081705), NSF Career Award (M.M., grant no. 0449088), AISIN/IMRA America Inc., and Thermo Fisher – CCG Collaborative Pilot Project Initiative.

References

1. Ali M, Schiedt B, Healy K et al. (2008) Modifying the surface charge of single track-etched conical nanopores in polyimide. Nanotechnology, 19: 085713
2. Ali M, Yameen B, Neumann R et al. (2008) Biosensing and supramolecular bioconjugation in single conical polymer nanochannels. Facile incorporation of biorecognition elements into nanoconfined geometries. J. Am. Chem. Soc., 130: 16351–16357
3. Blake S, Capone R, Mayer M et al. (2008) Chemically reactive derivatives of gramicidin a for developing ion channel-based nanoprobes. Bioconjug. Chem., 19: 1614–1624
4. Capone R, Blake S, Restrepo M R et al. (2007) Designing nanosensors based on charged derivatives of gramicidin a. J. Am. Chem. Soc., 129: 9737–9745
5. Chang H, Venkatesan B M, Iqbal S M et al. (2006) DNA counterion current and saturation examined by a mems-based solid state nanopore sensor. Biomed. Microdevices, 8: 263–269
6. Chun K Y and Stroeve P (2002) Protein transport in nanoporous membranes modified with self-assembled monolayers of functionalized thiols. Langmuir, 18: 4653–4658
7. Chun K Y, Mafe S, Ramirez P et al. (2006) Protein transport through gold-coated, charged nanopores: effects of applied voltage. Chem. Phys. Lett., 418: 561–564
8. Clarke J, Wu H C, Jayasinghe L et al. (2009) Continuous base identification for single-molecule nanopore DNA sequencing. Nat. Nanotechnol., 4: 265–270

9. Cockroft S L, Chu J, Amorin M et al. (2008) A single-molecule nanopore device detects DNA polymerase activity with single-nucleotide resolution. J. Am. Chem. Soc., 130: 818–820
10. Copeland R A (2000) Enzymes. John Wiley & Sons, New York
11. DeBlois R W and Bean C P (1970) Counting and sizing of submicron particles by the resistive pulse technique. Rev. Sci. Instrum., 41: 909–916
12. Ding S, Gao C L and Gu L Q (2009) Capturing single molecules of immunoglobulin and ricin with an aptamer-encoded glass nanopore. Anal. Chem., 81: 6649–6655
13. Fologea D, Ledden B, David S M et al. (2007) Electrical characterization of protein molecules by a solid-state nanopore. Appl. Phys. Lett., 91: 053901
14. Hammann C H, Hamnett A and Vielstich W (1998) Electrochemistry. Wiley-VCH, New York
15. Han A, Creus M, Schurmann G et al. (2008) Label-free detection of single protein molecules and protein-protein interactions using synthetic nanopores. Anal. Chem., 80: 4651–4658
16. Han A P, Schurmann G, Mondin G et al. (2006) Sensing protein molecules using nanofabricated pores. Appl. Phys. Lett., 88: 093901
17. Hille B (2001) Ion channels of excitable membranes. Sinauer Associates, Inc., Sunderland
18. Hladky S B and Haydon D A (1970) Discreteness of conductance change in bimolecular lipid membranes in presence of certain antibiotics. Nature, 225: 451–453
19. Hou X, Guo W, Xia F et al. (2009) A biomimetic potassium responsive nanochannel: G-quadruplex DNA conformational switching in a synthetic nanopore. J. Am. Chem. Soc., 131: 7800–7805
20. Howorka S and Siwy Z (2009) Nanopore analytics: sensing of single molecules. Chem. Soc. Rev., 38: 2360–2384
21. Howorka S, Nam J, Bayley H et al. (2004) Stochastic detection of monovalent and bivalent protein-ligand interactions. Angew. Chem. Int. Ed., 43: 842–846
22. Howorka S, Movileanu L, Lu X et al. (2000) A protein pore with a single polymer chain tethered within the lumen. J. Am. Chem. Soc., 122: 2411–2416
23. Jovanovic-Talisman T, Tetenbaum-Novatt J, McKenney A S et al. (2009) Artificial nanopores that mimic the transport selectivity of the nuclear pore complex. Nature, 457: 1023–1027
24. Kasianowicz J J, Henrickson S E, Weetall H H et al. (2001) Simultaneous multianalyte detection with a nanometer-scale pore. Anal. Chem., 73: 2268–2272
25. Ku J-R and Stroeve P (2004) Protein diffusion in charged nanotubes: "On-off" Behavior of molecular transport. Langmuir, 20: 2030–2032
26. Laitinen O H, Hytonen V P, Nordlund H R et al. (2006) Genetically engineered avidins and streptavidins. Cell. Mol. Life Sci., 63: 2992–3017
27. Macrae M X, Blake S, Jiang X et al. (2009) A semi-synthetic ion channel platform for detection of phosphatase and protease activity. ACS Nano, 3: 3567–3580
28. Majd S, Yusko E C, MacBriar A D et al. (2009) Gramicidin pores report the activity of membrane-active enzymes. J. Am. Chem. Soc., 131: 16119–16126
29. Majd M, Yusko E C, Billeh Y N et al. (2010) Applications of biological pores in nanomedicine, sensing and nanoelectronics. Curr. Opin. in Biotechnol., 21: 439–476
30. Makarov D E (2009) Computer simulations and theory of protein translocation. Acc. Chem. Res., 42: 281–289
31. Maxwell J C (1904) Treatise on Electricity and Magnetism. 3 ed. Clarendon: Oxford. Vol 1
32. Mayer M, Semetey V, Gitlin I et al. (2008) Using ion channel-forming peptides to quantify protein-ligand interactions. J. Am. Chem. Soc., 130: 1453–1465
33. Movileanu L, Howorka S, Braha O et al. (2000) Detecting protein analytes that modulate transmembrane movement of a polymer chain within a single protein pore. Nat. Biotechnol., 18: 1091–1095
34. Neher E and Sakmann B (1976) Single-channel currents recorded from membrane of denervated frog muscle-fibers. Nature, 260: 799–802
35. Nelson D L and Cox M M (2008) Lehninger principles of biochemistry. W. H. Freeman and Company, New York

36. Oukhaled G, Mathe J, Biance A L et al. (2007) Unfolding of proteins and long transient conformations detected by single nanopore recording. Phys. Rev. Lett., 98: 158101
37. Rokitskaya T I, Antonenko Y N and Kotova E A (1996) Photodynamic inactivation of gramicidin channels: a flash-photolysis study. Biochim. Biophys. Acta Bioenerg., 1275: 221–226
38. Saleh O A and Sohn L L (2003) Direct detection of antibody-antigen binding using an on-chip artificial pore. Proc. Natl. Acad. Sci. U.S.A., 100: 820–824
39. Schneider S W, Larmer J, Henderson R M et al. (1998) Molecular weights of individual proteins correlate with molecular volumes measured by atomic force microscopy. Pflugers Arch., 435: 362–367
40. Sexton L T, Horne L P, Sherrill S A et al. (2007) Resistive-pulse studies of proteins and protein/antibody complexes using a conical nanotube sensor. J. Am. Chem. Soc., 129: 13144–13152
41. Siwy Z, Heins E, Harrell C C et al. (2004) Conical-nanotube ion-current rectifiers: the role of surface charge. J. Am. Chem. Soc., 126: 10850–10851
42. Siwy Z, Gu Y, Spohr H A et al. (2002) Rectification and voltage gating of ion currents in a nanofabricated pore. Europhys. Lett., 60: 349–355
43. Siwy Z S (2006) Ion-current rectification in nanopores and nanotubes with broken symmetry. Adv. Funct. Mater., 16: 735–746
44. Smeets R M M, Keyser U F, Krapf D et al. (2006) Salt dependence of ion transport and DNA translocation through solid-state nanopores. Nano Lett., 6: 89–95
45. Storm A J, Storm C, Chen J H et al. (2005) Fast DNA translocation through a solid-state nanopore. Nano Lett., 5: 1193–1197
46. Talaga D S and Li J L (2009) Single-molecule protein unfolding in solid state nanopores. J. Am. Chem. Soc., 131: 9287–9297
47. Uram J D and Mayer M (2007) Estimation of solid phase affinity constants using resistive-pulses from functionalized nanoparticles. Biosens. Bioelectron., 22: 1556–1560
48. Uram J D, Ke K and Mayer M (2008) Noise and bandwidth of current recordings from submicrometer pores and nanopores. ACS Nano., 2: 857–872
49. Uram J D, Ke K, Hunt A J et al. (2006) Label-free affinity assays by rapid detection of immune complexes in submicrometer pores. Angew. Chem. Int. Ed., 45: 2281–2285
50. Uram J D, Ke K, Hunt A J et al. (2006) Submicrometer pore-based characterization and quantification of antibody-virus interactions. Small, 2: 967–972
51. Vlassiouk I, Kozel T R and Siwy Z S (2009) Biosensing with nanofluidic diodes. J. Am. Chem. Soc., 131: 8211–8220
52. Yameen B, Ali M, Neumann R et al. (2009) Synthetic proton-gated ion channels via single solid-state nanochannels modified with responsive polymer brushes. Nano Lett., 9: 2788–2793
53. Yusko E C, An R and Mayer M (2010) Electroosmotic flow can generate ion current rectification in nano- and micropores. ACS Nano., 4: 477–487
54. Yusko E C, Johnson J M, Majd S et al. (2011) Controlling protein translocation through nanopores with bio-inspired fluid walls. Nature Nanotech., 6: 254–260
55. Zhao Q T, de Zoysa R S S, Wang D Q et al. (2009) Real-time monitoring of peptide cleavage using a nanopore probe. J. Am. Chem. Soc., 131: 6324–6325

Chapter 10
Capture and Translocation of Nucleic Acids into Sub-5 nm Solid-State Nanopores

Meni Wanunu, Allison Squires, and Amit Meller

Abstract Nanopores have emerged as single-molecule analytic tools for fundamental biophysical characterization of nucleic acids as well as for future genomic applications. The enormous interest in single-molecule analysis has spurred the development of many different approaches to nanopore fabrication. Of these, ultrathin solid-state membranes are the most promising substrates, combining exceptional robustness and control over pore size and shape with an inherently planar geometry that enables parallel detection with nanopore arrays. Moreover, nanopores with diameters in the range of 1–5 nm represent an important size regime for studying nucleic acids, as these pores can translocate long DNA and RNA molecules in a linear fashion, enabling readout of local nucleic acid structure with unparalleled read-length. In this review, we focus on two fundamental aspects of nucleic acid analysis using nanopores, namely the process of DNA capture and the subsequent translocation dynamics. We compile here a multi-parametric study of double-stranded DNA molecules of lengths ranging from 50 to 50,000 bp, and discuss the influence of DNA length, applied voltage, temperature, and salt buffer concentrations on the capture and translocation processes.

Keywords Solid-state nanopores • Single-molecule detection • DNA translocation • DNA capture • DNA-pore interactions • Salt gradient focusing • Attomole detection

A. Meller (✉)
Department of Biomedical Engineering, Boston University,
44 Cummington St., Boston, MA 02215, USA
e-mail: ameller@bu.edu

S.M. Iqbal and R. Bashir (eds.), *Nanopores: Sensing and Fundamental Biological Interactions*, DOI 10.1007/978-1-4419-8252-0_10,

10.1 Introduction: Biopolymer Translocation Through Narrow Pores

Resistive sensing is based on monitoring changes in the ionic conductance through micrometer-scale apertures during the passage of colloidal particles (i.e. cells). This method, also known as "Coulter counting," has been broadly applied across the life sciences and biochemistry [10]. The nanopore method employs a similar detection principle, but uses an aperture that is only a few nanometers wide – suitable for probing individual molecules [11, 15, 48]. Another crucial difference between classical particle counters and nanopores is the ability of the latter to linearize long polymer coils during detection (Fig. 10.1) [49]. Consider a long biopolymer (e.g. DNA or RNA) with chain cross section a, and a nanopore of diameter d, where d is only slightly larger than a and is much smaller than the polymer's coil size, commonly estimated by its radius of gyration r_g $(a<d = r_g)$. When this biopolymer is threaded through the nanopore, its equilibrium coil structure must be deformed and linearized by the narrow constriction. Thus, the ionic conductance of the blocked nanopore can be directly related to local characteristics of the linearized biopolymer. This powerful feature sets the nanopore method apart from most other sensing methods, and makes it particularly useful for analysis of long biopolymers, particularly nucleic acids [1, 22, 33, 34]. It is therefore anticipated that nanopore-based platforms will be developed as future-generation tools for genomic profiling and for DNA sequencing, where the ability to linearly analyze long biopolymers ranging from thousands to hundreds of thousands of base pairs in length is particularly advantageous [5].

The ability to experimentally analyze individual molecules during transit through nanopores has greatly contributed to our basic understanding of the physical processes governing biopolymer transport. This, in turn, has led to the development of

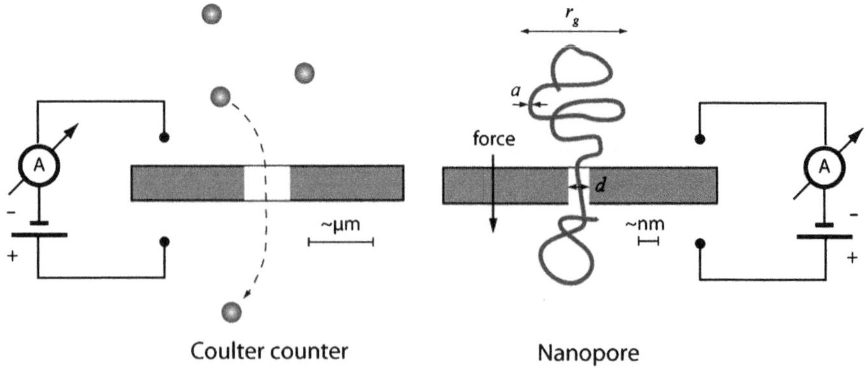

Fig. 10.1 The principle of resistive sensing. *Left panel*: Illustration of Coulter counter. An electric field draws charged particles through a micrometer-scale aperture in a membrane separating two salt buffer-filled chambers. Particles are detected and counted as they occlude this opening during their passage, blocking the flow of ions through the pore and producing a transient drop in current. *Right panel*: Illustration of nanopore setup. The applied voltage forces long, charged biopolymers to translocate linearly through a nanoscale pore. Molecules are both detected and characterized by the resulting transient changes in current through the nanopore

many novel approaches for nanopore-based detection of individual, unlabeled biopolymers. The impact of these methods is twofold: First, they expand the existing arsenal of single-molecule techniques by providing a wide variety of new means for analyzing biopolymers. For example, nanopores have been used to differentiate among nucleic acid (DNA or RNA) populations of varied length and sequence [34, 49], to probe secondary structure formation in nucleic acids [8, 32, 33, 40, 46], to detect specific or nonspecific interactions of nucleic acids with proteins or small molecules (i.e. drugs) [4, 14, 18, 49, 51], and to map individual transcription factors bound to genomic DNA molecules [24]. Second, the unique ability of the nanopore method to analyze unlabeled single biomolecules has been adapted to engineer many novel sensing approaches for a broad range of biomedical and biochemical applications, including genome profiling [41], single-molecule DNA sequencing [5, 9, 42, 43], and stochastic chemical sensing [20, 21]. Nanopores have just begun to impact scientific and technological discovery, and the next major breakthroughs in the development of this platform hinge upon progress made in two parallel avenues of research:

1. Development of a fundamental understanding of the dynamical processes governing biomolecules near and inside nanoscale devices, and of the interplay between various forces and interactions to which biomolecules are subject.
2. Availability of tools and methods allowing routine fabrication and manipulation of materials with nanometer resolution and accuracy.

Here, we will focus primarily on current progress toward a basic description and understanding of the fundamental physical rules governing voltage-driven biopolymer transport through nanoscale pores, because it is this knowledge that uniquely enables the development of future nanopore sensing methodologies. Over the past decade, both large biological membrane channels, like the toxin α-hemolysin, as well as man-made nanopores fabricated in ultra-thin solid-state membranes, have been used to detect, characterize, and manipulate biopolymers. The quest for next generation DNA sequencing methods has also sparked an interest in using nanopores for the analysis of nucleic acids, such as single- and double-stranded DNA and RNA molecules [1, 6, 22, 31, 34, 35, 40]. These studies shed light on the two equally important and essential molecular processes involved in nanopore sensing, namely: biopolymer *capture* and biopolymer *translocation*.

Capture is the process by which polymers (and more specifically, polyelectrolytes) arrive to the vicinity of the pore and are threaded into the pore. This process determines the overall throughput of nanopore sensing. A low capture rate translates to low throughput, which in turn limits the practicality of nanopore applications. On the other hand, an extremely high capture rate would have profound implications for biopolymer sensing, as this would enable the possibility for analysis of minute amounts of sample (e.g. number of DNA copies) [50], while circumventing the need for costly and time-consuming preamplification steps [5].

Translocation describes the dynamics of the biopolymer after it has already entered the pore, as it is progressively transported from the source chamber ("*cis*") to the receiving chamber ("*trans*"). This process dictates the ultimate sensitivity of nanopore detection; an extremely fast translocation speed may result in poor spatial resolution and the inability to detect fine details of any biopolymer side-groups [5]. In contrast, a sufficiently slow and steady translocation process would provide enough time for local averaging over the stochastic noise associated with single molecule sensing, resulting in superior device sensitivity.

Together, these two processes encompass the entirety of a molecule's progress to and through the nanopore. Moreover, as our understanding of these processes continues to develop, novel ways to manipulate and control the capture rate and the translocation speed can be realized. Focusing on solid-state nanopores made in inorganic membranes of silicon nitride, in this chapter we review the basic principles affecting voltage driven DNA capture and DNA translocation. More specifically, we focus here on nanopores roughly 5 nm in diameter, which exclusively allow linear passage of double-stranded DNA. The capture and passage of linearized biopolymers is distinctly different from transport of coiled or even partly folded biopolymers, as the latter may involve multiple conformations or folded states of the biopolymer, all of which are excluded from a purely linear passage. This distinction is extremely important for many prominent applications of nanopores, such as DNA sequencing, genome profiling, or mapping of DNA-bound molecules [5].

In a typical solid-state nanopore experiment, a pair of electrodes is used to apply an electrical potential ΔV far away from the thin insulating membrane separating the two chambers and containing the pore (Fig. 10.2a) [48]. Because the system is suspended in a high ionic-strength electrolyte solution (e.g. physiological conditions), even modest electrical potentials ($\Delta V \sim 0.1$ V) produce sufficiently large local electrical fields in the pore (~10^5 V/cm) to overcome the free energy barrier associated with stretching and threading extremely long, charged biopolymers into the pore [17, 37]. As a result, molecules like DNA and RNA will translocate from the negatively biased *cis* into the *trans* chamber.

As a macromolecule enters the pore, its presence physically excludes a substantial fraction of electrolytes from the pore, causing an abrupt blockade in the ionic current flowing through the pore. Figure 10.2b displays the ionic current just before and after the addition of 400 bp DNA molecules to the negatively biased *cis* chamber, using a 4 nm synthetic pore [49]. Upon addition of the DNA molecules (see arrow) a series of downwards spikes in the ionic current flowing through the pore are clearly visible. Closer inspection of these current "blockades" (Fig. 10.2c) reveals that a large fraction of the current spikes remain at the blocked state for a relatively long dwell-time (t_D). This timescale was found to scale with DNA length, and to reduce exponentially with the applied voltage [49]. Moreover, as we discuss in Sect. 10.4, t_D was also found to decrease exponentially with the pore diameter, d. Specifically, a small change in pore size (e.g. 1 nm or less) results in a considerable change in dwell time. These findings imply that t_D represents the passage time (or the "translocation" time) of each biopolymer from *cis* to *trans*. This was further confirmed by PCR measurements of material from the *trans* chamber [49], similar to previous results proving DNA translocation through protein pores [22].

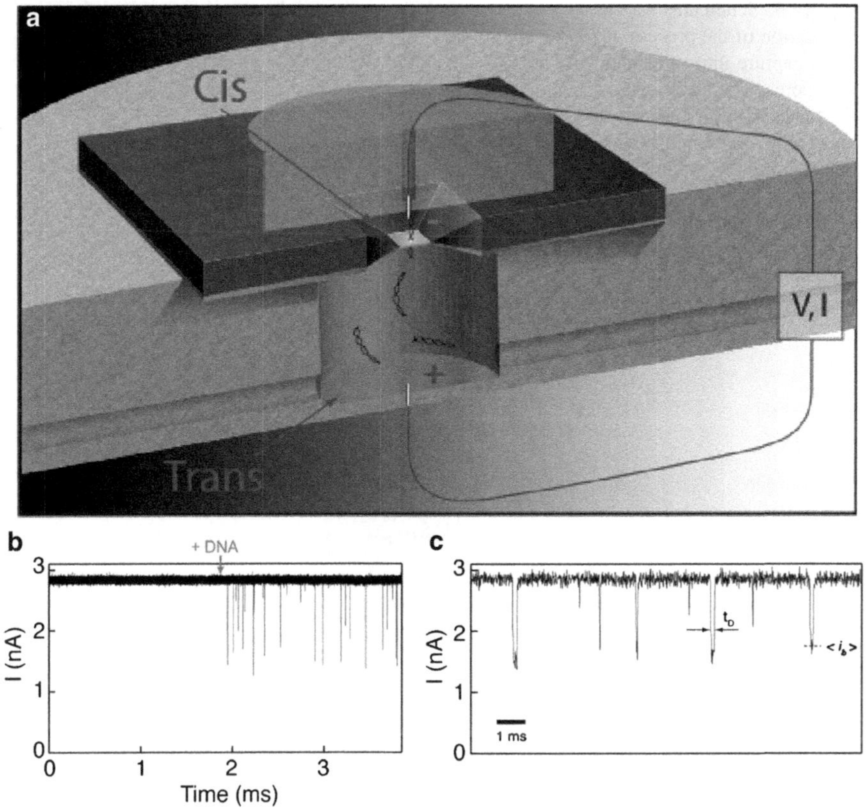

Fig. 10.2 (**a**) Schematic illustration of a solid-state nanopore setup. A silicon wafer contains a thin solid state membrane, where a single ~4 nm pore is fabricated. This membrane separates two tiny fluid chambers (*cis* and *trans*). A pair of electrodes immersed in these chambers are used to apply voltage and measure the ionic current through the pore. (**b**) Ion-current trace just before and after the addition of double-stranded DNA molecules to the negatively biased *cis* chamber. The linear passage of the DNA molecules through the pore causes *discrete current blockades*. (**c**) Magnified view of several current blockades. Reproduced with permission from Wanunu et al. [50], copyright Nature Publishing Group, and from Wanunu et al. [49], copyright Elsevier Inc

10.2 The Problem of DNA Capture

While nanopore sensing involves reading the properties of biopolymers as they move through the pore, the initial process of threading a molecule into the nanopore governs the overall throughput of the method, as described above. DNA capture consists of two steps: arrival of a molecule from the bulk to the mouth of the pore, and threading one end of that polymer into the pore. Studies investigating the capture of single-stranded DNA into lipid-embedded α-hemolysin channels indicate that there is a significant free energy barrier for DNA entry into the pore, which is associated with threading of the first few bases [17, 36, 37]. However, such a barrier was *not*

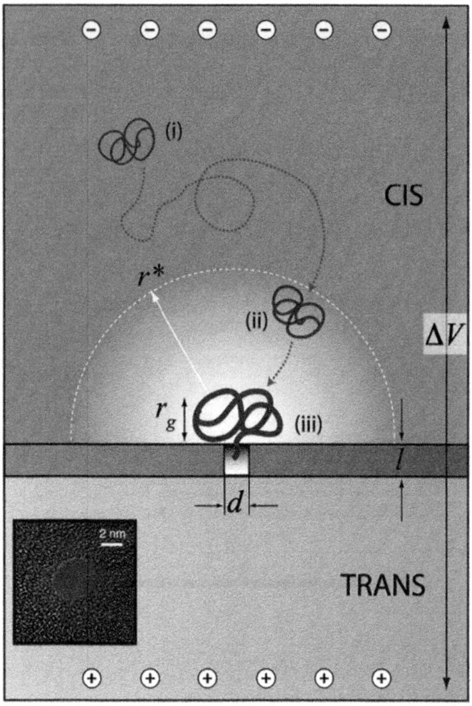

Fig. 10.3 Schematic illustration of the process of DNA capture into solid state nanopores. The electrical field near the pore (shaded area) plays a crucial role in determining the capture dynamics. Reproduced with permission from Wanunu et al. [50], copyright Nature Publishing Group

observed for capture of dsDNA of lengths 4–6 kbp and 48 kbp (λ-phage DNA) into large solid-state pores (diameter > 5 nm) [17, 36, 37, 52]. More recent studies of the capture mechanism of dsDNA capture into sub-5 nm solid-state nanopores have revealed two distinct steps in DNA capture, as illustrated in Fig. 10.3 [50]:

1. As a DNA coil approaches the pore from bulk (i) to some critical radius r^* larger than the coil size r_g, its motion transitions from purely diffusive motion to biased diffusive motion, driven by the decaying electric field outside the pore. This field is maintained by the ion current through the pore, which creates a potential profile $V(r)$ outside the pore mouth that attracts the DNA coil from a distance r^*, where r^* is orders of magnitude greater than Debye screening length scales (0.1–1 nm). Thus, a DNA coil experiencing this field is "funneled" towards the pore mouth (ii).
2. Once the DNA coil approaches the pore to within a distance of approximately one r_g, one DNA end must be threaded into the pore, a process that involves crossing a free-energy barrier (iii).

10.2.1 Theoretical Considerations

The arrival of the molecules from bulk to the pore mouth and the threading of one end into the pore involve independent physical mechanisms. The *rate-limiting step* (the slower of the two steps) determines the observed capture rate in experiments. It is therefore useful to consider the way in which these effects vary with experimentally accessible parameters such as DNA length, voltage and temperature, all of which may substantially affect these two processes. Before we consider these steps in detail, we must obtain a description of the spatial distribution of the electrical potential near the pore, as it plays a central role in both mechanisms.

Under the high ionic strength conditions typically used in nanopore experiments (e.g. >50 mM of monovalent salt), the characteristic Debye screening length is of the order of ~1 nm or smaller. In the absence of an external electric field and with the system in thermal equilibrium, this implies that static electrical charges in the system are *effectively screened*. Therefore, a highly negatively charged object (such as a DNA molecule) will not be electrically attracted to the area around the pore, even if this area is positively charged. However, when an external electrical potential ΔV is applied far away from the nanopore, a finite current density of ions flows through the pore, *drastically* changing the electrical profile of the system. Ohm's law states that at every point in space the electric field lines must follow the electric current lines, i.e. $\vec{j} = \sigma\vec{E}$, where $\vec{j}$ is the current density, $\vec{E}$ is the electric field, and σ is the conductivity (proportional to the local salt concentration). The steady-state ion-current flow through a cylindrical pore of diameter d and length l dictates a long-range power-law dependence of the electric potential in and around the pore surface (see the Supporting Information section in Wanunu et al., 2010 for derivation) [50]:

$$V(r) = \frac{d^2}{8lr}\Delta V. \tag{10.1}$$

This profile decays as $1/r$ with the distance r from the pore.

In the negatively-biased chamber, (10.1) describes an attractive, funnel-shaped potential landscape for the negatively-charged DNA coil. Far away from the nanopore the DNA motion is purely diffusive, because the electrostatic forces pulling the DNA towards the pore are negligible in comparison with the thermal forces randomizing its motion. Closer to the pore, DNA diffusion begins to be biased along the potential gradient lines by the increasingly strong electric field. The resulting motion may be described by the space-dependent drift velocity $v(r) = \mu\nabla V(r)$, where μ is the DNA electrophoretic mobility. It is well known that the electrophoretic mobility of DNA does not depend on its length for molecules longer than a few persistence lengths ($l_p \approx 50$ nm) [38, 39]. This is the result of an opposing electrophoretic drag force; while the charge of the DNA does scale linearly with the number of base pairs N, the corresponding electrostatic force $-Q^*\vec{E}$ is almost entirely canceled by the opposing force transmitted to the DNA through hydrodynamic drag of K^+ counterions flowing in the other direction through the DNA coil [27]. Therefore, there is a regime in the vicinity of the pore where short as well as long DNA molecules are expected to be equally mobilized by the long-range potential $V(r)$ [50].

To estimate the effective range where DNA motion is biased towards the pore, we evaluate the critical radius r^* from the nanopore where biased diffusion (for $r<r*$) dominates over unbiased diffusion (for $r>r*$), as shown in Fig. 10.3 [50]. If the DNA coil is positioned at a distance r from the pore, then pure diffusion over this distance would require time r^2/D, which corresponds to a speed $r/(r^2/D) \sim D/r$.

This must be compared with the electrophoretic speed $\mu E(r)$. Biased diffusion dominates when $\mu E(r) > D/r$; or more simply, when $\mu V(r)/D > 1$. Substituting this into (10.1) we obtain:

$$r^* = \frac{d^2 \mu}{8lD} \Delta V. \tag{10.2}$$

Note that r^* grows with DNA length because the diffusion constant D decreases with length. In other words, longer DNA molecules begin to sense the nanopore farther away from the pore than shorter DNA molecules. However, the rate R_{diff} at which DNA coils randomly and enters a hemisphere of radius r^* is proportional to their diffusion constant, and is given by Smoluchowski theory as $R_{diff} = 2\pi Dr$. Combining this with (10.2), we find:

$$R_{diff} = 2\pi D r^* = \frac{\pi d^2 \mu}{4l} \Delta V. \tag{10.3}$$

Thus, in the regime where diffusion to the pore is the rate-limiting step, the capture rate should be *independent of DNA length*. This result may seem counterintuitive, since free diffusion for longer DNA is inherently slower than diffusion for shorter DNA, and the Smoluchowski equation dictates that R_{diff} is proportional to D. However, the effective capture radius r^* grows with D^{-1}, which cancels out this length dependence. This result predicts the diffusion-limited "current density" of DNA translocating through the pore to be $J_{diff} = R_{diff}c$, where c is the bulk DNA concentration [50].

The DNA translocation rate is only diffusion-limited for a certain range of DNA lengths. For other lengths, the rate-limiting step occurs when the polymer is threaded into the pore (step (iii) in Fig. 10.3) [50]. The confinement of the DNA end, as well as possible unfavorable interactions of the highly charged DNA with the pore itself, creates a free energy barrier to capture. This barrier was experimentally observed for DNA transport through the 1.5 nm protein pore α-HL [17, 36], characterized by an exponential dependence of the capture rate on voltage that could be explained theoretically: When DNA capture is governed by an energy barrier, its rate, according to classical Kramers theory, can be written in the form $J = Rc = \omega \exp[(q\Delta V - U)/k_B T]$ [52], where U is the height of the threading barrier in the absence of applied voltage and q is the effective charge of a DNA end segment, which is DNA length-independent. The pre-factor ω in the expression above is usually interpreted as the threading attempt rate. Since in stage (iii) the DNA coil is at or very near to the mouth of the pore, the biasing effect of the local potential $V(r)$ on this attempt rate must be considered. Specifically, we find that the electrical bias leads to two capture enhancement mechanisms [50]:

1. An exponential attempt rate enhancement: The potential well $V(r)$traps the DNA a distance $\sim r_g$ from the pore mouth, where threading is repeatedly attempted. Due to the energy barrier, DNA molecules are delivered to the pore mouth multiple times before a successful translocation occurs. Therefore, while the probability of finding a DNA coil within a distance r_g from the pore mouth

is cr_g^3 in the absence of $V(r)$, the presence of $V(r)$ enhances this probability by the exponential factor $\exp(Ne\alpha/k_BT)$, where α is a constant related to the fact that not all phosphates on the DNA are ionized due to Onsager-Manning condensation.

2. When the DNA coil is placed at the pore mouth, the probability of successful end threading into the pore is determined by the internal dynamics of the coil. While the coil relaxation time may be estimated by its Zimm time (τ_{Zimm}), its dynamics are affected by the potential $V(r)$, which provides further bias of DNA segments towards the pore and thus, enhances the attempt rate of DNA end threading. Together, these two effects result in an increased capture rate R_{bar}:

$$R_{bar} = \frac{r_g^3}{\tau} \exp\left[\frac{q\Delta V - U}{k_BT} + \frac{e\Delta V}{k_BT}\frac{\alpha d^2}{al}\sqrt{\frac{N}{4N_p}}\right]. \tag{10.4}$$

Here, τ is proportional to $\tau_{Zimm}(kT/e\Delta V)(8la/d^2)$, $N_p \approx 150$ is the number of base pairs in a dsDNA persistence length, and a is the length per base pair of dsDNA. We consider (10.4) to be valid only for sufficiently long DNA, i.e. $N{>}4N_p$. Since $\tau_{Zimm} \propto r_g^3$, the only dependence on DNA length in (10.4) appears in the second term of the exponent, $R_{bar} \sim e^{C\sqrt{N/4N_p}}$. Thus in the threading-limited regime, the translocation rate should increase with DNA length [50].

To summarize, the dynamics determining capture rate of double-stranded DNA into small nanopores involves two steps: first, a transition from pure diffusion to biased diffusion that funnels DNA coils toward the pore region driven by $V(r)$ (10.1), and second, a threading stage occurring at the mouth of the pore, which involves crossing an energy barrier [50]. Equation (10.3) predicts that the first step is *DNA length-independent*, while (10.4) shows that in the second step the translocation rate *grows with increasing DNA length.* Since the DNA capture rate measured in experiments is limited by the slower process of the two steps, these equations predict that when the overall rate is limited by the diffusion, the capture rate will be length-independent and the rate will grow linearly with the applied voltage ΔV. In contrast, if the overall rate is limited by crossing the energy barrier associated with threading, then we expect to see a growing capture rate with length, and exponential growth with ΔV.

10.2.2 Experiments Measuring the DNA Capture Rate

DNA translocation data are typically acquired using custom software that either collects a continuous current recording, or detects and records only the current pulses in real time [3]. A DNA sample is typically characterized by statistical analysis of the square translocation pulse depths (current blockage level, I_B) and widths (dwell-times, t_D) for thousands of molecules, as shown in Fig. 10.4 [49]. Capture rates were calculated for each experiment from the mean time-delay between two successive events (δt in Fig. 10.4). To obtain a reliable measure of the capture rate, thousands of

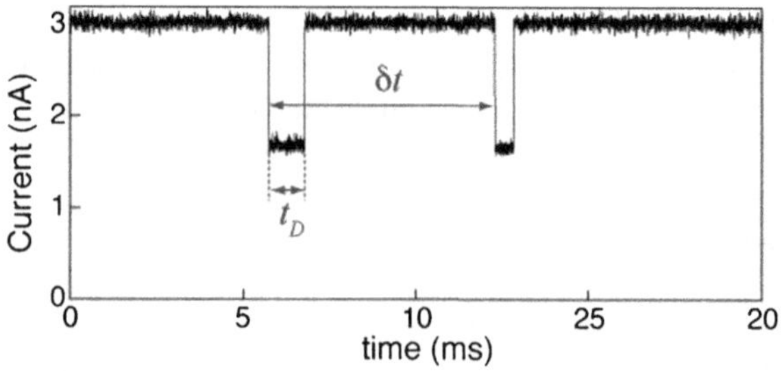

Fig. 10.4 Ion-current trace through a 4 nm pore, where two DNA translocation events are observed. The transport time of the molecule is denoted as t_D, whereas the time-delay between two successive events is denoted as δt. Reproduced with permission from Wanunu et al. [50], Copyright Nature Publishing Group

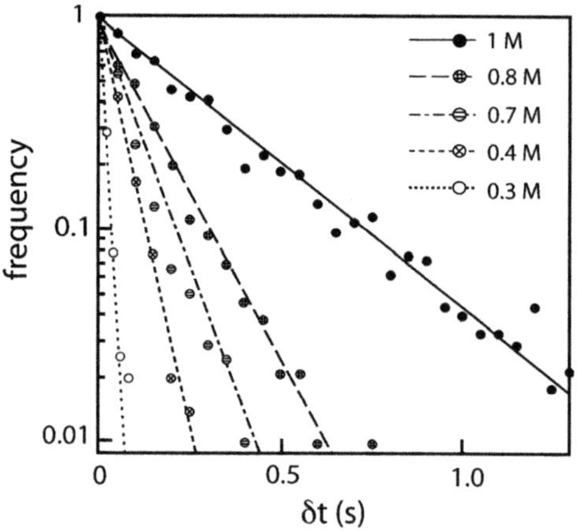

Fig. 10.5 Normalized capture rate distributions of 2,000 bp DNA as a function of the *cis* chamber KCl concentration as indicated ($[\mathrm{KCl}]_{\mathrm{trans}} = 1$ M for all cases). Reproduced with permission from Wanunu et al. [50], Copyright Nature Publishing Group

events were acquired and the distributions of their δt values evaluated. Due to the stochastic nature of the capture process, as long as the DNA sample is sufficiently dilute, these distributions should obey single-exponential decays, from which the slope R_c can be taken as the mean capture rate. Figure 10.5 displays a typical semi-log time-delay distribution measured for 2,000 bp DNA, where the KCl concentration in the *trans* chamber was kept at 1 M, and the *cis* chamber KCl concentrations were decreased to the indicated concentrations. For both symmetric and asymmetric salt gradients, we find that the capture rate scales linearly with DNA bulk concentration, *c*, as shown in Fig. 10.6 [50].

The dependence of the specific capture rate on DNA length using a 4 nm solid-state nanopore at an applied voltage of 300 mV is displayed in Fig. 10.7. The data clearly show two behavior regimes: For DNA molecules in the range ~400–8,000 bp, the capture rate per nM of DNA molecules *increases* with *N*, whereas for molecules longer than ~8,000 bp the capture rate is nearly *independent* of length [50]. These results support the hypothesis that for $N < 8,000$ bp the capture

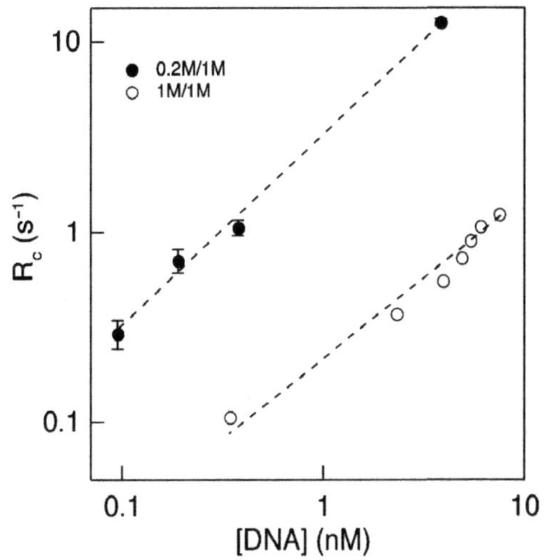

Fig. 10.6 Log–log plots of the capture rates of a 400 bp fragment into a 4 nm pore as a function of DNA concentration, under symmetric and asymmetric salt conditions (slopes of lines ~1). Reproduced with permission from Wanunu et al. [50], Copyright Nature Publishing Group

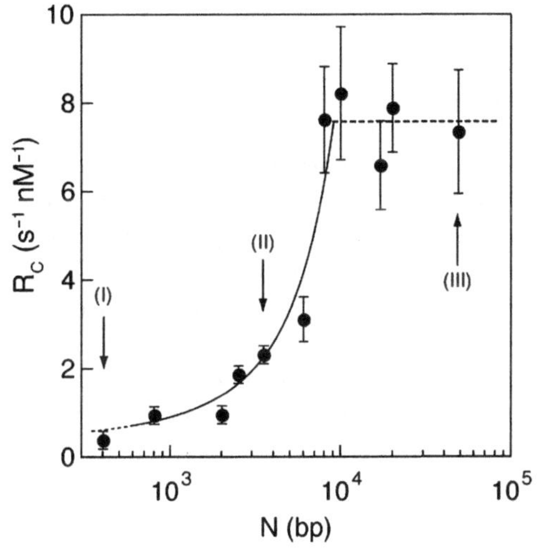

Fig. 10.7 The dependence of the specific capture rate of dsDNA into 4 nm solid-state nanopores, as a function of DNA length. A transition from length-dependent to length-independent regimes is observed at ~8,000 bp. Reproduced with permission from Wanunu et al. [50], Copyright Nature Publishing Group

process is dominated by a free energy barrier, described by (10.4). This is indeed confirmed by an excellent fit to the data (shown by the solid line). Similarly, in the range 8,000–48,000 bp, we observe length-independent behavior, indicating that this regime is diffusion-dominated, as described by (10.3) (dashed line).

The voltage dependence of the capture rate was also studied in these two regimes. Figure 10.8 displays the results for three DNA lengths: 400, 3,500 and 48,000 bp (I–III respectively). For molecules shorter than ~8,000 bp we observe an exponential dependence of R_c on V for voltages <450 mV [50]. These results support the hypothesis that in this length regime, the capture process is dominated by the threading energy barrier process, described by (10.4). In contrast, for both

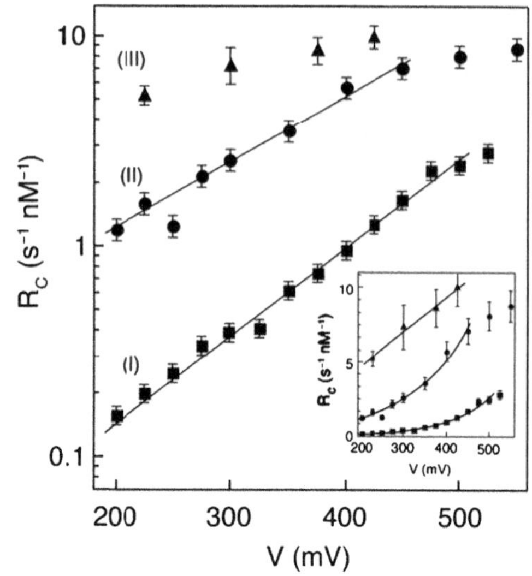

Fig. 10.8 The dependence of the specific capture rate of dsDNA on voltage. DNA shorter than ~8,000 bp displays an exponential dependence on *V* and longer DNA displays a linear dependence on *V*. A linear-linear plot of the data is shown in the *inset* to better illustrate these dependencies. Reproduced with permission from Wanunu et al. [50], Copyright Nature Publishing Group

longer DNA molecules we observe a weaker, linear dependence on voltage (III), which is predicted for a diffusion-limited process (10.2).

The dependence of capture rate on *N* for short molecules in sub-5-nm pores represents a unique example of capture selectivity towards longer DNA molecules. Although DNA mobility in free-flow electrophoresis is not length-dependent, here the rate-limiting step is R_{bar}. Thus, in this regime longer DNA molecules are more efficiently threaded than short ones, as their higher charge is advantageous for capture into the pore mouth. Moreover, the threading selectivity has a very steep dependence on DNA size for molecules in the range 800–8,000 bp [50]. For longer molecules, diffusion to the pore is the rate-limiting step, so the corresponding translocation rate shows no length dependence. Both regimes are also apparent in voltage studies, which show that R_c is exponentially dependent upon V for shorter molecules, but linearly dependent for very long molecules (Fig. 10.8).

10.2.3 Capture Enhancement Using Manipulated Fields

The strong impact of $V(r)$ on the capture rate suggests that manipulation of the voltage profile outside the pore is potentially useful for increasing throughput. One effective way to alter the potential profile is the application of an ion gradient across the pore. This results in selective "pumping" of the positive ions from the *trans* to the *cis* chamber, in the opposite translocation direction of the negatively charged DNA [50]. As it turns out, this phenomenon has two positive outcomes for nanopore-based platforms: First, the higher density of cations in the vicinity of the pore on the *cis* side increases the attraction of DNA to the pore (it increases the magnitude of $V(r)$), resulting in a significantly higher capture rate. Second, additional hydrodynamic

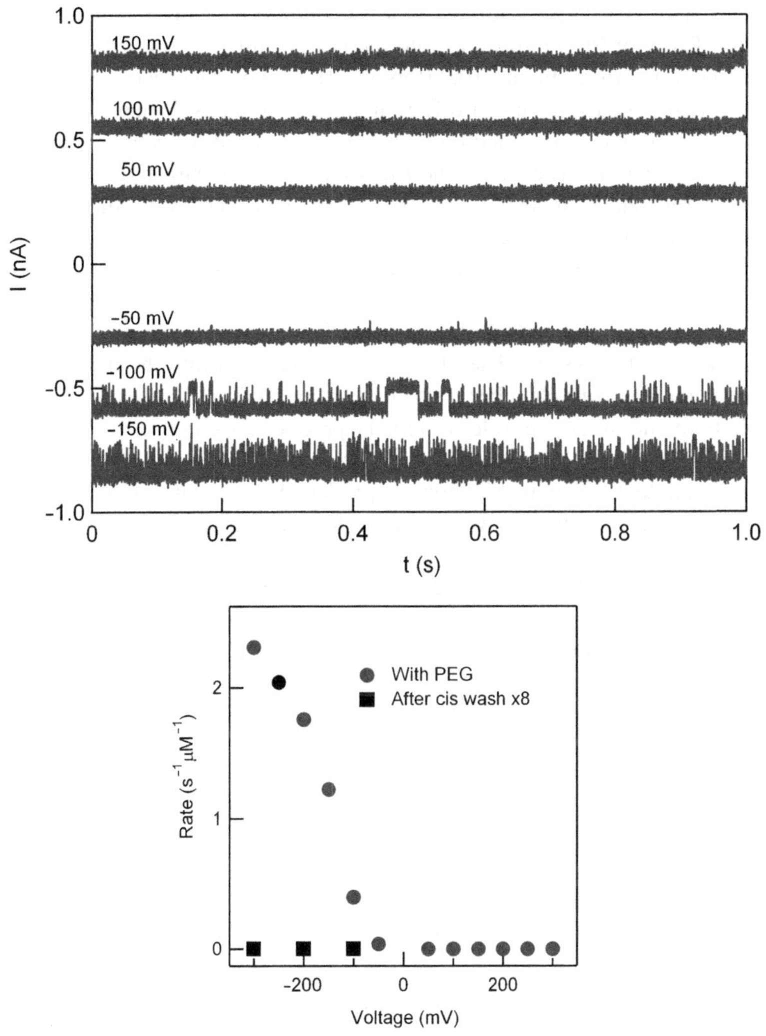

Fig. 10.9 Ion-current traces showing PEG entry into a 4 nm solid-state pore at varying voltage V applied to the *trans* chamber (*top*). Entry rate as a function of voltage (*bottom*). Reproduced with permission from Wanunu et al. [50], Copyright Nature Publishing Group

drag during translocation, which is associated with cation flow in the opposite direction to DNA movement, slows the DNA and increases the translocation time.

Experiments using the uncharged polymer polyethylene glycol (PEG) were performed in order to verify that the direction of electroosmotic flow (EOF) in SiN nanopores is governed by the cations, and therefore is directed towards the negatively biased electrode [50]. In Fig. 10.9, we show a set of ion current traces after PEG (MW = 12,000) was added to the *cis* chamber at a final concentration of 415 µM (small enough to avoid creating an appreciable osmotic flow), under symmetrical 1 M/1 M KCl buffer (pH 8.5) conditions, using a nanopore 4 nm in diameter. Entry of PEG molecules into the pore resulted in brief stochastic blockades of the ion current. Figure 10.9 (top panel) displays one second of typical gap-free recorded ion currents under different applied voltages. When positive voltage was applied to the *trans* electrode, no ion current blockades were observed. Reversing the voltage (i.e. when the *trans* chamber was negatively biased), clear stochastic blockades appeared, the frequency of which grew sharply with the voltage magnitude. The lower panel of Fig. 10.9 displays the blockade event rate as a function of voltage applied to the *trans* electrode over a range from -300 to +300 mV. Clearly, PEG entry to the pore is only observed at negative voltages. Entry of the uncharged PEG molecules into the nanopore is driven by the EOF of cations (K^+) in this system. Therefore entry of the PEG into the nanopore is only feasible when the *trans* chamber is negatively biased. These measurements suggest that EOF in the silicon nitride system is produced by *cations*, and is directed towards the negative electrode.

Given that the EOF in the SiN pores is produced by the cations, we next examine the effect of salt gradients (which can either enhance or suppress the flow of cations) on the capture rate of negatively charged DNA. Using one nanopore, the capture rate R_c of 400 bp DNA was measured using three different *trans/cis* salt concentrations and the same *cis* DNA concentration (3.8 nM), shown in Fig. 10.10a [50]. The top trace corresponds to 1 M/1 M *trans/cis* KCl concentrations, the middle trace to 1 M/0.2 M, and the lower trace to 0.2 M/1 M. Continuous data streams were collected for each gradient, and representative 2-s snapshots are shown. Strikingly, when the *cis* concentration was lowered from 1 to 0.2 M, the capture rate increased ninefold from ~0.4 to ~3.7 s^{-1} nM^{-1}. Reversal of the conditions (to 0.2 M/1 M) effectively suppressed DNA capture. These results were completely reversible, i.e. returning to 1 M/1 M yielded the same capture rate as was previously obtained for those concentrations. Figure 10.10b shows the enhancement in R_c for several other DNA lengths (400, 2,000, 3,500 and 8,000 bp), over a large range of asymmetries (C_{trans}/C_{cis} from 1 to ~32). Clearly, above a threshold value of roughly $C_{trans}/C_{cis} \sim 1.5$, a roughly linear, length-independent increase in the capture rate is observed. In addition to providing rate enhancement, these asymmetric salt conditions are particularly advantageous because they allow experiments to be performed where DNA is kept under physiological conditions (i.e. 125 mM KCl), while still preserving the signal-to-noise ratio of an experiment performed at high salt concentration.

Under asymmetric salt conditions, the electrical potential near the pore is proportional, to first approximation, to the ratio of bulk ionic concentrations in the *cis* and *trans* chambers: $V(r) = V_{sym}(r)C_{trans}/C_{cis}$, where $V_{sym}(r)$ is the potential under symmetric conditions. Accordingly, the value of r^* is also modified to $r^* = r^*_{sym}C_{trans}/C_{cis}$ [50]. The rationale for this is that the voltage drop in the solution of lower conductance is higher, and therefore a high/low *trans/cis* salt gradient will yield an asymmetric potential profile, with more of the voltage drop occurring within the *cis* chamber than in the *trans* chamber. Finite-element Comsol simulations confirm that salt-dependent charge imbalance at the pore causes the potential to protrude significantly farther into the *cis* chamber for

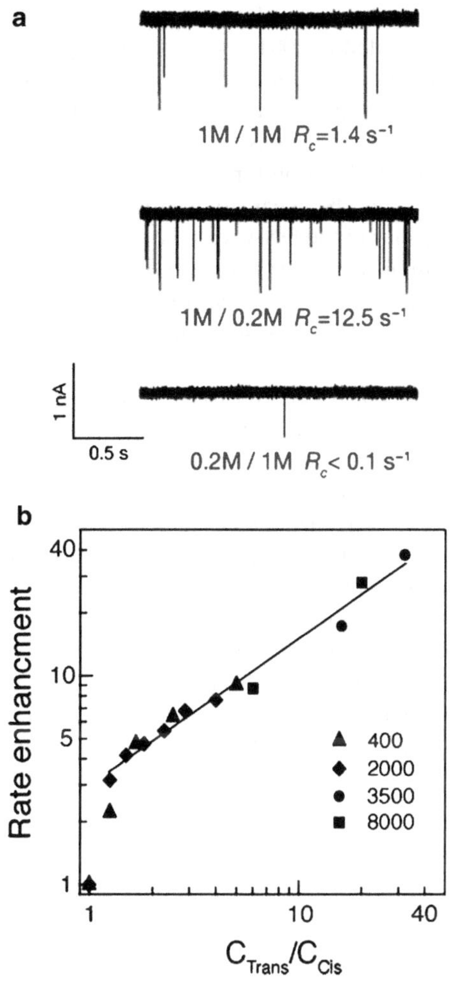

Fig. 10.10 Ion-current vs. time traces showing translocation frequency for different salt concentration ratios between the *cis* and *trans* chambers (*top*). Capture rate as a function of salt concentration ratio (*bottom*). Reproduced with permission from Wanunu et al. [50], Copyright Nature Publishing Group

large C_{trans}/C_{cis} ratios than for symmetric salt conditions [50]. Notably, for $C_{trans}/C_{cis} = 20$ (4 M/0.2 M KCl), $V(r)$ extends *hundreds of nanometers* away from the pore before dropping below 50% of its maximum.

10.3 Voltage Driven DNA Translocation

The process of DNA transport through small pores is intriguing both from a fundamental perspective of investigating polymer behavior in confined volumes, as well as from a more practical side, which translates the ion current collected during polymer passage to its local structure and properties. Since solid-state nanopores were first reported in 2001, the vast majority of publications have been focused on DNA translocation through large nanopores (10–20 nm), in which typical transport times correspond to average DNA speeds of 10–50 ns / bp, largely governed by hydrodynamics of the DNA polymer [7, 13, 26, 45]. Here we focus on the dynamics of DNA translocation through nanopores with sub-5 nm diameters, where dsDNA transport must be linear (folded molecules cannot enter small pores). In this regime, interactions of the DNA with the pore walls significantly slow down its average translocation speed, (0.1–1 μs/bp) [2, 12, 16], thus enhancing the resolution of the method (electrical measurements of low ion current levels are not practical below ~1 μs due to noise), allowing the investigation of a broader range of DNA lengths.

As shown in Fig. 10.1, our system consists of a single solid-state nanopore in a SiN membrane [23, 25, 44]. DNA translocation is characterized by recording ion current as a function of time, and by analysis of the distributions of two main parameters, t_D and I_B, which are the transport time and fractional pore current, respectively ($I_B = \langle i_b \rangle / \langle i_o \rangle$, where i_b is the mean blocked current and i_o is the mean open pore current). In Fig. 10.11, we display I_B vs. t_D scatter plots measured for 8,000 bp DNA through nanopores with two indicated diameters ($V = 300$ mV, $T = 21$ °C) [49]. Three salient features emerge upon decreasing the pore size from 8 to 4 nm: (1) An increase in the contrast of the signal in the 4 nm pore (highlighted by decreasing values of I_B from 0.9 to 0.5). (2) For the 8 nm pore, DNA can enter in both folded and unfolded configurations, whereas in the 4 nm pore, only unfolded events are observed. The ion current distributions of the 8 nm pore exhibit multiple peaks, whereas the 4 nm pore exhibits a single peak (Fig. 10.11 right-hand panel).

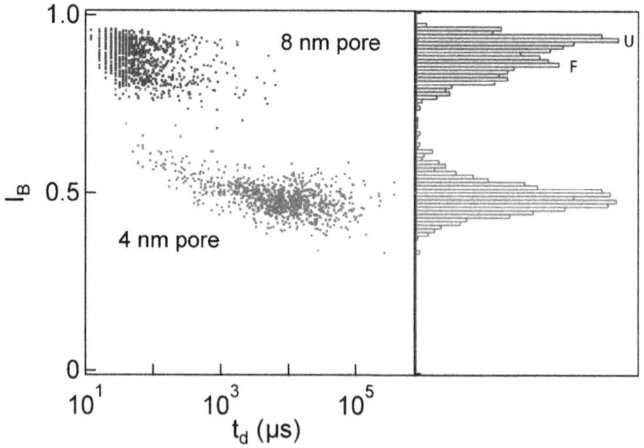

Fig. 10.11 Translocation scatter plots of 8 kbp dsDNA through 8 and 4 nm solid-state nanopores ($V = 300$ mV). The I_B histograms are given on the *right hand panel*, where for the 8 nm pore, both folded (F) and unfolded (U) events are shown. Reproduced with permission from Wanunu et al. [49], Copyright Elsevier Inc

(3) The mean values of t_D are nearly two orders of magnitude *longer* for the 4 nm pore, suggesting that the DNA experiences slower translocation through the smaller pores.

The last feature is of great interest for nanopore applications in DNA analysis: while in the 8 nm pore the transport timescales approach the minimum measurable dwell time allowed by the electrical bandwidth, in the 4 nm case, the longer dwell times results in significantly larger total ion current flux during DNA transport, and a proportional reduction in the measurements' shot noise [49]. This section is divided into several parts: First, we discuss the nature of the dwell-time distribution shapes obtained for DNA translocation through sub-5 nm pores. We then discuss in more detail the effect of pore size, applied voltage, and temperature on the DNA transport times. Finally, we examine some possible explanations for the observed dynamics, and conclude with the effect of salt concentration on the transport dynamics.

10.3.1 Transport Time Distributions: The Effect of DNA–Pore Interactions on the Distribution Shape

DNA translocation through nanopores has been a subject of numerous theoretical studies. Lubensky and Nelson have predicted for transport of single-stranded DNA through 1.5 nm protein pores that the first-passage time distribution, or the probability density function for polymer translocation, is given by a non-Gaussian form for a strongly interacting pore [29]. This form is characterized by a peak at short times (the most probable translocation time) and a tail for longer times (Fig. 10.12a).

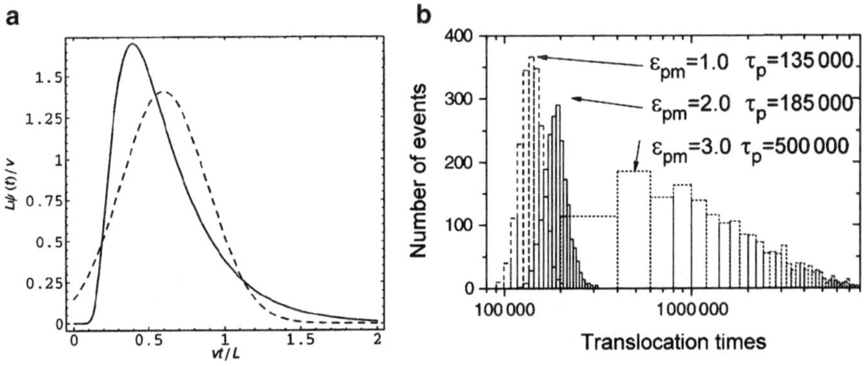

Fig. 10.12 (**a**) Analytically-derived first-passage dwell-time distribution $\psi(t)$ (dimensioned by L/v) for a polymer of length L (time axis is re-dimensioned by v/L) translocating through a small pore (*solid line*), and a Gaussian distribution (*dashed line*) with the same mean and variance for comparison. (Reprinted from Lubensky and Nelson [29], copyright 1999, with permission from Elsevier.) (**b**) Simulated dwell-time distributions for a polymer translocating through a pore (*solid line*), at different values of polymer–pore interactions (ε_{pm}). (Reproduced with permission from Luo et al. [30], Copyright (2007) by the American Physical Society)

This asymmetric distribution shape is the result of a series of potential wells that the polymer must overcome during its passage through the pore (i.e. from a random or periodic set of interaction sites with the pore along the polymer). In large pores, where the DNA is less confined, the contribution of interactions with the pore is reduced, and the DNA undergoes fast transport. In smaller pores, the DNA driven through the pore is configurationally restricted, thereby forcing it to repeatedly make contact with the inner pore structure during its passage. These interactions dominate the biopolymer dynamics through small pores. Another effect that effectively reduces the driving force of the DNA through small pores is the electroosmotic ion current along the DNA backbone (in the direction opposite to DNA motion), which slows DNA by generating hydrodynamic drag on the translocating polymer [27].

To investigate the effect of polymer–pore interactions on the width of the distribution tail, Luo et al. used Langevin dynamics to model a 2D polymer translocating through a nanopore [30]. In their work, the DNA chains were represented as bead-spring chains held together by a Lennard-Jones potential, with additional constraints accounting for the polymer's elasticity and for excluded volume interactions. Polymer–pore interactions were modeled by placing a Lennard-Jones potential between the pore and the beads. In Fig. 10.12, we show their numerically-obtained translocation distributions for this polymer as a function of the pore–monomer interaction parameter ε_{pm}. As may be seen in these distributions, increasing interactions result in increased peak dwell times (τ_p). Even more strikingly, increasing interactions also increase the width of the dwell-time distribution. Increasing pore–polymer interactions reduce the effective driving force of the DNA in the pore, which in turn increases its dwell time. However, the increasing number of possible conformations the polymer can assume as it resides in the pore for a longer time results in broad distributions. In fact, for $\tau_p = 3.0$, while the most probable dwell time is $\tau_p = 500{,}000$, the mean dwell time is much higher than this value. In the limit of very strong pore-polymer interactions, this timescale can be estimated by fitting the tail of the dwell-time distribution, i.e. $P(t, t>\tau_p)$ to a mono-exponentially decaying function $P(t) \sim A\exp(-t/t_d)$, where, t_d is the mean translocation timescale.

In Fig. 10.13 we show the experimentally measured dwell-time distributions for a 6,000 bp DNA sample through a 4 nm pore [49]. To fully characterize the translocation process, statistical analysis of thousands of DNA translocation events is necessary. The shape of the distribution resembles the first-passage time distribution shown in Fig. 10.12: the probability of translocation is close to zero at the very shortest times, then sharply increases for slightly longer times, reaching a clear peak at $t_\mathrm{p} \sim 200$ μs. Unlike the theoretical model, the experimental distribution exhibits a broad decay at the longer times. A systematic study of different DNA lengths revealed that the distributions clearly follow bi-exponential distributions, with two distinct timescales [49]. The solid line illustrates the typical features of the translocation distribution, with the sharp increase followed by an exponential decay with a characteristic timescale, here $t_1 = 1.4$ ms. We have not shown the actual fit here for clarity, despite the fact that the t_2 population becomes the dominant one for DNA lengths greater than 3,500 bp (see Wanunu et al. [49] for further discussion of t_2).

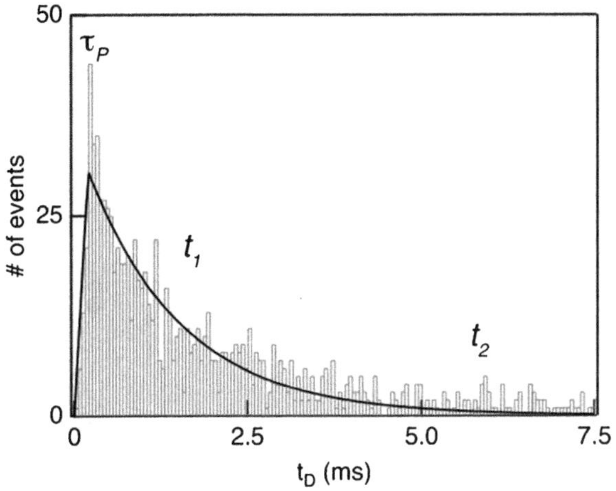

Fig. 10.13 Dwell-time distribution, shown in a linear-linear scale for translocations of a 6,000 bp DNA fragment (for a 4-nm pore at 21°C, 300 mV applied voltage). The distributions obey first-passage time distributions in the limit of strong pore–polymer interactions. The mean transport time can be well-approximated by fitting an exponentially decaying function to the tail of the distribution (i.e. $t > \tau_P$). The emergence of a second dominant timescale ($t2_2$) is visible for longer DNA molecules. Reproduced with permission from Wanunu et al. [49], Copyright Elsevier Inc

10.3.2 Effect of Nanopore Size on DNA Transport

One obvious way to control the polymer–pore interactions is to fine-tune the pore size. Control over the nanopore size can be achieved with the sub-nm precision fabrications capabilities of a TEM [23, 25, 44]. This facilitated an investigation into the effect of nanopore size on its transport dynamics. This investigation involved preparing a set of nanopores in the range of ~2.5–5 nm in diameter, and measuring the typical dwell times of double-stranded DNA in each pore, as explained above [49]. As shown in Fig. 10.14, a decrease of over an order of magnitude in dwell time was observed when the pore diameter was increased by a mere ~2.3 nm. This increase is much larger than the expected increase based on viscous drag alone, illustrating the dominance of DNA–pore interactions in determining the biopolymer's dynamics. Additionally, the collision timescale appears to be only mildly affected by the pore size, as shown by the open circles in Fig. 10.15. Since colliding molecules do not make extended contact with the pore, logically pore size should not affect their dynamics as much as it affects actual translocations.

10.3.3 DNA Length Dependence on the Transport Dynamics

As explained above, the dwell-time distributions show two distinct timescales (t_1 and t_2), obtained by fitting bi-exponential functions to the distributions. Notably,

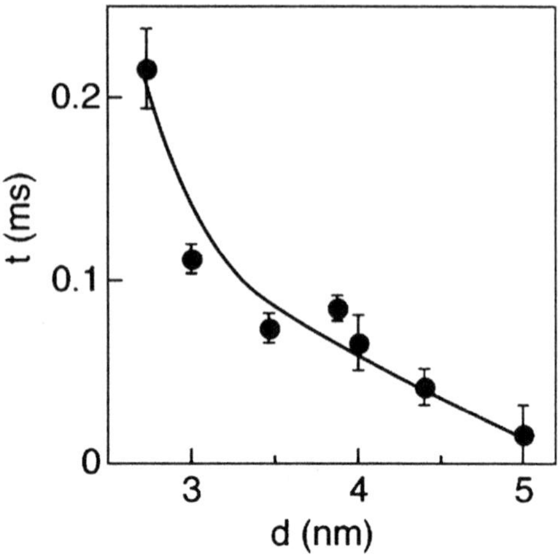

Fig. 10.14 Plots of the translocation timescale (*solid circles*) for 400-bp DNA as a function of nanopore diameters (*d*) in the range 2.7–5 nm. The line is a guide to the eye. Reproduced with permission from Wanunu et al. [49], Copyright Elsevier Inc

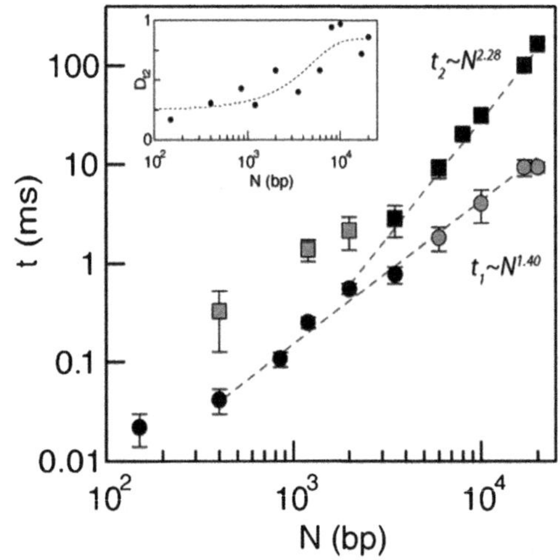

Fig. 10.15 Log–log plot of DNA translocation timescales as a function of DNA length (*N*) measured using a 4-nm pore. Bi-exponential fits to the dwell-time distributions yield two timescales: t_1 (*circles*) and t_2 (*squares*) which follow power laws with $\alpha_1 = 1.40$ and $\alpha_1 = 2.28$ respectively. The inset displays the dependence of the relative weight (the fraction of events in the t_2 population vs. the t_1 population) as a function of DNA length. Reproduced with permission from Wanunu et al. [49], copyright Elsevier Inc

at the limits of both short and long DNA molecules, the dwell-time distributions are mostly mono-exponential, with t_1 dominating the short molecules, and t_2 dominating for long molecules. Intermediate DNA lengths require the use of two exponentials. Figure 10.15 displays a summary of characteristic translocation times for 11 different DNA lengths from ~150 to 20,000 bp, measured using a 4 nm pore at room temperature with an applied bias of 300 mV [49]. For a DNA molecule shorter than

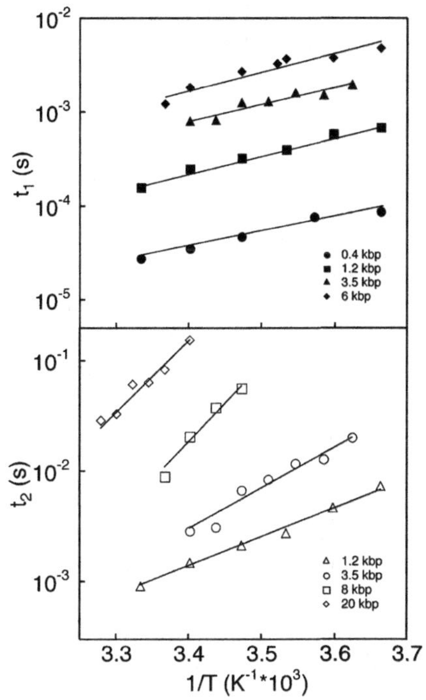

Fig. 10.16 The dependence of bath temperature on DNA transport dynamics through 4 nm pores. The *top graph* shows a semi-log plot of t_1 for selected DNA lengths <6 kbp. The Arrhenius slope of the t_1 timescale (~12 K_BT) is length independent, suggesting that the dynamics are governed by interactions within the pore. In contrast, the t_2 timescale (*lower graph*) exhibits increasing temperature dependence for increasing DNA lengths, suggesting external interactions of the DNA coil. Reproduced with permission from Wanunu et al. [49], Copyright Elsevier Inc

3,500 bp, the prominent timescale (t_1) follows a power-law dependence on N, where the power is equal to 1.40. For longer polymer lengths where t_2 dominates, a much stronger power-law dependence emerges: $t_2 < N^{2.28}$. The striking shift from t_1-dominated to t_2-dominated transport dynamics can be attributed to additional interactions external to the pore that occur for longer DNA coils, either with the silicon nitride membrane or with itself.

Temperature dependence studies of biopolymers' dynamics provide further evidence for the dominant role of interaction in determining translocation timescales. A study of the temperature dependence of translocation time for selected DNA lengths revealed nearly a tenfold increase in the characteristic t_1 and t_2 timescales when the temperature was varied from 30 to 0°C [49]. Figure 10.16 shows a semi-log plot of their mean transport time as a function of T^{-1}. A simplified Arrhenius model for the temperature dependence $t_D = Ae^{\Delta G/k_BT}$ yields similar effective energy barriers for all DNA lengths studied: ΔG ~ 12.0 $\pm$ 0.5 k_BT, or 7.1 $\pm$ 0.3 kcal/mol for t_1. The invariance of ΔG with N for the t_1 timescale suggests that interactions within the pore dominate its dynamics, since these are not expected to depend on the external coil length. In contrast, the t_2 timescale displays increasing ΔG values for increasing N; with ΔG ~ 18 $\pm$ 1, 25.5 $\pm$ 1, 48 $\pm$ 4, and 45 $\pm$ 2 k_BT for 1200, 3500, 8000 and 20,000 bp, respectively. Notably, the slowing down observed with reduced temperature in both t_1 and t_2 *cannot* be attributed to increased fluid viscosity, as the viscosity was expected to increase by only a factor of ~2.7 over the same temperature range.

We observed that for the longer DNA molecules, the increased Arrhenius slopes for increasing DNA lengths in the t_2 regime can be qualitatively explained by considering the environment outside the pore [49]. Recall that the potential outside the pore exerts a significant force on the DNA coil outside the pore, effectively squeezing it towards the membrane. The force on the coil results in stronger interactions with the membrane that grow with the biopolymer length, increasing the effective barrier for DNA translocation. We note, however, that although this qualitative explanation is consistent with empirical observations, it has yet to be modeled theoretically.

In addition to interactions with the interior walls of the pore, another important factor which alters DNA dynamics arises from hydrodynamic interactions of the confined DNA with counterion electroosmotic flow. The same field which electrophoretically drives negatively-charged DNA through the pore also produces cation flow in the opposite direction, which exerts a drag force on the biomolecule. Luan and Aksimentiev have recently employed molecular dynamics (MD) simulations to address this question [28]. The main features of their MD studies are schematically shown in Fig. 10.17. In bulk electrolyte solution, as well as in nanopores, motion and forces of DNA are modeled by considering 20 bp DNA [poly(dA)20-poly(dT)20] in 0.1 M KCl solution under an applied electric field. The electrical force applied to the DNA was balanced by connecting the DNA to a weak harmonic spring with a known spring constant, simulating optical tweezers. Bead displacement was used to estimate the overall force, revealing that force varied linearly with the applied electric field, with a proportionality constant of 0.25e.

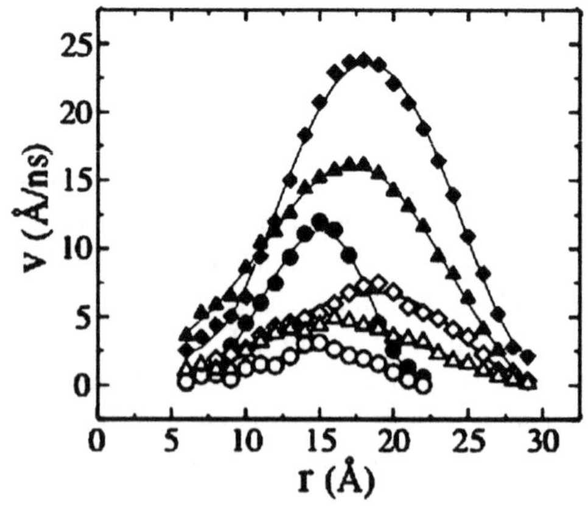

Fig. 10.17 Molecular dynamics simulation of DNA in nanopore under applied electric field. Water flow profile in three types of nanopores as a function of distance from the centre of DNA. *Circles*: 2.25 nm radius pore, *triangles*: 3 nm radius pore with rough surface, and *rhombus*: 3 nm radius pore with smooth walls. Open and closed symbols are simulations done at external electric field values of 0.125 and 0.5 V/6.4 nm respectively. Reproduced with permission from Luan and Aksimentiev [28], Copyright (2008) by the American Physical Society

Luan and coworkers also found that although the counterion distribution around DNA is independent of solution viscosity, the stall force itself is viscosity-dependent, in apparent contradiction to classical Manning theory. It was noted that the redistribution of ions around the negatively charged DNA surface results in an unbalanced total electrolyte charge, which is driven by the electric field, generating an electro-osmotic flow of water near the DNA. The direction of the additional drag on the DNA due to this flow is opposite to that of the electrical field, thus reducing the effective force sensed by the spring when the DNA is at rest [28].

10.3.4 Effect of Salt Gradients on the Transport Dynamics

After identifying the importance of counterion electroosmosis in DNA/small pore systems, we expect that increasing the hydrodynamic profile of the counterions around the DNA will reduce the effective force acting on it in the pore. In Fig. 10.18, we show the normalized translocation times under asymmetric salt conditions for two DNA lengths [50]. In these experiments, we have systematically reduced the salt concentrations in the *cis* chamber and recorded DNA translocations at each indicated C_{cis} value. We then compare the mean transport time of the DNA under a salt gradient to its mean transport time under symmetric 1 M/1 M salt conditions. We have observed a systematic retardation for both DNA lengths (400 and 2,000 bp) when C_{cis} is reduced: for the shorter DNA, we find that translocation is slowed by a factor of ~50% when the salt concentration is reduced by a factor of 5, whereas translocation is slowed by nearly fourfold for the longer DNA in the same salt conditions. It is not surprising that the shorter DNA molecules slow down less than the longer molecules; given the 1.40 power law that we find for DNA <3 kbp,

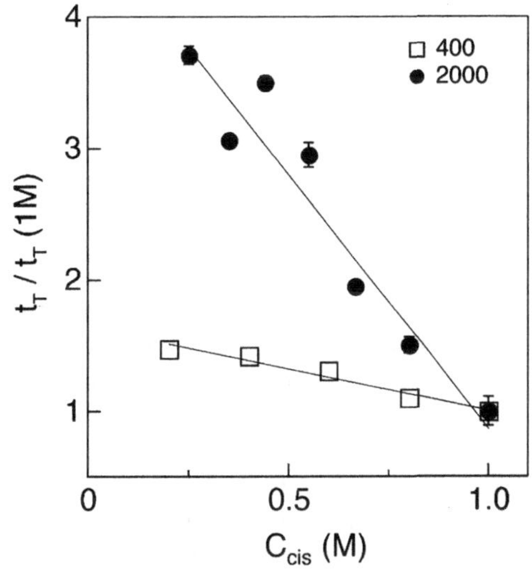

Fig. 10.18 Dependence of applied salt gradient on DNA transport for 400 and 2,000 bp DNA molecules. The quantity t_T/t_T(1 M) is the ratio of the mean transport time at the indicated salt gradient (C_{trans} = 1 M) to the transport time under symmetric 1 M/1 M KCl conditions. DNA slows down upon reduction of salt concentrations, suggesting counterion electroosmosis further damps the electrical force on the DNA. Reproduced with permission from Wanunu et al. [50], Copyright Nature Publishing Group

the actual ratio of speeds between 400 and 2,000 bp should be $(2{,}000/400)^{1.4} = 9.5$. This value is similar to our experimental result, which shows a slope ratio of ~7 for these two DNA samples (see line). These results are also consistent with the simulations performed by Luan and Aksimentiev, in which electroosmotic drag functions as an additional stalling force on the DNA [28]. However, it is not clear what the quantitative effect of a salt gradient on transport speed should be, primarily because the actual salt gradient profile inside the nanopore is complex and, to date, experimentally unquantified.

While the dynamic profile of DNA translocating through a nanopore is a subject of great importance for nanopore applications, a comprehensive model that accounts for biopolymer interactions with the membrane and pore walls, electrophoretic effects, and biopolymer hydrodynamics, has yet to be developed. Empirically, we know that the electric field in a nanopore exerts forces of tens of pN on the dsDNA, sufficient to overcome the entropic free-energy barrier associated with linearizing the biopolymer, as well as interactions with surfaces and any EOF which acts to slow the translocation process. A quantitative prediction of the actual instantaneous velocity or dwell time distributions will require a more sophisticated understanding of the complex salt gradient profile within the pore.

10.4 Conclusion

The studies reviewed here assess current progress toward understanding the dynamics of biopolymer capture and translocation through small solid-state nanopores, as illustrated by the case of double-stranded DNA. A theoretical model for DNA capture has been outlined, composed of two distinct steps: diffusion to the vicinity of the pore, and threading of one end into the pore. DNA coils diffusing randomly throughout the *cis* chamber are electrically focused towards the pore when they enter a hemisphere of influence described by the critical radius r^*. Once a coil reaches the pore mouth, the threading process is governed by an energy barrier that is highly biased by the local electrical potential near the pore. Investigation of DNA length dependence on the capture rate revealed that for shorter DNA, threading is the rate-limiting step, in which regime capture rate varies exponentially with square root of DNA length. For long DNA, diffusion to the pore is the rate-limiting step, and thus the overall capture rate is insensitive to DNA length. These two features are extremely useful, as they imply that small nanopores are capable of capturing even long biopolymers with the same efficiency as shorter biopolymers, despite the fact that longer molecules naturally diffuse slower. Observations of capture rate for a single DNA length as a function of voltage further confirmed theoretical predictions for these two dynamical regimes.

The dynamics of translocation are currently accessible primarily by examining the distribution of dwell times, t_D, of molecules in the nanopore under different conditions. The size of the nanopore is a particularly critical parameter; in large pores DNA molecules are driven through the pore with forces that are comparable to those in free electrophoresis, with both folded and unfolded configurations, but in sub-5 nm pores translocation proceeds only in an unfolded manner, where the net electrophoretic force on the DNA is greatly reduced by hydrodynamic interactions and friction with the nanopore surface. The distribution of dwell-times (t_D) in small pores was both analytically and numerically predicted to resemble a Poissonian distribution, due to these interactions of the DNA with the wall of the nanopore. Similar results are obtained from experimental data, and the characteristic timescales related to collisions and translocations may be extracted using exponential fits to the distribution tails. Small nanopores produce translocations that have a mean dwell time t_D that is orders of magnitude longer than that for larger pores, which is advantageous for many practical applications. The dwell time is reduced by a combination of factors: friction with the membrane, net electrophoretic force, and the electroosmotic drag of condensed counterions on the DNA.

We find that the application of salt gradients across the pore permits manipulation of the electric field profile in and around the pore. This implies that (1) The capture rate can be increased by altering the critical radius for biased diffusion and increasing the local potential near the pore that governs the threading efficiency. (2) The translocation time can be extended, presumably due to the increased EOF of cations acting to slow down the velocity of the negatively charged DNA during its translocation process.

Yet there are many aspects of translocation dynamics that remain unexplored. One particularly important question is that of a biopolymer's velocity profile as it passes through the pore. For many applications, it would be advantageous if the translocating biopolymer maintained a constant velocity throughout, since this would simplify the signal interpretation. Even given a non-uniform velocity profile, understanding the interactions and dynamics responsible would enable adjustment of materials and fabrication to exert some measure of control over this parameter. The question of velocity is very closely linked to the ultimate resolution of a nanopore. At present, solid-state nanopores are tens of nanometers thick, limiting the ultimate sensing resolution to a minimum of ~30 bases. This restricts the resolution of some practical applications, such as DNA sequencing. In order to improve the nanopore resolution, advances in materials fabrication are required in order to provide stable, geometrically well-defined nanopores that may ultimately enable differentiation of single base-pairs (~0.5 nm), or even of short sequences. Finally, control over the selectivity of nanopores and dynamics of biopolymer transport may be achieved by chemically modifying the internal and external surfaces of nanopores [19, 47]. Progress in material design, fabrication, and sub-nanometer pore control, coupled with a better understanding of the physical parameters that govern voltage-induced biomolecular motion through small pores, will be essential for opening new horizons in single biomolecule sensing.

Acknowledgements We acknowledge stimulating discussions contributing to this chapter with B. McNally, A. Singer, Y. Rabin, A. Grosberg, D. Nelson, A. Kolomeisky and W. Morrison. A.M. acknowledges support from NIH award HG-004128, and NSF award PHY-0646637.

References

1. Akeson M, Branton D, Kasianowicz JJ, Brandin E, Deamer DW (1999) Microsecond time-scale discrimination among polycytidylic acid, polyadenylic acid, and polyuridylic acid as homopolymers or as segments within single RNA molecules. Biophys J **77**(6):3227–3233
2. Aksimentiev A, Heng ZB, Timp G, Schulten K (2004) Microscopic kinetics of DNA translocation through synthetic nanopores. Biophys J **87**:2086–2097
3. Bates M, Burns M, Meller A (2003) Dynamics of DNA molecules in a membrane channel probed by active control techniques. Biophys J **84**(4):2366–2372
4. Benner S, et al. (2007) Sequence-specific detection of individual DNA polymerase complexes in real time using a nanopore. Nat Nanotechnol **2**(11):718–724
5. Branton D, et al. (2008) The potential and challenges of nanopore sequencing. Nat Biotechnol **26**(10):1146–1153
6. Butler TZ, Gundlach JH, Troll MA (2006) Determination of RNA orientation during translocation through a biological nanopore. Biophys J **90**(1):190–199
7. Chen P, et al. (2004) Probing single DNA molecule transport using fabricated nanopores. Nano Lett **4**(11):2293–2298
8. Chen P, Li CM (2007) Nanopore unstacking of single-stranded DNA helices. Small **3**(7):1204–1208
9. Clarke J, et al. (2009) Continuous base identification for single-molecule nanopore DNA sequencing. Nat Nanotechnol **4**(4):265–270
10. Coulter WH. 1953: US 2,656,508.
11. Dekker C (2007) Solid-state nanopores. Nat Nanotechnol **2**(4):209–215
12. Fologea D, Uplinger J, Thomas B, McNabb DS, Li J (2005) Slowing DNA translocation in a solid-state nanopore. Nano Lett **5**:1734–1737
13. Fologea D, Brandin E, Uplinger J, Branton D, Li J (2007) DNA conformation and base number simultaneously determined in a nanopore. Electrophoresis **28**(18):3186–3192
14. Fologea D, Ledden B, McNabb DS, Li JL (2007) Electrical characterization of protein molecules by a solid-state nanopore. App Phys Lett **91**(5)
15. Healy K (2007) Nanopore-based single-molecule DNA analysis. Nanomed **2**(4):459–481
16. Heng ZB, et al. (2004) Sizing DNA using a nanometer-diameter pore. Biophys J **87**:2905–2911
17. Henrickson SE, Misakian M, Robertson B, Kasianowicz JJ (2000) Driven DNA transport into an asymmetric nanometer scale pore. Phys Rev Lett **85**:3
18. Hornblower B, et al. (2007) Single-molecule analysis of DNA-protein complexes using nanopores. Nat Methods **4**(4):315–317
19. Iqbal SM, Akin D, Bashir R (2007) Solid-state nanopore channels with DNA selectivity. Nat Nanotechnol **2**(4):243–248
20. Jayawardhana D, Crank J, Zhao Q, Armstrong D, Guan X (2009) Nanopore stochastic detection of a liquid explosive component and sensitizers using boromycin and an ionic liquid supporting electrolyte. Anal Chem **81**(1):460–464
21. Kang X, Cheley S, Guan X, Bayley H (2006) Stochastic detection of enantiomers. J Am Chem Soc **128**(33):10684–10685
22. Kasianowicz JJ, Brandin E, Branton D, Deamer DW (1996) Characterization of individual polynucleotide molecules using a membrane channel. Proc Natl Acad Sci USA **93**(24): 13770–13773
23. Kim MJ, Wanunu M, Bell DC, Meller A (2006) Rapid fabrication of uniformly sized nanopores and nanopore arrays for parallel DNA analysis. Adv Mater **18**(23):3149–3153
24. Kowalczyk S, Hall A, Dekker C (2010) Detection of local protein structures along DNA using solid-state nanopores. Nano Lett **10**(1):324–428

25. Li J, et al. (2001) Ion-beam sculpting at nanometre length scales. Nature **412**(6843):166–169
26. Li JL, Gershow M, Stein D, Brandin E, Golovchenko JA (2003) DNA molecules and configurations in a solid-state nanopore microscope. Nat Mater **2**(9):611–615
27. Long D, Viovy J-L, Ajdari A (1996) Simultaneous action of electric fields and nonelectric forces on a polyelectrolyte: motion and deformation. Phys Rev Lett **76**:3858–3861
28. Luan B, Aksimentiev A (2008) Electro-osmotic screening of the DNA charge in a nanopore. Phys Rev E **78**(2):021912
29. Lubensky DK, Nelson DR (1999) Driven polymer translocation through a narrow pore. Biophys J **77**:1824–1838
30. Luo K, Ala-Nissila T, Ying S-C, Bhattacharya A (2007) Influence of polymer-pore interactions on translocation. Phys Rev Lett **99**(14)
31. Mathe J, Aksimentiev A, Nelson DR, Schulten K, Meller A (2005) Orientation discrimination of single-stranded DNA inside the alpha-hemolysin membrane channel. Proc Natl Acad Sci USA **102**(35):12377–12382
32. Mathe J, Arinstein A, Rabin Y, Meller A (2006) Equilibrium and irreversible unzipping of DNA in a nanopore. Europhys Lett **73**(1):128–134
33. McNally B, Wanunu M, Meller A (2008) Electro-mechanical unzipping of individual DNA molecules using synthetic sub-2 nm pores. Nano Lett **8**(10):3418–3422
34. Meller A, Nivon L, Brandin E, Golovchenko J, Branton D (2000) Rapid nanopore discrimination between single polynucleotide molecules. Proc Natl Acad Sci USA **97**(3):1079–1084
35. Meller A, Nivon L, Branton D (2001) Voltage-driven DNA translocations through a nanopore. Phys Rev Lett **86**:3435–3438
36. Meller A, Branton D (2002) Single molecule measurements of DNA transport through a nanopore. Electrophoresis **23**(16):2583–2591
37. Meller A (2003) Dynamics of polynucleotide transport through nanometre-scale pores. J Phys Condens Matter **15**:R581–R607
38. Nkodo AE, et al. (2001) Diffusion coefficient of DNA molecules during free solution electrophoresis. Electrophoresis **22**:2424–2432
39. Olivera BM, Baine P, Davidson N (1964) Electrophoresis of the nucleic acids. Biopolymers **2**:245–257
40. Sauer-Budge AF, Nyamwanda JA, Lubensky DK, Branton D (2003) Unzipping kinetics of double-stranded DNA in a nanopore. Phys Rev Lett **90**(23)
41. Singer A, et al. (2010) Nanopore based sequence specific detection of duplex DNA for genomic profiling. Nano Lett **10**:738–742
42. Soni GV, Meller A (2007) Progress toward ultrafast DNA Sequencing using solid-state nanopores. Clin Chem **53**(11):1996–2001
43. Soni GV, et al. (2010) Synchronous optical and electrical detection of biomolecules traversing through solid-state nanopores. Rev Sci Instrum **81**(1):014301–014307
44. Storm AJ, Chen JH, Ling XS, Zandbergen HW, Dekker C (2003) Fabrication of solid-state nanopores with single-nanometre precision. Nat Mat **2**(8):537–540
45. Storm AJ, et al. (2005) Fast DNA translocation through a solid-state nanopore. Nano Lett **5**(7):1193–1197
46. Vercoutere W, et al. (2001) Rapid discrimination among individual DNA hairpin molecules at single-nucleotide resolution using an ion channel. Nat Biotechnol **19**(3):248–252
47. Wanunu M, Meller A (2007) Chemically modified solid-state nanopores. Nano Lett **7**(6):1580–1585
48. Wanunu M, Meller A (2008) Single-molecule analysis of nucleic acids and DNA-protein interactions using nanopores. In: Selvin P, Ha TJ (eds) Single-Molecule Techniques: A Laboratory Manual. Cold Spring Harbor Laboratory Press: Cold Spring Harbor, New York
49. Wanunu M, Sutin J, McNally B, Chow A, Meller A (2008) DNA translocation governed by interactions with solid state nanopores. Biophys J **95**(10):4716–4725

50. Wanunu M, Morrison W, Rabin Y, Grosberg A, Meller A (2010) Electrostatic focusing of unlabelled DNA into nanoscale pores using a salt gradient. Nat Nanotechnol **5**:160–165
51. Wanunu M, Sutin J, Meller A (2009) DNA profiling using solid-state nanopores: detection of DNA-binding molecules. Nano Lett **9**:3498–3502
52. Zhang J, Shklovskii BI (2007) Effective charge and free energy of DNA inside an ion channel. Phys Rev E **75**:021906

Chapter 11
Nanopore-Based DNA Sequencing and DNA Motion Control

Hongbo Peng, Binquan Luan, and Gustavo Stolovitzky

Abstract Compared to traditional Sanger's DNA sequencing methods or currently commercialized next-generation-sequencing solutions (454, Roche, Basel; Solexa, Illumina, San Diego; SOLiD, Applied Biosystems, Foster City, CA, USA/Agencourt, Beverly, MA, USA; HelioScope, Helicos, Cambridge, MA, USA), nanopore-based DNA sequencing proposals have a number of advantages that fueled intense research efforts both in the industry and academia. If these efforts are successful, nanopore-based DNA sequencing will enable real-time single molecular DNA sequencing methods with little to no sample preparation. Nanopore sequencing has the potential to reduce the cost of sequencing of a whole human genome to less than $1,000. However, the road to conquer this technology is not without serious challenges. Two key issues in the field are to control the DNA translocation through the nanopore and to sense different DNA bases that compose the DNA molecule being sequenced. In this chapter, we present an overview of some commercial DNA sequencing technologies that will provide the context for our discussion of various nanopore DNA sequencing approaches. We will focus our discussion on the efforts to control the DNA translocation through the pore, as a step to realize the objective of nanopore-based DNA sequencing.

Keywords Blockade • Blockage • Genome • Ionic current • Nanopore • Nanopore – capacitor • Pyrosequencing • DNA ratcheting • DNA sequencing • DNA transistor • DNA translocation control • Single stranded DNA • Sequencing method • Solid-state nanopore • Translocation time • Transverse electronic transport • Trapping field • Trapping energy • Unzipping DNA

H. Peng (✉)
IBM Thomas J. Watson Research Center, 1101 Kitchawan Road,
Yorktown Heights, NY 10598, USA
e-mail: pengho@us.ibm.com

S.M. Iqbal and R. Bashir (eds.), *Nanopores: Sensing and Fundamental Biological Interactions*, DOI 10.1007/978-1-4419-8252-0_11,

11.1 DNA Sequencing

The information to produce many of the components of the cell such as RNAs and proteins is encoded in the sequence of nucleotides of a cell's DNA. Some alterations to this sequence can contribute to diseases such as cancer, cystic fibrosis, Huntington disease, to name a few. DNA also contains some of the determinants of an individual's propensity for some diseases, response to a treatment and prognosis of clinical course. Determining the DNA sequence is therefore fundamental to molecular biology and biomedicine, and could potentially open a door to a medicine tailored for each individual, the ambitious but achievable goal of personalized medicine. In this section we discuss existing techniques for DNA sequencing.

11.1.1 Sanger's Sequencing Method

The key principle of the Sanger method [1, 2] is the use of dideoxynucleotides triphosphates (ddNTPs, where the N stands for A, C, G or T) as DNA chain terminators. As shown in Fig. 11.1a, during the DNA elongation, the 5′-phosphate group of the dNTPs is condensed with the 3′-hydroxyl group at the end of the extending DNA strand to form a phosphodiester bond by releasing a pyrophosphate. If the oxygen atom from the 3′-OH group is removed to form a dideoxynucleotide, as shown in Fig. 11.1b, the DNA strand extension will be terminated since the phosphodiester bond between two nucleotides cannot be formed under this condition. The chain terminator dideoxynucleotide triphosphates are used to convert the DNA sequence information into length information of DNA fragments by polymerase chain reaction (PCR). As shown in Fig. 11.2, the target DNA

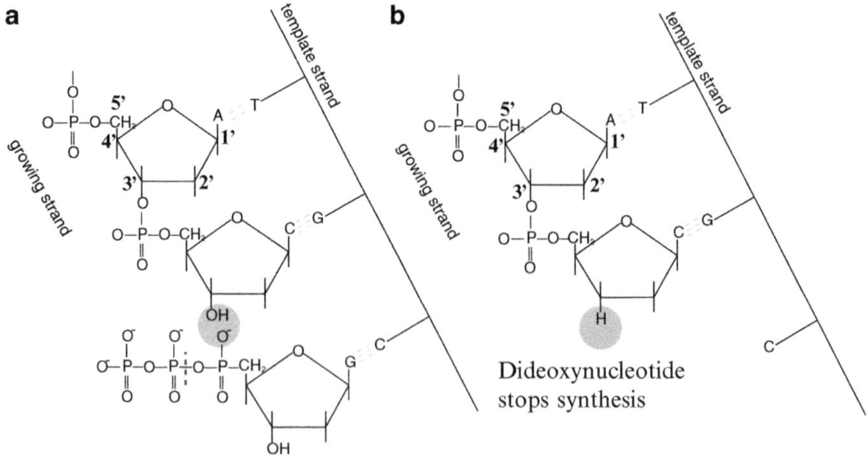

Fig. 11.1 Diagram showing standard DNA synthesis (**a**) and a chain terminated by the addition of a dideoxynucleotide chain terminator (**b**)

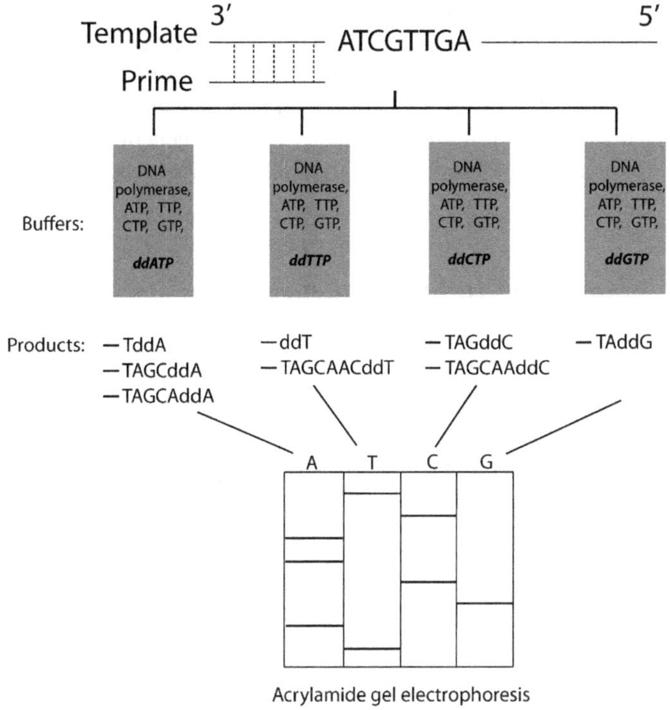

Fig. 11.2 Principle of the chain-terminating method. Besides having the all the dNTPs, each of four lanes has one of the four ddNTP, which, when incorporated, terminates the polymerase chain reaction at different lengths. Adapted from [2]

sample with annealed template is loaded into four separate reaction buffers, each containing the four standard radioactively or fluorescently labeled deoxynucleotides (dATP, dGTP, dCTP and dTTP) and the DNA polymerase. Only one of the four dideoxynucleotides (ddATP, ddGTP, ddCTP, or ddTTP) is added to each reaction buffer at lower concentration than the standard deoxynucleotides. Since the dideoxynucleotides can terminate DNA strand extension, the random incorporation of dideoxynucleotides instead of the corresponding standard deoxynucleotides results in various DNA fragments of varying lengths. The length information of the DNA fragments is then extracted by gel or capillary electrophoresis with a resolution of a single nucleotide. The length of each DNA fragment in a reaction buffer corresponds to the distance between the primer to the DNA base complementary to the added dideoxynucleotide in the target DNA sample.

Sanger sequencing has been an extremely successful technique for DNA sequencing, and continues to be the gold standard when it comes to data quality of DNA sequences. Through parallelization, automation and refinement of the established Sanger's sequencing method, the Human Genome Project (HGP) achieved a sequencing cost of \$0.01/base in 2003 [3], which resulted in a cost of about \$30,000,000 to sequence the three billion bases of a human genome.

Much-lower-cost methods for DNA sequencing will be required to make genome sequencing feasible for routine healthcare practice. Aggressive research and development during the past decade resulted in a decrease of both cost and time of sequencing by at least five orders of magnitude. The current faster and cheaper sequencing methods are referred generically as "next-generation sequencing technologies" to distinguish them from the traditional Sanger sequencing.

11.1.2 Fluorescent In Situ Sequencing

Fluorescent in situ sequencing (FISSEQ), a method faster and cheaper than Sanger Sequencing, is based on detecting the fluorescent signals of the added deoxynucleotide triphosphates in real time. As demonstrated by Mitra et al. [4], and illustrated in Fig. 11.3, the target DNA molecules are first PCR amplified in a polyacrylamide gel [5]. One strand of the amplified DNA is then covalently

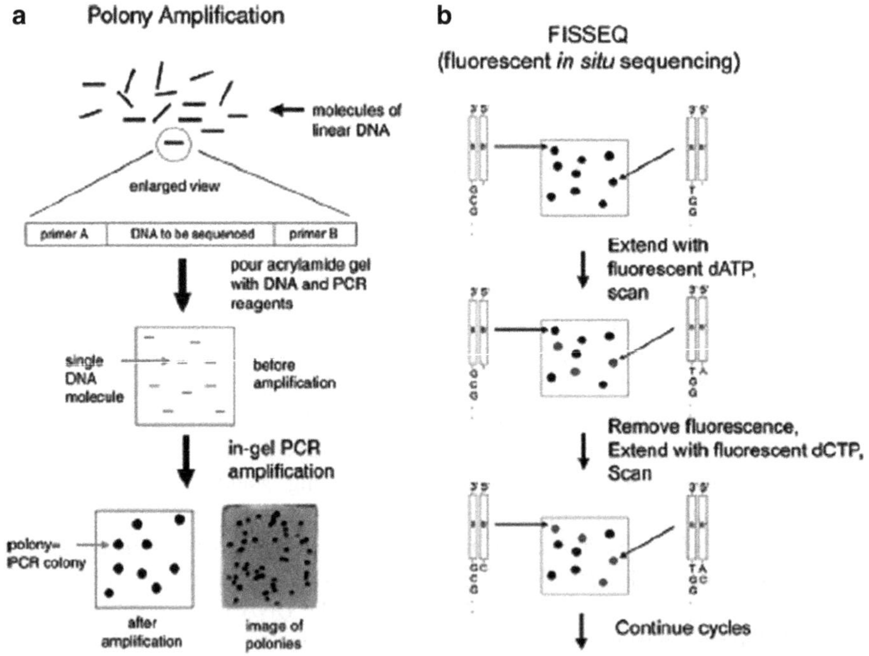

Fig. 11.3 Fluorescent in situ sequencing. (**a**) Polony amplification. A library of linear DNA molecules with universal priming sites is PCR amplified in a polyacrylamide gel. A single template molecule gives rise to a polymerase colony or polony. (**b**) Fluorescent in situ sequencing. Polonies are denatured, and a sequencing primer is annealed. Polonies are sequenced by serial additions of a single fluorescent nucleotide. Adapted from [4]

attached to the polyacrylamide matrix by acrydite modification [6, 7] of the 5′ end. To perform the sequencing, the immobilized DNA is denatured, hybridized with a primer and the complementary strand is elongated with deoxynucleotides that are reversibly coupled to fluorescent groups. A scanning fluorescence microscope is used to check if each type of deoxynucleotides has been incorporated or not, telling the sequencing information of the complementary strand. After checking one type of deoxynucleotides, the fluorescence is removed by cleaving the linker between the fluorophore and the nucleotide. Thc cycle is repeated by adding a different type of fluorescently labeled deoxynucleotide and scanning the gel. In this way, the sequence of every target DNA molecule (represented by a PCR colony on the gel) can be determined in parallel.

Based on the FISSEQ concept, but modifying the DNA amplification in gel polonies by solid substrate "bridge amplification," Solexa (now part of Illumina) developed a DNA sequencing technology which is presently commercialized by Illumina as the "Genome Analyzer System."

11.1.3 Pyrosequencing

Pyrosequencing is a cyclic-array sequencing method based on detecting the released pyrophosphate during incorporation of nucleotide triphosphate. The principle of pyrosequencing [8] can be described as follows: (1) the target single strand DNA molecules are fixed on solid surfaces, hybridized with a primer and then incubated with DNA polymerase, ATP sulfurylase, firefly luciferase, and a nucleotide-degrading enzyme apyrase; (2) the pyrophosphate (PPi) released during the synthesis of DNA is converted to ATP by ATP sulfurylase and the concentration of ATP is then sensed by the luciferase, which utilizes the energy of ATP to emit light; (3) the amount of light can be estimated by a CCD (charge-coupled device) camera; (4) unincorporated deoxynucleotides and the residue ATP are then degraded by the nucleotide-degrading enzyme apyrase after each cycle; (5) repeated cycles of deoxynucleotide addition are performed until the whole target DNA is sequenced.

An example of the light signals resulting from the pyrosequencing reactions are shown in Fig. 11.4. The intensity of the signal at each cycle is proportional to the number of nucleotides that are incorporated to each template during the cycle. The technique has been further developed by a company named 454 Life Sciences (now owned by Roche diagnostics) into a technology known as 454 Pyrosequencing.

For more details and other DNA sequencing technologies such as SOLiD platform (Applied Biosystems; Foster City, CA, USA) and the HeliScope Single Molecule Sequencer technology (Helicos; Cambridge, MA, USA), interested readers may refer to references [9, 10]. The remaining of this chapter will focus on DNA sequencing approaches based on nanopore technology.

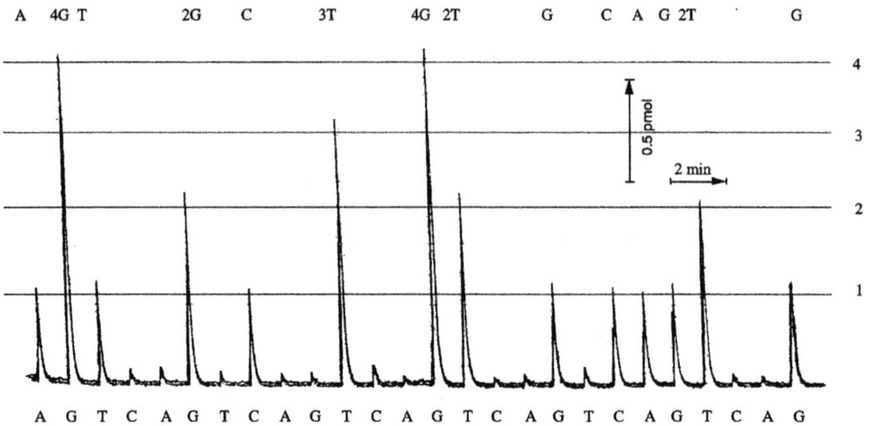

Fig. 11.4 Pyrosequencing performed on a 25-base-long oligonucleotide template. Proportional signals are obtained for one, two, three and four base incorporations. Repeated cycles of A, G, T, C incorporations are indicated at the bottom part of the figure, while the top part of the figure illustrates the obtained sequence based on the intensity of the fluorescence signal. Adapted from [8]

11.1.4 Nanopore-Based DNA Sequencing: Challenges and Opportunities

11.1.4.1 KBBD Proposal

Nanopore-based DNA sequencing is a non-cyclical, single-molecule, real-time DNA sequencing method. It was first proposed in 1996 by Kasianowicz et al. in their seminal paper [11] on DNA translocation. A biological protein channel (alpha-hemolysin, or alpha-HL, with 1.5 nm pore diameter) can form in a diphytanoyl phosphatidylcholine lipid bilayer (Fig. 11.5), which is across a horizontal, conical aperture at the end of a Teflon tube (Fig. 11.6). The ionic current through the nanopore can be detected using the Patch-Clamp electronics when a voltage is applied across the membrane. Due to the volume exclusion, passage of molecules will decrease the number of ions inside the nanopore, thus decreasing the ionic current. As shown in Fig. 11.7, the passage of single-stranded DNA molecules through this membrane-embedded α-hemolysin channel, for example, causes a ~90% blockage of the ionic current at 1 M KCl solution. This fact led to the KBBD proposal (for Kasianowicz, Brandin, Branton and Deamer), that it may be possible to use the ionic current passing through the nanopore to directly determine the sequence of a polynucleotide.

One obvious experiment is to check the current blockades caused by homopolymers of polycytidylic acid (poly (C)), polyadenylic acid (poly (A)). A single adenylic base is larger than a single cytidylic base, as shown in Fig. 11.8, one would expect that the current blockage of poly (A) is larger than that of poly (C). But the experiment done by Akeson et al. shows the opposite effect [13]. As in Fig. 11.9, there are three current blockage levels for poly (A) and all of them are actually smaller than those of poly (C).

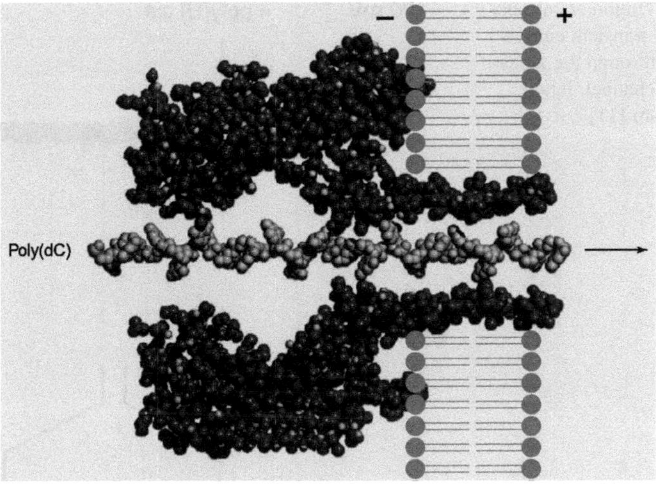

Fig. 11.5 A single alpha-HL channel is shown in cross section embedded in a lipid bilayer. Image adapted from [12]

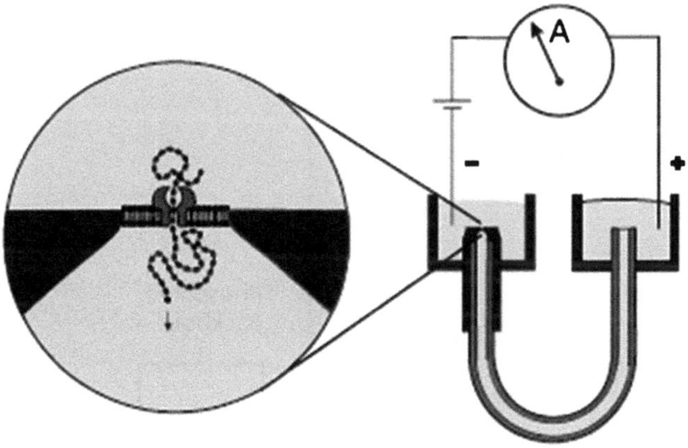

Fig. 11.6 Supporting apparatus for horizontal bilayer experiments. Image adapted from [13]

To figure out the reason for this counter intuitive effect, the authors of [13] examined blockades caused by RNA of composition A(30)C(70)Gp [12]. As shown in Fig. 11.9, there are two levels of ionic current blockages: always 95–85%, indicating that the 3′ poly (C) end of A(30)C(70)Gp always entered the pore first. The explanation is that poly (C) traverses through the α-hemolysin channel as a helix, while helical segments along a poly (A) strand have to be unwound and extended to permit translocation through the channel, which is also the reason of this opposite effect that the current blockage of poly (A) is smaller than that of poly (C) (Fig. 11.10).

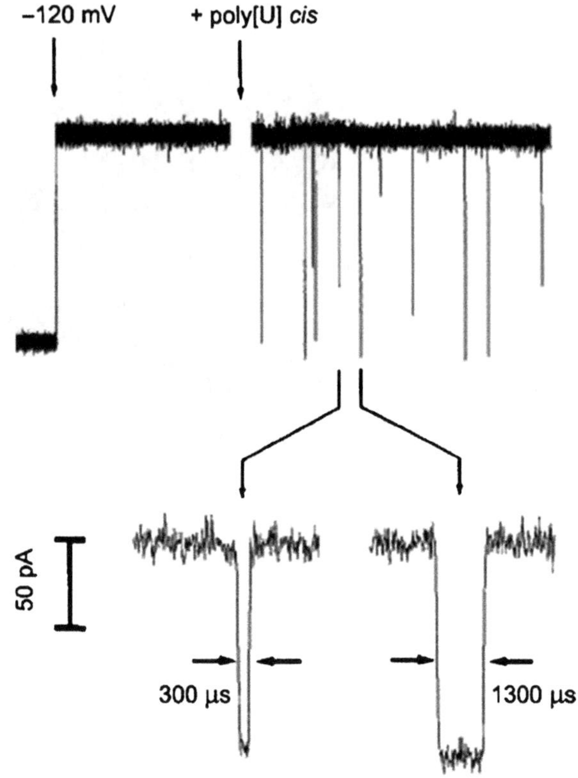

Fig. 11.7 Oligomers of poly (U) caused transient current blockades through the alpha-HL single-channel. Image adapted from [11]

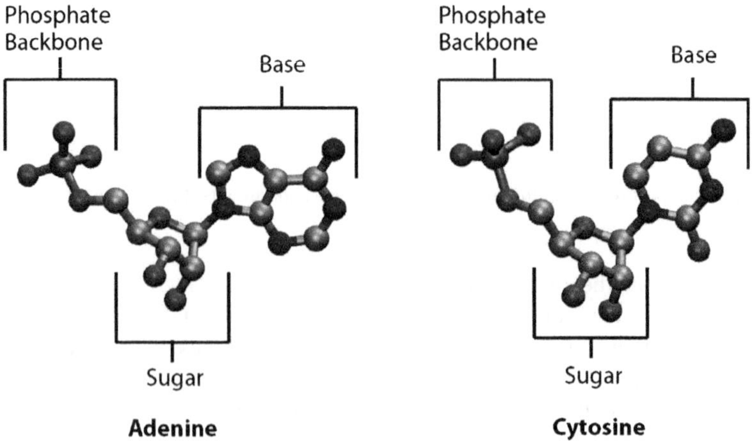

Fig. 11.8 Schematic of one unit of polyadenylic and polycytocylic acids

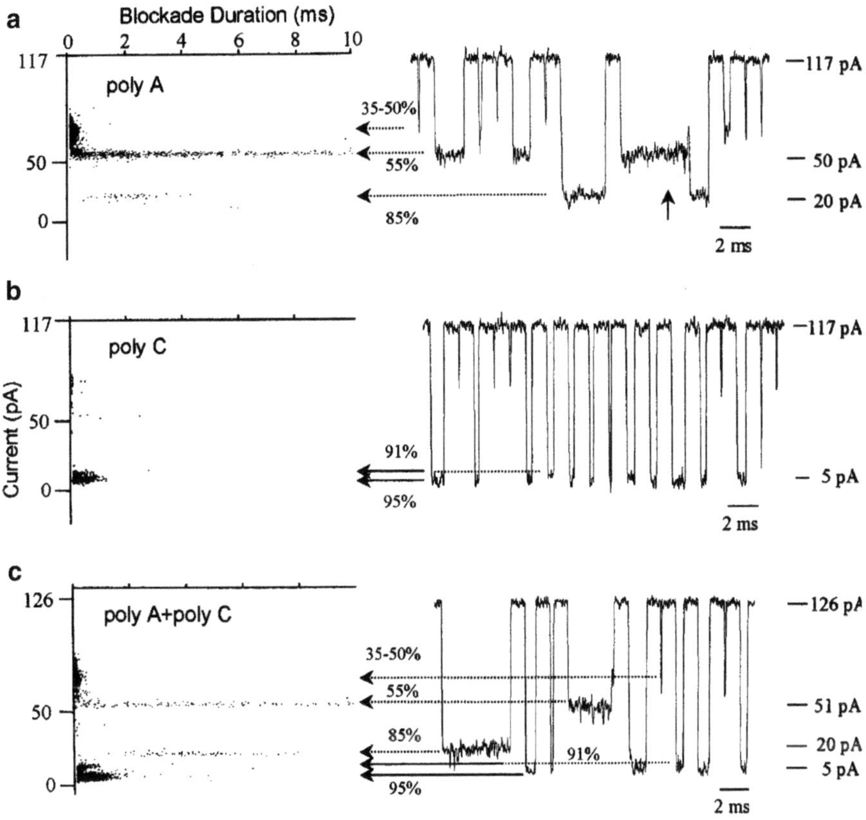

Fig. 11.9 Scatter plots of amplitude and duration of current blockades caused by RNA homopolymers passing through the alpha-HL pore. Typical blockade events comprising each plot are shown on the right. (*Top*) Blockades caused by poly(A) (150 nucleotide nominal length, 200 mg ml^{-1}). (*Middle*) Blockades caused by poly(C) (125 nucleotide nominal length, 200 mg ml^{-1}). (*bottom*) Blockades caused by a mixture of poly(C) (125 nucleotide nominal length) and poly(A) (175 nucleotide nominal length). Quiescent periods between events were spliced out of the plots on the right so that numerous blockades could be presented in one figure. Image adapted from [13]

The first lesson one can learn from this experiment is that the secondary structure of DNA has to be eliminated for the purpose of sequencing DNA. As suggested by Kasianowicz et al. [11], at least four conditions have to be met for this nanopore DNA sequencing method to be possible:

1. The channel and membrane must be sufficiently robust to withstand whatever temperature and chemical treatments are required to eliminate interference from polynucleotide secondary structures;
2. The length of the nanopore channel has to be comparable with the inter base distance of 0.4 nm;
3. Each nucleotide must produce a characteristic transient current blockage;
4. The rate of nucleotide movement through the channel must be slower than the time resolution of ionic current measurement system.

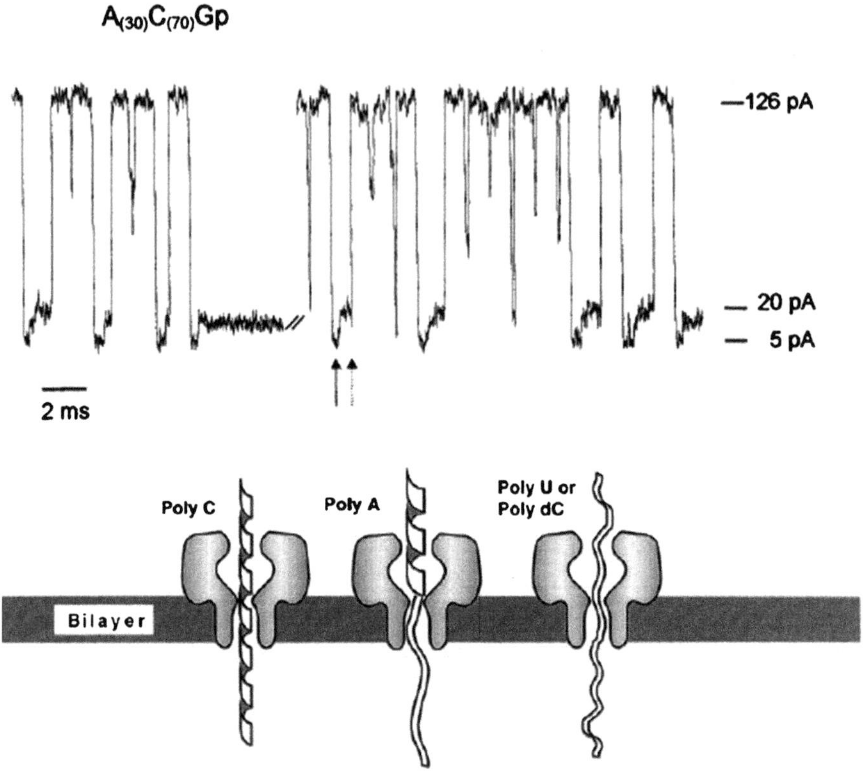

Fig. 11.10 Typical blockades of ion current in the alpha-HL pore caused by A(30)C(70)Gp RNA. Image adapted from [13]

Solid state nanopores [14–18] have been developed to satisfy condition (1). For identification of bases (basically at the root of conditions (2) and (3)) there are proposals based on "Transverse electronic transport" [19, 20], "Semiconductor nanopore–capacitor" [21], "unzipping designed DNAs" [22] and "Hybridization-assisted Nanopore DNA Sequencing" [23]. Issues on condition (4) will be discussed in Section 11.2.

11.1.4.2 Nanopore-Based DNA Sequencing via Transverse Electronic Transport

The principle of nanopore-based DNA sequencing via transverse electronic transport method is shown in Fig. 11.11. If one puts two nanometer-scale electrodes around the nanopore, the tunneling current through the two electrodes will be modulated as the DNA bases are translocating through the nanopore one at a time. Using carbon nanotubes as electrodes one can make condition (2) satisfied by carefully breaking the nanotube to get a sharp point electrode with dimensions

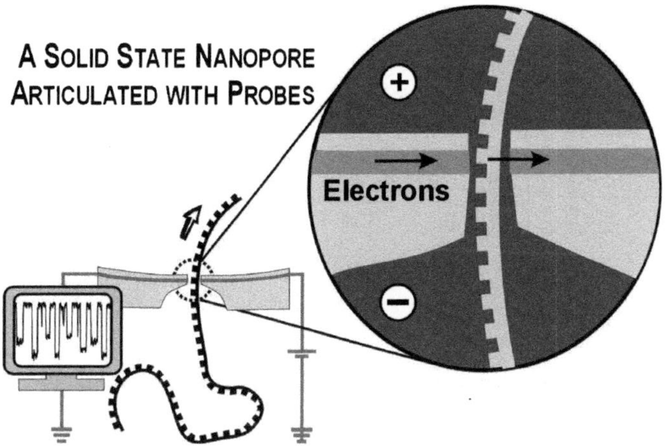

Fig. 11.11 Principle of DNA sequencing via transverse electronic transport. A nanopore is in an insulating membrane that separates two ionic solution-filled compartments. In response to a voltage bias (labeled "−" and "+") across the membrane, ssDNA molecules in the "−" compartment are driven, one at a time, into and through the nanopore. The probes across the nanopore serve as emitter and collector of a tunneling "microscope." Elevated temperatures and denaturants maintain the DNA in an unstructured, single-stranded form. (Courtesy of D. Branton, Harvard)

smaller than the diameter of the carbon nanotube. It is also possible for condition (3) to be satisfied, as shown in the theoretical calculation by Lagerqvist, J. et al. [19] in Fig. 11.12. As pointed out by Zhang X. G. et al. [20], electron tunneling depends critically on the distance between the DNA base and the electrodes, which is a challenge from an experimental point of view. Controlling the motion of DNA through the nanopore with single base resolution is also a great challenge, which we will discuss in Section 11.2.

11.1.4.3 Nanopore-Based DNA Sequencing via Semiconductor Nanopore–Capacitor

A semiconductor nanopore–capacitor device is fabricated from a metal–oxide–semiconductor (MOS) capacitor, as shown in Fig. 11.13 [21]. Voltage signals on the MOS capacitor is induced as the charged DNA molecule is driven through the nanopore-capacitor by an external voltage bias across the nanopore. Simultaneous measurements of the current through the pore, the voltages on the poly and c-Si electrodes, and the voltage on the MOS capacitor are shown in Fig. 11.14. The voltage signals on the MOS capacitor are nearly coincident with an event observed in the ionic current through the pore, which is a proof that these voltage signals are due to the translocation of a single DNA molecule through the pore.

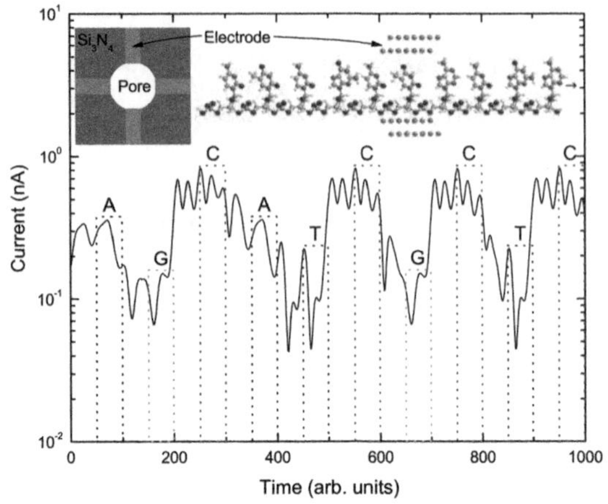

Fig. 11.12 Transverse current vs. time (in arbitrary units) of a highly idealized single strand of DNA translocating through a nanopore with a constant motion. The sequence of the single strand is AGCATCGCTC. The *left inset* shows a top-view schematic of the pore cross section with four electrodes. The *right inset* shows an atomistic side view of the idealized single strand of DNA and one set of gold electrodes across which electrical current is calculated. The *boxes* show half the time each nucleotide spends in the junction. Within each box, a unique signal from each of the bases can be seen. Image adapted from [19]

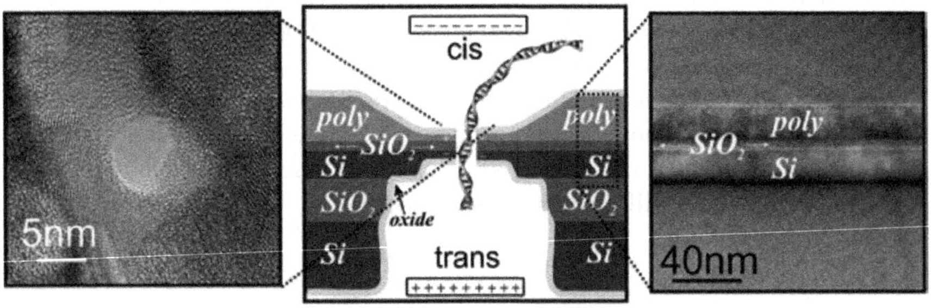

Fig. 11.13 *Center panel*: schematic diagram of a novel biosensor consisting of a nanopore in a capacitor membrane. *Right panel*: high-resolution TEM image through the capacitor membrane structure. *Left panel*: high-resolution lattice image of a nanopore sputtered into the capacitor membrane. Image adapted from [21]

The high spatial resolution can be achieved by using a thin gate oxide, such as 0.7 nm in the standard MOSFET technology, so condition (2) can be satisfied. For condition (3), MD simulation was done by Gracheva et al. [24] showing that it is possible to resolve individual nucleotides as well as their types in the absence of conformational disorder (Fig. 11.15). However, avoiding conformational disorders and controlling the position of the DNA (against thermal agitation) remains a serious challenge from an experimental point of view. Some of these issues will be discussed in Section 11.2.

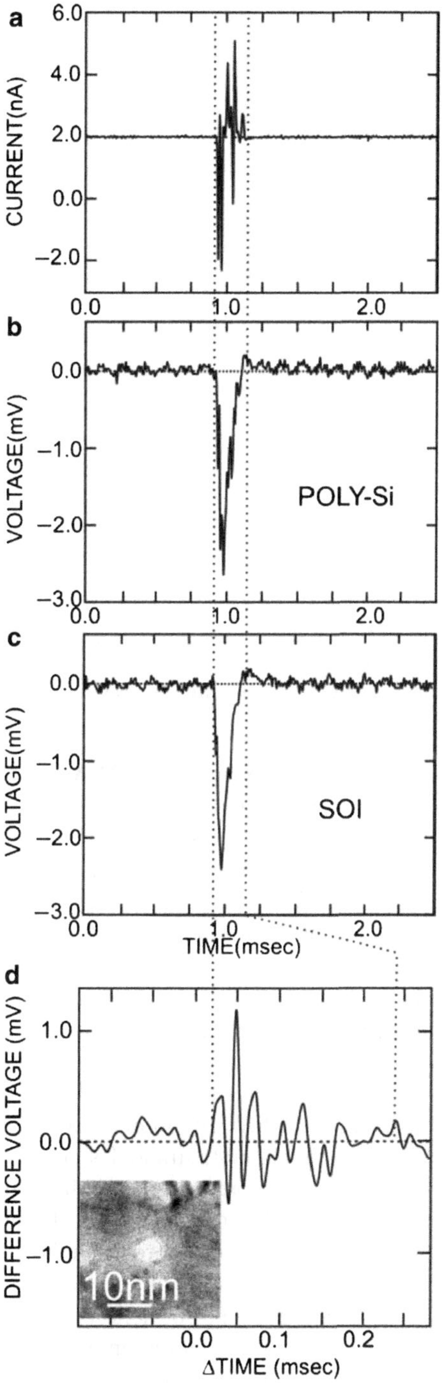

Fig. 11.14 DNA translocation through a 7 nm diameter pore in a capacitor. Corresponding to a transient in the current (**a**), voltage signatures on the SOI and poly-silicon electrodes are shown in (**b**) and (**c**) respectively. (**d**) shows the difference between the poly and SOI electrodes. A STEM image of the 7 nm diameter pore is shown in the inset of (**d**). Image adapted from [21]

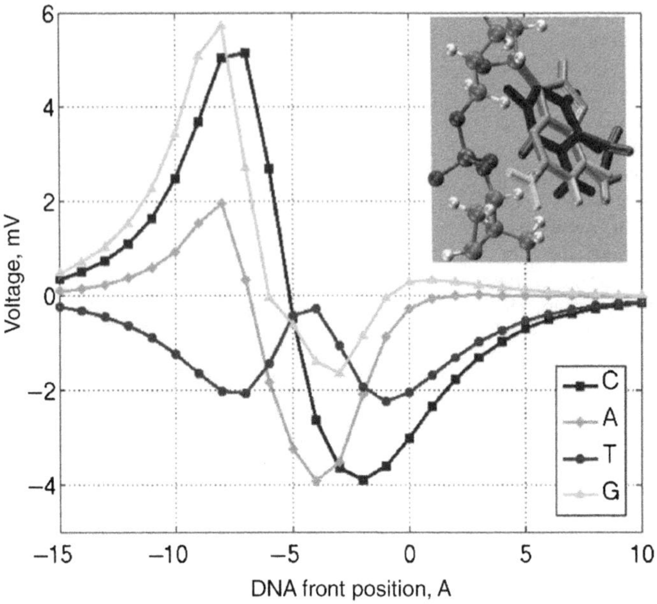

Fig. 11.15 Voltage traces due to translocation of single bases C, A, T and G. The *inset* illustrates the conformations of the replaced nucleotides. The orientations of their aromatic rings as well as the backbone structures are identical. Image adapted from [24]

11.1.4.4 Nanopore-Based DNA Sequencing via Unzipping Designed DNAs

Quite different from the above electronic detection approaches, this proposal [22] is based on the detection of optical signals while a designed DNA is unzipped by a nanopore, as shown in Fig. 11.16. A biochemical process is employed to create the designed DNA [22]. The DNA molecule to be sequenced is recoded in such a way that each nucleotide in the original DNA is represented by 20-base sequences in the designed DNA. The 20-base sequences are two blocks of 10-base strands and each 10-base strand has two variations, which can be used to code the four different bases A, T, C, G in the original DNA. The two variations of the 10-base strands are each labeled with a different fluorescent tag on one end and a quenching molecule at the other end. In a double strand designed DNA, a quenching molecule is always close enough to a fluorescent tag, which prevents detecting the fluorescence from the background fluorescent tags. The fluorescence signal can only be detected at the entrance of the nanopore when the quencher is pulled away from the fluorescent tag during unzipping of the ds-DNA by the nanopore [25–27], as the oligomers complementary to the 20-mers coding each nucleotide form a stem-loop structure in which the quencher/fluorescent pairs are in close proximity (Fig. 11.16). Although this method requires some biochemical process, it is still a very promising near-term DNA sequencing technology, as it can be parallelized by using nanopore arrays.

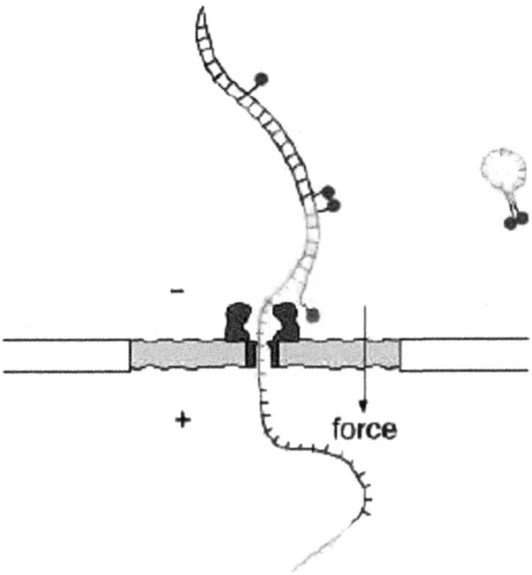

Fig. 11.16 Schematic of the concept of nanopore DNA sequencing via unzipping designed DNAs. (Courtesy of A. Meller)

11.1.4.5 Hybridization-Assisted Nanopore DNA Sequencing

To overcome two of the current difficulties in nanopore DNA sequencing: fabricating a nanopore with length comparable with the length of a single base and detecting the characteristic transient current blockages of single DNA bases, Ling et al. [23] proposed the Hybridization-assisted Nanopore DNA Sequencing (HANS) method. The principle of this proposal is shown in Fig. 11.17. An insulating solid state nanopore membrane (shown in gray) separates two ionic buffers. The unknown single strand DNA is shown as the top strand in the figure. Short known oligonucleotides (DNA probes, shown as the bottom strands in the figure) are annealed onto their complimentary part of the single stranded template DNA. The single stranded DNA with hybridized probes is then driven through the nanopore by applying a voltage bias across the nanopore. The expected ionic current trace due to the blockage of the single strand part and the double strand part of the DNA is shown in the lower panel of the figure. In this way, one only needs to detect the ionic current differences between the single strand part and double strand part of the DNA, thus bypassing the difficulty to detect the transient current differences between single DNA bases in the original KBBD proposal. Additionally, one only needs to detect ionic current signals of oligonucleotides instead of single bases, thus bypassing the difficulty of fabricating a nanopore with length comparable with the length of a single base.

Sequencing by hybridization (SBH) [29, 30] is currently not competitive for sequencing as it cannot detect the repeats of DNA probes, which limits the ability to have the complete information needed for reconstruction of the DNA sequence. By combing SBH with the nanopore technology [4], one can detect the repeating DNA probes as well as the location of the probes from the ionic current trace. For probe length n, one simply tries all 4^n possible probes separately and records the

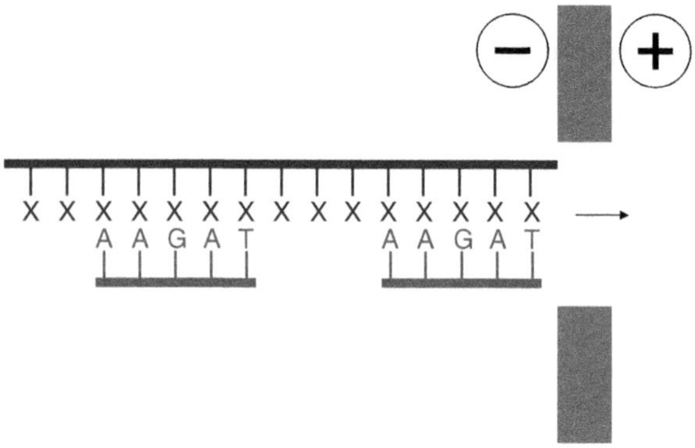

Fig. 11.17 Principle of Hybridization-assisted Nanopore DNA Sequencing. The *upper panel* shows a single stranded DNA with two DNA probes translocating through a nanopore. The *lower panel* shows the corresponding ionic current (not real data). (Courtesy of X. Ling)

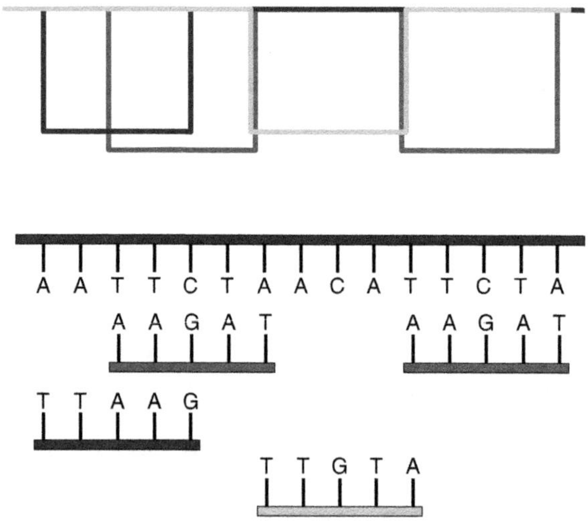

Fig. 11.18 Sequence reconstruction from ionic current traces. The *upper panel* shows the ionic current trace when the unknown DNA with different DNA probes translocates through the nanopore. The *lower panel* shows reconstructing DNA sequence from the ionic current trace. Three different probes "TTAAG," "AAGAT" and "TTGTA" and their corresponding ionic current traces are shown in the figure. (Courtesy of X. Ling)

corresponding ionic current traces, then reconstructs the DNA sequencing from the location information of the DNA probes. An example is shown in Fig. 11.18: from the ionic current trace of three different probes "TTAAG," "AAGAT" and "TTGTA," one can reconstruct the original sequence of the target DNA "AATTC-TAACATTCTA." This proposal is currently under development by a company called Nabsys (http://www.nabsys.com/).

11.1.4.6 Oxford Nanopore DNA Sequencing

The Oxford nanopore DNA sequencing approach [28] employs an exonuclease enzyme to cleave individual nucleotide molecules from the target DNA and drives them through a protein nanopore, which is a hemolysin mutant illustrated as WT-(M113R/N139Q)6(M113R/N139Q/L135C)1. The protein nanopore is covalently attached to an adaptor molecule, heptakis(6-deoxy-6-amino)-6-N-mono (2-pyridyl)dithiopropanoyl-β-cyclodextrin ($am_6amPDP_1\beta CD$), which will slow down the motion of nucleotide molecules through the pore [28]. By cleaving the target DNA molecule into individual nucleotide, one does not need to worry about the secondary structure of DNA or the length of the nanopore channel (previously mentioned conditions (1) and (2)). By putting an adaptor molecule

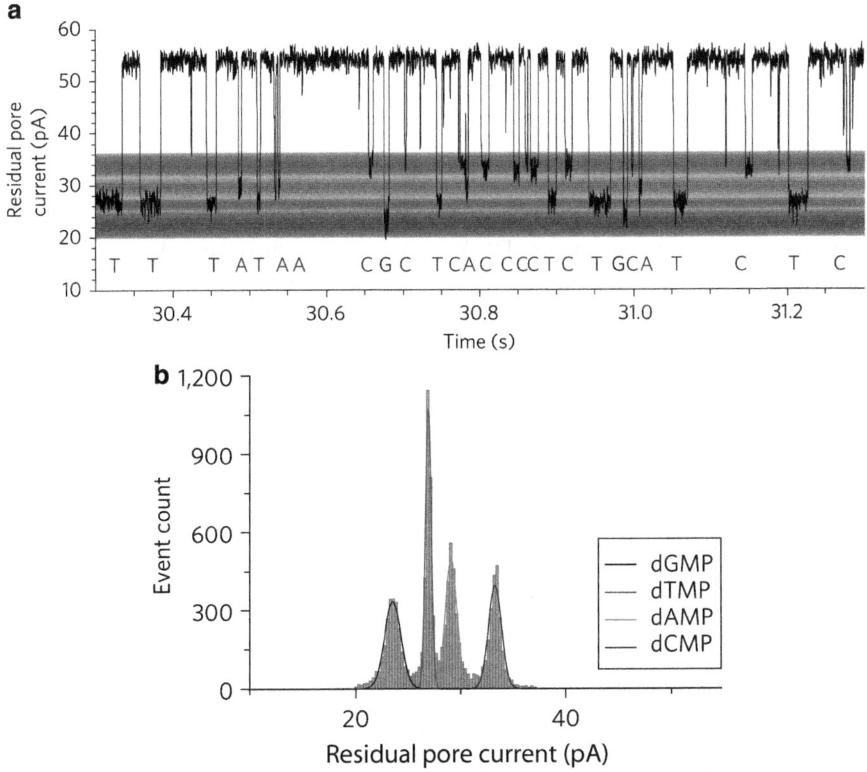

Fig. 11.19 Nucleotide event distributions with the adapter. (**a**) Single-channel recording from the WT-$(M113R/N139Q)_6(M113R/N139Q/L135C)_1$-am6amDP$_1\beta$CD pore showing dGMP, dTMP, dAMP and dCMP discrimination, with shaded bands (three standard deviations from the centre of the individual Gaussian fits) added to represent the residual current distribution for each nucleotide. (**b**) Corresponding residual current histogram of nucleotide binding events, including Gaussian fits. Data acquired in 400 mM KCl, 25 mM Tris HCl, pH 7.5, at 180 mV in the presence of 10 mM dGMP, 10 mM dTMP, 10 mM dAMP and 10 mM dCMP. Adapted from [28]

inside the protein nanopore, one also solves the concerns of previously mentioned condition (3) and (4): characteristic transient current signatures for each nucleotide molecules and enough dwelling time of each nucleotide molecule inside the nanopore.

Figure 11.19a shows the ionic current trace of 10 mM dGMP, 10 mM dTMP, 10 mM dAMP and 10 mM dCMP in 400 mM KCl, 25 mM Tris HCl, pH 7.5, at 180 mV voltage bias across the hemolysin mutant pore with the adaptor molecule. Individual nucleotide molecules dGMP, dTMP, dAMP and dCMP can be discriminated based on the reduced ionic current while they bind in the nanopore [28], as illustrated by the shaded bands. Residual current histogram of nucleotide binding events, including Gaussian fits are shown in Fig. 11.19b, with clear separation of Gaussian peaks for each type of nucleotide molecules [28]. Demonstration of differentiating individual nucleotide molecules is a significant achievement and the remaining challenge of this approach is to guarantee that the cleaved nucleotide molecules will be transported to the pore in the original sequence of the target DNA molecule.

11.2 DNA Motion Control in the Nanopore

Among the various nanopore DNA sequencing approaches, electrical sensor approaches [19, 20, 24] require no modifications on the target single stranded DNA molecules, and thus have the potential to attain the lowest cost. For electrical sensor approaches, simulations [19, 21, 24] have suggested that the detection of different electrical signals for different DNA bases is plausible assuming that the position of DNA is well controlled against thermal agitation. Whatever electrical sensor is proposed for differentiating individual DNA bases while DNA is moved through a nanopore, the motion of the DNA has to be controlled at single base resolution. Otherwise the readings from the sensor will not be able to tell the "sequence" of the bases. Thus, controlling the motion of DNA translocation through a nanopore is a key challenge in the nanopore DNA sequencing field, and is the topic of this section.

11.2.1 Viscosity, Voltage, Ionic Concentration and Temperature Control

As demonstrated by Fologea et al. [31], the translocation speed of DNA through the nanopore is affected by the viscosity of the buffer (Fig. 11.20), the applied voltage bias (Fig. 11.21), the salt concentration (Fig. 11.22) and the temperature of the buffer (Fig. 11.23). DNA translocation speed can be considerably slowed down by increasing viscosity, decreasing applied voltage bias, decreasing the ionic

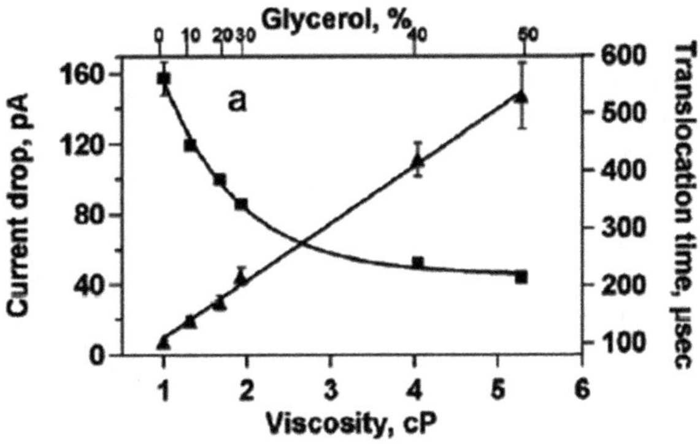

Fig. 11.20 Current blockage (*closed squares*) and translocation time (*closed triangles*) vs. viscosity for 3-kbp DNA in 1.5 M KCl-TE solution at 120 mV bias voltage. The *solid curves* are fits for $\Delta I_t \approx 1/\eta$ and $\Delta t_d \approx \eta$, where η is the viscosity of the buffer. Image adapted from [31]

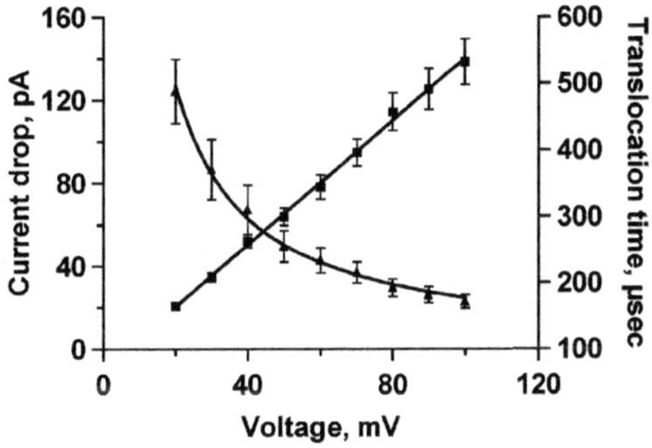

Fig. 11.21 Current blockage (*closed squares*) and translocation time (*closed triangles*) vs. applied voltage measured in 1.6 M KCl-TE buffer containing 20% glycerol. The solid curves are fits for $\Delta I_b \approx V$ and $t_d \approx 1/V$. Image adapted from [31]

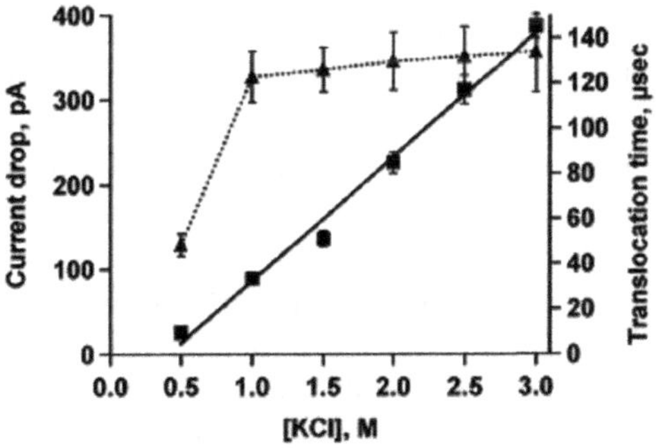

Fig. 11.22 Current blockage (*closed squares*) and translocation time (*closed triangles*) vs. applied voltage change as a function of KCl concentration. The *solid curves* are linear fits of current blockage and translocation time to KCl concentration. Image adapted from [31]

concentration and decreasing the temperature. Note that these approaches also decrease the ionic current signal by the same factor as that for the speed of DNA. Thus the signal to noise ratio for DNA sequencing proposals based on the ionic current measurement may degrade if using these methods for slowing down the translocation speed of DNA.

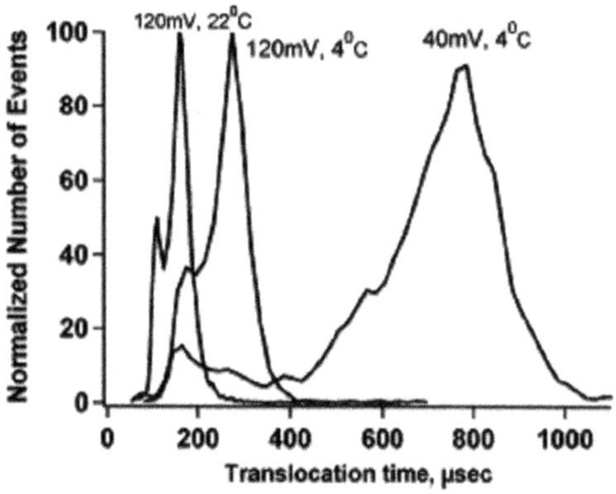

Fig. 11.23 Translocation time is increased when the temperature is decreased. At low voltage (40 mV) and low temperature (4 °C) in 1.6 M KCl-TE buffer containing 20% glycerol, the translocation time is about 800 μs. Image adapted from [31]

11.2.2 Effective Electric Driving Force on DNA in the Nanopore and Optical Tweezers Approach for Motion Control

Optical tweezers use a focused laser beam to trap, as well as control the motion of a micrometer-sized particle. After characterizing the stiffness k of the interaction between a laser beam and the particle, the optical tweezers can be used to measure forces from a few pico-Newtons to about 100 pico-Newtons. As a single molecule technique, optical tweezers are widely used to manipulate biological molecules and interrogate biological processes, such as stretching DNA [32] and probing the stepping of a kinesin molecule on micortubules [33]. Recently, optical tweezers have been used to measure forces on DNA in a solid-state nanopore [34, 35]. To facilitate a DNA sequencing process, ideally, optical tweezers can drive DNA through a solid-state nanopore at an arbitrarily slow speed [36].

In the experiment reported in [34], one end of a DNA molecule was attached to a bead via a standard streptavidin-biotin bond while the other end of DNA was electrically driven through a solid state nanopore. The effective electric driving force F_{el} that is balanced by the force from the optical tweezers can be computed from the measured displacement z of a bead, i.e. $F_{el} = -kz$ (see Fig. 11.24a). When increasing the biasing electric field across the solid membrane, the bead was dragged towards the membrane until the effective electric driving force was balanced by the restorative force from the optical tweezers. Figure 11.24b shows that the effective electric driving force increases linearly with the voltage drop across the membrane. Interestingly, the slope that characterizes the effective charge q_{eff} of DNA does not vary with the ion concentration in an electrolyte. This experiment elegantly demonstrated the successful combination of two nanotechnologies to obtain mechanical and electrical interactions between DNA and a nanopore. The nature of q_{eff}, however, was not determined, as it could result from either counterion condensation or hydrodynamic friction or both.

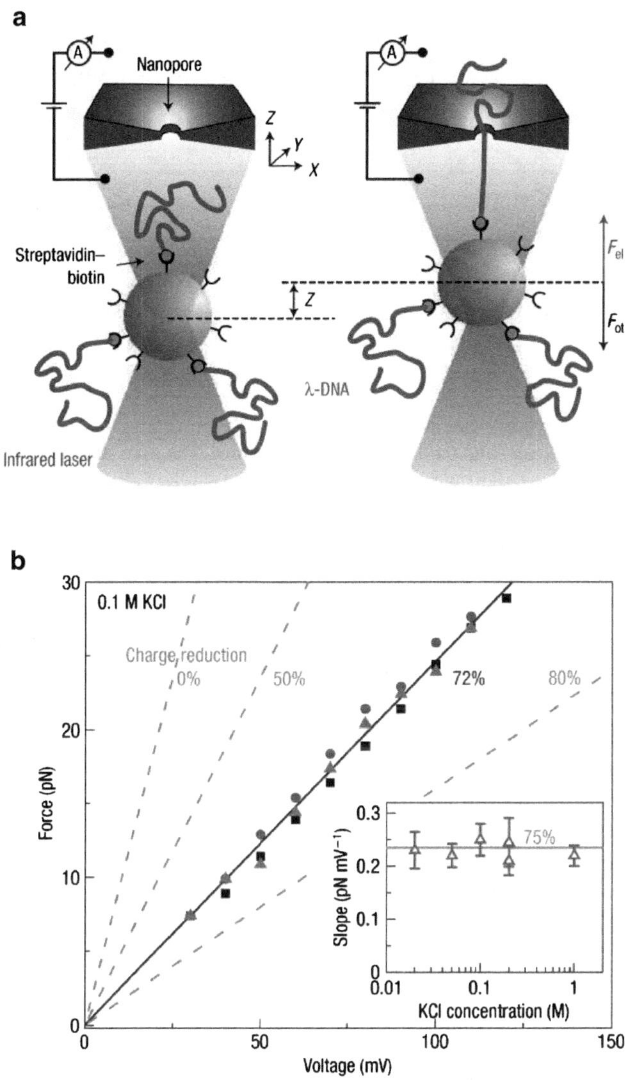

Fig. 11.24 (**a**) Experimental set-up. (**b**) Effective electric driving forces vs. voltage drops across the solid membrane for different salt concentrations. (Adapted from [34]. Figures permitted by Nature Physics)

Molecular dynamics simulations that can capture atomic-level details were performed to characterize the effective charge q_{eff} of DNA in a solid-state nanopore [37]. Simulation results show that the effective driving force on DNA is independent of the counterion-screened charge of DNA. Because of non-specific binding between mono-valent ions and DNA, a typical residence time of such ions is only about 10 ps. In an electric field, counterions near the DNA surface move in the field direction, causing an electro-osmotic flow inside a nanopore (Fig. 11.25a). Thus, the electric

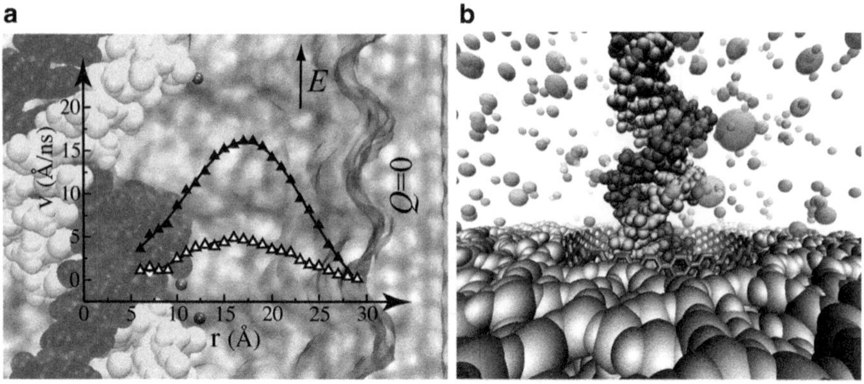

Fig. 11.25 (**a**) Water flow velocity profiles between DNA and pore (3-nm-radius) surfaces, when E = 0.02 (*open triangles*) and 0.08 (*filled triangles*) V/nm. The pore surface is neutral. (**b**) DNA translocation through a solid-state nanopore coated with a single-wall carbon nanotube

driving force was reduced by the hydrodynamic drag of the electro-osmotic flow along the DNA surface. Simulations further demonstrate that the effective electric driving force is $\xi\mu$, where ξ is the friction coefficient and μ the mobility. This proposed hydrodynamic effect was confirmed in experiments that showed different effective driving forces on DNA in pores with different sizes [38], indicating a hydrodynamic effect.

Besides the pore size, the hydrodynamic drag force on DNA could be affected by other boundary conditions of the flow, such as surface charge density and roughness. It was shown in simulations that the effective driving force decreases when the surface charge density of a pore changes from a negative to a positive value [39] and increases when the surface becomes rougher [37]. More importantly, microscopic interactions between DNA and a nanopore surface could result in irregular motion of DNA or even immobilize DNA. However, a steady and controllable motion of DNA inside a nanopore is critical for the ultimate success of the nanopore-based DNA sequencing method. To avoid or reduce such an unfavorable interaction, a solid-state nanopore might be coated with a single-walled carbon nanotube (SWCNT) that provides a smooth surface, reducing surface imperfections of a solid (Fig. 11.25b). Potentially, SWCNT modification of a pore surface could reduce the adhesive and frictional forces between DNA and a nanopore, yielding a steadier translocation of DNA. Note that DNA translocation through a SWCNT has been demonstrated experimentally [40], but the dynamics of that motion is still not clear and deserves further study.

11.2.3 Magnetic Tweezers

Although optical tweezers can drive DNA through a solid-state nanopore at an arbitrarily slow speed [36], it is difficult to scale such assay to large number of

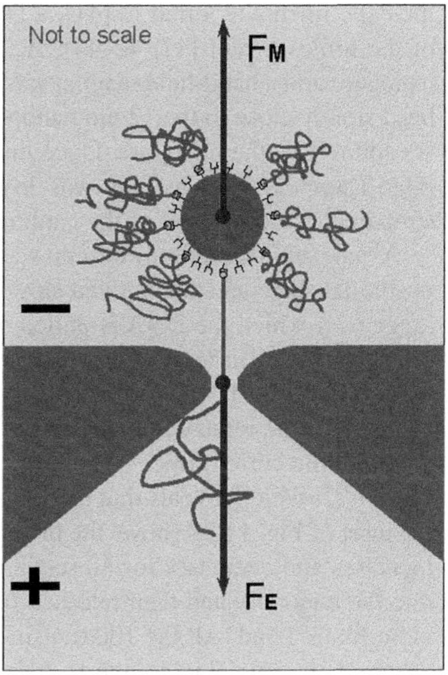

Fig. 11.26 Schematic of the reverse DNA translocation using magnetic tweezers. Two reservoirs, filled with ionic buffer, are separated by a nanopore chip (shown in *gray*). A voltage bias is applied across the chip. DNAs (shown as *wiggly lines*) attach to the magnetic bead via standard streptavidin-biotin bonds. Electrical force F_E on the DNA and magnetic force F_M on the magnetic bead are drawn as thick *solid arrows*. Adapted from [41]

nanopores. The basic concept of the magnetic tweezers approach [41] is shown in Figure 11.26. DNA molecules are attached to magnetic beads via the standard streptavidin-biotin bonds. The free end of the DNA can be captured into the nanopore by the applied electric field. Subsequently, one can apply a precisely controlled magnetic force on the magnetic bead to balance the electrical force on the trapped DNA, i.e., the DNA is in a tug-of-war between the magnetic bead and the nanopore. By increasing the magnetic force further, or reducing the bias voltage, until the magnetic force exceeds the electrical force, the DNA can be pulled out of the nanopore from the *cis* side of the nanopore. In this way, the minimized net force on the DNA and the hydrodynamic drag on the micron-sized bead will slow down the motion of DNA while it is pulled out from the nanopore. Since one can construct a magnetic field gradient over a large space, this technique is inherently applicable to large number of addressable nanopores. By ramping the magnetic field slowly, the DNAs in all the nanopores can be pulled out slowly during one ramping step.

The experiment reported in [41] was done on a 12 nm size nanopore at 0.1 M KCl in Tris-EDTA buffer at pH 8. Tween 20 was used to prevent the sticking of magnetic bead to the surface of a nanopore chip. It is known [42–44] that the presence of a DNA in a nanopore has two competing effects for the nanopore conductance: the physical volume of the DNA leads to a reduction in total ion population in the nanopore, thereby reducing nanopore conductance; the negatively charged DNA brings in extra counterions, leading to a conductance enhancement. The net effect of a translocating DNA on the nanopore conductance depends on the ionic strength of the buffer solution. In the case of 0.1 M KCl that was used in the experiments reported in [41], a DNA capture event is indicated as an increase of the ionic current [17, 42–44]. A 200 mV voltage bias was applied across the nanopore and a hand-held magnet was employed to drag one DNA-coated magnetic bead slowly close to the 12 nm nanopore. The ionic current increases when a DNA is captured by the nanopore. Once this occurs, the magnetic bead becomes immobile. Then the voltage bias was lowered to 50 mV to avoid breakage of the streptavidin-biotin bond of the captured DNA.

A pair of magnetic tweezers was mounted on a Burleigh Inchworm nanopositional stage and was moved slowly towards the nanopore chip. In the distance range over which the DNA is pulled out, the force increase rate is less than 0.2 fN per step on the Burleigh stage. Once the magnetic force exceeds the electric force (at 50 mV), the DNA was pulled out from the nanopore, as indicated from the decrease of the ionic conductance to the baseline value.

The ionic current trace of a DNA being pulled out of the nanopore is shown in Fig. 11.27, which reveals that the transitional time is about 19 ms. For comparison, the inset of Fig. 11.27 shows the histogram of the transitional times of ionic current increases and decreases for 50 sequential DNA events, in which DNA is captured into the nanopore and then released from the nanopore due to breakage of streptavidin-biotin bond. All the 100 transitional times are less than 2 ms. (The histogram peaks at 1.2 ms due to the 1 kHz cutoff frequency set in the ionic current measurement system.) So the 19 ms is the actual time that it takes for the end of the DNA to pass through the 60 nm long nanopore channel by the pulling of the

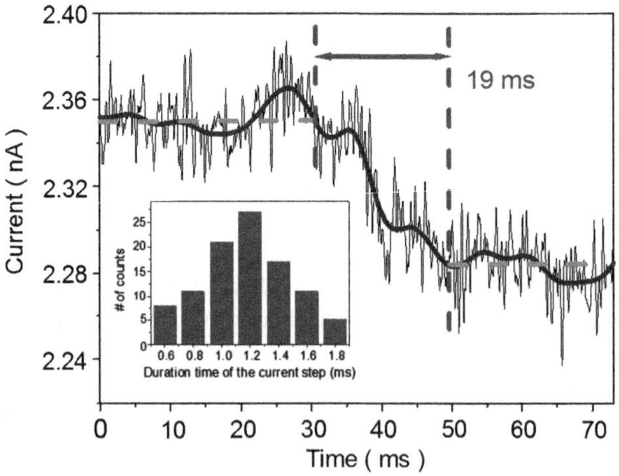

Fig. 11.27 The time trace of the nanopore ionic current as the DNA is being pulled out of the nanopore by magnetic force. (Note $t = 0$ is re-defined.) The thick *solid line* is a smoothed curve of the same data. For comparison, the inset shows the histogram of the transitional times of ionic current increases and decreases from 50 sequential events (DNA is captured into the nanopore and then released from the nanopore due to breakage of streptavidin-biotin bond). Note that the peak at 1.2 ms in the *inset* is not the actual time that the end of the DNA passes through the nanopore. It is due to the 1 kHz cutoff frequency set in the ionic current measurement system. Adapted from [41]

magnetic bead. As such, the speed of the end of the DNA moving out of the nanopore can be estimated to be about 0.00316 nm/μs, or 0.0096 bases/μs. This is so because when the end of the DNA is in the nanopore, the net force on the DNA is largest (the magnetic force is largest as the bead is closest to the magnetic tweezers and the entropic force exerted by the random coil part of the DNA is smallest as the DNA has been stretched), and in consequence the instant speed of the DNA is the largest: thus the average speed of the DNA during the "pulling out" process is actually slower than the instant speed of the end of DNA moving out of the nanopore. By comparing to the average speed of the standard DNA translocation [14, 18], the average speed of this *reverse DNA translocation* (by magnetic tweezers) is more than 2,000-fold slower than that of the standard DNA translocation [14, 18].

11.2.4 "DNA Transistor"

As experimentally demonstrated by Keyser et al. [34], the electrical force on double-stranded DNA in a nanopore is proportional to the voltage bias across the nanopore, about 0.24 pN/mV. This should allow the designing of a device to electrically trap a single DNA in a nanopore against the thermal agitation, which is exactly the idea of "DNA transistor" [45] of putting a stack of metal/dielectric/metal/dielectric/metal layers into the membrane with the nanopore. As shown in Fig. 11.28a, the "shaded" layers (thickness *W*) are metal while the non-shaded layers (thickness *S*) are dielectric. The discrete negative charges of a DNA

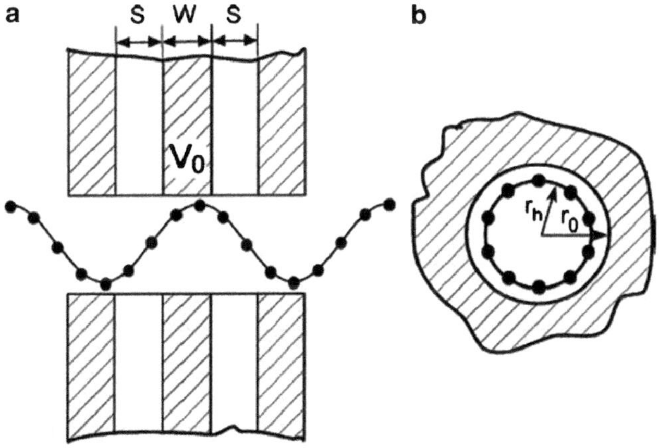

Fig. 11.28 A schematic of the DNA transistor. The *shaded regions* in the sketch are metal layers, while the non-shaded regions are dielectric layers. The negative charges of the ssDNA are represented by the *solid circles*: (**a**) cross section and (**b**) top view. (Adapted from [45])

molecule are represented by solid circles. The thicknesses W and S are in units of the intercharge distance of the DNA, d (assumed to be ~0.7 nm for stretched ssDNA). Voltage bias V_0 is applied on the middle metal layer while the two other metal layers are grounded. If we ignore the distortion of the electrical field inside the nanopore, then the electric field inside the nanopore is 0 at the area with metal wall (metal zone) and takes a constant value $E_{TR} = -V_0/(Sd)$ or $V_0/(Sd)$ at the area with dielectric wall (dielectric zone). The net force on the DNA will be $F_{net} = q^*(N_R\text{-}N_L)^*V_0/(Sd)$, q^* is the effective charge on each phosphate group of the DNA backbone (about 0.5 e) according to [34], N_R is the number of charges on the DNA that are in the right dielectric zone and N_L is the number of charges on the DNA that are in the left dielectric zone.

As shown in Fig. 11.29a, assuming the dielectric layer thickness S is a half integer, and the metal layer thickness W is slightly (ε) larger than integer length, the

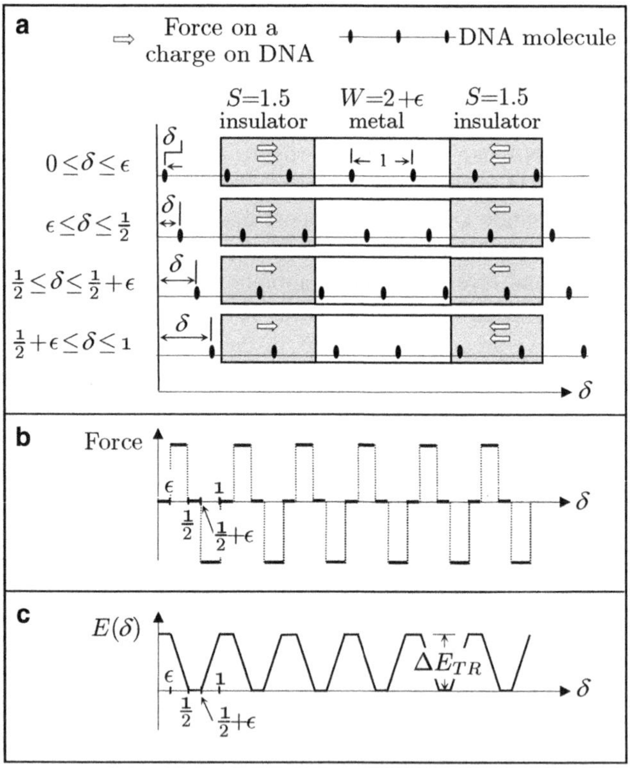

Fig. 11.29 For the case that the thickness of the dielectric layer S is a half integer and the thickness of the metal layer W is slightly larger than an integer: (**a**) Schematic of forces on the charges of the DNA at different displacements of the DNA (relative to the nanopore). The force on each charge of the DNA is represented by a *white arrow*, pointing the direction of the force. (**b**) Schematic of the net force on the DNA vs. the displacement of the DNA δ (relative to the nanopore). (**c**) The potential associated with the force depicted in (**b**). ΔE_{TR} is the effective trapping barrier. (Adapted from [45])

net force on the DNA will be dependent on the displacement of the DNA relative to the nanopore (δ). The net force on the DNA vs. the displacement of DNA is summarized in Fig. 11.29b and the potential associated with the net force on the DNA is summarized in Fig. 11.29c. The potential is a trapping potential with effective trapping barrier $\Delta E_{TR} = q^* E_{TR}(1/2\text{-}\varepsilon)d$. The maximum attainable ΔE_{TR} is dependent on electrical breakdown field of the dielectric material. For a reasonable estimation of trapping energy on a single-stranded DNA, assuming unscreened DNA charges, the dielectric thickness 1.5 d (1.05 nm), the voltage $V_0 = 1$ V on the order of the electrical breakdown voltage 1 V for 1 nm thick SiO_2), ΔE_{TR} is 13.6 $k_BT \gg$ thermal energy k_BT, k_B is Boltzmann constant and T is room temperature 300 K.

To move the single-stranded DNA through the DNA transistor base by base, one will need to adjust the gate voltages and voltage bias across the nanopore. Referring to Fig. 11.29c, for the case $\varepsilon = 0$ under the condition that a voltage bias is applied across the nanopore to break the symmetry, the trapping potential of the DNA vs. the displacement of the DNA is represented in Fig. 11.30a. The peak to peak distance is the inter-charge-space or inter-base-space. The DNA center of mass is initially trapped at the position indicated with the solid circle and labeled "0". If one reverses the polarity of the gate voltage, the minimum and maximum trapping potential locations are swapped, as shown in Fig. 11.30b. Due to the broken symmetry, the DNA will move to the new potential minimum point to the right of the original potential minimum position, as shown by the solid circle labeled "1". As the polarity of the gate voltage is switched back and forth cyclically, the DNA will be moved to the right at a rate of one base per cycle.

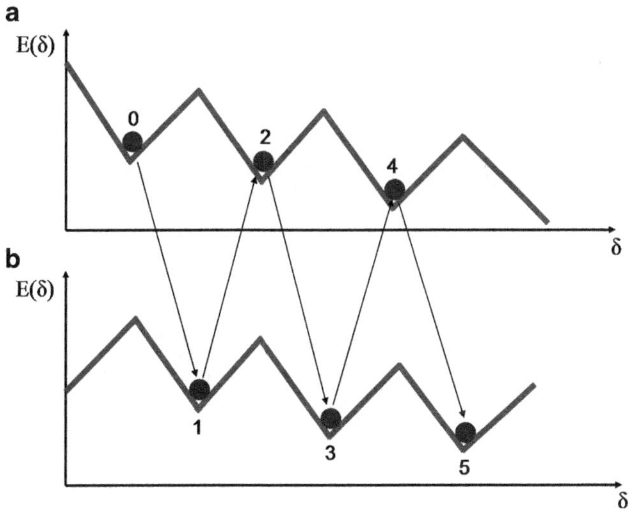

Fig. 11.30 (**a**) A schematic of trapping potential vs. the displacement of the DNA in the DNA transistor when a voltage bias is applied across the DNA transistor. (**b**) A schematic of trapping potential vs. the displacement of the DNA in the DNA transistor when the polarity of the gate voltage in (**a**) is reversed. The *solid circle* labeled "*0*" shows the initial position of DNA. *Circles* "*1*," "*2*," "*3*," "*4*," and "*5*" show the sequential positions of the DNA when the polarity of the gate voltage is switched back and forth

Further realistic molecular dynamics simulations [46] confirmed that single DNA molecules can be driven through the nanopore in a ratchet-like fashion, with a step size equal to the spacing between neighboring phosphate groups in the ssDNA backbone during the simulation time scale of tens of nanoseconds. Fig. 11.31a illustrates the DNA transistor system in the simulations. In order to characterize the field of the electric force in the DNA transistor, a fragment of a single stranded DNA (ssDNA) containing 20 adenine deoxynucleotides, poly (dA_{20}), was simulated as being pulled by a spring while submerged in a NaCl electrolyte solution confined in a 4 nm diameter nano-channel. The nanochannel was assumed to be an opening through multiple layers of amorphous SiO_2 (regions where the electric field indicated with the arrows is present) and metal (light grey). As for the trapping electric field, we used the maximum value that SiO_2 can take before dielectric breakdown ($E = 108$ mV/Å). This field was applied in the dielectric regions, as shown in the figure. The thickness of each dielectric region is $2.5d$, and the thickness of middle metal region is $2d$, where d is the spacing between neighboring phosphate groups. After equilibration in the NVT (T = 300 K)

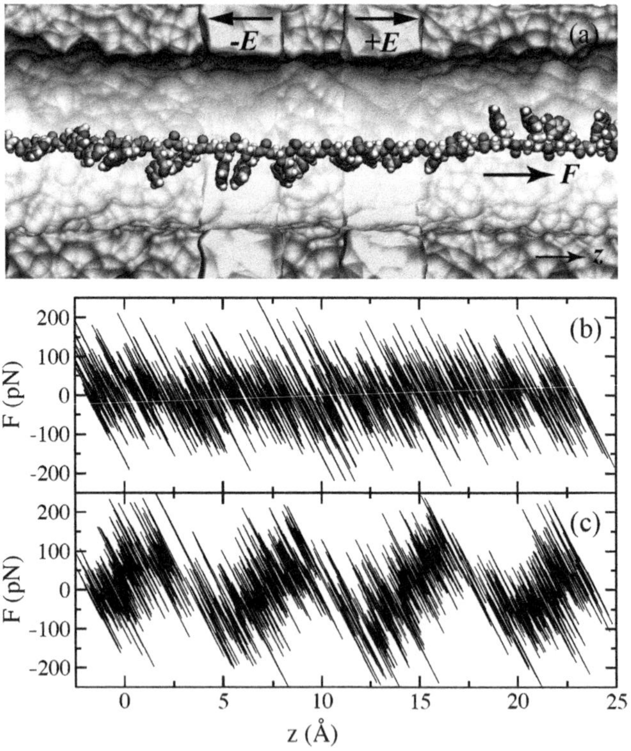

Fig. 11.31 Simulation of DNA's motion in a DNA transistor when DNA is pulled by harmonic spring. (**a**) Illustration of the simulation system. (**b**) Measured force exerted by the spring vs. DNA position relative to the DNA transistor when $E = 0$. (**c**) Measured force exerted by the spring vs. DNA position relative to the DNA transistor when $E = 108$ mv/Å. Adapted from [46]

ensemble for 2 ns, DNA was pulled by a harmonic spring (mimicking optical tweezers) of constant k and at a constant velocity of 1 Å/ns along the nano-channel. When the trapping fields are off (Fig. 11.31b), the pulling force exerted by the spring can be characterized by $F = \xi v$, where v is the pulling velocity and ξ the friction coefficient. The large fluctuation in force can be explained by thermal fluctuations$\sqrt{k_B T k}$. When the trapping fields are on (with value 108 mV/Å), the force exerted by the spring vs. the position of DNA relative to the DNA transistor is shown in Fig. 11.31c, indicating the uphill and downhill motions of DNA on the landscape of a periodic potential of period d. The maximum trapping force is about 90 pN, corresponding to a trapping potential of about 100 meV, much larger than the thermal energy 26 meV.

This "harmonic spring" approach can be used to characterize the electrical trapping force inside the DNA transistor. But for real applications, rather than using optical tweezers, one will use a biasing electric field across the nanochannel to drive the DNA through the DNA transistor base by base. As shown in Fig. 11.32, when the biasing field is strong (9.38 mV/Å), DNA moves through the transistor at a constant velocity. However, when decreasing the biasing electrical field, the motion of DNA consists of a stick state and a slip state, as shown in Fig. 11.32. The stick–slip motion of DNA in a transistor can also be seen in http://www.youtube.com/watch?v=aLXw2pebWPg. This movie was made from a trajectory of a 10 ns molecular dynamics simulation. The simulation setup is same as described above. The biasing electric field is 0.625 mV/Å and the trapping field is 108 mV/Å. This stick–slip motion can be extremely useful for controlling the DNA

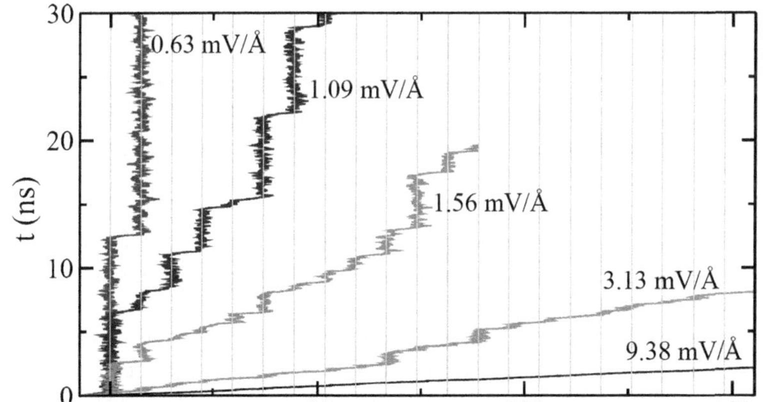

Fig. 11.32 Simulation results (time vs. the position of the DNA center of mass) for the stick–slip motion of DNA in a DNA transistor. DNA is driven by a biasing electric field across the whole channel while the trapping fields are on. The applied biasing fields vary from 0.625 to 9.375 mV/Å. The spacing of *vertical lines* is d, the spacing between neighboring phosphate groups. Adapted from [46]

motion inside nanopore. In a typical slip event, ssDNA advances one nucleotide spacing d (as clearly seen in Fig. 11.32).

11.3 Discussion: Beyond Sequencing

The technological progress made from the early chain termination methods [1] of the late 1970s to the present day sequencing methods is truly remarkable. The ingenuity of the methods presented in this chapter, and of other methods left out for consideration of length (e.g., [47] and [48]), have pushed the boundaries of DNA sequencing in terms of speed, cost and reduced complexity. Indeed, the cost per finished base pair has decreased seven orders of magnitude in the two decades that elapsed since onset of the Human Genome Project (1990), when the cost was about $10 per base [49]. Likewise the time required to sequence a human genome decreased between three to four orders of magnitude. In the words of Fred Sanger, one of the protagonists of this saga [50]: "When we started working on DNA I don't believe we were thinking about sequencing the entire human genome—perhaps in our wildest dreams but certainly not within the next 30 years. After all, the human genome was probably the ultimate goal of all our work."

Indeed, the *raison d'etre* of all the elegant technologies described earlier in this chapter is that they will enhance our knowledge of biology in general and of human biology in particular. We have started to see tremendous progress in that direction. The ENCyclopedia Of DNA Elements (ENCODE) project, an effort funded by the National Human Research Institute, has started to produce unforeseen new vistas into the structural and functional organization of the human genome. The finding of complex patterns of regulation and transcription together with conservation of non-coding regions and the abundance of non-coding RNA genes have led some researchers to postulate new definitions for the concept of gene [51]. Other pioneering effort designed to bridge the gap between human genetic variation and human health is the Personal Genome Project (PGP) [52]. By recruiting volunteers willing to share both their genome sequence and their clinical information with the research community, the PGP will produce a dataset of personal genome sequences that can be evaluated along with the medical data, to enable the discovery of associations prior to more rigorous data collection.

Eventually, the hope is that by enabling a phase of intense research in the association between genomic features and clinical characteristics, fast and cheap sequencing will enable the era of personalized medicine. The race and efforts to find the best sequencing technology that we are witnessing is an important part of the history of the medicine of the future.

Acknowledgments The authors acknowledge useful discussions with members of the IBM DNA-transistor team: Ali Afzali, Arjang Hassibi, George Walker, Glenn Martyna, Philip Waggoner, Stanislav Polonsky, Stefan Harrer and Stephen Rossnagel. This work was supported in part by a grant from the National Institutes of Health (R01-HG05110-01).

References

1. Sanger F, Nicklen S, Coulson, AR (1977) DNA sequencing with chain-terminating inhibitors Proc. Natl. Acad. Sci. USA 74(12): 5463–5467.
2. Sanger F (8 December. 1980) Determination of nucleotide sequences in DNA, Nobel lecture.
3. Collins FS, Morgan M, Patrinos A (2003) The human genome project: lessons from large-scale biology. Science 300: 286–290.
4. Mitra RD, Shendure J, Olejnik J, Olejnik EK, Church G.M (2003) Fluorescent in situ sequencing on polymerase colonies. Analyt. Biochem. 320: 55–65.
5. Mitra RD, Church GM (1999) In situ localized amplification and contact replication of many individual DNA molecules, Nucleic Acids Res. 27.
6. Rehman FN, Audeh M, Abrams ES, Hammond PW, Kenney M, Boles TC (1999) Immobilization of acrylamide-modified oligonucleotides by co-polymerization, Nucleic Acids Res. 27: 649–655.
7. Vasiliskov AV, Timofeev EN, Surzhikov SA, Drobyshev AL, Shick VV, Mirzabekov AD (1999) Fabrication of microarray of gelimmobilized compounds on a chip by copolymerization, Biotechniques 27: 592–594, 596–598, 600.
8. Ronaghi M (2001) Pyrosequencing sheds light on DNA sequencing. Genome Res. 11: 3–11.
9. Shendure J, and Ji H (2008) Next-generation DNA sequencing. Nat. Biotechnol. 26: 1135–1145.
10. Mardis ER (2008) Next-Generation DNA Sequencing Methods. Annu Rev Genomics Hum Genet 9:387–402.
11. Kasianowicz JJ, Brandin E, Branton D, Deamer DW (1996) Characterization of individual polynucleotide molecules using a membrane channel. Proc. Natl. Acad. Sci. USA 93: 13770–133773.
12. Deamer DW, Akeson M (2000) Nanopores and nucleic acids: prospects for ultrarapid sequencing Trends Biotechnol. 18: 147–51.
13. Akeson M et al. (1999) Microsecond time-scale discrimination among polycytidylic acid, polyadenylic acid, and polyuridylic acid as homopolymers or as segments within single RNA molecules Biophys. J. 77: 3227–33.
14. Li J, Gershow M, Stein D, Brandin E, Golovchenko JA (2003) DNA molecules and configurations in a solid state nanopore microscope Nature Mat. 2: 611.
15. Heng JB et al. (2003) The detection of DNA using a silicon nanopore IEDM Tech. Digest 767–70.
16. Heng JB et al. (2004) Sizing DNA Using a Nanometer-Diameter Pore Biophys. J. 87: 2905.
17. Chang H et al. (2004) DNA-mediated fluctuations in ionic current through silicon oxide nanopore channels Nano Lett. 4: 1551–1558.
18. Storm AJ et al. (2003) Fast DNA translocation through a solid-state nanopore Nat. Mater. 2: 537–40.
19. Lagerqvist J, Zwolak M, Di Ventra M (2006) Fast DNA sequencing via transverse electronic transport. Nano Lett. 6: 779–782.
20. Zhang XG et al. (2006) First-principles transversal DNA conductance deconstructed. Biophys. J. 91: L04–L06.
21. Gracheva ME, Xiong A, Aksimentiev A, Schulten K, Timp G, Leburton JP (2006) Nanotechnology 17: 622–633.
22. Lee JW, Meller A (2007) Perspectives in Bioanalysis, edited by K. Mitchelson (Elsevier).
23. Ling XS, Bready B, Pertsinidis A (2006) Hybridization-Assisted Nanopore Sequencing of Nucleic Acids, USPTO Patent Application No. 20070190542.
24. Gracheva ME, Aksimentiev A, Leburton JP (2006) Electrical signatures of single-stranded DNA with single base mutations in a nanopore capacitor Nanotechnology 17: 3160–3165.
25. Sauer-Budge AF, Nyamwanda JA, Lubensky DK, Branton D (2003) Unzipping Kinetics of Double-Stranded DNA in a Nanopore Phys. Rev. Lett. 90: 238101.
26. Mathé J, Viasnoff HVV, Rabin Y, Meller A (2004) Nanopore unzipping of individual DNA hairpin molecules Biophys. J. 87: 3205.
27. Mathé, J, Arinstein A, Rabin Y, Meller A (2006) Equilibrium and irreversible unzipping of DNA in a nanopore Europhys. Lett. 73: 128.

28. Clarke J, Wu HC, Jayasinghe L, Patel1 A, Reid S, Bayley H (2009) Continuous base identification for single-molecule nanopore DNA sequencing Nature Nanotechnology 4: 265–270.
29. Drmanac R, Drmanac S, Labat I, Crkvenjakov R, Vicentric A, Gemmell A (1992) Sequencing by hybridization: towards an automated sequencing of one million M13 clones arrayed on membranes. Electrophoresis 13: 566–573.
30. Southern EM (1996) DNA chips: analyzing sequence by hybridization to oligonucleotides on a large scale Trends Genet. 12(3): 110–115.
31. Fologea D, Uplinger J, Thomas B, McNabb DS, Li J (2005) Slowing DNA Translocation in a Solid-State Nanopore Nano Lett. 5: 1734–1737.
32. Bustamante C, Bryant Z, Smith SB (2003) Ten years of tension: single-molecule DNA mechanics. Nature 421: 423–427.
33. Yildiz A, Tomishige M, Gennerich A, Vale RD (2007) Intramolecular Strain Coordinates Kinesin Stepping Behavior along Microtubules. Cell 134: 1030–1041.
34. Keyser UF, Koeleman BN, van Dorp S et al. (2006) Direct force measurements on DNA in a solid-state nanopore. Nat Phys 2: 473–477.
35. Trepagnier EH, Radenovic A, Sivak D et al. (2007) Controlling DNA capture and propagation through artificial nanopore. Nano Letts 7: 2824–2830.
36. Dekker C (2007) Solid-state nanopore. Nat Nanotech 2: 209–215.
37. Luan B, Aksimentiev A (2008) Electroosmotic screening of the DNA charge in a nanopore. Phys Rev E78: 021912–021915.
38. van Dorp S, Keyser UF, Dekker NH, Dekker C, Lemay SG (2009) Origin of the electrophoretic force on DNA in solid-state nanopores. Nat Phys 5: 347–351.
39. Luan B, Aksimentiev A (2010) Control and reversal of the Electrophoretic force on DNA in a charged nanopore. J Phys Condens Matter 22: 44123.
40. Liu H, He J, Tang J et al. (2010) Translocation of Single-Stranded DNA Through Single-Walled Carbon Nanotubes. Science 327: 64–67.
41. Peng HB, Ling XS (2009) Reverse DNA translocation through a solid-state nanopore by magnetic tweezers. Nanotechnology 20: 185101.
42. Fan R, Karnik R, Yue M, Li D, Majumdar A, Yang P (2005) DNA Translocation in Inorganic Nanotubes Nano Lett. 5: 1633–1637.
43. Smeets RMM, Keyser UF, Krapf D, Wu MY, Dekker NH, Dekker C (2006) Salt dependence of ion transport and DNA translocation through solid-state nanopores Nano Lett. 6: 89–95.
44. Chang H, Venkatesan BM, Iqbal SM, Andreadakis G, Kosari F, Vasmatzis G, Peroulis D, Bashir R (2006) DNA counterion current and saturation examined by a MEMS-based solid state nanopore sensor Biomed Microdevices 8: 263–269.
45. Polonsky S, Rossnagel S and Stolovitzky G (2007) Nanopore in metal-dielectric sandwich for DNA position control Appl. Phys. Lett. 91: 153103.
46. Luan B, Peng H, Polonsky S, Rossnagel S, Stolovitzky G, Martyna G (2010) Base-By-Base Ratcheting of Single Stranded DNA through a Solid-State Nanopore Phys. Rev. Lett. 104: 238103.
47. John Eid et al. (2009) Real-Time DNA Sequencing from Single Polymerase Molecules. Science 323: 133.
48. Radoje Drmanac et al. (2010) Human Genome Sequencing Using Unchained Base Reads on Self-Assembling DNA Nanoarrays. Science 327: 78.
49. Service RF (2006) Science. 311: 1544–1546.
50. Fred Sanger (2001) The early days of DNA sequences, Nature Medicine 7: 267.
51. Gerstein MB, Bruce C, Rozowsky JS, Zheng D, Du J, Korbel JO, Emanuelsson O, Zhang ZD, Weissman S, Snyder M (2007) What is a gene, post-ENCODE? History and updated definition, Genome Res. 17: 669.
52. http://www.personalgenomes.org/

Chapter 12
Third Generation DNA Sequencing with a Nanopore

Gregory Timp, Utkur Mirsaidov, Winston Timp, Jiwook Shim, Deqiang Wang, Valentin Dimitrov, Jan Scrimgeour, Chunchen Lin, Jeffrey Comer, Anthony H. Ho, Xueqing Zou, Aleksei Aksimentiev, and Klaus Schulten

Abstract With the advent of Next-Generation-Sequencing (NGS) technologies, an enormous volume of DNA sequencing data can be generated at low cost, placing genomic science within the grasp of everyday medicine. However, mired in this voluminous data, a new problem has emerged: the assembly of the genome from the short reads. In this chapter we examine the prospects for sequencing DNA using a synthetic nanopore. Nanopore sequencing has the potential for very long reads, reducing the computational burden posed by alignment and genome assembly, while at the same time eliminating logistically challenging and error-prone amplification and library formation due to its exquisite single molecule sensitivity. On the other hand, long high fidelity reads demand stringent control over both the DNA configuration in the pore and the translocation kinetics. We examine the prospects for satisfying these specifications with a synthetic nanopore.

Keywords Sequencing DNA • Trapping DNA • Comparison between biological and synthetic nanopores • Molecular dynamics simulations • Hydrodynamic focusing • λ-DNA

12.1 Introduction

The Sanger method of DNA sequencing has transformed biology – it has given us the first draft of the human genome at a price tag of about $2.7 billion [1, 2]. The basis for Sanger sequencing is the polymerase chain reaction (PCR), which is used in combination with dideoxy-terminated nucleotides to prematurely terminate the elongation reaction. By mixing fluorescently labeled dideoxynucleotides with deoxynucleotides, PCR is prematurely terminated, leading to fragmentary single-stranded copies of the

---

G. Timp (✉)
Stinson-Remick Hall, University of Notre Dame, Notre Dame, IN 46556, USA
e-mail: gtimp@nd.edu

S.M. Iqbal and R. Bashir (eds.), *Nanopores: Sensing and Fundamental Biological Interactions*, DOI 10.1007/978-1-4419-8252-0_12,

template that differ in length with the last base fluorescently labeled with a characteristic fluorescent moiety. Separating these fragments by size through electrophoresis, the sequence can be determined from the color of fluorescence produced at a given fragment length. Though serviceable, this procedure is problematic for several reasons. The template read length using this method is limited to ~800 bp with base-calling quality dropping as read length increases. This introduces significant challenges, especially for de novo sequencing, requiring that either chromosome walking or shotgun sequencing be used, which are logistically challenging and require computationally intensive assembly of the completed sequence. The chain termination reaction itself is also time consuming, as is electrophoretic separation [3]. However, the overarching problems with Sanger sequencing are the relatively large amounts (~100 μg) of DNA required – amplification leads to errors – and the expense due to reagents for labeling and separation.

Lately, Sanger sequencing has been superseded by next generation sequencing (NGS) technologies such as Illumina's Genome Analyzer, Roche's 454 sequencer and Applied Biosystems' SOLiD System, which can resequence a human genome for less than $100,000. NGS produces an enormous volume of DNA sequencing data – in excess of one billion short reads per instrument per day [4]. NGS methods generally involve randomly shearing genomic DNA into smaller fragments from which sequencing libraries are created. Many NGS technologies, including cyclic reversible termination, single-nucleotide addition and single molecule real time (sic), can be categorized as sequencing-by-synthesis that involves DNA polymerase. These short templates are immobilized in a flow cell allowing billions of sequencing reactions to be performed simultaneously. The immobilized libraries are subjected to either sequencing-by-synthesis techniques (cyclic reversible termination, single nucleotide addition) or sequencing-by-ligation. Finally, NGS uses either fluorescent or luminescent detection of nucleotide incorporation in combination with imaging to readout the sequence one reaction at a time.

Libraries are prepared either through clonally amplified templates or single DNA molecule templates. In both cases the nucleic acid is first fragmented, then adaptor primers are annealed to the end to make templates. In the case of clonal amplification, individual fragments are then amplified using emulsion PCR, isothermal bridge amplification, or another technique to cluster a group of clonal amplicons. This amplification produces greater signal as each of the amplicons will contribute signal to the sequencing process. For example, Illumina relies on the amplification of clonal clusters and a sequencing-by-synthesis reversible terminator method that uses reversible versions of dye-terminators, adding one nucleotide at a time and then detecting fluorescence at each position by repeated removal of the blocking group to allow polymerization of another nucleotide. Pyrosequencing as used by 454 also uses clonal amplification along with DNA polymerization, adding one nucleotide at a time and then detecting the nucleotides added to a location through the light emitted through release of the attached pyrophosphates. Since nucleotides are added to the templates in a given cycle, each cycle demands a high fidelity addition process. Incomplete extension or misincorporation of the template results in signal de-phasing, where bases become out of step with the rest in the cluster, which increases fluorescence noise and causes base-calling errors forcing shorter reads [4].

12.1.1 Single Molecule Sequencing

On the other hand, de-phasing is not an issue with single-molecule templates and so the requirement for cycling efficiency is relaxed. However, single molecule templates are still susceptible to multiple nucleotide additions in a cycle. Deletion errors occur due to either quenching effects between adjacent dye molecules or incorporation of dark nucleotides. And finally, achieving the signal-to-noise ratio (SNR) required for single molecule detection still remains a challenge. *Pacific Biosystems* is developing a SMRT (single molecule real time) sequencing method that relies on enzymatic incorporation of a fluorescently labeled nucleotide through a DNA-polymerase, applying zero-mode waveguide (ZMW) technique to suppress the ambient radiation so that one molecule can be identified. While this strategy boasts the potential for 5–10 Gbp/s from an array of ZMWs, it is still error prone and suffers polymerase-dependent shortcomings [5].

The polymerase-dependent methods favor short reads due to the nature of a processive sequencing process; because errors accumulate over time, the quality of the base-calling deteriorates with read length. As a result of the short reads, a bioinformatic bottleneck develops, which makes alignment and assembly of the genome an especially vexing proposition [6]. It has been shown that the assembly quality deteriorates rapidly as the read length decreases [7, 8]. For Sanger sequencing (750 bp) an assembly of *Neisseria meningitidis* resulted in 59 contigs, 48 of which were >1 kbp, whereas at 70 bp, the assembly consisted of >1,800 contigs. The assembly was still fragmented (296 contigs) even for relatively long reads (200 bp) within reach of NGS [7]. Irrespective of the sequencing technology, de novo assembly becomes difficult whenever the length of the read is shorter than the repetitive segments in the sequence because the sequences cannot be aligned unambiguously. For such cases assembly falls into a class of NP-hard problems with no efficient computational solution [6]. This is especially problematic with a large number of repeats – which can be used for genotyping or relevant in mobile genetic element location [9]. It becomes exponentially harder to assemble a genome as the number of repeats grows.

Single molecule DNA sequencing represents the logical, end-of-the-line in development sequencing technology, which extracts the maximum amount of information from a minimum of material and pre-processing. The low material requirement coupled with quick results would allow for easy sequencing of precious primary samples from human patients, e.g. allowing doctors to look for rare diseases in the pre-clinical stage [10]. Among the emerging third generation technologies, sequencing a single molecule of DNA with a nanopore seems to have the brightest prospects [11], because it has the potential for very long read lengths (>1 kbp), thereby reducing the time required for alignment of the reads and the subsequent assembly of the genome, while eliminating logistically challenging and error-prone amplification and library formation due to the exquisite single molecule sensitivity. Moreover, electrical detection of DNA using a nanopore has several other advantages over luminescent or fluorescent detection – it has more dynamic range and requires less preprocessing.

12.1.2 The Prospects for Single Molecule Sequencing with a Nanopore

The nanopore sequencing concept uses a radically new approach to detection that relies on the electrical signal that develops when a DNA molecule immersed in electrolyte translocates through a pore in a membrane. By applying an electric field to a nanopore in a thin membrane, it is possible to force the highly charged polyanionic DNA molecule to move through it with nucleotides in single-file sequential order. The electric potential of the nucleotides in a pore presents an energy barrier to the passage of ions and blocks the current in a characteristic way. If each nucleotide has a characteristic electrical signature, then a pore could be used to analyze the sequence, reporting all the signatures in a single long read without resorting to multiple DNA copies.

The prospect for sequencing DNA with a nanopore is being carefully scrutinized right now [11–20]. To sequence DNA using a nanopore requires a robust, porous structure of an appropriate size – comparable to or even less than the size of DNA – to maximize the electrical signal. The prototypes being explored fall into two categories generally: (1) proteinaceous pores such as α-hemolysin (α-HL) shown in cross-section in Fig. 12.1a and its mutants; or (2) nanopores in solid-state membranes like that illustrated in Fig. 12.1b. Since the translocation velocity through the pore can be very high – about 1 bp/10 μs for α-HL [12] and ~1 bp/10 ns

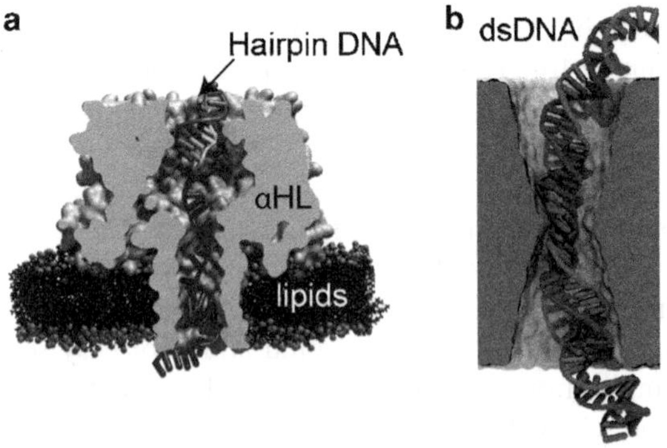

Fig. 12.1 Nanopores for sequencing DNA. (**a**) A cross-section through an α-HL pore taken from a simulation showing a trapped DNA hairpin. The pore diameter (1.3 nm) is smaller than the stem of the hairpin. (**b**) A cross-section through a solid-state pore in a silicon nitride membrane showing a double stranded DNA molecule trapped in a 2 nm diameter pore. The diameter of the solid-state pore can be any size. Adapted from reference [33]

[21] for a pore in a silicon nitride membrane – it may be possible to sequence a DNA molecule quickly, driving the cost to practically nothing if the bases can be discriminated electrically. *But single base resolution on a translocating strand has not been demonstrated yet.*

High fidelity reads demand stringent control over both the molecular configuration in the pore and the translocation kinetics. Control of the molecular configuration determines how the ions passing through the pore come into contact with the nucleotides in the constriction, while the translocation kinetics affect the time interval in which the same nucleotides are held in the constriction and data is acquired. *Until now, none of the nanopore prototypes proffered for sequencing has shown any prospect of satisfying both of these specifications at the same time.*

Bayley et al. [19, 20, 22] recently engineered α-HL in such a way to improve the SNR, i.e. to hold a nucleotide in place for a longer period of time, in order to perform more averaging. By modifying the α-HL such that a cyclodextrin is placed in the β-barrel, the time that the pore is occluded by a single nucleotide can be extended. This allows for more accurate determination (>90%) of which nucleotide is in the pore based on blockade current. This method was used in combination with an exonuclease to determine the composition of single-stranded DNA (*ssDNA*) by cleaving off individual nucleotides, then measuring them in an α-HL pore. This scheme could potentially be used on raw, genomic DNA. However, it suffers from the crippling problem of logistics: i.e. how to transport the cleaved nucleotides from the exonuclease to the pore, ensuring that they arrive in the same sequence as found in the original DNA, that none escape (missing a base), and that the exonuclease does not outpace the pore. Tethering the exonuclease to the α-HL has been proffered as a solution, but this scheme is a nontrivial extension to the original concept.

There are other limitations to using α-HL for sequencing. First are the obvious structural limitations – the protein structure is difficult to change in a predictable way. Though it is possible to introduce subtle mutations into the protein, gross structural changes are inordinately difficult. Chief among these structural limitations in α-HL are the length of the nanopore, and hence the thickness of the membrane, and the diameter of the pore. For example, the α-HL channel limiting aperture is only 1.5 nm in diameter and several nucleotides long. The secondary structure of *ssDNA* along with the shape and length of the nanopore obfuscate the interpretation of a long read, making it practically impossible to measure only one base at a time. The lipid bilayer presents yet another limitation. The lipid bilayer is typically 25–100 μm in diameter and only 5 nm thick. It ruptures after a few hours of use or after cycling the electrolyte a few times and the large size of the membrane produces a capacitance that adversely affects the frequency and noise performance

In contrast to α-HL, in a solid-state membrane, the pore geometry, the thickness and composition of the membrane can all be controlled with sub-nanometer precision using semiconductor nanofabrication practices. This aspect represents a promising new approach to the creation of a pore. This is important because the precision translates directly into control of the distribution of the electric field. The control of the electric field on this scale has already been used to produce the most sensitive electronic device – the single-electron transistor [23] – and now the same techniques are being leveraged to measure the charge distribution in a single DNA molecule.

So far, the focus of attention has been on sequencing strategies that rely on electric signals such as the blockade current that develops when DNA translocates through a pore in a membrane. Alternatively, using electrodes which are placed extremely close to the DNA molecule on either side of a nanopore orthogonal to the DNA backbone, each nucleotide might be probed directly as it translocates passed through the pore. Theoretical studies have shown that with such electrodes, the electron tunneling current should be significantly different between the different nucleotides. There is an overwhelming advantage to this approach – e.g., the nucleotide is directly probed, rather than indirectly measured using the the blockage of ionic current. Lagerqvist et al. suggests that 10^7 current measurements per second should be sufficient to identify individual bases – or 10 MHz – which is possible with proper electrode design [24]. However, differences in tunneling current have not yet been experimentally shown to differentiate individual bases, or base pairs, from each other using typical STM probes [25, 26]. Moreover, the tunneling current is exquisitely sensitive to position, which can be problematic especially considering thermal fluctuations that occur in an oversized constriction [27].

Another method of potentially detecting the nucleotide present in the nanopore is through capacitive detection. By placing electrodes in the nanopore, the electrostatic potential in the pore can be measured. If DNA is cycled back and forth through the pore (with an amplitude of oscillation of ~1 nm), an effect due in part to the dipole moment of the base present in the pore constriction and in part to the velocity difference for different bases is measured. The combination of the dipole moment and the velocity are characteristic of the base in the pore, allowing for identification of the DNA sequence [28].

All of these implementations suffer from the same limitations: i.e. lack of control over the translocation kinetics and the molecular configuration. Prior work [21, 29, 30] has indicated that the velocity of DNA through a solid-state nanopore can be large, exceeding 1 bp/10 ns even at low voltage (200 mV). On the one hand, the high velocity promises high throughput. On the other hand, it also makes it difficult to resolve the electrical transients associated with a single nucleobase, especially if the bandwidth is narrow enough to minimize noise. Sequencing with single base resolution also demands sub-nanometer control of the molecular configuration in the pore because the configuration affects the signal derived from the ion current. This stringent condition follows because B-form double-stranded DNA (*dsDNA*) has a twisting, propeller-like, helical structure ~2 nm in diameter with an axial rise of 0.34 nm per base-pair. Repeated measurements made using either multiple pores with multiple copies of DNA [31], or multiple passes with a single molecule [32] might address throughput, but they don't offer control of the molecular configuration in the pore.

12.1.3 Chapter Overview

Nanopores promise a new paradigm not only for DNA sequencing, but for polynucleotide synthesis, and even information processing and computing. However, that promise won't be realized unless the translocation kinetics and the molecular configuration in the pore are stringently controlled. In this chapter, following several recent reviews, [33–35] we will narrowly focus on recent developments [21] concerning the prospects for sequencing double-stranded DNA using a synthetic nanopore with a diameter less than the double helix. Our plan is to use this example to illuminate both the challenges and some of the available solutions. First, we illustrate how state-of-the-art in semiconductor nanofabrication can be applied to produce a sub-nanometer diameter nanopores in heterostructure membranes, and tailor the geometry to the propeller-like structure of DNA. Thus, semiconductor nanotechnology plays a key role in our plan for sequencing a single DNA molecule.

A nanopore is sensitive to single molecules DNA. But to take full advantage of the minimal material requirements, we have to somehow convey the DNA to the pore. We show two schemes that use microfluidics as a molecular conveyance for DNA: hydrodynamic focusing and integrated valves in nanofluidic circuits. Working separately or together, these strategies could be used to force the DNA to within the capture radius of the electric field extending outside the nanopore. Once it is captured by the electric field, the translocation kinetics through the pore has to be controlled to ensure signal fidelity. We have developed a new method for trapping a single molecule of *dsDNA* in a nanopore, smaller in diameter than the double helix. We show that it is possible to trap a single *dsDNA* molecule in a nanopore <3 nm in diameter by first applying a voltage larger than the stretching threshold, forcing the molecule to translocate through the pore. According to molecular dynamics (MD) simulations, this leaves the *dsDNA* stretched in the pore constriction with the base-pairs tilted, while the B-form canonical structure is preserved outside the pore. If the electric field is rapidly switched to a value below the threshold during the translocation, a single *dsDNA* molecule becomes trapped in this configuration, in what is effectively a harmonic potential. In principle, the molecule can subsequently pushed through the pore one base at a time, provided the voltage forcing the DNA through the pore is switched fast enough. If the duration in the trap is commensurate with the bandwidth, we further assert that we can discriminate distinct signatures of C-G and A-T base-pairs, under less than optimal conditions, by simply measuring the pore current.

We envision using this type of trap in a sequencing protocol whereby the translocation kinetics of *dsDNA* in a pore are stringently controlled and measurements are performed to extract the identities of the nucleotides from the pore current. One obvious problem with sequencing this way is determining which nucleotide is on which strand, e.g. distinguishing A-T from T-A. And so, finally we consider the limitations imposed on the SNR by synthetic nanopores currently in production. We show that it should be possible to call base-pairs with sufficient resolution, provided that the membrane capacitance is minimized using conventional nanofabrication practices to reduce the area.

12.2 Single Molecule Sequencing with a Synthetic Nanopore

12.2.1 Nanopore Fabrication

We have developed methods for producing nanometer diameter pores in robust membranes as illustrated in Fig. 12.2. Pores in solid-state membranes offer vastly improved chemical stability over phosphor-lipid layers. They are resilient in the

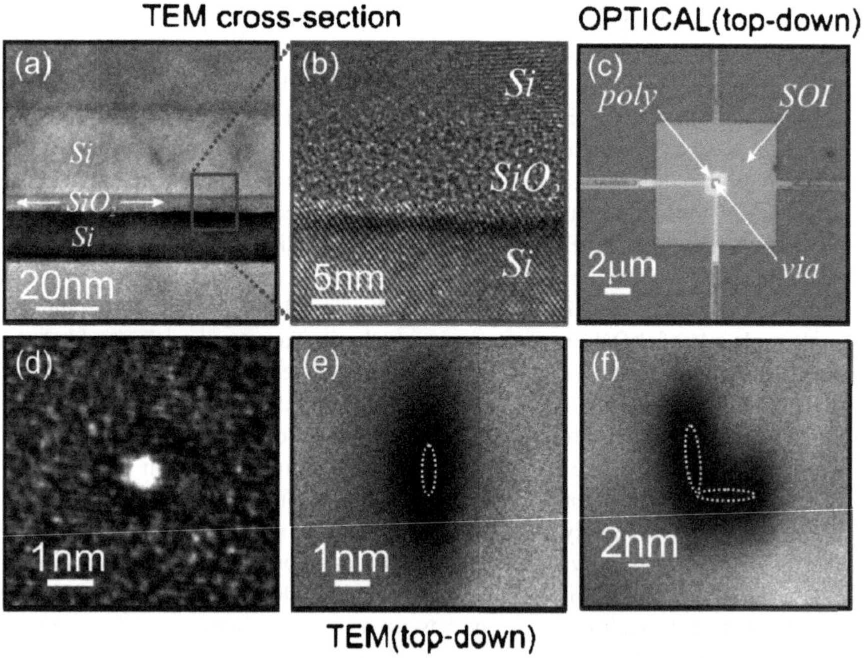

Fig. 12.2 Silicon nanotechnology is a key innovation for nanopore sequencing. (**a**) A TEM cross-section through the membrane structure. SiO_2 membranes are formed by depositing a sacrificial silicon layer on a substrate. A thin oxide <5 nm thick shown in (**b**) is grown by RTO and subsequently a thin layer of polysilicon is deposited on top and polished back using CMP. With DUV lithography and a combination of wet and dry etching a membrane is revealed. (**c**) Optical micrograph of a membrane showing a $(2\ \mu m)^2$ top silicon layer with a 1 μm via through it and a $(10\ \mu m)^2$ bottom Si layer. The nitride layer on the top and bottom are transparent. The top and bottom Si layers act as etch stops. (d) After revealing the membrane, a pore is sputtered using a tightly focused, high energy electron beam. The TEM micrograph shows a ~0.7 nm diameter pore. (**e**) Scanning TEM (STEM) micrograph of a $0.9 \times 2.4\ nm^2$ slit produced with Titan at 300 kV. (**f**) An "L" shaped pattern indicating the stability of the 1 nm beam (<1 nm drift/30 min). Adapted from references [33–35]

harsh chemical and thermal environments useful for denaturing the DNA, as well as allowing for easier integration with other electrical or microfluidic components. Solid-state membrane materials range from polymers [36], solid-state dielectrics [37, 38], and semiconductors [35, 39] to metal films [40, 41]. Different membrane materials allow for tailoring different electrical properties, such as surface charge density and capacitance, and they are susceptible to a wide variety of surface treatments useful for controlling the pore charge and hydrophilicity. Most importantly, a solid-state membrane can be reduced to sub-micrometer scale using conventional semiconductor nanofabrication techniques, mitigating parasitic capacitance effects and improving electrical performance [38, 39, 44]. There is also a panoply of methods for fabricating pores – ion-beam milling [42], ion-track etching [43], reflow [45] or electron-beam ablation [37] to name a few. In particular, we were the first to report the use of a tightly focused, high energy electron beam to sputter a nanopore [37]. Figure 12.2a–d are examples of the devices we have produced: i.e. a 0.7 nm nanopore sputtered into a 4.5 nm thick silicon oxide membrane – this is the smallest synthetic pore in the thinnest membrane ever reported. Leveraging this capability and the phenomenal electron beam stability and brightness below 2 nm, we can use a scanning TEM to tailor the shape of the pore making slits like that shown in Fig. 12.2e or even irregular patterns like the "L" in (f).

12.2.2 DNA *Conveyance to the Nanopore*

While a nanopore is the ultimate analytical tool with single molecule sensitivity, there is a shortcoming in its application to sequencing DNA that is related to the diffusion equivalent capacitance [46, 47]. When an electric field is applied across a membrane with a $d < 3$ nm bi-conical pore in it that is immersed in electrolyte, the voltage is effectively focused near the center of the membrane over a region about 1–3 nm wide [48, 49]. This means that *dsDNA* has to first diffuse within range of the pore to be driven through it by the electric field. The rate of DNA capture is roughly given by $R = 2\pi CDr$, with R the capture rate, C the concentration of DNA, D the DNA diffusion coefficient in free solution, and r the radius of probable capture by the pore, which is on the scale of microns at high voltage [50]. The diffusion capacitance governs the time required to capture a molecule, which is about 1 s for the 10^9 molecules/μL concentration used routinely to test nanopore performance. This capacitance leads to a trade-off between response time and detectable concentration. So, the key to single molecule operation of a pore is conveying a small volume of material within the capture radius.

To maintain a response time of ~1 s while utilizing single molecule fragments, we have to focus the molecules into the capture volume over the pore. We can accomplish this one of two ways: (1) by adapting a technique used prevalently in flow cytometry to concentrate cells [51, 52] – hydrodynamic focusing or (2) by trapping the molecule in a small volume over the pore. Figure 12.3 demonstrates the principle of hydrodynamic focusing using fluorescent molecules in a laminar flow

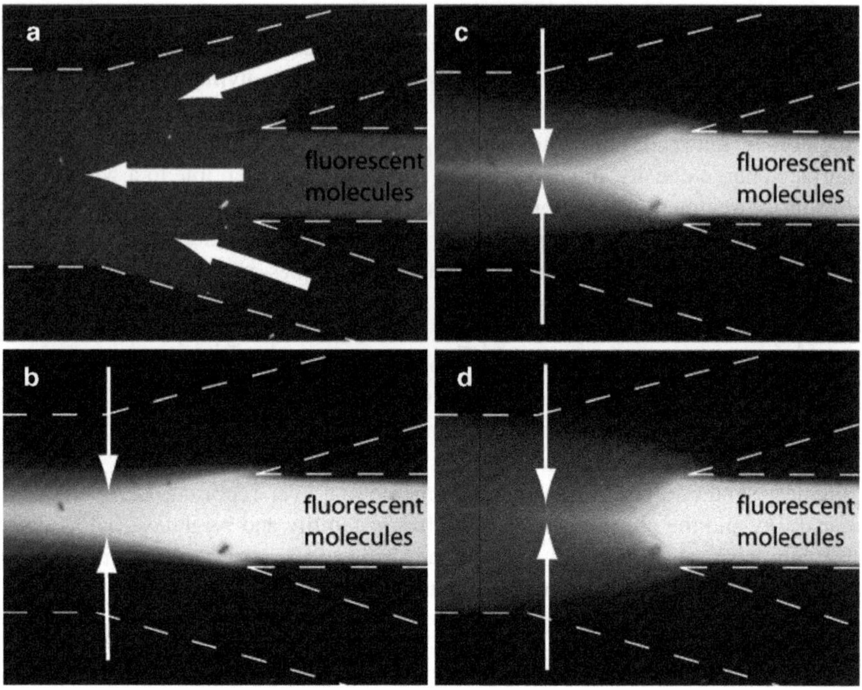

Fig. 12.3 Fluorescent images taken from a microfluidic device consisting of three channels that merge into a single channel at a 15° angle. Fluorescent molecules indicate the flow lines. The microfluidic is highlighted by *white dashed* lines. The flow is right to left in the channel. Four different flow conditions are used to illustrate hydrodynamic focusing of fluorescent molecules: (**a**) 0.03 μL/min; (**b**) 10 μL/min; (**c**) 5 μL/min; and (**d**) 2.5 μL/min

(from right to left) in a microfluidic device. By altering the ratio of the sheath (top and bottom channels) to the input (center channel) flow rate, the flow in the center channel can be focused into a thin jet – as evident from the fluorescent peak in the center. Using this approach, we expect to convey single DNAs within the pore capture radius without compromising the response time.

Alternatively, we designed a microfluidic chip with integrated micromechanical valves enabling us to concentrate in the volume over the pore. This approach was motivated by microfluidic applications such as genetic analysis of single cell [53, 54]. Figure 12.4a shows the plan of the chip, which we implemented with soft lithography in polydimethylsiloxane (PDMS) using 12 μm line-rules. This design utilizes two-layer PDMS push-down microfluidic valves to control fluid flow in a 10-input microchannel array that is bonded to a silicon nitride membrane (either at high temperature >76 °C for 3 h or using plasma oxidation). An elastomeric membrane is formed where the flow channel is positioned orthogonal to the control channel directly above it. Magnified top-down views of two pressurized valves are shown in Fig. 12.4c, d. According to this design, a single nanopore in a membrane $< (10\ \mu m)^2$ is located in the active volume at the intersection between the horizontal and vertical microchannels, and independent fluid flows containing

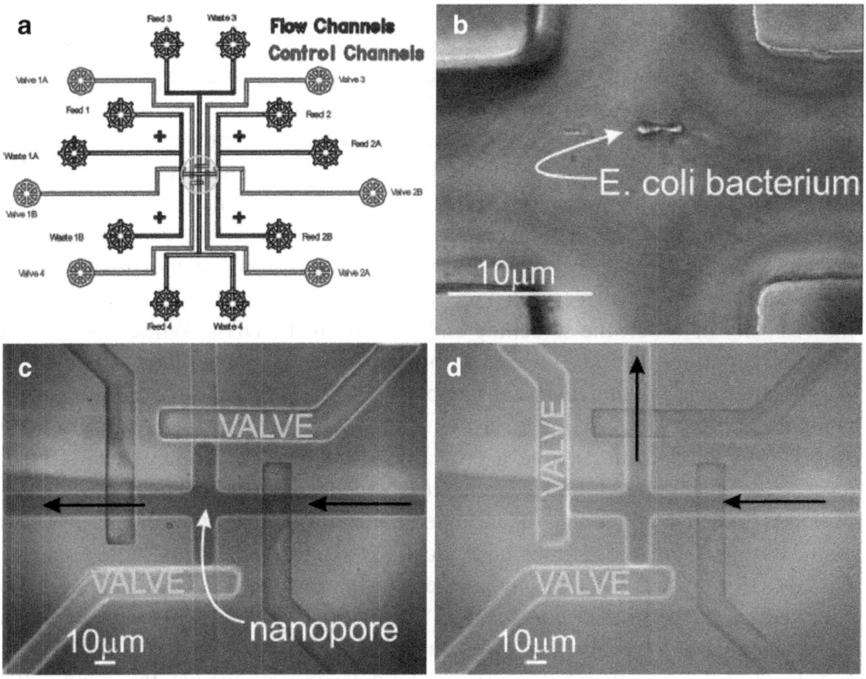

Fig. 12.4 Nano-fluidics Integrated Circuits. Using the same line-rules used for single cell analysis we implemented with soft lithography in PDMS a 6-valve, 10 input microfluidic circuit shown in (**a**). (**b**) A magnified view of the region highlighted in (**a**) shows a single cell is trapped in the microfluidic. (**c**, **d**) Magnified view of the same region highlighted by the circle in (**a**) illustrating the operation of the valves. By applying pressure to the valves (*lightly colored dye*) in this circuit we can control the flow of small DNA (*dark colored dye*) to the pore in the cross

DNA are guided to the pore via the valves. The operation of the valves (highlighted with dye) illustrated in Fig. 12.4c, d allows us to direct a small volume of material (represented by dark colored dye fluid) toward the pore. Because of the line-rules we are able to manipulate very small volumes of material (9 pL) and at the same time gate three different flows into the volume over the pore. For example, using the valves to gate the flow, a single *E. coli* bacterium can be positioned in the active area (Fig. 12.4b). Using 12 μm line-rules with a concentration of 10^9 molecules/μL, the volume over the pore shown in Fig. 12.4c contains ~1,000 molecules. Thus, by leveraging this integrated microfluidic circuitry, a nanopore can be connected to the products of a single cell analysis.

12.2.3 *Trapping a* DNA *in a Synthetic Nanopore*

Once inside the pore, there are three main forces that affect the DNA according to MD simulations [55]. The first and strongest force is the electric field, acting primarily on the negatively charged phosphate backbone of DNA, which drives the molecule forward into the pore while the positively charged ion cloud surrounding it is driven back. There are also electrostatic and nonpolar (van der Waals) interactions with the pore walls, and finally a drag force associated with the movement of the DNA polymer in solution.

When pore diameter is large (~10 nm), *dsDNA* may translocate through the pore in folded configuration making single base pair analysis impracticable [18, 56]. Therefore, pores with diameters comparable to a double helix (<3 nm) are of special interest. The smaller the pore diameter; the larger the blockade current and therefore the larger the signal. However, the electromechanics involved in a translocation are very different in pores with smaller diameters (<3 nm) because of the viscosity of water, the screening and the size of the DNA. While only a small voltage is required to force *dsDNA* through a pore >3 nm in diameter, when the pore diameter is smaller than the double helix the leading edge of the *dsDNA* penetrates into a constriction to a diameter of about ~2.5 nm and stalls there. Applying a voltage bias above a critical threshold provides enough differential force, exceeding that required to stretch *dsDNA* (~60 pN) [48, 49], and the molecule is pulled towards the center of and eventually through the membrane. The two strands comprising the double helix do not pass through pores with diameters 1.6<d<2.5 nm in the same way as they do through larger pores [57, 58]. The confinement of the smaller pores causes the base-pairs to tilt. Due to the activation energy required to begin this stretching transition, DNA will not be able to translocate below the threshold. The threshold depends on the pH, the composition of the strand and the methylation profile [49, 59].

We have thoroughly characterized the electromechanics of *dsDNA* in a nanopore [48, 49, 57–59]. As illustrated in Fig. 12.5a, we find that a 2.0 nm pore in a nominally 10 nm thick silicon nitride membrane exhibits a threshold $U = 2.9$ V for permeation of *dsDNA*, while a 2.0 nm pore in the 20 nm membrane shows a threshold voltage $U \approx 5.5$ V because of the change in the field profile associated with the thickness of the membrane. The estimated differential tensile force on the leading nucleotides in the strand is $F = qE > 60$ pN where E is the electric field in the pore, which is large enough to stretch *dsDNA*. It must be that the origin of the sharp field threshold for permeation is due to the stretching transition.

Recently, we established that the voltage threshold for permeation of *dsDNA* through a nanopore with a diameter smaller than the double helix, depends on the methylation level and pattern too, and it can be substantially smaller than for an unmethylated variant of the same DNA [58]. These observations could have a bearing on sequencing the epigenome with a nanopore without bi-sulfite treatment of the DNA first. We investigated the permeability of MS3 with different methylation levels and profiles through two pores with similar (~1.8 nm) diameters as show in Fig. 12.5b, c. Figure 12.5b represents the results of three *qPCR* analyses – one for unmethylated, one for hemi- and another for fully methylated DNA – showing the number of DNA copies permeating the pore as a function of the applied potential. The threshold voltages, U, for fully- and hemi-methylated MS3 are easily resolved and fall below that observed for the unmethylated variant. For example, the

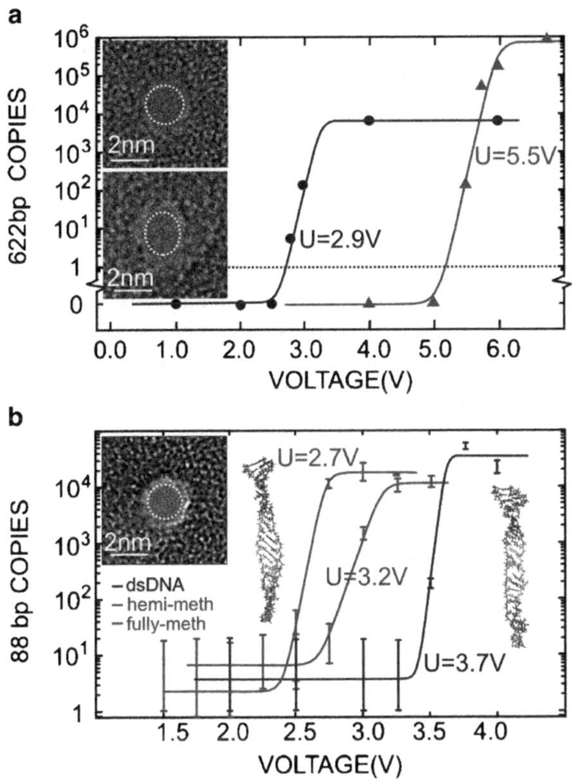

Fig. 12.5 Stretching DNA in a nanopore. (**a**) qPCR results obtained for the 2 nm pores showing the copy number vs. voltage. 622 bp *dsDNA* permeates the 2 nm pore in 10 nm thick Si_3N_4 membrane (*top inset*) for $V > 2.5$ *V*, and the 1.8 × 2.2 nm pore in 20 nm thick Si_3N_4 membrane (*bottom inset*) for $V > 5.5$ *V*. (**b**) qPCR results indicating the number of MS3 DNA copies that permeate through the 1.8 ± 0.2 nm pore shown in the *inset* as a function of the membrane voltage. *Insets* on the *left* and *right* are snapshots of methylated and unmethylated MS3 translocating through the 1.8 nm pore respectively. Both DNA exhibit an ordered B-DNA form, but there is a significant degree of disorder for unmethylated DNA. The highlighted region of the strand shows the portion of the DNA where methylated cytosines are located. The same region is also highlighted in the unmethylated strand for comparison. Adapted from references [48, 49, 59]

threshold for unmethylated MS3 in Fig. 12.5b is $U = 3.6$ V while hemi- and fully methylated MS3 show $U = 3.2$ V and $U = 2.7$ V, respectively.

The large shifts in the thresholds with methylation are surprising because the leading nucleotides in the strand that are important to stretching in a pore are separated by more than 18 bp (~6 nm) from a methylation site. The structure of methylated DNA, inferred from X-ray diffraction and NMR, indicates that the effect of methylation on the conformation of DNA is very subtle, and localized near the methylation site [59]. On the other hand, MD simulations indicate that the methyl groups reduce the DNA flexibility because of the steric hindrance by bulky methyl groups and because the DNA folds around methyl groups [60]. Snapshots of the DNA in the pore shown in the insets to Fig. 12.5b reveal the molecular structure with atomic detail, indicating that the methylated DNA is more ordered and stiffer [61]. This is also evident in the root-mean-square (rms) deviation in the helix diameter. At 4 V, the interior segments of methylated and unmethylated DNA (shown in yellow in the insets to Fig. 12.5b) have an rms-deviation 0.29 and 0.49 nm, respectively.

When λ-DNA is injected into the electrolyte at the negative (cis) electrode and a voltage is applied across the membrane 15.0 ± 2.2 nm thick membrane with a pore 2.5 × 2.0 ± 0.2 nm cross-section – smaller than the DNA double helix – in it like that shown in Fig. 12.6a, current transients such as those shown in Fig. 12.6b are observed [21]. The current transients occur randomly as a function of time as illustrated in the figure, but the inter-arrival time decreases with increasing concentration of DNA. Such blockades ostensibly represent the reduction of the electrolytic pore current due to the translocation of DNA.

Figure 12.6b illustrates threshold behavior presumably due to stretching as alluded to above, showing a dearth of transients found in a current trace measured at 200 mV compared to an 800 mV trace. Figure 12.6c summarizes the voltage dependence of the frequency of blockade events over the range from 100 mV to 1 V. Generally, we observe an abrupt rise in the number of blockades over a range of ~200 mV near the critical threshold. If we assume each blockade corresponds to *dsDNA* permeating the pore, then the permeation rate can be described by the transition-state relation of the Kramers type: $R = R_0 V/(1 + exp[q^*(U\text{-}V)/kT]$, where R_0 is a frequency factor, q^*U is the effective barrier height, q^*V is the reduction in the energy barrier due to the applied potential, and kT is the thermal energy. Using this relation, the data was fit and the results were overlaid on the scatter plot in Fig. 12.6c. We deduce a threshold of $U = 0.46 \pm 0.02$ V with $q^* = 0.8 \pm 0.2e$, which presumably corresponds to the force required to stretch the leading nucleotides in the pore.

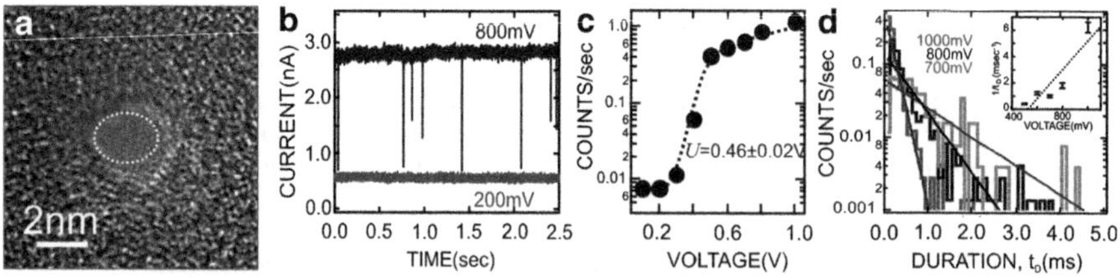

Fig. 12.6 (**a**) A TEM image of a 2.5 × 2.0 nm cross-section pore in a silicon nitride membrane 15 nm thick. (**b**) Electrolytic current measured in 100 mM KCl at 800 mV (*top*) and 200 mV (*bottom*) through the pore shown in (**a**) as a function of time. The frequency of blockades decreases dramatically with voltage; at 0.2 V practically no transients are observed. (**c**) The frequency of blockades observed with the 2.5 nm pore as a function of membrane voltage, illustrating the frequency drop as voltage decreases below 0.5 V. The *dotted line* represents a fit to the data. (**d**) Distributions illustrating the frequency as a function of the duration of a current blockade, t_D, above threshold at 1.0 V (*left*), 800 mV (*center*) and 700 mV (*right*). The distribution depends sensitively on the voltage. *Inset*: The reciprocal of the duration, t_D^{-1}, as a function of the applied voltage. Adapted from reference [21]

The dependence of the blockade duration on the membrane voltage offers more support to the interpretation of the blockade current as a translocation across the membrane. Figure 12.6d shows the frequency of current transients associated with λ-DNA as a function of duration with the voltage as a parameter. If the blockade duration corresponds with the interval that DNA blocks the pore, then the average transient width t_D signifies the time required for 48.502 kbp λ-DNA to translocate through the pore. We find that $t_D = 0.16 \pm 0.01$ ms, 0.53 ± 0.06 ms, 1.1 ± 0.1 ms, 0.82 ± 0.07 ms and 2.49 ± 0.25 ms for voltages of 1.0 V, 800 mV, 700 mV, 600 mV and 500 mV, respectively. The corresponding translocation velocity is 1 bp/3.3 ns at 1 V and 1 bp/11 ns at 800 mV to 1 bp/50 ns at 500 mV. The inset to Fig. 12.6d shows a plot of the voltage dependence of the reciprocal of the average transient width, i.e. $1/t_D$, measured above threshold. $1/t_D$ vanishes near the threshold value. The line in the inset is a least-square fit to the data. It has a slope of 11 V^{-1} s^{-1} with a voltage-intercept of 0.53 V, which is comparable to the threshold voltage inferred from Fig. 12.6c.

So far, no one has been able to use current blockades in a nanopore to sequence a continuous strand of DNA. Poor SNR is part of the reason. We have shown elsewhere (and recapitulate in what follows) that the effect of thermal, *1/f*, and dielectric noise can be successfully mitigated by minimizing the bandwidth and controlling the membrane geometry [44]. However, the motion of a DNA base relative to the detector inevitably introduces "translocation noise" if a continuous strand is driven electrophoretically through a pore [62]. Translocation noise is captured succinctly in a one-dimensional transport model by the ratio of the drift to diffusion velocities: i.e. $v_{drift}/v_{diff} = \mu E/(D/L_m) = V/(kT/q)$ where D and μ are the DNA diffusivity and the mobility in a pore, respectively, which are related by the Einstein relation, and the electric field is given by $E = V/L_m$, where V is the voltage applied across a membrane of thickness, L_m. From this relation, we infer that voltages large compared to kT/q are desirable to offset diffusion of the DNA in the pore. But large voltages can adversely affect membrane reliability and increase the translocation velocity forcing high frequency operation to electrically read each base. To obviate the need for continuous high voltage operation, while still suppressing translocation noise, the base should be trapped in the pore, read, and then impelled to the next base in the sequence.

To sequence *dsDNA* we plan to trap a molecule by stretching it in pore geometry smaller than the double helix, then use low-noise measurements of the blockade current to identify the base-pairs in the trap. Trapping stabilizes the position of the molecule for measurement and minimizes the translocation noise (if the trap is deep enough). We have already demonstrated the efficacy of this kind of trap [21]. The voltage-intercept for $1/t_D$ near the threshold in Fig. 12.6d suggests that *dsDNA* can be trapped in a nanopore that is smaller in diameter than the double helix. To test this hypothesis, we forced *dsDNA* into the 2.6×2.0 nm pore shown in Fig. 12.7a using a 600 mV transmembrane bias, which is above threshold value of $U = 0.30 \pm 0.04$ V according to Fig. 12.7b, then reduced the bias to 100 mV, which is below the threshold while the DNA was still in the pore. During this procedure, the pore current is constantly monitored. Eventually, the current returns

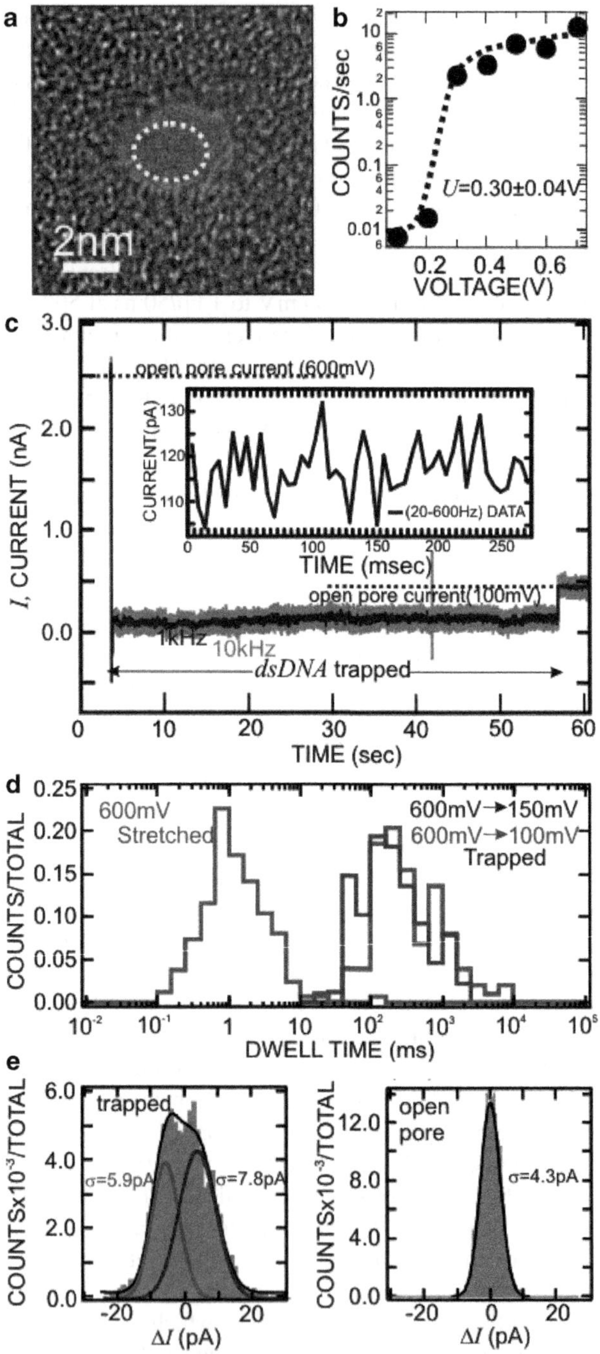

Fig. 12.7 Trapping a λ-DNA molecule in a nanopore. (**a**) A TEM micrograph of a 2.6 × 2.0 nm cross-section pore in a silicon nitride membrane 15 nm thick. (**b**) The frequency of blockades observed with this pore as a function of membrane voltage illustrating the frequency drop as voltage

to the open pore value, near $t = 58.8$ s in Fig. 12.7c. We assume that the transient at $t = 58.8$ s is indicative of the molecule exiting the pore after 56 s, corresponding to a translocation velocity of >1 bp/1.8 ms – about *97,000× slower* than the peak value obtained at a constant bias of 600 mV shown in Fig. 12.7d. Thus, when the bias is reduced below the stretching threshold, the pore acts as a trap resisting the molecular motion.

MD simulations corroborate our interpretation of the experiments by showing that the motion of the *dsDNA* can be slowed or effectively stopped when the driving voltage is turned off. Figure 12.8a illustrates a simulated system that includes a *dsDNA* fragment, 100 mM KCl, and a 2.0 nm diameter pore in a silicon nitride membrane. The *dsDNA* molecular conformation within the pore was stretched from 0.34 to a 0.41 nm rise per base-pair even at 0 V. The distortion from the B-form *dsDNA* structure can be clearly seen in Fig. 12.8a. With a 500 mV bias applied, the *dsDNA* was stretched further to 0.44 nm. As illustrated in Fig. 12.8b, at 250 and 500 mV, the *dsDNA's* motion is arrested: only by applying 1 V could the *dsDNA* coax out of the trap. By analyzing the *dsDNA* displacements at a 0 V (inset of Fig. 12.8b), we determined that the pore acts as a harmonic trap with an effective spring constant of 7.2 ± 0.8 nN/nm. The probability of escape from a trap depends sharply on the force applied, explaining the threshold. The force required to restart the motion is essentially determined by the product of the spring constant, k, and the distance over which escape of a base-pair from the trap occurs x_0: i.e. $q^*\mathrm{E} \sim kx_0$ where $x_0 = L/2$ with L the length / base-pair. Thus, we find that the force required to move the *dsDNA* one base-pair in a 2.0 nm diameter pore is $kx_0 = 1.4$ nN.

Our estimates also suggest that this trap mechanism is robust with respect to the stochastic forces that act on the *dsDNA* coils above and below the trap. To estimate the magnitude of a stochastic force, we approximate the *dsDNA* coil outside the pore as a bead with the hydrodynamic radius, r_h, of λ – DNA, which is about 750 nm [63, 64]. The stochastic force from the solvent on a bead of that size, $\Phi(\tau) = (1/\tau)\int \Phi(\mathrm{t})\mathrm{dt}$ follows from the second moment of the stochastic force: $<F(t_1)\, F(t_2)> = 6\, k_B\, T\, \xi\, \delta(t_2 - t_1)$, where ξ is the friction coefficient of the bead.

Fig. 12.7 (continued) decreases below 0.3 V. The *dotted line* represents a fit to the data. (**c**) Triggered by the onset of a current blockade indicating that λ-*dsDNA* is translocating through the pore of (**a**), the voltage is switched from 600 mV (above the stretching threshold) to 100 mV (below threshold). As a result the molecule is trapped in the pore till $t = 58.8$ s. *Inset* is a magnified view of the current fluctuations observed near $t = 58.8$ s. (**d**) Histograms of the dwell times observed at a constant voltage of 600 mV(*grey*) and the distribution of elapsed time spanning the instant when a blockade event triggers the voltage switch from 600 mV to 150 mV or 100 mV to the return of the current to the open pore value seconds later. The peak in the distribution of current blockade durations increases from about 900 μs to about 200 ms, increasing 200×. (**e, f**) Histograms showing the distribution of the current during the blockade in the interval 14–14.5 s, when λ-DNA is trapped (**e**), and the open pore for $t > 58.8$ s (**f**). The distribution for the trapped molecule must be fit to at least two Gaussians: one (*right*) offset from the median ($\Delta I = 0$) by $\Delta I = +2.9$ pA with a width of $\sigma = 7.8$ pA; and another (*left*) offset by −6.51 pA with a width of $\sigma = 5.9$ pA. The *black line* represents the sum. In contrast, the data in (**f**) representing the open pore can be fit by a single Gaussian with a width $\sigma = 4.3$ pA. Adapted from ref. [21]

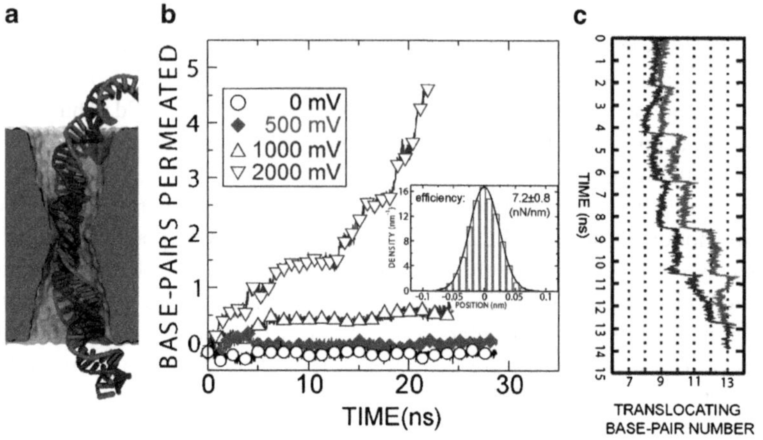

Fig. 12.8 MD simulation of the nanopore trap. (**a**) Snapshot of the system that includes *dsDNA*, water and ions (not shown) as well as a 2.0 nm diameter pore. The molecular conformation within the constriction is stretched beyond 0.34 nm/base-pair by 20–30%. (**b**) The number of base-pairs permeating through the larger or the smaller pore in 4 MD simulations done at different biases. The simulations predict a threshold between 500 mV and 1.0 V. (*Inset*) Histogram of the displacement of the base-pair nearest to the membrane's center at a 0 V bias. The *solid line* shows the distribution expected for a harmonic trap with a 7.2 nN/nm spring constant. (**c**) MD simulation showing stepwise transport of *dsDNA*. A square-wave bias was applied to produce 1 and 2 base pair step transport separated by a 2 ns rest interval through a $2.3 \times 2.3\ \text{nm}^2$ pore. Adapted from reference [21]

Using the Stokes' formula $\xi = 6\pi\eta r_h$, where η is the viscosity of water, and expressing the mean force as the time average of the instantaneous force, we find $<|\Phi|^2> = 36\pi\eta k_B T r_h/\tau = 0.034\ \text{nN}^2$, where $\tau = 10$ ns is the time interval of the average. Thus, the rms-force is ~180 pN. Even if all of this force is transferred to the DNA residing in the pore, which is unlikely due to the finite speed of the stress propagation, the magnitude of the force will still be insufficient to displace the DNA out of the trap shown in the MD simulations.

This potential could be used to control the motion of the *dsDNA* in the pore as illustrated by the preliminary analysis given in Fig. 12.8c. In these MD simulations, with the application of a 0.11 ns duration 2 V square-wave pulse followed by 2 ns at 0 V (blue), the *dsDNA* is clearly conveyed through the pore in base-pair steps, although sometimes it slides back and skips steps, presumably because the voltage profile is not quite optimal. Two base-pair step are produced using a 0.12 ns pulse. Thus, it seems feasible that *dsDNA* can be stepped through the pore in single base-pair step under optimal conditions.

The current fluctuations associated with the trapped λ-DNA observed at low voltage shown in the inset to Fig. 12.7c and tallied in the histogram of Fig. 12.7e support the hypothesis that the pore current can be used to detect the sequence of base-pairs. We observed that for $t < 58.8$ s the amplitude of the current fluctuations increases relative to the open pore value found for longer times. With the molecule trapped (i.e. for $t < 58.8$ s in Fig. 12.7c), we filtered the current data and formed histograms of the current fluctuations like that shown in Fig. 12.7e, which can be represented by the superposition of two Gaussian distributions: one (solid blue) offset from the median ($\Delta I = 0$) by $\Delta I = +2.9$ pA with a width of $\sigma = 7.8$ pA; and another (solid red) offset by $\Delta I = -6.5$ pA with a width of $\sigma = 5.9$ pA. The goodness of fit, measured by the reduced χ^2-statistic, improves from $\chi^2 = 5.04$ for a single Gaussian fit to $\chi^2 = 0.71$ for two Gaussians, indicating that two Gaussians represent a superior model. This is in contrast to the open pore current data shown in Fig. 12.7f, which can be fit by a single Gaussian with a width $\sigma = 4.3$ pA with $\chi^2 = 0.67$ measured at the same voltage.

We tentatively attribute the separate peaks in Fig. 12.7e at $\Delta I = +2.9$ pA and $\Delta I = -6.5$ pA to partially resolved signals associated with C-G/G-C and A-T/T-A base-pairs, respectively. This identification is supported by the observation that λ-DNA has a nearly uniform distribution of base-pairs T-A, A-T, C-G, G-C. Thus, the currents corresponding to each type of base-pair should translate to approximately equal, distinct distributions in the blockade current convolved with the electrical noise. The difference between C-G and A-T base-pairs can be resolved in this case because of the long (~2 ms) time each base-pair spends in the constriction. However, apparently the SNR is inadequate for discriminating C-G from G-C or A-T from T-A, despite the slowing. This assertion is also corroborated by even longer duration measurements of blockade currents associated with streptavidin bound, 100 bp long, C-G and A-T biotinylated duplexes trapped by the electric field in a pore in a configuration described elsewhere [65].

One obvious problem with sequencing this way is determining which nucleotide is on which strand. However, our simulations show that the base-pair tilt, caused by the confinement, is maintained during a translocation with the nucleotides of one strand always lagging their partners on the other. At low bias, the electric potential of the nucleotides in the tilted configuration presents a peculiar energy barrier to the ions with a passage rate that is exponentially related to the height. The differences in the heights for different sequences could therefore have substantial effects on the current–voltage relation.

12.2.4 *Noise in a Nanopore*

Thus, the current measurement shown in Fig. 12.7c apparently corresponds to a single long read of >10,000 bp! While trapping a DNA molecule facilitates the high fidelity reads required for sequencing it, the electrical noise associated with the pore current introduces ambiguity that adversely affects base-calling. Both the high frequency and noise performance of the pore current are critical for applications like sequencing. High throughput demands a high translocation velocity of the DNA, and while a solid-state nanopore can satisfy this specification, the noise is proportional to the bandwidth so that the fidelity of base-calling may be compromised.

According to the model of Smeets et al. [66], the frequency response is essentially determined by the series combination of of the membrane capacitance, C_m,

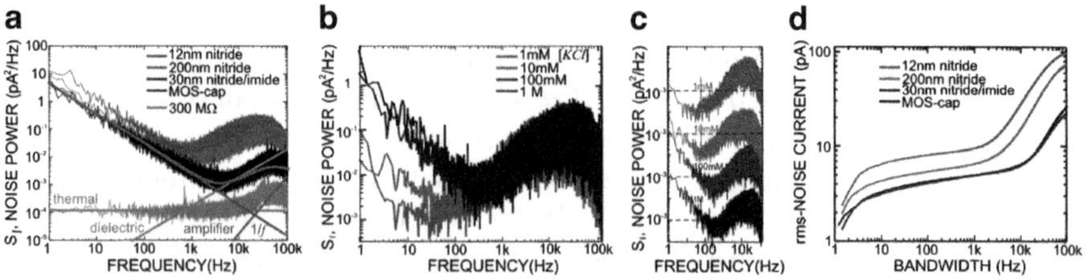

Fig. 12.9 Current noise spectra measured in nanopores. (**a**) Noise power spectra of nanopores in the four membranes measured in 1 M KCl with different effective capacitance. From *bottom to top*, a 300 MΩ resistor, a 7.1 × 7.3 ± 0.3 nm pore in a MOS membrane, a 1.7 × 2.8 ± 0.2 nm pore in a polyimide coated Si_3N_4 membrane, a 3.0 × 3.8 ± 0.2 nm pore in ~200 nm Si_3N_4 membrane, and a 3.3 × 4.8 ± 0.2 nm in 12 nm Si_3N_4 membrane. The low frequency *1/f* noise, the high frequency dielectric noise along with the amplifier noise are analyzed for the ~2.4 nm pore in polyimide coated membrane. The fit to the total noise is shown as a light colored continuous curve. (**b**) Noise spectra of a 2.1 × 2.3 ± 0.2 nm pore in 12 nm nitride membrane for different electrolyte concentrations. (**c**) The same noise spectra shown in (**c**) but offset to show the high frequency noise without overlap. (**d**) The rms-current noise vs. bandwidth for the membranes of (**a**); the largest contributor to the rms-current noise is dielectric noise prevalent above 1 kHz frequency. Taken from reference [44]

and the electrolyte resistance, R_{el}, so that $R_{el}C_m$ >1–10 μs for the 2.2 nm pore in a 15 nm thick nitride membrane in 100 mM KCl, corresponding to a bandwidth of $\Delta f = 1/2\pi R_{el}\ C_m \sim 100$ kHz. If the translocation velocity is high, it becomes impossible to resolve that portion of the blockade associated with a single base; a sampling frequency >16 MHz would be required. More gain can't resolve this problem due to the concomitant increase in electrical noise.

Motivated to call bases with high fidelity, we along with others have investigated the noise associated with current measurements in a nanopore. Fig. 12.9a shows the noise power spectra of four nanopores – with diameters comparable to the DNA double helix – in different types of membranes: two associated with different nitride thicknesses 12 nm and 200 nm on a silicon substrate; a third associated with a composite 30 nm nitride membrane coated with a 4 μm polyimide film on a silicon substrate; and a fourth associated with a membrane formed from a MOS capacitor; all measured in 1 M *KCl* along with the spectrum of a 300 MΩ resistor, a value comparable to the resistance of the 2.2 nm pore. The measured noise spectra of all the nanopore in electrolyte can be analyzed into four components: thermal, *1/f*, dielectric and amplifier noise contributions. (The light colored continuous trace represents a sum of all the components fit to a 2.4 nm pore in polyimide coated membrane.)

We expect that thermal noise associated with pore $S_t = 4k_BT/R$, resistance, will be negligible over the band <100 kHz since the noise from the resistor is below all of the other spectra. We observed at low frequencies that the noise power density is inversely proportional to the frequency indicative of *1/f* noise, i.e.: $S_{1/f} = I^2A/f^{\beta} = I^2(\alpha/N_c)f^{\beta}$ where I is the pore current, α is the Hooge parameter (an empirically determined proportionality constant that depends on the type and concentration of charge carriers, N_c is the total number of current carriers, f is the frequency, β is an exponent that is typically unity and $A \sim R_p^\gamma$ Consistent with these observations, Hooge suggested that *l/f* noise occurs in bulk conductors due to the fluctuating mobility of charge carriers that produces current fluctuations [44, 66, 67]. Generally, associated with the coefficient A we find that $\gamma = 1.03 \pm 0.44$ over a factor of 10,000$\times$ in pore resistance, in support of Hooge's phenomenological picture.

As illustrated in Fig. 12.9b, c, we find that the noise spectrum for $f < 100$ Hz is sensitive to the electrolyte concentration, while the high frequency noise is not. *1/f* noise becomes negligible at frequencies $f > 1$ kHz, and the spectrum exhibits linear frequency dependence up to about 50 kHz. The linear frequency dependence, coupled with the lack of a dependence of this part of the spectrum on the electrolyte concentration, which is evident from Fig. 12.9b, c, indicates dielectric noise with a spectrum of the form: $S_D = 4k_BTDC_m(2\pi f)$, where D and C_m are the loss tangent and the effective capacitance of the dielectric membrane.

Finally, according to the Fig. 12.9d, the noise in the range 100–50 kHz is the dominant contribution to the rms-current noise – it is exponentially larger than the *1/f* component. Above 50 kHz, the spectrum is strongly affected by the bandwidth of the amplifier and interconnections to it. Thus, reducing parasitic capacitances, electrolyte resistance and amplifier noise are all key elements for improving both the frequency and noise performance.

12.3 Conclusion

A nanopore is an analytical tool with single molecule sensitivity. It operates in a way that is reminiscent of Coulter's original idea of using dielectric objects within a constricted current path to alter the electrical resistance [68]. In this chapter we have explored both the promise and limitations of using a nanopore in a solid state membrane for one especially compelling application: sequencing double-stranded DNA.

It may now be possible to control both the translocation kinetics and the configuration of double stranded DNA in a pore by controlling the pore geometry. Semiconductor nanofabrication is a key factor, enabling the creation of structures with sub-nanometer precision. Leveraging the precision to produce nanopores in a solid-state membrane smaller in diameter than the double helix,, a single molecule of double-stranded DNA can be trapped in a pore by applying an electric force larger than the stretching threshold. Once a current blockade associated with a translocating molecule is detected, the electric field in the pore is switched in an interval less than the translocation time to a value below the threshold for stretching. This leaves the *dsDNA* stretched in the pore constriction with the base-pairs tilted, while the B-form canonical structure is preserved outside the pore. In this configuration, the translocation velocity is substantially reduced to ~1 bp/2 ms in the extreme, which could allow discrimination between A-T and C-G base-pairs just by measuring the ionic current.

While trapping a DNA molecule this way may facilitate the high fidelity reads required for sequencing it, the electrical noise associated with the ionic pore current introduces ambiguity that adversely affects base-calling. If dielectric noise associated with the membrane predominates for $f > 1$ kHz, then $I_{rms}^2 = 4kT\,DC_m\pi\Delta f^2$ Thus, the membrane capacitance, C_m, should be minimized to promote better SNR. But even if the dielectric noise contribution is minimized, it is likely that thermal and/or $1/f$ noise will remain problematic. Thus, it is likely that base-calling will be accomplished in a noisy environment, extracting the current associated with a sequence of bases from a signal buried in the noise, which will ultimately affect throughput of signal fidelity.

To make this more concrete, we can estimate noise and the affect on the throughput. The data shown above indicates that the relative change in current associated with λ-DNA blockading a pore is typically $\Delta I/I < 0.8$, which translates to a ΔI ~2–3 nA for a 2.2 nm diameter pore in 1 M KCl at 0.5 V, near the stretching threshold. Therefore, to detect a molecule with SNR > 2, we need peak-to-peak noise <1.5 nA or an rms value of $\Delta I_{rms} \sim 1.5$ nA/8 = 190 pA. For a bandwidth of $\Delta f < 100$ kHz, we estimate that $DC_m \sim 70$ pF is required to detect a current signature [44], which is easy to do. On the other hand, according to MD simulations [65], to detect the difference between a C-G and A-T base-pair in a pore like this, we must resolve a difference signal of ΔI ~20 pA or smaller so that $\Delta I_{rms} \sim 2$ pA. It is realistic to suppose that the dielectric noise can be forced to satisfy this specification by minimizing the capacitance [44] even for a bandwidth of 16 MHz. However, if thermal noise predominates then $I_{rms} = \sqrt{4kT\,\Delta f/R} \sim 2pA$. Thus, the rms-current noise specification forces a bandwidth no greater than ~40 kHz, which limits the throughput. From these extrapolations, we conclude that a single, solid-state nanopore in a membrane engineered with state-of-the-art fabrication techniques would have adequate frequency and noise performance for sequencing DNA, but the noise is likely to compromise the throughput and base-calling fidelity.

Acknowledgments We gratefully acknowledge numerous contributions and our close collaboration with Jiunn Heng, Chuen Ho and Greg Sigalov. This work was funded by grants from National Institutes of Health [R01 HG003713A, PHS 5 P41-RR05969], the Large Resource Allocation Committee [MCA05S028], the Petroleum Research Fund (48352-G6), and the National Science Foundation [TH 2008–01040 ANTC, PHY-0822613 and DMR-0955959].

References

1. Lander ES, Linton LM, Birren B, *et al* (2001) Initial sequencing and analysis of the human genome. Nature, **409**(6822), 860–921.
2. Venter JC, Adams MD, Myers E, Li PW, Mural RJ, *et al* (2001) The sequence of human genome. Science **291**, 1304–1351.
3. Mardis ER (2008) The impact of next-generation sequencing technology on genetics," Trends Genetics **24**(3), 133–141.
4. Metzker ML (2010) Sequencing technologies—the next generation. Nature Rev. Genetics **11**, 31–46.

5. Eid J, Fehr A, Gray J, Luong K, Lyle J, *et al* (2009) Real-time DNA sequencing from single polymerase molecules. Science **323**, 133–138.
6. Pop M, and Salzberg SL (2008) Bioinformatics Challenges in New sequencing Technologies," Trends in Genetics **24**(3), 142–149.
7. Chaisson M, *et al* (2004) Fragment assembly with short reads. Bioinformatics **20**, 2067–2074.
8. Whiteford N, *et al* (2005) An analysis of the feasibility of short read sequencing. Nucleic Acids Res. **33**, e171.
9. Voelkerding KV, Dames SA, and Durtschi JD (2009) Next-generation sequencing: from basic research to diagnostics. Clin. Chem. **55**(4), 641–58.
10. Anker P, Mulcahy H, Chen XQ, and Stroun M (1999) Detection of circulating tumour DNA in the blood (plasma/serum) of cancer patients. Cancer and Metastasis Reviews **18**(1), 65–73.
11. Branton D, Deamer DW, Marziali A, Bayley H, *et al* (2008) The potential and challenges of nanopore sequencing. Nature biotechnology **26**(10), 1146–1153.
12. Akeson M, Branton D, Kasianowicz JJ, Brandin E, and Deamer DW (1999) Microsecond time-scale discrimination among polycytidylic acid, polyadenylic acid, and polyuridylic acid as homopolymers or as segments within single RNA molecules. Biophys. J. **77**(6), 3227–3233.
13. Chen P, *et al* (2004) Probing single DNA molecule transport using fabricated nanopores. Nano Lett. **4**(11), 2293–2298.
14. Heng JB, *et al* (2004) Sizing DNA using a nanometer-diameter pore. Biophys. J. **87**(4), 2905–2911.
15. Kasianowicz JJ, Brandin E, Branton D, and Deamer DW (1996) Characterization of individual polynucleotide molecules using a membrane channel. Proc. Natl. Acad. Sci. U.S.A. **93**(24), 13770–13773.
16. Li JL, Gershow M, Stein D, Brandin E, and Golovchenko JA (2003) DNA molecules and configurations in a solid-state nanopore microscope. Nat. Mater. **2**(9), 611–615.
17. Meller A, Nivon L, Brandin E, Golovchenko JA, and Branton D (2000) Rapid nanopore discrimination between single polynucleotide molecules. Proc. Natl. Acad. Sci. U.S.A. **97**(3), 1079–1084.
18. Storm AJ, Chen JH, Zandbergen HW, and Dekker C (2005) Translocation of double-strand DNA through a silicon oxide nanopore. Phys. Rev. E **71**(5), 10.
19. Clarke J, *et al* (2009) Continuous base identification for single-molecule nanopore DNA sequencing. Nat. Nanotechnol doi: 10.1038/NNANO.2009.12.
20. Cockroft S, Chu J, Amorin M, Bayley H, and Ghadiri M (2008) A single-molecule nanopore device detects DNA polymerase activity with single-nucleotide resolution. J. Am. Chem. Soc. **130**(3), 818.
21. Mirsaidov U, Comer J, Dimitrov V, Aksimentiev A, and Timp G (2010) Slowing the Translocation of Double-Stranded DNA Using a Nanopore Smaller than the Double Helix. Nanotechnology **21**, 395501.
22. Stoddart D, Heron AJ, Mikhailova E, Maglia G, and Bayley H (2009) Single-nucleotide discrimination in immobilized DNA oligonucleotides with a biological nanopore. Proceedings of National Academy of Sciences **106**(19), 7702–7707.
23. Likharev KK (1999) Single-Electron Devices and Their Applications. Proc. IEEE **87**, 606–632.
24. Lagerqvist J, Zwolak M, and Di Ventra M (2006) Fast DNA sequencing via transverse electronic transport," Nano Letters **6**(4), 779.
25. Chang SA, He J, Lin LS, *et al* (2009) Tunnel conductance of Watson-Crick nucleoside-base pairs from telegraph noise. Nanotechnology **20**(18), 7.
26. He J, Lin LS, Liu H, *et al* (2009) A hydrogen-bonded electron-tunneling circuit reads the base composition of unmodified DNA. Nanotechnology **20**(7), 8.
27. Meunier V, and Krstic PS (2008) Enhancement of the transverse conductance in DNA nucleotides. Journal of Chemical Physics **128**(4), 4.
28. Sigalov G, Comer J, Timp G, *et al* (2008) Detection of DNA Sequences Using an Alternating Electric Field in a Nanopore Capacitor, Nano Letters **8**(1), 56–63.

29. Fologea D, Uplinger J, Thomas B, McNabb DS and Li JL (2005) Slowing DNA translocation in a solid-state nanopore. Nano Lett. **5**, 1734–1737.
30. Storm AJ, Storm C, Chen JH, Zandbergen H, Joanny JF, and Dekker C (2005) Fast DNA translocation through a solid-state nanopore. Nano Lett. **5**, 1193–1197.
31. Soni GV and Meller A (2007) Progress toward ultrafast DNA Sequencing using solid-state nanopores. Clinical Chemistry **53**(11), 1996–2001.
32. Gershow M and Golovchenko JA (2007) Recapturing and trapping single molecules with a solid-state nanopore. Nat. Nanotechnol. **2**(12), 775–779.
33. Timp W, Mirsaidov UM, Wang D, Comer J, Aksimentiev O, and Timp G (2010) Nanopore Sequencing: Electrical Measurements of the Code of Life. IEEE Trans. Nanotechnology **9**, 281–294.
34. Mirsaidov UM, Wang D, Timp W, and Timp G (2010) Molecular Diagnostics for Personal Medicine Using a Nanopore," WIRES Review Nanomedicine Nanobiotechnology, **2**, 367–381.
35. Heng JB, Aksimentiev A, Ho C, Dimitrov V, Sorsch T, Miner J, Mansfield W, Schulten K, and Timp G (2005) Beyond the Gene Chip. Bell Labs Tech. J. **10**(3), 5–22.
36. Cruz-Chu ER, Ritz T, Siwy ZS, Schulten K (2009) Molecular control of ionic conduction in polymer nanopores. Faraday Disc. **143**, 47–62.
37. Ho C, Qiao R, Heng JB, Chatterjee A, Timp R, Aluru NR, and Timp G (2005) Electrolytic transport through a synthetic nanometer-diameter pore. Proceedings of National Academy of Sciences **102**(30), 10445–10450.
38. Venkatesan BM, Dorvel B, Yemenicioglu S, Watkins N, Petrov I, and Bashir R (2009) Highly sensitive, mechanically stable nanopore sensors for DNA analysis Advanced Materials, **21**(27), 2771–2776.
39. Dimitrov V, Aksimentiev A, Schulten K, Heng JB, Sorsch T, *et al* (2006) Exploring the Prospects for a Nanometer-scale Gene Chip. IEDM Proceedings 169–172.
40. Fischbein MD and Drndić M (2007) Sub-10 nm Device Fabrication in a Transmission Electron Microscope. Nano Lett. **7**(5), 1329–1337.
41. Martin CR, Nishizawa M, Jiarge K, Kang M, and Lee SB (2001) Controlling Ion Transport Selectively in Gold Nanotubule Membranes. Adv. Mater. **13,** 1351–1362.
42. Li J, Stein D, McMullan C, *et al* (2001) Ion-beam sculpting at nanometre length scales. Nature **412**, 6843, 166–169.
43. Siwy Z and Fulinski A (2002) Fabrication of a synthetic nanopore ion pump. Physical Review Letters **89**(19), 4.
44. Dimitrov V, Mirsaidov UM, Wang D, Sorsch T, *et al* (2010) Nanopores in Solid-State Membranes Engineered for Single-Molecule Detection. Nanotechnology **21**, 065502.
45. Storm AJ, Chen JH, Ling XS, *et al* (2003) Fabrication of solid-state nanopores with single-nanometre precision. Nature Materials **2**(8), 537–540.
46. Nair PR and Alam MA (2006) Performance limits of nanobiosensors. Appl. Phys. Lett. **88**, 233120.
47. Berg HC (1993) Random walks in biology, Princeton University Press, Princeton, N.J.
48. Heng JB, Aksimentiev A, Ho C, Marks P, Grinkova YV, Sligar S, Schulten K, and Timp G, (2006) The Electromechanics of DNA in a Synthetic Nanopore. Biophys. J. **90**, 1098–1106.
49. Heng JB, Aksimentiev A, Ho C, Marks P, Grinkova YV, Sligar S, Schulten K, and Timp G (2005) Stretching DNA using the Field in a Synthetic Nanopore. Nano Let. **5**(10), 1883–1888.
50. Nakane J, Akeson M, and Marziali A (2002) Evaluation of nanopores as candidates for electronic analyte detection. Electrophoresis **23**(16), 2592–2601.
51. Durack G (2003) *Cell Sorting Techniques and Technologies* in Emerging Tools for Single Cell Analysis: Advances in optical measurement technologies. Wiley - Liss.
52. Scott R, Sethu P, Harnett CK (2008) 3D hydrodynamic focusing in a microfluidic. Rev. Sci. Instru. **79**, 046104.
53. Marcus JS, Anderson WF, and Quake SR (2006) Microfluidic single-cell MRNA isolation and analysis. Anal. Chem. **78**(9), 3084–3089.

54. Melin J and Quake SR (2007) Microfluidic Large-Scale Integration: The Evolution of Design Rules for Biological Automation. Annu. Rev. Biophys. Biomol. Struct. **36**, 213–231.
55. Luan B and Aksimentiev A (2008) Strain softening in stretched DNA. Physical Review Letters **101**, 11.
56. Li JL, Gershow M, Stein D, *et al* (2003) DNA molecules and configurations in a solid-state nanopore microscope. Nature Materials **2**(9), 611–615.
57. Zhao Q, Comer J, Dimitrov V, *et al* (2008) Stretching and unzipping nucleic acid hairpins using a synthetic nanopore. Nucleic Acids Research **36**(5), 1532–1541.
58. Comer J, Dimitrov V, Zhao Q, *et al* (2009) Microscopic Mechanics of Hairpin DNA Translocation through Synthetic Nanopores. Biophysical Journal **96**(2), 593–608.
59. Mirsaidov UM, Timp W, Zou X, *et al* (2009) Nanoelectromechanics of Methylated DNA in a Synthetic Nanopore. Biophysical Journal **96**(4), L32–L34.
60. Heinemann U and Hahn M (1992) CCAGGC-m5C-TGG, "Helical fine structure, hydration, and comparison with CCAGGCCTGG. Journal of Biological Chemistry **267**, 7332–7341.
61. Derreumaux S, Chaoui M, Tevanian G, and Fermandjian S (2001) Impact of CpG methylation on structure, dynamics and solvation of cAMP DNA responsive element. Nucleic Acids Research **29**, 2314–2326.
62. Golovchenko JA, private communication.
63. Dawson JR and Harpst JA (1971) Light Scattering and Hydrodynamic Properties of Linear and Circular Bacteriophage Lambda DNA. Biopolymers **10,** 2499–2508.
64. Moffitt JR, Chemla YR, Izhaky D and Bustamante C (2006) Differential detection of dual traps improves the spatial resolution of optical tweezers. Proc. Natl. Acad. Sci, U.S.A. **103**(24), 9006–9011.
65. Shim J, Timp W, Comer J, Wang D, Mirsaidov U, Aksimentiev A, and Timp G, unpublished.
66. Smeets RMM, Keyser U, Dekker N, *et al* (2008) Noise in solid-state nanopores. Proceedings of the National Academy of Sciences **105**(2), 417.
67. Smeets RMM, Dekker NH, and Dekker C (2009) Low-frequency noise in solid-state nanopores. Nanotechnology **20**, 095501.
68. Coulter W (1953) Means for Counting Particles Suspended in a Fluid, USPTO.

Chapter 13
Stochastic Detection of Terrorist Agents and Biomolecules in a Biological Channel

Xiyun Guan, Ranulu Samanthi S. de Zoysa, Dilani A. Jayawardhana, and Qitao Zhao

Abstract Stochastic sensing can detect analytes at the single-molecule level, in which a biological ion channel embedded in a lipid bilayer or a nano-scale sized pore fabricated in a solid-state membrane is used as the sensing element. By monitoring the ionic current modulations induced by the passage of the target analyte through the single pore, both the concentration and the identity of the analyte can be revealed. In this chapter, we highlight recent advances in the stochastic detection of terrorist agents and biomolecules, and in real-world sample analysis using alpha-hemolysin protein ion channels.

Keywords Stochastic sensing • Ion channel • Lipid bilayer • Alpha-hemolysin • Artificial nanopore • Single-channel recording • Terrorist agents • Peptides • DNA • Real-world sample analysis

13.1 Introduction

Nanopore stochastic sensing is currently an active research area. In large part this growing interest is driven by the discovery that nanopore sensors can successfully detect analytes at the single-molecule level, potentially offering a highly sensitive, rapid, and multi-functional sensing system [1, 2]. Nanopore detection is achieved by monitoring the ionic current flowing through a single pore at a fixed applied potential. Typically a buffer solution containing a high salt concentration (e.g., 1 M NaCl or KCl) at or near pH 7.4 (i.e., physiological pH) is used to produce the open channel

X. Guan (✉)
Department of Chemistry and Biochemistry, The University of Texas at Arlington, 700 Planetarium Place, Arlington, TX 76019-0065, USA
e-mail: xguan@uta.edu

S.M. Iqbal and R. Bashir (eds.), *Nanopores: Sensing and Fundamental Biological Interactions*, DOI 10.1007/978-1-4419-8252-0_13,

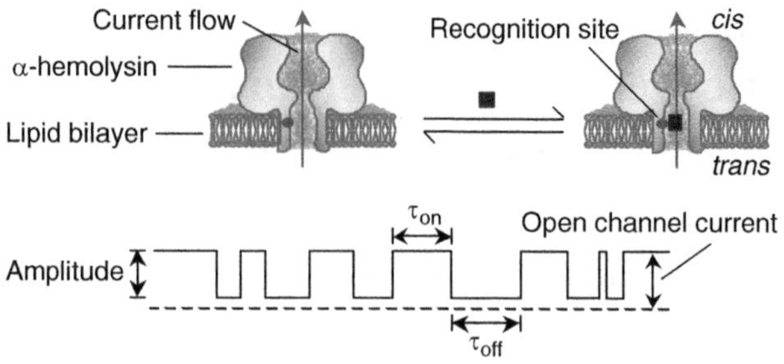

Fig. 13.1 Schematic representation of nanopore stochastic sensing

current which is monitored. In the absence of compounds, the channel is always open and a constant ionic current (called open-channel current) could be observed. In contrast, when a target molecule enters the pore, it will physically block the channel, thus resulting in a decrease in the ionic current flowing through the pore; when the molecule leaves the channel, the pore re-opens and the ionic current will increase (back to the open-channel state). In this way, a sequence of individual single-molecule binding events can be detected as transient modulations in the recorded current. Although each individual current blockage event is random, the statistical mean values of the residence time (τ_{off}) and amplitude of the events are reproducible and are also unique for different analytes. Hence, τ_{off} and amplitude can serve as a characteristic current signature to reveal the identity of an analyte (Fig. 13.1). Furthermore, the concentration of the analyte can be obtained from the frequency of occurrence ($1/\tau_{on}$) of the binding events. In stochastic sensing, since each analyte produces a characteristic signature, the sensor element itself need not be highly selective. Theoretically, this allows several analytes to be quantitated concurrently using a single sensor element (Fig. 13.2) [3], as long as the sensor itself can provide enough resolution.

The most often used nanopore stochastic sensor element is a single transmembrane protein α-hemolysin (αHL) channel embedded in a planar lipid bilayer. α-Hemolysin is a spontaneous pore-forming toxin secreted by *Staphylococcus aureus*. The wild-type αHL (Fig. 13.3) forms a mushroom-shaped pore, which consists of seven identical subunits arranged around a central axis [4]. The opening of the channel on the *cis* side of the bilayer measures 29 Å in diameter and broadens into a cavity of ~41 Å across. The cavity is connected to the trans-membrane domain, a 14-stranded β-barrel with an average diameter of 20 Å. The αHL pore has several properties, which make it unique as a sensor element in stochastic sensing. First, compared with other protein channels, such as porin [5] and leukocidin [6], the open αHL channel is quiet without transient background current modulation events. Thus, the αHL pore is an ideal sensor element for sensitive detection of trace amounts of analytes. Second, since the three-dimensional structure of the αHL pore is known [4], it can be modified with a variety of new functions, which greatly enhance its potential sensor application. Furthermore, the transmembrane portion (β-barrel) of the protein pore is sufficiently stable to

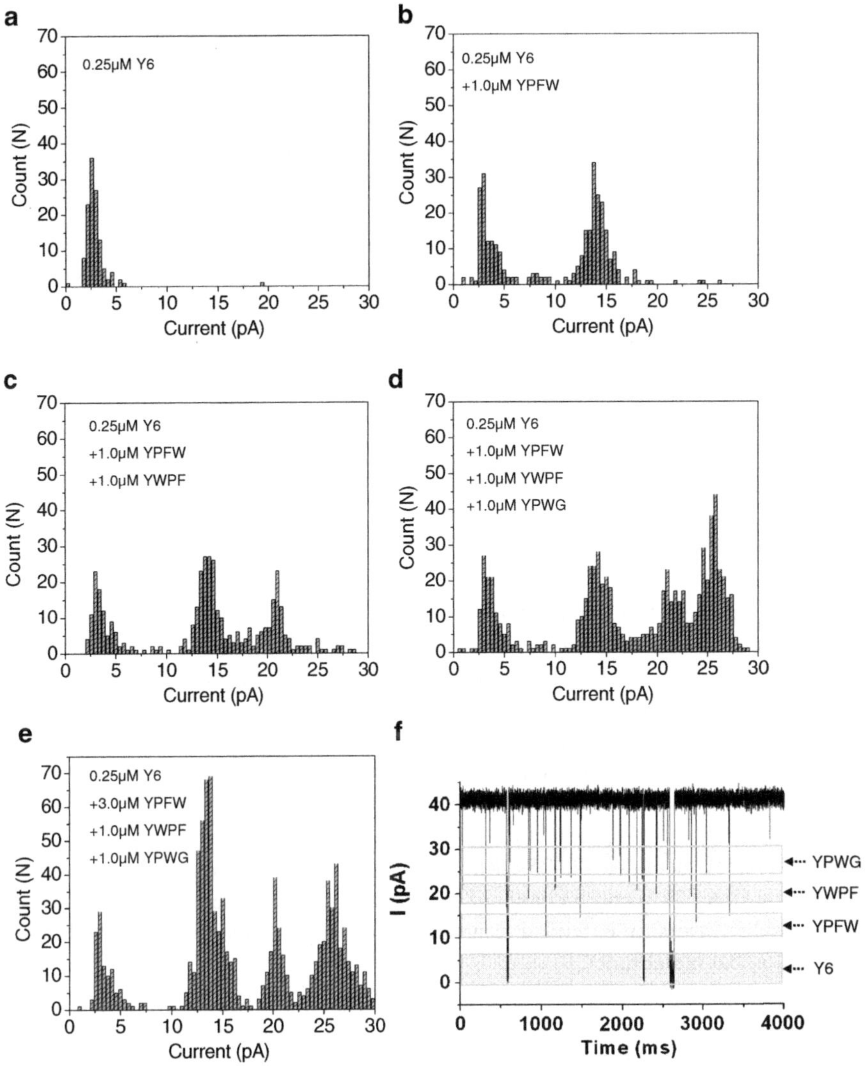

Fig. 13.2 Simultaneous detection of a mixture of peptides. Event histograms of (**a**) peptide Y6; (**b**) a two-peptide mixture containing Y6 and YPFW; (**c**) a mixture of three peptides containing Y6, YPFW and YWPF; (**d**) a mixture of four peptides containing Y6, YPFW, YWPF and YPWG; (**e**) a mixture of Y6, YPFW, YWPF and YPWG, where the concentration of YPFW was three times larger than that of (**d**); and (**f**) the corresponding representative single-channel current trace of (**d**), in which the individual Y6, YPFW, YWPF and YPWG events were arrow-marked at different levels. The experiments were performed at +30 mV (*cis* at ground) under a series of symmetrical buffer conditions with a 2.0 mL solution comprising 2 M NaCl, and 10 mM Tris–HCl (pH 7.5). αHL protein $(M113F)_7$ was added to the *cis* compartment, while peptides Y6, YPFW, YWPF, and YPWG were sequentially added to the *trans* compartment of the chamber with concentrations as shown in the figure. Figure reprinted with permission from [3], Copyright 2009 American Chemical Society

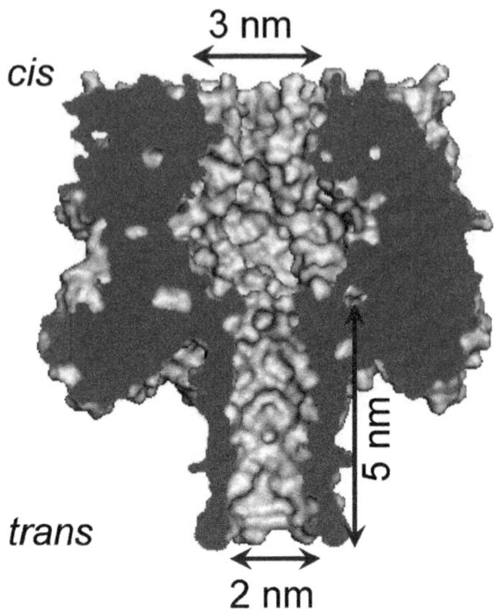

Fig. 13.3 α-hemolysin structure

permit the protein to tolerate modifications without losing functionality (i.e., still having the ability to form a channel in a lipid bilayer after the protein is engineered). Additionally, the relatively large αHL pore, and hence, the large single-channel conductance facilitates current recording. For example, the wild-type and engineered versions of the transmembrane protein pore in αHL have been used as stochastic sensing elements for the identification and quantification of a wide variety of substances [1], including metal ions [7, 8], anions [9], organic molecules [10], reactive molecules [11], enantiomers [12], terrorist agents [13–16], DNA [17–24], peptides [3, 25–28], and proteins [29–32]. Whereas this work is of obvious interest, a major challenge to the transition of protein pore technology to deployable sensors for extended usage is the fragility and the long-term stability of the biological membranes.

In light of the above limitation of natural pores, efforts are underway to construct robust pores in other materials, e.g., polymer and solid inorganic membranes [33–37]. Although the fabricated artificial nanopores improve the fragility aspect of the protein pore and they can function in a variety of extreme conditions, including voltage, temperature, and solvent variations, the following two issues have, to date, severely limited the practical application of the artificial pore approach. First, unlike the protein pore, it is very difficult, time-consuming, and requires specialized expensive equipment to prepare artificial pores with reproducible pore size. Therefore, the current signatures for a given analyte (e.g., amplitude and residence time) may change significantly from pore to pore, thus making the statistical analysis of the data extremely difficult and non transferable from lab to lab and even one sensor to another in the same lab. Second, the currently available artificial pore technology provides a very poor resolution due to the lack of surface functionalities [2], although it should be noted that the most recent study showed that it was possible to impart solid-state nanopores with selectivity and chemical functions [38–40]. This allows only large molecules such as DNA to be detected by using the electrophoretic effect to drive these molecules through the pore [33]. Furthermore, the poor resolution provided by the artificial pore makes it extremely difficult to characterize the current signatures of different large molecules, thus not permitting differentiation of molecules that differ slightly in composition.

There are also other efforts to overcome the fragility aspect of the protein pore. For example, carbon nanotubes have been tested as an alternative to the protein pore [41, 42]. However, in addition to the surface functionality problem noted above, it is currently not possible to reproducibly fabricate single-walled nanotubes of specified length and pore size. Another nanopore approach, called hybrid nanopore system, attempts to combine the advantage of the protein pore technology (i.e., the ease of engineering the nanopore with numerous functions, and having an identical pore size) and that of artificial pore approach (i.e., robustness). For example, it has been suggested that natural protein pores could be placed in artificial membranes [43], or located in etched inorganic nanotubes [44]. Although inherently appealing, the relative lack of results to date makes it impossible to gauge the long-term effectiveness of this approach at this point in time.

The recent advance in the protein-based nanopore technology has shown great promise to transition the nanopore sensor from pure academic curiosity to deployable use as a laboratory or clinical tool for routine sensor applications. For example, it has been demonstrated that a single αHL pore embedded in a planar phospholipid bilayer could be sandwiched between two agarose gel layers in situ, which may be transported, stored, and used repeatedly [45]. Furthermore, it was reported that the lipid bilayer supported by a glass nanopore membrane was very stable and could last for at least 2 weeks [46]. Thus, the issue regarding the fragility and the long-term stability of the lipid bilayer, i.e., the major hurdle of utilizing the protein pore method for extended usage could be overcome. More recently, Zhao and co-workers have demonstrated that an array of protein pores modified with a variety of surface functions could be used to construct a pattern-recognition stochastic sensor [47]. This nanopore sensor array technique allows identification of a target analyte from a mixture and the potential for simultaneous detection of multiple analytes.

A thorough overview of various aspects of nanopore stochastic sensing technology was recently provided in an excellent review article by Howorka and Siwy [48]. In this chapter, we will first briefly introduce planar bilayer recording, and then summarize the recent advances pertinent to the stochastic detection of terrorist agents and biomolecules in a biological channel.

13.2 Planar Bilayer Recording

Protein pore-based stochastic sensing experiments are usually performed in a two-compartment chamber device similar to that as shown in Fig. 13.4, where the *cis* and *trans* compartments are separated by a hydrophobic film (e.g., Teflon) having

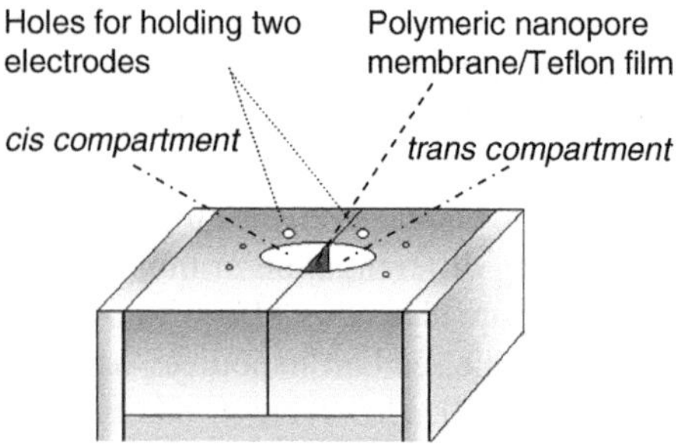

Fig. 13.4 Diagram of the nanopore stochastic sensing chamber device

an aperture of ~120 μm. As mentioned in the introduction section, in stochastic sensing, the protein pore is usually supported by a phospholipid bilayer. Thus far, there are a number of methods for the formation of planar lipid bilayer, including painting, tip dipping, apposition, and microfluidics. Among them, the apposition approach (also called the Montal–Mueller or monolayer folding method [49]) is the most popular one because the method is easy to use and also the bilayer formed in this way is very stable, which usually can last for at least several hours. To form a lipid bilayer by using the Montal–Mueller method, the aperture in the film is first pretreated with an organic solvent, e.g., hexadecane. Then, the two chamber compartments are filled with an appropriate volume of an electrolyte (e.g., 1 M NaCl/KCl) solution to make sure that the solution levels are below the aperture. Next, one or two drops of phospholipid solution such as 1,2-diphytanoylphosphatidylcholine in *n*-pentane are added to the surface of electrolyte solutions in both the compartments. After waiting for ~1–2 min to allow the pentane to evaporate and leave the lipid monolayers on the surface of the electrolyte solutions, certain volume of the same electrolyte solution is added to both the chamber compartments to raise the solution levels past the aperture. In such a way, a bilayer of phospholipid is formed on the aperture. The formation of the lipid bilayers can be monitored by using a function generator. In general, the bilayer recording experiments are performed under symmetrical buffer conditions with each compartment containing a ~1.0 mL solution of 1 M NaCl/KCl and 10 mM Tris–HCl (pH 7.4). Proteins (e.g., αHL) are added to the *cis* compartment, which is connected to "ground." In such a way, after insertion of a single αHL channel, its mushroom cap is located in the *cis* compartment, while the β-barrel of the αHL inserts into the lipid bilayer and connect with the *trans* of the chamber device. An appropriate transmembrane potential is applied to the chamber device using two Ag/AgCl electrodes with 3% agarose bridges containing 3 M KCl. Currents are recorded with a patch clamp amplifier. To shield against ambient electrical noise, a metal box is used to serve as a Faraday cage, inside which the bilayer recording amplifying headstage, stirring system, chamber, and chamber holder are enclosed.

It should be noted that, whether the observed ionic current is attributed to the insertion of a single αHL channel into the lipid bilayer or not can be conveniently judged from the planar bilayer recording experiment. Since the bilayer lipid membrane typically has a sealing resistance on the order of 100–200 GΩ [50], it can isolate the electrolyte solutions in the two chamber compartments very well. Under an applied voltage bias of e.g., 40 mV, the current will be 0 pA even if the proteins have been added to the *cis* compartment, but have not yet inserted into the bilayer. In contrast, after one αHL protein channel inserts into the bilayer, the two chamber compartments (*cis* and *trans*) are connected. Thus, the cations and/or anions in the electrolyte solutions will be electrophoretically driven toward the other compartment via the protein nano-channel under the applied potential, and a sudden current jump could be observed. The single channel currents for the wild-type and various mutant αHL protein pores have been well documented, usually about 21–32 pA in 1 M NaCl solution under an applied potential of 40 mV depending on the protein pore used [51]. If two αHL channels insert into the bilayer, the current will be doubled, around 42–64 pA.

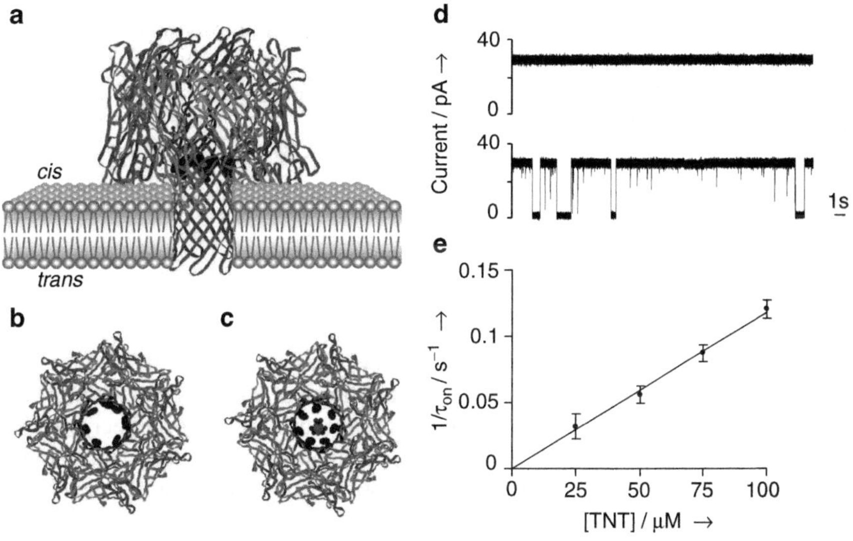

Fig. 13.5 Stochastic sensing of TNT in an engineered protein pore. (**a**) Side view of the Met113Phe7 pore highlighting position 113, where the naturally occurring Met residue has been substituted with Phe. (**b**) View into the Met113Phe7 pore from the *cis* side of the lipid bilayer. (**c**) View into the Met113Phe7 pore showing a TNT molecule at the same scale. The Phe side chains are shown in extended conformations and the TNT molecule has been placed with the plane of the ring perpendicular to the central axis of symmetry in the pore. (**d**) Current traces in the absence of TNT (*upper panel*) and in the presence of 50 μm TNT (*lower panel*). The TNT was added to the chamber (1.52 mL) in acetonitrile (17.2 mL). The same volume of acetonitrile was included in the experiment shown in the upper trace. (**e**) Plot of $1/\tau_{on}$ as a function of TNT concentration. The calculation of $1/\tau_{on}$ (n = 3) was based on the long blocking events. Experiments shown in Figs. 13.5d and 13.5e were performed at +50 mV (*cis* at ground) in NaCl (1 M), Tris–HCl (10 mM, pH 7.5) with the Met113Trp7 pore. Both the protein and TNT were added to the *cis* chamber. Figure reprinted with permission from [13], Copyright 2005 Wiley-VCH Verlag GmbH & Co. KGaA

13.3 Detection of Explosive and Chemical Warfare Agents

13.3.1 2,4,6-Trinitrotolune

2,4,6-Trinitrotolune (TNT) is an explosive used in military and industrial applications and also a priority pollutant listed by the US Environmental Protection Agency [52–54]. To detect TNT, aromatic surface function groups could be introduced inside the αHL protein pore (Fig. 13.5). For example, the Met amino acid residues at position 113 of the wild-type αHL pore can be replaced with Phe, Tyr, or Trp side chains [13]. Note that position 113 is close to the constriction area of the αHL protein channel and this position has been employed for the development of numerous nanopore sensors for a variety of analytes [3, 9]. When a TNT molecule enters the mutant protein pore, it will bind near that position (containing a cluster of seven aromatic residues) due to the aromatic-aromatic interaction, thus causing current blockage (Fig. 13.5). Since the event frequency was linearly related to the concentration of the added TNT, and other related nitroaromatic compounds had different current signatures (e.g., residence times and amplitudes) from those of TNT [13], the method should be able to find a potential application in identifying and quantifying TNT in an aqueous environment.

13.3.2 Liquid Explosives

Liquid explosives are usually binary mixtures where the two individual components alone are nonexplosives [55]. Since they can be transported easily and without being noticed, liquid explosives may be smuggled on the airplane to be the secret terrorist weapon. In fact, recently, there have been some reports about the failed plot to attack airplanes with liquid explosives [56]. Nanopore sensors have been developed as an effective platform to detect the liquid explosive component and sensitizers. For example, hydrazine (HZ) is a component of the liquid explosive, Astrolite. When hydrazine is mixed with ammonium nitrate, a powerful explosive is made. Diethylamine (DEA), triethylamine (TEA), and morpholine (CM) are liquid explosive sensitizers for nitromethane. The identification of amine-type liquid explosive components and the associated sensitizers was achieved by using boromycin as a molecular adaptor [14]. The cleft formed by the boromycin structure can accommodate monovalent cations such as potassium, ammonium, amine compounds, etc. (Fig. 13.6). When the host–guest complex enters the pore, current blockage events could be observed. Since different analyte–boromycin complexes showed significantly different event mean residence times and/or amplitudes (Fig. 13.6), the differentiation and even simultaneous detection of liquid explosive components in aqueous environments could be conveniently achieved.

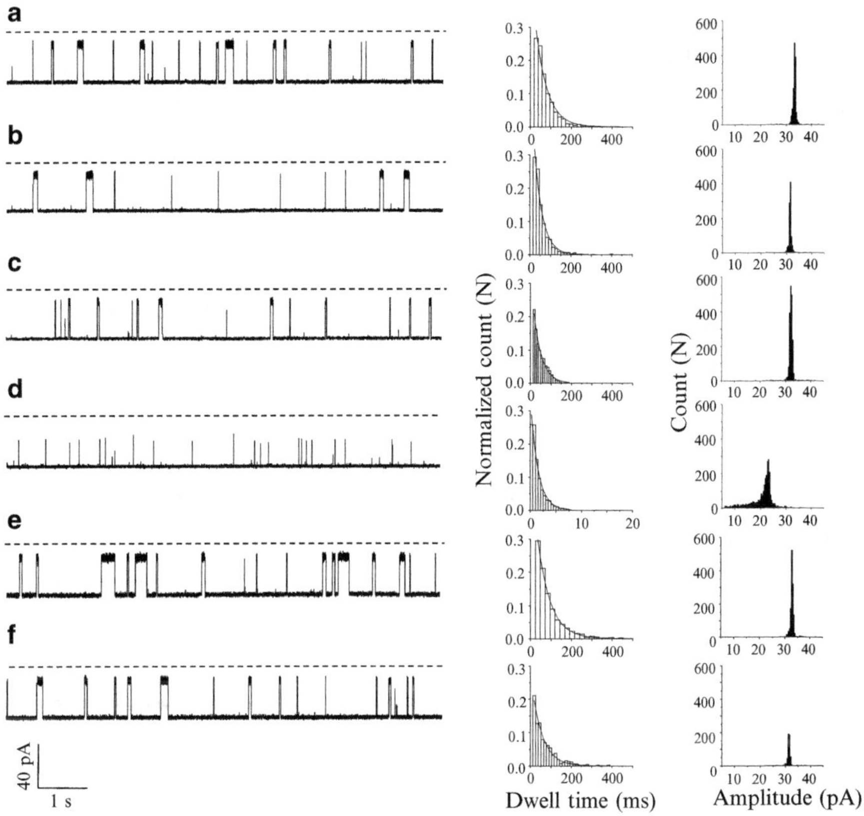

Fig. 13.6 Stochastic sensing of analytes in 1 M BMIM-Cl solution and using boromycin as a host. (*Left*) Typical single channel current recording traces. (*Middle*) Dwell time histograms. (*Right*) Amplitude histograms. (**a**) TEA (τ_{off} = 60.6 ± 1.9 ms, amplitude = 33.7 ± 0.4 pA, n = 7); (**b**) DEA (τ_{off} = 40.5 ± 0.5 ms, amplitude = 33.6 ± 0.8 pA, n = 4); (**c**) CM (τ_{off} = 34.7 ± 0.6 ms, amplitude = 33.5 ± 1.2 pA, n = 4); (**d**) HZ (τ_{off} = 1.71 ± 0.06 ms, amplitude = 22.6 ± 0.7 pA, n = 6); (**e**) NH_4^+ (τ_{off} = 75.0 ± 1.6 ms, amplitude = 33.7 ± 0.2 pA, n = 3); and (**f**) K^+ (τ_{off} = 52.1 ± 0.3 ms, amplitude = 33.6 ± 0.3 pA, n = 3). *Dashed lines* represent the levels of zero current. The *solid lines* in the dwell time histograms are fits of the normalized event distributions to single exponential functions. This normalization eliminates the event variation between experiments with different analytes. Figure reprinted with permission from [14]. Copyright 2009 American Chemical Society

13.3.3 Nerve Agent Hydrolytes

Organophosphorus chemical agents are the most toxic group in chemical warfare agents [57]. They are also known as nerve gases because they can disrupt the nervous system due to their irreversible binding to acetylcholine esterase, an enzyme which affects the accumulation of the neurotransmitter acetylcholine [57]. The same host–guest interaction approach as adopted in the detection of liquid explosives

Fig. 13.7 Detection of nerve agent hydrolytes. (**a**) Side view of the $(M113F/K147N)_7$ pore; and (**b**) View into the $(M113F/K147N)_7$ pore from the *cis* side of the lipid bilayer, highlighting positions 113 and 147, and the lodged βCD molecule. (**c**) Structures of PMPA and CMPA. Typical single channel current recording traces: (**d**) Without PMPA/CMPA; (**e**) 2 μM PMPA; and (**f**) 2 μM CMPA. The experiments were performed at −80 mV in 1 M NaCl and 10 mM Tris–HCl (pH 7.5), and in the presence of 40 μM βCD. Figure reprinted with permission from [15], Copyright 2009 Elsevier B.V.

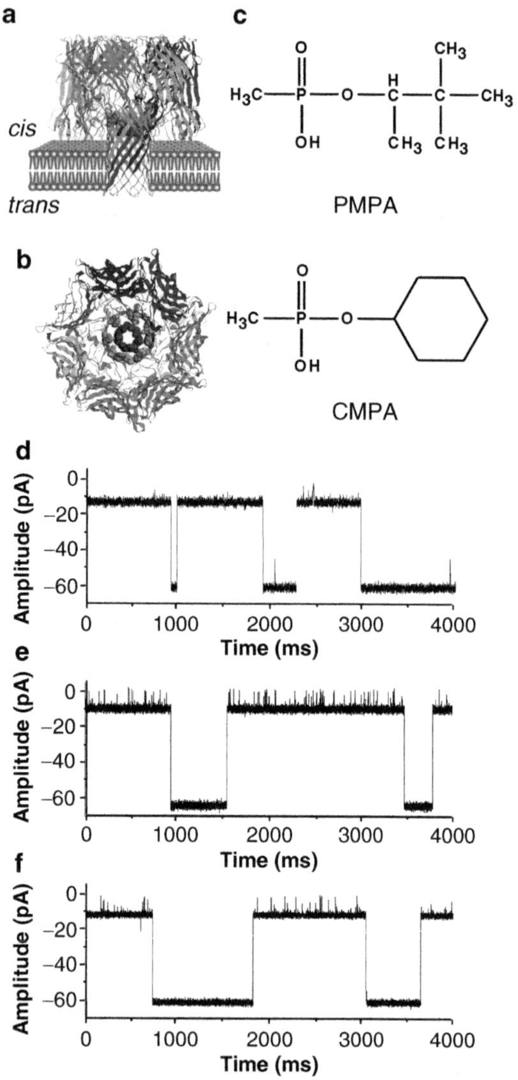

with a boromycin host can be employed for the analysis of organophosphorus chemical agents. For example, a host beta-cyclodextrin (βCD) molecule can be lodged in the lumen of the channel of an engineered αHL $(M113F/K147N)_7$ pore (as depicted in Fig. 13.7a, b) as a molecular adapter to detect pinacolyl methylphosphonic acid (PMPA) and cyclohexyl methylphosphonic acid (CMPA, Fig. 13.7c), the hydrolyzed products of the nerve agents, GD and GF respectively [15]. Note that in the two nanopore sensing systems (i.e., detection of liquid explosives and nerve agent hydrolytes), the host compounds boromycin and βCD behaved significantly different. In the case of boromycin, since it formed a stable host–guest complex with various amine-based compounds, the long duration blockage events were in fact attributed to the interaction of the host–guest complex with the protein pore. Although the un-complexed boromycin also bound to the pore, the interaction was much weaker, leading to a much shorter duration event. And hence, the unlabeled boromycin events and the host–guest complex events could be differentiated based on the residence time although these two types of events had similar amplitudes. In contrast, βCD bound to the mutant $(M113F/K147N)_7$ protein very tightly, so that the channel was partially blocked for a long time after addition of βCD to the buffer solution. If PMPA or CMPA compounds are additionally added to the mutant $(M113F/K147N)_7$ αHL pore, they will be further captured by the βCD host, thus causing a further current drop (Fig. 13.7b, c) [15].

13.4 Detection of Biomolecules

13.4.1 *Peptides*

Peptides play important roles in a variety of physiological processes in living systems. Accordingly, assay of these biologically active molecules are of paramount importance in diagnostics and therapeutics [58–60]. Since the amino acid components of a peptide can be categorized into four major groups: aromatic, hydrophobic, positive charge, and negative charge, it is likely that various peptides may be identified and differentiated by using engineered protein pores with different surface functions via weak non-covalent bonding interactions. To demonstrate this concept, a series of short peptides consisting of mainly aromatic amino acids and with various lengths was analyzed with a $(M113Y)_7$ pore, which contains an aromatic binding site with seven aromatic Tyr side chains. The experimental results showed that with an increase in the length of the peptide, both the mean residence time and the current blockage amplitude of the events increased (Fig. 13.8). Furthermore, it was found out that the composition of the peptide would also affect the event residence time and amplitude [3]. Take the protein channel modified with aromatic binding sites for example, aromatic amino acid components of a peptide contributed more to the residence time and current blockage of the events than other types of amino acids, although the van der Waals volumes of amino acids also affected the event signatures. Further experiments showed that with a properly engineered protein pore, not only could peptides that differed by only an amino acid be successfully differentiated, but also a mixture of peptides could be simultaneously detected (Fig. 13.2). This weak non-covalent interaction nanopore sensing approach could even permit differentiation of peptide sequences. For example, a series of short peptides with the same length and identical composition but with different sequences such as PYWF, YPWF, YWPF, and YPFW were able to be

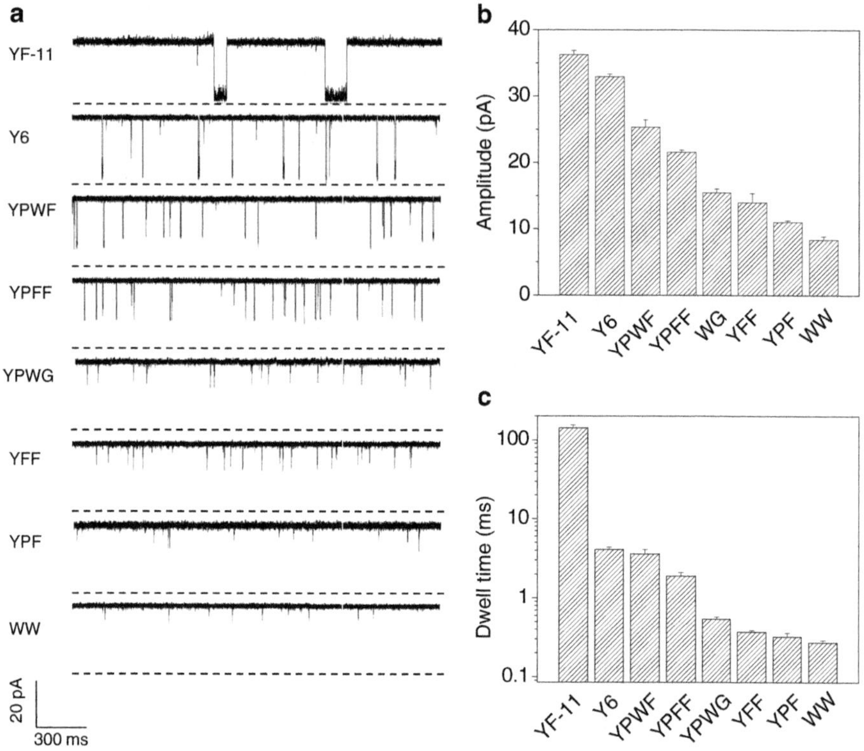

Fig. 13.8 Effect of peptide length and structure on the transport of peptides through a single $(M113Y)_7$ pore. (**a**) Representative single-channel traces; (**b**) Current blockage amplitudes; and (**c**) Event mean dwell times. The experiments were performed under a series of symmetrical buffer conditions with a 2.0 mL solution comprising 1 M NaCl, and 10 mM Tris–HCl (pH 7.5) at +50 mV (*cis* at ground). Peptides were added to the *trans* compartment, while the $(M113Y)_7$ protein was added from *cis* of the chamber device. The final concentrations of peptides in the buffer were 1.0 μM each. Figure reprinted with permission from [3]. Copyright 2009 American Chemical Society

differentiated from each other based on the different blockage residual currents and mean residence times of their events produced in the $(M113F)_7$ pore [3].

The finding is significant in that a promising peptide/protein sequencing method could be visualized. For example, long peptides or protein molecules can be cleaved into short fragments by enzymatic digestion or chemical cleavage. And then, the identities (e.g., structure and sequence) of these short peptide fragments can be determined by examining their translocation in the engineered protein pores with weak non-covalent binding sites. This concept has been demonstrated by real-time monitoring trypsin cleavage of the amyloid-β peptide (A-β) in the $(M113F)_7$ protein pore [61]. As shown in Fig. 13.9, before addition of trypsin to the peptide A-β (10–20) solution, only a single type of current blockage events was observed. In contrast, after addition of trypsin, two new types of current modulation events with different residence times and amplitudes from those of peptide A-β

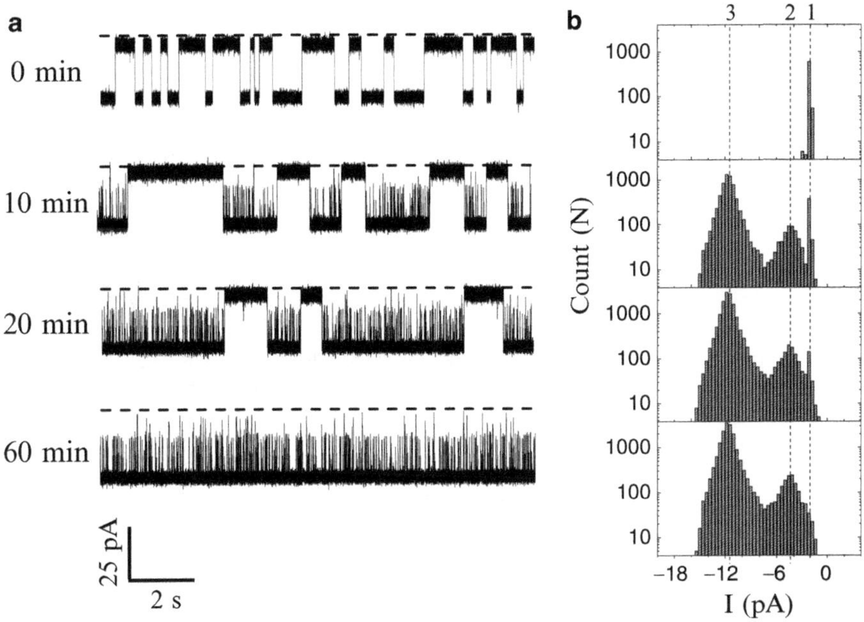

Fig. 13.9 Monitoring of A-β (10–20) cleavage by trypsin. (**a**) Representative segments of a single channel recording trace at various times. *Dashed lines* represent the levels of zero current; (**b**) The corresponding time-dependent event amplitude histograms. Dashed lines 1, 2, and 3 represent the mean residual current levels for peptides YEVHHQKLVFF, YEVHHQK, and LVFF, respectively. The experiment was performed at −40 mV with 10 μM A-β (10–20) and 0.025 μM trypsin. Figure reprinted with permission from [61], Copyright 2009 American Chemical Society

(10–20) were clearly observed. The identities of these two new types of events, which were believed to be attributed to the two breakdown products of A-β (10–20), were further confirmed by investigating the translocation of individual peptide standards. It is worth mentioning that in addition to the determination of the identities of short peptide fragments, this approach could be also employed as a real-time label-free method to study enzyme kinetics, e.g., to obtain V_{max} and K_m. This may be very useful since incorporation of a label may alter the structure and function of the target biomolecule.

13.4.2 DNA

Nanopore DNA sequencing, first introduced in the mid-1990 s [17], has emerged as one of the most promising sequencing technology among the other sequencing methods [62, 63]. A good overview of the current status of nanopore DNA

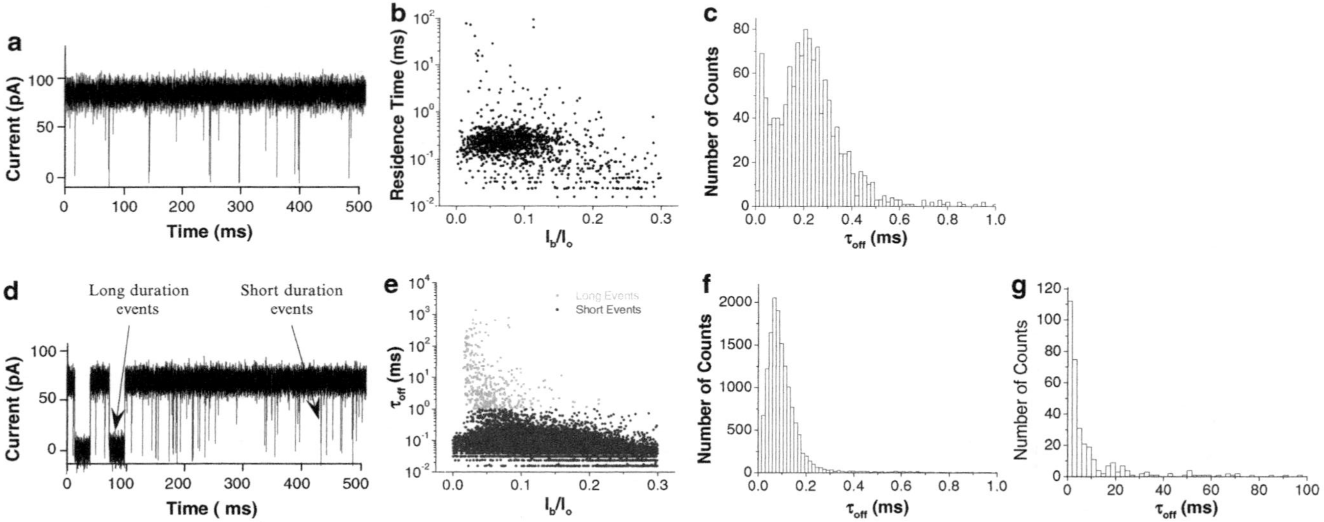

Fig. 13.10 *Upper panel*: translocation of $(dA)_{20}$ in the mutant $(M113F)_7$ αHL pore in 1 M NaCl solution. (**a**) Representative single channel current recording trace; (**b**) scatter plot of event amplitude vs. residence time; and (**c**) event residence time histogram. *Lower panel*: Translocation of $(dA)_{20}$ in the mutant $(M113F)_7$ αHL pore in 1 M BMIM-Cl solution. (**d**) Representative single channel current recording trace; (**e**) scatter plot of event amplitude vs. residence time; (**f**) residence time histogram of short-lived events; and (**g**) residence time histogram of long duration events. I_b/I_o in (**b**) and (**e**) is normalized blockage residual current, which was obtained by dividing the average blockage residual current of an event by the average open channel current. Figure reprinted with permission from [70]. Copyright 2009 American Chemical Society

sequencing was recently provided by Branton et al. [64]. One key advantage of utilizing a nanopore to sequence DNA molecules is that it can read the linear sequence of nucleotides from the native DNA sample without copying the DNA and without incorporating labels. Optimal implementation of this method would include: "sequencing without amplification or modification, and would provide very long sequence reads (tens of thousands to millions of bases) rapidly and at sufficiently high redundancy to produce assembled sequence of high quality" [62]. However, as pointed out by Branton and co-workers, one of the major hurdles of utilizing nanopores to sequence single-stranded DNA (ssDNA) molecules is the rapid DNA translocation velocity (~1–3 μs/base) through the nanopore so that the accurate detection of single nucleotide bases could not be achieved with the currently available single-channel recording technique [63]. In the past decade, various approaches have been used to slow ssDNA translocation, including immobilization of DNA polynucleotides with streptavidin [65], formation of DNA-hemolysin rotaxane [20], and formation of double-strand DNA [19]. Furthermore, single nucleotide bases could be captured by using a host compound such as β-cyclodextrin [66]. Using this host–guest interaction approach, single nucleotide base A, T, G, C were successfully differentiated. In addition, experimental physical conditions such as temperature, voltage bias, viscosity, and application of an alternating electric field was also found useful to control the DNA translocation rate [67–69].

Recently, it was reported that the use of ionic liquid solutions as the background electrolyte instead of the commonly used KCl/NaCl solution increased the values of the residence time of liquid explosives and monovalent cations in the αHL pore [14]. The feasibility of utilizing organic salt solutions to slow the translocation of ssDNA in the αHL pore was further investigated [70]. It was found that in addition to the rapid DNA translocation events with the mean residence time similar to that in the NaCl/KCl solution, another type of events with the translocation rates on the order of hundreds of microseconds per nucleotide base were identified (Fig. 13.10). Since the mean residence time of these long-lived events increased linearly with an increase in the DNA length [70], they were suitable for the analysis of the length and structure of a polynucleotide molecule. More recently, the effect of alkaline pH on DNA translocation was also systematically investigated by Maglia and co-workers [21]. It was reported that the frequency of ssDNA translocation through the wild-type αHL pore decreased with an increase in the pH of the buffer solution. When pH was larger than 10.7, no events could be identified. Further experiment showed that ssDNA translocation events were observed in the mutant $(M113R)_7$ pore even at pH 11.7. These results suggest that the structure of αHL pores is stable enough to tolerate high pH. Furthermore, ssDNA translocation is dependent upon the charge distribution within the lumen of the αHL channel [17]. The finding is significant in that the translocation of ssDNA through the nanopore could be manipulated via introduction of various surface functions into the nanopore, and double-stranded DNA was able to be studied using the αHL pore approach at high pH (Fig. 13.11).

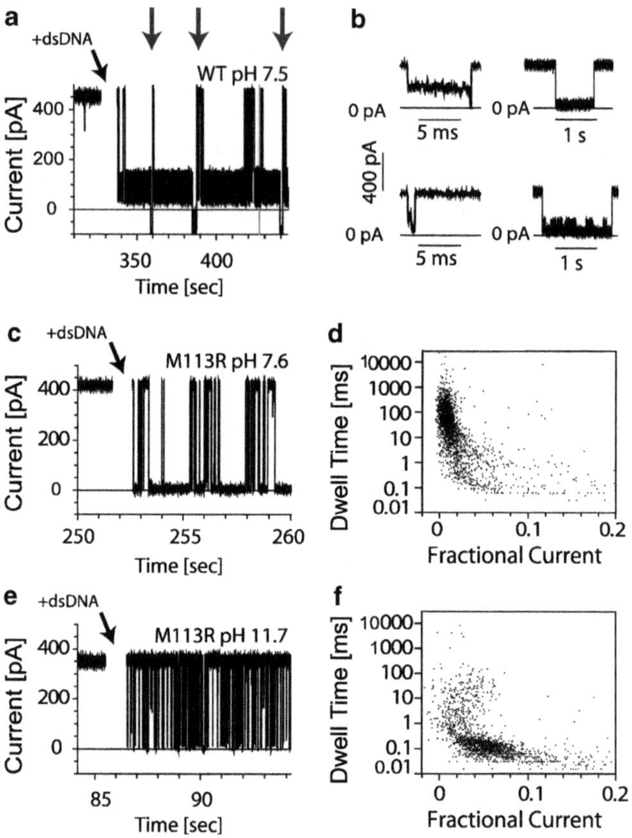

Fig. 13.11 Interaction of dsDNA with WT and M113R-αHL pores. (**a**) Current blockades observed after the addition of dsDNA (1.0 μM) to the *cis* side of a WT pore at pH 7.6. The very long current blockades most likely do not correspond to DNA translocation events (see the text). The occluded pore was reopened by ramping the potential to negative and then positive potentials (*vertical arrows*). (**b**) Selection of shorter current blockades caused by the interaction of dsDNA with WT nanopores at pH 7.6. (**c**) Current blockades observed after the addition of dsDNA (1.0 μM) to M113R pores at pH 7.6. The relatively long current blockades (but shorter than the very long events seen with WT-αHL) suggest that M113R promotes the unzipping of duplexes. ssDNA then traverses the pore (see the text). (**d**) Fractional residual current vs. dwell time of the current blockades for the interaction of dsDNA with M113R nanopores at pH 7.6. (**e**) dsDNA-induced blockades through M113R pores at pH 11.7. The very short current blockades represent direct translocation of ssDNA after denaturation of the dsDNA (2.0 μM) at the alkaline pH of the solution. (**f**) Fractional residual current vs. dwell time of the current blockades for the interaction of (dissociated) dsDNA with M113R nanopores at pH 11.7. The buffer was 10 mM Tris–HCl, 10 mM CAPS, 15 mM K phosphate, 1.0 M KCl, 100 μM EDTA, pH 7.6. The pH in panels (**e**) and (**f**) was adjusted to 11.7 by adding small aliquots of 1.0 M KOH directly to the *cis* and *trans* chambers with a pore in the bilayer. Figure reprinted with permission from [21]. Copyright 2009 American Chemical Society

13.5 Nanopore Analysis of Real World Samples

In addition to the development of various nanopore sensors for a variety of analytes, efforts are under way to transition the protein-based nanopore technology to deployable sensor applications. As overviewed in the introduction section, one key challenge to nanopore stochastic sensing using protein channels is the fragility and the long-term stability of the lipid bilayers. The recent advance in the protein pore technology demonstrated that this limitation could be potentially overcome by using a protein nanopore chip [45], or employing lipid bilayer supported by a glass nanopore membrane [46]. Another challenge to nanopore (including both protein pore and synthetic nanopore) stochastic sensing is how effective this approach is used for the analysis of a real-world sample. It is likely that a number of components in the sample matrix could potentially interfere with the target analyte detection since the nanopore sensor is semi-selective. To address this issue, recently, simulated contaminated water samples were analyzed by nanopore sensor for the first time [15]. Among them, one water sample served as the false positive, which was obtained by spiking methyl phosphonic acid, ethyl methylphosphonic acid, and 2-(dimethylamino) ethanethiol in tap water. Other water samples served as real alarms. In addition to the three organophosphate compounds, these samples also contained different concentrations of pinacolyl methylphosphonic acid (PMPA). The experimental results showed that if the sample only contained false positives, no current blockage events were observed. In contrast, current modulations were identified in all of the samples that contained the target compound (i.e., PMPA) [15]. Furthermore, the concentrations of pinacolyl methylphosphonic acid in the real alarm samples obtained by using the nanopore sensor were in agreement with their corresponding theoretical values (Table 13.1). The results showed that the nanopore sensing method could indeed differentiate the target compound from false positives, and even allowed the accurate identification and quantification of PMPA in the presence of a mixture of structure-similar compounds, thus demonstrating the feasibility of utilizing nanopore sensors for real-world sample analysis.

Table 13.1 Recovery of PMPA from liquid samples[a] by use of the nanopore stochastic sensing method

Sample number	Theoretical value (μM)	Experimental value ± SD (μM)
1	0	not detected
2	10.0	10.6 ± 0.4
3	15.0	14.2 ± 0.9
4	20.0	18.8 ± 1.7

Each experimental value represents the mean of three replicate analyses ± one standard deviation. The experiments were performed at −80 mV with mutant αHL $(M113K/F147N)_7$ pore in the presence of 40 μM βCD. Table reprinted with permission from [15], Copyright 2009 Elsevier B.V.

[a]All of the four liquid samples contained additional 100 μM of methyl phosphonic acid, ethyl methylphosphonic acid, and 2-(dimethylamino) ethanethiol.

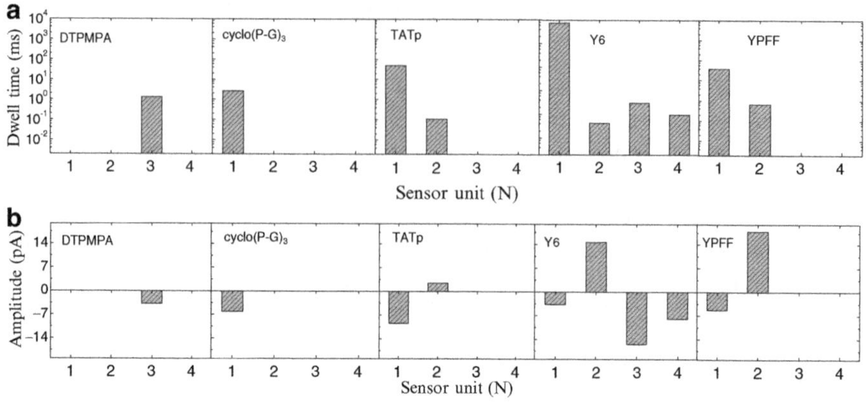

Fig. 13.12 Pattern-recognition differentiation of a variety of molecules. (**a**) Dwell time plot, and (**b**) amplitude plot. The experiments were performed at +40 mV or −40 mV (*cis* at ground) with 1 M NaCl and 10 mM Tris–HCl (pH 7.5). Figure reprinted with permission from [47], Copyright 2008 IOP Publishing Ltd

To enhance the resolution and/or selectivity of nanopore sensors for multi-analyte analysis, pattern-recognition nanopore sensor array technique was developed [47]. The proposed sensing device consisted of an array of at least two nanopores, each of which had different functional groups (i.e., non-covalent bonding sites), ranging from super hydrophobic to ultra hydrophilic, as well as having both positive and negative charged surfaces. Thus, each individual pore of the sensor array was different and reacted differently toward a compound. Furthermore, a parallel electric circuit of on/off switches was employed to control which channel(s) to be monitored. In this system, the collective responses of each individual component nanopore to a compound produced a diagnostic pattern, which can serve as an analyte's signature (Fig. 13.12). This sensing principle is similar to the mammalian olfactory system [71], and has been utilized in a variety of chemical sensors, including electronic nose [72, 73], to increase their differentiation resolution. With an increase in the dimensionality of the sensing system, the nanopore pattern-recognition technology has proven that it indeed provides an enhanced resolution for the differentiation of analytes compared to a single-pore configuration, and allows identification of a target analyte from a mixture or the potential for simultaneous detection.

13.6 Conclusions

Protein pore-based stochastic sensing technique has many advantages. These include high sensitivity, rapid response, low cost, instrumental simplicity, ease of use, and low false positive rate. Although the transition of protein pore technology to deployable sensors for extended usage has been hindered by the fragility and the long-term stability of the biological membranes in the past, recent studies demonstrate that this limitation could be overcome by using glass nanopore membrane or stable protein pore chips. Furthermore, the resolution of nanopore stochastic sensing can be further improved by using a pattern-recognition nanopore sensor array and with an organic salt solution as the background electrolyte. The protein-pore based stochastic sensing method is envisioned for further development as a high-throughput, portable technique for real-time, on-site analysis of chemical, biological, and explosive agents, as well as compounds of environmental and biological importance.

References

1. Bayley, H.; Cremer, P. S. Stochastic sensors inspired by biology. *Nature* **2001**, *413*, 226–230.
2. Schmidt, J. Stochastic sensors, *J. Mater. Chem.* **2005**, *15*, 831–840.
3. Zhao, Q.; Jayawardhana, D. A.; Wang, D.; Guan, X. Study of peptide transport through engineered protein channels. *J. Phys. Chem. B* **2009**, *113*, 3572–3578.
4. Song, L.; Hobaugh, M. R.; Shustak, C.; Cheley, S.; Bayley, H.; Gouaux, J. E. Structure of staphylococcal alpha-hemolysin, a heptameric transmembrane pore. *Science* **1996**, *274*, 1859–1866.
5. Conlan, S.; Zhang, Y.; Cheley, S.; Bayley, H. Biochemical and biophysical characterization of OmpG: A monomeric porin. *Biochemistry* **2000**, *39*, 11845–11854.
6. Miles, G.; Cheley, S.; Braha, O.; Bayley, H. The staphylococcal leukocidin bicomponent toxin forms large ionic channels. *Biochemistry* **2001**, *40*, 8514–8522.
7. Braha, O.; Gu, L.-Q.; Zhou, L.; Lu, X.; Cheley, S.; Bayley, H. Simultaneous stochastic sensing of divalent metal ions. *Nat. Biotechnol.* **2000**, *17*, 1005–1007.
8. Braha, O.; Walker, B.; Cheley, S.; Kasianowicz, J. J.; Song, L.; Gouaux, J. E.; Bayley, H. Designed protein pores as components for biosensors. *Chem. Biol.* **1997**, *4*, 497–505.
9. Cheley, S.; Gu, L.-Q.; Bayley, H. Stochastic sensing of nanomolar inositol 1,4,5-trisphosphate with an engineered pore. *Chem. Biol.* **2002**, *9*, 829–838.
10. Gu, L.-Q.; Braha, O.; Conlan, S.; Cheley, S.; Bayley, H. Stochastic sensing of organic analytes by a pore-forming protein containing a molecular adapter. *Nature* **1999**, *398*, 686–690.
11. Shin, S.-H.; Luchian, T.; Cheley, S.; Braha, O.; Bayley, H. Kinetics of a reversible covalent-bond-forming reaction observed at the single-molecule level. *Angew. Chem. Int. Ed.* **2002**, *41*, 3707–3709.
12. Kang, X. F.; Cheley, S.; Guan, X.; Bayley, H. Stochastic Detection of Enantiomers. *J. Am. Chem. Soc.* **2006**, 128, 10684–10685.
13. Guan, X.; Gu, L. Q.; Cheley, S.; Braha, O.; Bayley, H. Stochastic sensing of TNT with a genetically engineered pore. *ChemBioChem*, **2005**, *6*, 1875–1881.
14. Jayawardhana, D. A.; Crank, J. A.; Zhao, Q.; Armstrong, D. W.; Guan X. Nanopore stochastic detection of a liquid explosive component and sensitizers using boromycin and an ionic liquid supporting electrolyte. *Anal. Chem.* **2009**, *81*, 460–464.
15. Wang, D.; Zhao, Q.; Guan, X. Detection of nerve agent hydrolytes in an engineered nanopore. *Sens. Actuators B Chem.* **2009**, *139*, 440–446.
16. Wu, H. C.; Bayley, H. Single-molecule detection of nitrogen mustards by covalent reaction within a protein nanopore. *J. Am. Chem. Soc.* **2008**, *130*, 6813–6819.
17. Kasianowicz, J. J.; Brandin, E.; Branton, D.; Deamer, D. Characterization of individual polynucleotide molecules using a membrane channel. *Proc. Natl. Acad. Sci. U. S. A.* 1996, *93*, 13770–13773.
18. Meller, A.; Nivon, L.; Brandin, E.; Golovchenko, J.; Branton, D. Rapid nanopore discrimination between single polynucleotide molecules. *Proc. Natl. Acad. Sci. U.S.A.* **2000**, *97*, 1079–1084.

19. Howorka, S.; Cheley, S.; Bayley, H. Sequence-specific detection of individual DNA strands using engineered nanopores. *Nat. Biotechnol.* **2001**, *19*, 636–639.
20. Sanchez-Quesada, J.; Saghatelian, A.; Cheley, S.; Bayley, H.; Ghadiri, M. R. Single DNA rotaxanes of a transmembrane pore protein. *Angew. Chem. Int. Ed. Engl.* **2004**, *43*, 3063–3067.
21. Maglia, G.; Henricus, M.; Wyss, R.; Li, Q.; Cheley, S.; Bayley, H. DNA strands from denatured duplexes are translocated through engineered protein nanopores at alkaline pH. *Nano Lett.* **2009**, *9*, 3831–3836.
22. Stoddart, D.; Heron, A. J.; Mikhailova, E.; Maglia, G.; Bayley, H. Single-nucleotide discrimination in immobilized DNA oligonucleotides with a biological nanopore. *Proc. Natl. Acad. Sci. U.S.A.* **2009**, *106*, 7702–7707.
23. Clarke, J.; Wu, H. C.; Jayasinghe, L.; Patel, A.; Reid, S.; Bayley, H. Continuous base identification for single-molecule nanopore DNA sequencing. *Nat. Nanotechnol.* **2009**, *4*, 265–270.
24. Maglia, G.; Restrepo, M. R.; Mikhailova, E.; Bayley, H. Enhanced translocation of single DNA molecules through alpha-hemolysin nanopores by manipulation of internal charge. *Proc. Natl. Acad. Sci. U.S.A.* **2008**, *105*, 19720–19725.
25. Stefureac, R.; Long, Y. T.; Kraatz, H. B.; Howard, P.; Lee, J. S. Transport of alpha-helical peptides through alpha-hemolysin and aerolysin pores. *Biochem.* **2006**, *45*, 9172–9179.
26. Movileanu, L.; Schmittschmitt, J. P.; Scholtz, J. M.; Bayley, H. Interactions of peptides with a protein pore. *Biophys. J.* **2005**, *89*, 1030–1045.
27. Wolfe, A. J.; Mohammad, M. M.; Cheley, S.; Bayley, H.; Movileanu, L. Catalyzing the translocation of polypeptides through attractive interactions. *J. Am. Chem. Soc.* **2007**, *129*, 14034–14041.
28. Mohammad, M. M.; Movileanu, L. Excursion of a single polypeptide into a protein pore: simple physics, but complicated biology. *Eur. Biophys. J.* **2008**, *37*, 913–925.
29. Movileanu, L.; Howorka, S.; Braha, O.; Bayley, H. Detecting protein analytes that modulate transmembrane movement of a polymer chain within a single protein pore. *Nat. Biotechnol.* **2000**, *18*, 1091–1095.
30. Howorka, S; Nam, J.; Bayley, H.; Kahne, D. Stochastic detection of monovalent and bivalent protein-ligand interactions. *Angew. Chem. Int. Ed. Engl.* **2004**, 43, 842–846.
31. Xie, H.; Braha, O.; Gu, L.-Q.; Cheley, S.; Bayley, H. Single-molecule observation of the catalytic subunit of cAMP-dependent protein kinase binding to an inhibitor peptide. *Chem. Biol.* **2005**, *12*, 109–120.
32. Cheley, S.; Xie, H.; Bayley, H. A genetically encoded pore for the stochastic detection of a protein kinase. *Chembiochem* **2006**, *7*, 1923–1927.
33. Li, J.; Stein, D.; McMullan, C.; Branton, D.; Aziz, M. J.; Golovchenko, J. A. Ion-beam sculpting at nanometre length scales. *Nature* **2001**, *412*, 166–169.
34. Storm, A. J.; Storm, C.; Chen, J.; Zandbergen, H.; Joanny, J. F.; Dekker, C. Fast DNA translocation through a solid-state nanopore. *Nano Lett.* **2005**, *5*, 1193–1197.
35. Storm, A. J.; Chen, J. H.; Ling, X. S.; Zandbergen, H. W.; Dekker, C. Fabrication of solid-state nanopores with single-nanometre precision. *Nat Mater.* **2003**, *2*, 537–540.
36. Heins, E. A.; Siwy, Z. S.; Baker, L. A.; Martin, C. R. Detecting single porphyrin molecules in a conically shaped synthetic nanopore. *Nano Lett.* **2005**, *5*, 1824–1829.
37. Siwy, Z.; Apel, P.; Dobrev, D.; Neumann, R.; Spohr, R.; Trautmann, C.; Voss, K. Ion transport through asymmetric nanopores prepared by ion track etching. *Nucl. Instrum. Methods Phys. Res., Sect. B* **2003**, *208*, 143–148.
38. Iqbal, S. M.; Akin, D.; Bashir, R. Solid-state nanopore channels with DNA selectivity. *Nature Nanotechnol.* **2007**, *2*, 243–248.
39. Wanunu, M.; Meller, A. Chemically modified solid-state nanopores. *Nano Lett.* **2007**, *7*, 1580–1585.
40. Vlassiouk, I.; Kozel, T. R.; Siwy, Z. S. Biosensing with nanofluidic diodes. *J. Am. Chem. Soc.* **2009**, *131*, 8211–8220.

41. Sun, L.; Crooks, R. M. Single Carbon Nanotube Membranes: A well-defined model for studying mass transport through nanoporous materials. *J. Am. Chem. Soc.* **2000**, *122*, 12340–12345.
42. Yeh, I. C.; Hummer, G. Nucleic acid transport through carbon nanotube membranes. *Proc. Natl. Acad. Sci. U.S.A.* **2004**, *101*, 12177–12182.
43. Nardin, C.; Meier, W. Hybrid materials from amphiphilic block copolymers and membrane proteins. *J. Biotechnol.* **2002**, *90*, 17–26.
44. Bayley, H.; Martin, C. R. Resistive-pulse sensing-from microbes to molecules. *Chem. Rev.* **2000**, *100*, 2575–2594.
45. Kang, X. F.; Cheley, S.; Rice-Ficht, A. C.; Bayley, H. A storable encapsulated bilayer chip containing a single protein nanopore. *J. Am. Chem. Soc.* **2007**, *129*, 4701–4705.
46. White, R. J.; Ervin, E. N.; Yang, T.; Chen, X.; Daniel, S.; Cremer, P. S.; White, H. S. Single ion-channel recordings using glass nanopore membranes. *J. Am. Chem. Soc.* **2007**, *129*, 11766–11775.
47. Zhao, Q.; Wang, D.; Jayawardhana, D. A.; Guan, X. Stochastic sensing of biomolecules in a nanopore sensor array. *Nanotechnology* **2008**, *19*, 505504.
48. Howorka, S.; Siwy Z. Nanopore analytics: sensing of single molecules. *Chem. Soc. Rev.* **2009**, *38*, 2360–2384.
49. Montal, M.; Mueller, P. Formation of bimolecular membranes from lipid monolayers and a study of their electrical properties. *Proc. Natl. Acad. Sci. U.S.A.* **1972**, *69*, 3561–3566.
50. Matsuno, Y.; Osono, C.; Hirano, A.; Sugawara, M. Single-channel recordings of gramicidin at agarose-supported bilayer lipid membranes formed by the tip-dip and painting methods. *Anal. Sci.* **2004**, *20*, 1217–1221.
51. Gu, L. Q.; Cheley, S.; Bayley, H. Prolonged residence time of a noncovalent molecular adapter, beta-cyclodextrin, within the lumen of mutant alpha-hemolysin pores. *J. Gen. Physiol.* **2001**, *118*, 481–494.
52. Pinnaduwage, L. A.; Gehl, A.; Hedden, D. L.; Muralidharan, G.; Thundat, T.; Lareau, R. T.; Sulchek, T.; Manning, L.; Rogers, B.; Jones, M.; Adams, J. D. Explosives: a microsensor for trinitrotoluene vapour. *Nature* **2003**, *425*, 474–474.
53. Looger, L. L.; Dwyer, M. A.; Smith, J. J.; Hellinga, H. W. Computational design of receptor and sensor proteins with novel functions. *Nature* **2003**, *423*, 185–190.
54. Meagher, R. B. Pink water, green plants, and pink elephants. *Nat. Biotechnol.* **2001**, *19*, 1120–1121.
55. http://www.globalsecurity.org/military/systems/munitions/explosives-liquid.htm.
56. http://www.howstuffworks.com/liquid-explosives.htm.
57. Hooijschuur, E. W.; Kientz, C. E.; Brinkman, U. A.Analytical separation techniques for the determination of chemical warfare agents. *J. Chromatogr. A* **2002**, *982*, 177–200.
58. Rathert, P.; Dhayalan, A.; Murakami, M.; Zhang, X.; Tamas, R.; Jurkowska, R.; Komatsu, Y.; Shinkai, Y.; Cheng, X. D.; Jeltsch, A. Protein lysine methyltransferase G9a acts on non-histone targets. *Nat. Chem. Biol.* **2008**, *4*, 344–346.
59. Asara, J. M.; Schweitzer, M. H.; Freimark, L. M.; Phillips, M.; Cantley, L. C. Protein sequences from mastodon and Tyrannosaurus rex revealed by mass spectrometry. *Science* **2007**, *316*, 280–285.
60. Baker, D.; Sali, A. Protein structure prediction and structural genomics. *Science* **2001**, *294*, 93–96.
61. Zhao, Q.; de Zoysa, R. S.; Wang, D.; Jayawardhana, D. A.; Guan, X. Real-time monitoring of peptide cleavage using a nanopore probe. *J. Am. Chem. Soc.* **2009**, *131*, 6324–6325.
62. National Human Genome Research Institute (2004) Revolutionary Genome Sequencing Technologies—The $1000 Genome. (http://grants1.nih.gov/grants/guide/rfa-files/RFA-HG-04-003.html)
63. Bayley, H. Sequencing single molecules of DNA. *Curr. Opin. Chem. Biol.* **2006**, *10*, 628–637.
64. Branton, D.; Deamer, D. W.; Marziali, A.; Bayley, H.; Benner, S. A.; Butler, T.; Di Ventra, M.; Garaj, S.; Hibbs, A.; Huang, X.; Jovanovich, S. B.; Krstic, P. S.; Lindsay, S.; Ling, X. S.; Mastrangelo, C. H.; Meller, A.; Oliver, J. S.; Pershin, Y. V.; Ramsey, J. M.; Riehn, R.; Soni,

G. V.; Tabard-Cossa, V.; Wanunu, M.; Wiggin, M.; Schloss, J. The potential and challenges of nanopore sequencing. A. *Nat. Biotechnol.* **2008**, *26*, 1146–1153.
65. Purnell, F. R.; Mehta, K. K.; Schmidt, J. J. Nucleotide identification and orientation discrimination of DNA homopolymers immobilized in a protein nanopore. *Nano Lett.* **2008**, *8*, 3029–3034.
66. Astier, Y.; Braha, O.; Bayley, H. Toward single molecule DNA sequencing: direct identification of ribonucleoside and deoxyribonucleoside 5'-monophosphates by using an engineered protein nanopore equipped with a molecular adapter. *J. Am. Chem. Soc.* **2006**, *128*, 1705–1710.
67. Meller, A.; Branton, D. Single molecule measurements of DNA transport through a nanopore. *Electrophoresis* **2002**, *23*, 2583–2591.
68. Meller, A.; Nivon, L.; Branton, D. Voltage-driven DNA translocations through a nanopore. *Phys. Rev. Lett.* **2001**, *86*, 3435–3438.
69. Sigalov, G.; Comer, J.; Timp, G.; Aksimentiev, A. Detection of DNA sequences using an alternating electric field in a nanopore capacitor. *Nano Lett.* **2008**, *8*, 56–63.
70. de Zoysa, R. S.; Jayawardhana, D. A.; Zhao, Q.; Wang, D.; Armstrong, D. W.; Guan, X. Slowing DNA translocation through nanopores using a solution containing organic salts. *J. Phys. Chem. B* **2009**, *113*, 13332–13336.
71. Dickinson, T. A.; White, J.; Kauer, J. S.; Walt, D. R. Current trends in 'artificial-nose' technology. *Trends Biotechnol.* **1998**, *16*, 250–258.
72. Turner, A. P.; Magan, N. Electronic noses and disease diagnostics. *Nat. Rev. Microbiol.* **2004**, *2*, 161–166.
73. Thaler, E. R.; Kennedy, D. W.; Hanson, C. W. Medical applications of electronic nose technology: review of current status. *Am. J. Rhinol.* **2001**, *15*, 291–295.

Chapter 14
Nanopore Force Spectroscopy: Insights from Molecular Dynamics Simulations

Jeffrey Comer and Aleksei Aksimentiev

Abstract Nanopore force spectroscopy (NFS) has emerged as a convenient method to characterize the behavior of single biomolecules and biomolecular assemblies under force. NFS has many advantages over conventional single molecule techniques, such as being label-free and high throughput; however, NFS lacks direct control over the force applied to the biomolecules and registers the conformational transitions induced by the force only indirectly, by monitoring changes in the ionic current passing through the pore. In this chapter, we describe how all-atom molecular dynamics simulations can complement NFS experiments by providing information inaccessible to experiment. The chapter illustrates applications of the molecular dynamics (MD) method to interpret the results of NFS measurements, characterize the forces involved and determine the microscopic origin of the observed phenomena. Important technical aspects of the method, as well as its pitfalls and limitations are briefly discussed.

Keywords Molecular dynamics • Force spectroscopy • Computer simulation • Nucleic acid • Restriction enzyme • Endonuclease • Protein–DNA recognition • Electrostatic tweezers • DNA unzipping • DNA hairpin • Nanobiotechnology • Biomolecular complex • Single-molecule analysis • Molecular rupture

14.1 Introduction

Single-molecule techniques have fomented a revolution in biology by allowing precise measurements of the nanoscale processes that give rise to the qualities that we ascribe to life – growth, motion, reproduction, assimilation of nutrients, etc. Determination of the mechanical properties of biomolecules has been instrumental

A. Aksimentiev (✉)
Department of Physics, Beckman Institute for Advanced Science and Technology, University of Illinois at Urbana-Champaign, 1110 W Green Street Urbana, Urbana, IL 61801, USA
e-mail: aksiment@illinois.edu

S.M. Iqbal and R. Bashir (eds.), *Nanopores: Sensing and Fundamental Biological Interactions*, DOI 10.1007/978-1-4419-8252-0_14,

in understanding many fundamental biological interactions. For example, knowledge of the mechanical properties of DNA gives insight into how packaging, transcription, and replication of genetic material occur.

Mechanical probing of single biomolecules has been accomplished using a variety of techniques [1] including optical tweezers [2, 3], magnetic tweezers [4, 5], atomic force microscopy [6], biomembrane probe [7, 8], and subjection to flow [9]. These techniques require the attachment of molecules by linkers to some larger object, such as an Atomic Force Microscope (AFM) tip/cantilever, or a micrometer-scale bead as required in optical and magnetic tweezers experiments. Attachment of linkers can change the behavior of the system, and reloading the apparatus with a new biomolecule after each measurement can be time-consuming and require prodigious effort.

Nanopore Force Spectroscopy (NFS), on the other hand, is a technique in which mechanical forces can be applied to single biomolecules without chemical modification or linker attachment. Instead, NFS works with biomolecular assemblies having a relatively thin portion, which can thread through the pore, and a second bulkier portion, which cannot. The basic design of a typical NFS apparatus is illustrated in Fig. 14.1a. A small chamber is separated into two sections by a membrane no more than a few tens of nanometers thick. Both sides of the chamber

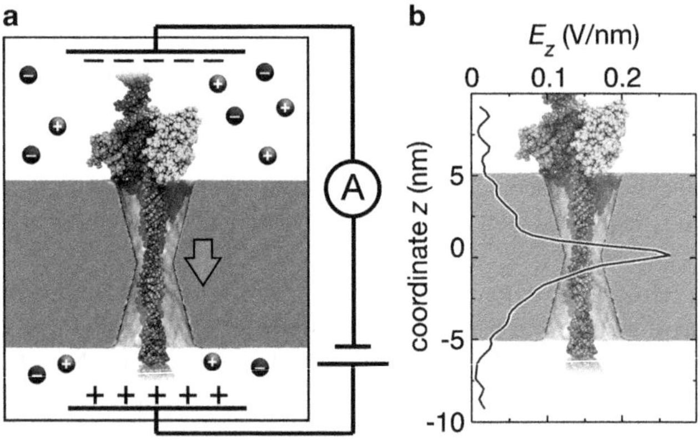

Fig. 14.1 Nanopore force spectroscopy. (**a**) Cartoon of a typical nanopore force spectroscopy experiment. Force is applied to double-stranded DNA threaded through the pore. An enzyme bound to the DNA is too large to pass through the pore. Therefore, the force on the DNA stresses the intermolecular bond between the enzyme and DNA, allowing the interaction between the two to be observed under the application of force. The top and bottom chambers (not to scale) are separated by a membrane having a thickness of several nanometers. The entire system is filled with electrolyte. Electrodes in each chamber are used to apply a voltage across the membrane. The transmembrane voltage determines the force on the analyte and also causes a flow of ions through the pore, which is measured using a sensitive ammeter. (**b**) Focusing of electrostatic force by the nanopore. The component of the electric field directed along the pore axis (E_z) is plotted as a function of position along this axis. An image of the pore faithfully overlays the plot, although the electric field was derived from a simulation in which no analyte was present in the pore. Because the solution filling the system is a good conductor, the magnitude of electric field near the pore constriction is many orders of magnitude larger than outside the pore. The data plotted here was derived from an MD simulation of a pore in the absence of an analyte

are filled with an electrolyte solution. A pore having a diameter from one to tens of nanometers has been drilled in the membrane. Solutes injected on one side of the chamber can only reach the other side by passing through the nanopore. In Fig. 14.1a, force is applied to a molecule of double-stranded DNA (dsDNA), which threads through the pore. A relatively bulky enzyme is bound to the DNA which has too large a cross-section to pass through the pore. The walls of the pore apply a force to the enzyme opposite to that applied to DNA, stressing the intermolecular bond between the two. Thus, in NFS we can determine the behavior of molecular assemblies under force, revealing the thermodynamic and kinetic properties of the intermolecular bond [10–12] as well as structural changes necessary for biological functionality and technological applications.

14.1.1 Nanopore Force Spectroscopy Experiments

In NFS experiments, the size of the nanopore can be tailored to the problem of interest. To probe enzyme–DNA interactions, such as those binding the complex shown in Fig. 14.1a, a pore having a diameter of 3 nm is sufficiently large to permit unfettered passage of dsDNA (which has a diameter of ~2.5 nm), but too small to accommodate the enzyme. Similarly, to probe interactions between DNA strands, a smaller pore (~2.0 nm) can be used that permits passage of unstructured DNA but excludes the passage of portions of the molecule having bulky secondary structure, such as double helices. Using focused beams of charged particles, nanopores having diameters as large as 20 nm and as small as 1.0 nm are routinely fabricated [13], permitting NFS to be applied to a broad range of molecular assemblies.

A majority of NFS experiments have involved DNA in some way, due to the importance of the molecule in biology and biotechnology. In addition to the extensive use of NFS for probing the forces involved in unfolding of single DNA molecules [14–20], protein–DNA interactions have also been explored [21–23]. DNA is an especially good subject for NFS because it is highly negatively charged ($-e$ per nucleotide at physiological pH) and, therefore, can easily be captured by the electric field of the nanopore. Despite this bias toward DNA, Goodrich et al. [24] used NFS to study unfolding of peptide molecules, while experiments by Siwy and colleagues [25], in which chemically modified nanopores were used to detect specific proteins, suggest the prospect of studying protein–protein interactions with NFS. Indeed, NFS is an extremely general method that can be used to probe many types of interactions at the single-molecule level when at least one of the analytes carries a nonzero charge.

A reader unfamiliar with NFS might be left with several questions: (i) How does one load an analyte into the nanopore? (ii) How does one apply force to the analyte once it gets there? (iii) How does one observe the effect of the force on the analyte?

Questions (i) and (ii) have the same answer: the nanopore geometry focuses an applied electric field, capturing charged analytes and allowing sustained forces to be applied. Electrodes are placed on both sides of the pore, which allow an electrical potential difference to be applied between the two chambers. Because the solution is conductor, a majority of this applied potential difference is dropped across the membrane, leading to a focusing of the electric field in and around the pore. Articles describing NFS experiments often quote the applied potential difference, or transmembrane voltage, which gives another experimenter sufficient information to reproduce the electrostatic force given the same membrane thickness and pore geometry. As shown in Fig. 14.1b, the electric field is sharply focused at the tightest constriction of the pore, with a magnitude on the order of 10^8 V/m for a typical nanopore under a transmembrane voltage of 1 V. Analytes in the surrounding solution of the appropriate charge are drawn to the pore by the electric field or an electro-osmotic flow. Loading of more than one analyte can be prevented by tuning the size of the pore to sterically exclude more than one analyte.

There are two common protocols for the application of the transmembrane voltage in NFS experiments. First, one can use a constant transmembrane voltage, which applies a constant force assuming a constant pore–analyte conformation, and measure the distribution of the time interval between the rupture and escape of the analyte. A second common protocol involves applying a linearly increasing transmembrane voltage, or a constant loading rate if one again assumes a constant pore–analyte conformation.

The answer to question (iii) above is that measurements of the ion current through the pore, which is driven by the transmembrane voltage, serve as an indirect means to determine the effect of the force on the analyte. This current, which typically ranges from a few picoamperes to many nanoamperes, can be measured by sensitive ammeters as suggested in Fig. 14.1a. Because the opening of the nanopore is near the size of the analyte, the ion current through it can change dramatically when an analyte is captured and loaded into the pore. For example, a drop (or increase) in the ion current can be observed when DNA enters a nanopore. The exit of the DNA from the nanopore is therefore accompanied by a return of the current to its original value. Furthermore, the ion current can change in response to conformational transformations of the analyte. Although protocols for nanopore assays have been constructed which employ fluorescent markers [26] or measurements of the potential on electrodes embedded in the walls of the nanopore [27], the ion current is invariably a convenient and useful source of information on events occurring in the pore during NFS experiments.

The unique properties of NFS also lead to some disadvantages with respect to other single-molecule force probing techniques. In a typical NFS experiment, there is just one input signal (the transmembrane voltage) and just one output signal (the measured electrolytic current). In contrast to other force spectroscopy methods, in pure NFS the force on the analyte cannot be directly measured (although hybrid experiments including both optical tweezers and nanopores have been performed [28] to directly measure the force). In NFS, the experimenter has precise control over the transmembrane voltage, which only indirectly determines the force on the analyte. This force depends on the position of the analyte within the pore as well as the pore geometry, the electrical properties of the membrane and electrolyte, and the positions of any other molecules that may be associated with the analyte.

Another difficulty of NFS is interpretation of the measured ion current signatures in terms of microscopic events in the pore. While some ion current signatures may be easy to interpret, such as a reduction in the current caused by the presence of DNA threading through the pore, a rich variety of current signatures have been observed in nanopore experiments. The ion current through nanopores containing DNA has been heavily studied [29]. It is understood that the effect of the DNA on the current is a competition between two effects: steric blockage of the pore by DNA, which decreases the current, and enhancement of the density of charge carriers due to the DNA's counterions, which increases the current [30, 31]. At high bulk ion concentrations, the first effect dominates and the current is reduced by the DNA. At low bulk ion concentrations, the second effect dominates and the current is enhanced. However, in some cases, both enhancements and reductions are seen under the same macroscopic conditions [18, 20, 31]. This suggests that the ion current depends on microscopic details of the DNA conformation in the pore.

14.1.2 Molecular Dynamics Simulation

Molecular Dynamics (MD) simulation, by providing an atomic resolution view of the processes occurring in the pore, has the ability to fill some of the gaps in the information furnished by NFS experiments. MD simulations can yield predictions of how the analyte interacts with the pore, what conformation the analyte is likely to have during different stages of the experiment, and sequences in which events such as dissociation are likely to occur. For example, if one were studying unzipping of DNA duplexes, one could estimate how deeply the duplex penetrates into the pore during the experiment, how much the duplex stretches at various transmembrane voltages, and which basepairs of the duplex are likely to break first under the applied force. Although NFS experiments rarely allow direct measurement of the forces applied to analyte, these forces can be calculated in MD simulations. Furthermore, MD allows us to observe the microscopic details of processes occurring in the nanopore, while simultaneously obtaining estimates of the ion current through the nanopore. We can therefore associate ranges of current values with specific molecular configurations of the analyte in the pore. The topic of interpretation of ion current signatures has been reviewed elsewhere [29]; however, considering its importance to NFS experiments, we describe some of the pertinent aspects here.

This chapter focuses on classical all-atom MD, which we will refer to simply as MD, although ab initio MD, which employs approximations to quantum mechanical laws, and coarse-grain MD, in which groups of atoms are represented by single simulated particles, could also be useful. MD simulations typically begin with the creation of a computational model of a small key portion of the experiment. For simulations of NFS in synthetic nanopores, we begin with a three-dimensional simulation cell having linear dimensions of a few tens of nanometers to comfortably accommodate the structures of interest. Atomic-resolution models of biomolecules, often constructed using data from X-ray crystallography or nuclear magnetic resonance, are added, along with ions, molecules of water, and the atoms of the synthetic membrane housing the nanopore. Such systems usually comprise tens of thousands to millions of atoms. In classical all-atom MD, the acceleration of each atom of the system – including the water – is calculated using Newton's Second Law: $\ddot{\mathbf{x}}_i(t) = \mathbf{F}_i(t)/m_i$, where $\mathbf{F}_i$ is the force on atom i and m_i is its mass. The force is calculated using a number of simple mathematical functions that include electrostatic forces, van der Waals forces, and various empirical forces that approximate the behavior of covalently bonded structures as well as external forces, such as those due to applied electric fields [32]. The particular forms of the mathematical functions and values of the parameters used for each type of atom constitute a force field. A number of force fields have been developed over the past few decades for the purpose of accurate simulations of biomolecular systems, notably the AMBER [33, 34] and CHARMM force fields [35]. Using forces calculated in this way, the trajectory of the entire system can be generated by numerical integration of Newton's Second Law in time. Using the best algorithms available today, it is possible to obtain tens or hundreds of nanoseconds of such a trajectory in a 24 h period. By parallelizing the computation over hundreds or thousands of individual processing units, systems containing as many as millions of atoms can be simulated at this rate. The MD simulations presented in this chapter were designed and executed following the protocols described by Comer et al. [36].

14.1.3 Chapter Overview

Through examples given in this chapter, we illustrate how MD simulations can assist in the design and interpretation of NFS experiments. First, we describe the application of the MD method to NFS of enzyme–DNA complexes. We show how the method is used to estimate the force required to rupture the enzyme–DNA complex as a function of the loading rate and to make qualitative predictions of the strength of enzyme binding to sequences that differ from the enzyme's target sequence. Next, we describe the use of MD simulations to interpret NFS experiments of DNA duplex unfolding. Here, the simulations have elucidated how the mode by which the duplex unfolds might be affected by the size of the nanopore and how various values of the ion current could correspond to particular conformations of the DNA in the pore. Finally, we describe how the MD method is used to determine the force experienced by DNA in a nanopore and has resolved the contributions of counterion screening and electro-osmotic flow to the effective reduction of the DNA charge.

14.2 Rupture of Protein–DNA Assemblies

Protein–DNA interactions are essential to many biological processes including transcription, packaging, and repair of genetic material. Protein–DNA interactions have also been exploited in biotechnology. For example, restriction enzymes, which are proteins that bind to double-stranded DNA at specific sequences and cleave the molecule in two, are indispensable tools in recombinant DNA technology [38] as well as many DNA sequencing and genotyping methods [39–41].

The type II restriction enzyme *Eco*RI binds to dsDNA and is thought to diffuse linearly, pausing near sequences resembling the cognate sequence GAATTC [42]. In the presence of Mg^{2+}, the *Eco*RI cleaves both DNA strands at particular points along the cognate sequence with high specificity. In the absence of Mg^{2+}, *Eco*RI does not cleave the DNA, but still binds strongly to sites having the cognate sequence.

The extraordinary specificity of type II restriction enzymes for their respective cognate sequences combined with the high throughput potential of NFS has been proposed as the basis of a method of genotyping genomic DNA [22]. In short, a solution of DNA molecules decorated with bound enzymes (*Eco*RI or similar enzymes such as *Bam*HI) can be introduced into the chamber on one side of the membrane. The DNA is captured by the pore and threads through it until the enzyme reaches the pore opening as illustrated in Fig. 14.2a. If the pore is sufficiently small (having a diameter < 5 nm) and the transmembrane voltage is sufficiently low, the enzyme will block the pore and halt translocation of the DNA. However, for larger transmembrane voltages, the force on the enzyme–DNA bond dramatically increases the probability for the rupture of the complex. Rupture of the complex permits the DNA to continue threading through the pore until the next enzyme is encountered. The value of the transmembrane voltage at which rupture occurs shows a strong dependence on the sequence to which the enzyme is bound, being nearly zero for nonspecific binding and a few volts for the cognate sequence. Therefore, by measuring the transmembrane voltage required to dissociate an enzyme from DNA, one can identify particular subsequences within a given DNA molecule.

Quantitative polymerase chain reaction has been used to count the number of DNA molecules passing through the pore at various transmembrane voltages [22]. When enzymes are specifically bound to the DNA molecules, this number increases sharply over a small range of transmembrane voltage. For example, in a nanopore having a minimum elliptical cross section of 3.4 nm $\times$ 4.5 nm, the molecule count increases from near zero to 10^8 over a range between 1.8 and 2.0 V [22]. One can therefore define a threshold value of the transmembrane voltage above which nearly all enzymes reaching the pore dissociate, and below which the mean time until rupture is much greater than the timescale of the experiment. When *Eco*RI was bound to its cognate sequence GAATTC, the complex was shown to have a much higher threshold voltage than when it was bound to a sequence that differed by one basepair [22, 23]. The sequence GGATCC, which differs from the cognate sequence by two basepairs, showed no threshold at all. Even the basepairs just next to the cognate sequence were shown to measurably affect the threshold voltage [23]. Furthermore, it was shown that another restriction enzyme (*Bam*HI) bound to its cognate sequence (GGATCC) could be distinguished from *Eco*RI bound to its cognate sequence due to the difference in the threshold voltage [22].

Despite the many insights yielded by the experiments, there were a number of questions about the behavior of the complex in the pore. How deeply did the enzyme

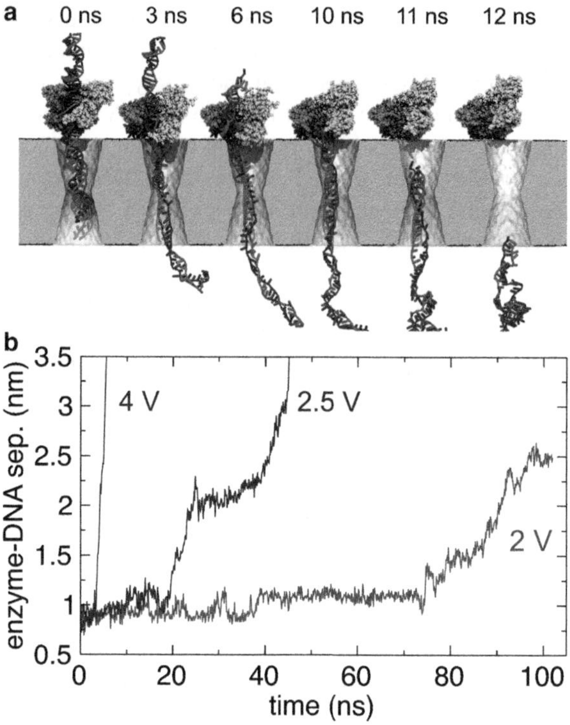

Fig. 14.2 MD simulations of a restriction enzyme–DNA complex interacting with a nanopore. (**a**) Snapshots from an MD simulation in which an external voltage of 4.0 V is applied across the membrane. The membrane has a thickness of 10 nm and the minimum diameter of the pore is 2.9 nm. The elapsed time since the application of the voltage is shown above each snapshot. (**b**) Rupture of the *Eco*RI–DNA molecular bond at various applied voltages. To characterize the separation between the enzyme and DNA, we plot the distance between the phosphate of the DNA adjacent to the bond cleaved by *Eco*RI (when Mg^{2+} is present) and the glutamic acid in the active site of *Eco*RI [37]. Due to the symmetry of the *Eco*RI dimer, there are two active sites; the data plotted here are for the site nearest the pore. Abrupt increases in this distance are observed when the enzyme and DNA dissociate. In all cases, a force was applied to reduce the affinity of the DNA for the pore walls [20]; however, for the 2 V simulation this force was not applied until $t = 56$ nm.

penetrate into the pore and how was it positioned in the pore when force was applied? How did the conformation of the enzyme and DNA change under force? How did rupture occur? Did the DNA simply slip through the binding site? Or did the conformation of the protein change considerably? What was the force at rupture? Which parts of the DNA decoupled from the protein first? MD simulations, one of which is illustrated in Fig. 14.2a, provided answers to all these questions.

The simulations immediately yielded qualitative insights. As can be seen from Fig. 14.2a, the enzyme did not completely enter the pores, even under a high force. Under load, however, the enzyme showed a tendency to tilt, allowing one lobe to penetrate the pore opening, as shown clearly in the 3 and 6 ns snapshots shown in

Fig. 14.2a. The tilt of the protein was accompanied by a bend in the DNA greater than 60°, which rotated the portion of the molecule above the enzyme towards the membrane. Such a sharp bend over just a few tens of basepairs would be difficult with B-form dsDNA, which has a persistence length of about 50 nm; however, at 2.0 V, the DNA below the enzyme was stretched into an extended form with only loose coupling between the strands. The simulations also showed that, during rupture, one strand leaves the binding site first, followed shortly by the other.

Our definition of the threshold voltage as the voltage at which the mean time until rupture is much greater than the timescale of the experiment make its estimation difficult for MD because the timescale accessible to MD is much less than that accessible to experiments. However, MD can give an upper bound for the threshold voltage that, due to the sharpness of the dependence of the rupture rate on the threshold voltage, comes quite close to the experimental value. Figure 14.2b shows that the time until rupture rises sharply as the transmembrane voltage approaches 2.0 V. At 4.0 V, rupture occurs in only 3.4 ± 0.1 ns. At 2.5 V, 20 ± 1 ns were required for rupture to occur. Decreasing the transmembrane voltage by just 0.5 V (to 2.0 V) increased the rupture time to 75 ± 1 ns. These results bear some resemblance to experiments, where threshold voltage for rupture of an *Eco*RI–DNA complex in a pore of a similar geometry was shown to lie between 1.5 and 2.0 V [22].

MD also permits qualitative comparison of the rupture kinetics between different systems. For example, the rupture kinetics of different enzymes bound to DNA or the same enzyme bound to different sequences could be compared. Note that comparisons of the kinetics among enzymes bound to different DNA sequences cannot be accomplished by simply mutating bases in the computational model. Crystallography revealed that the conformation of *Bam*HI (an enzyme with an active site very similar to that of *Eco*RI) is markedly different when it is specifically bound to its cognate sequence than when it is nonspecifically bound to a sequence that differed by a single basepair [43]. Hence, an experimentally derived structure containing the DNA sequence of interest is required because the necessary changes to the enzyme conformation are unlikely to occur on the timescale accessible to MD simulations if the simulation is begun from a generic structure.

In NFS experiments one can measure the transmembrane voltage required for rupture, but not the force on the complex at rupture. Thus, using MD, we embarked on two approaches to determine the force on the complex during the experiments. In the first approach, we extracted the force directly from simulations that mimicked experimental conditions [22]. This approach was complicated by the presence of solvent mediated forces, hydrodynamic effects, and artificial forces due to the thermostat. However, rupture forces ranging from 1,600 to 2,800 pN were recorded for simulations performed at 3.0 and 4.0 V.

In the second approach, we applied a methodology more akin to that used in optical tweezers, although with a number of conveniences not available in experiment. Here, a portion of the DNA was attached to one end of a virtual spring while the other end was displaced at various rates. The restriction enzyme was also restrained to its initial position by virtual springs; thus, the displacement of the spring end eventually led to rupture of the complex. Protocols such as these are often referred to as Steered Molecular Dynamics (SMD) [44, 45]. A diagram of how the forces were applied is shown in Fig. 14.3a. Figure 14.3b shows that the time until rupture decreased with the loading rate. An advantage of these SMD simulations is that the force on the virtual springs can be directly and unambiguously extracted. Figure 14.3c shows the force as a function of time for each of the SMD simulations. In all cases, the maximum value of the force was observed within a few hundred picoseconds of the time at which the displacement between the enzyme and DNA rapidly increased. The SMD simulations enabled us to determine the maximum force as a function of the loading rate, which is plotted in the inset of Fig. 14.3c.

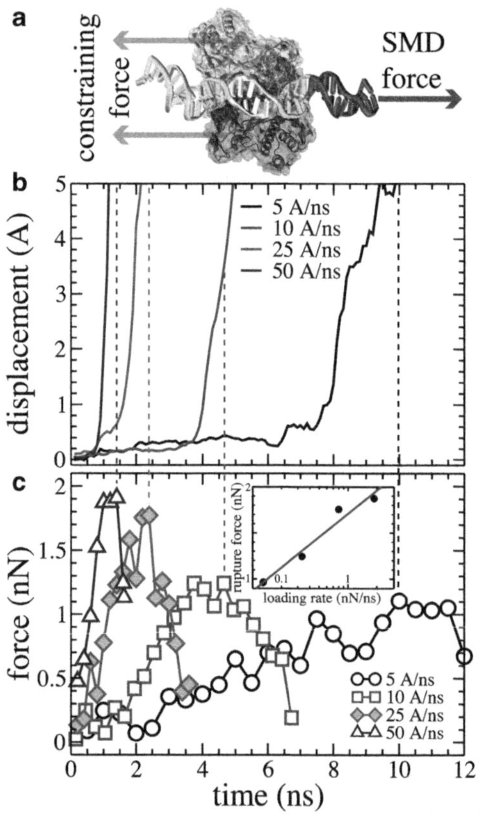

Fig. 14.3 SMD simulations of restriction enzyme–DNA rupture at different loading rates. (**a**) Schematic of the forces applied in the MD simulations. The center of mass of the 10 basepair fragment of DNA is attached to one end of a virtual spring, whose other end is moved to the right at a constant velocity. The α-carbons of the enzyme are also attached to virtual springs (constraining force) to keep the enzyme fixed. The portion of the DNA to which the *Eco*RI binds specifically is shown in *white*. (**b**) The distance between the *Eco*RI and the DNA vs. time for SMD simulations at different pulling rates. (**c**) Force restraining the enzyme as a function of time. The forces are averaged over 0.5, 0.3, 0.2, and 0.2 ns intervals for loading rates of 0.5, 1.0, 2.5, and 5.0 nm/ns. The *inset* shows the force at rupture f^* as a function of the loading rate r. The data can be fit by $f^*/(1\,\text{nN}) = 1.710 + 0.364\ln(r)$. Figure adapted from [22] by permission of the American Chemical Society

An approximately logarithmic dependence of the rupture force on the loading rate was found, consistent with theoretical expectations [14, 15].

*Eco*RI binds most strongly to the sequence GAATTC; however, some variations of this sequence still lead to different degrees of binding. Can MD predict the strength of binding for sequences differing by one or two basepairs from the cognate sequence?

The atomic-scale detail of the SMD simulations permitted us to identify the order in which each nucleotide of the DNA within the enzyme dissociated during rupture of the enzyme, thereby obtaining a qualitative estimate of the contribution of each basepair to the strength of the *Eco*RI–DNA bond [23]. Although there was a tendency for nucleotides nearest to the portion of the DNA to which force was applied to dissociate first, the first and last basepairs of the cognate sequence (GAATTC) were shown to be especially quick to dissociate. In accord with this, mutation of one of these two basepairs was shown to have only a small effect on the bulk free energy of dissociation (ΔG) – weakening it from the cognate value of – 15.2 to – 13.2 kcal/mol. The middle basepairs were slightly slower to dissociate (GAATTC), in agreement with the bulk ΔG of – 8.6 kcal/mol associated with the mutation of one of them. The underlined basepairs of GAATTC had a significant tendency to dissociate later than the others and were therefore predicted to have the greatest effect on *Eco*RI–DNA binding. Indeed, mutating just one of these two basepairs weakens the binding dramatically: $\Delta G = -4.9$ kcal/mol. Furthermore, no specific binding at all was observed in experiments in which both were mutated. Thus, the relative importance of each basepair to the binding predicted by simulation was in agreement with experimental measurements of the binding free energy.

The results presented here emphasize the sharpness of the threshold voltage. In fact, experiments seem to show that the dependence of the rupture rate on the transmembrane voltage is much stronger than exponential and apparently stronger than what is expected from theories of escape from a single well in the high-barrier limit [11, 12, 46]. Interaction of DNA with surfaces in and around pores have been implicated in modifying rupture kinetics from expected forms [47]. Furthermore, simulations have shown that, in the presence of strong attraction between the DNA and the pore walls, a threshold voltage for DNA translocation can exist even without an enzyme. Much work remains to be done to understand the origin of the sharpness of the threshold voltage and the small variance of the observed rupture times.

14.3 Unfolding of Hairpin DNA

Just as a bound enzyme can block passage of dsDNA through a nanopore unless a sufficiently large force is applied to dissociate the enzyme, regions having helical secondary structure on a DNA molecule can halt its translocation unless a sufficiently large force is applied to unfold these regions. Hairpin DNA, having both a double-helical portion and an overhanging coil portion, has been extensively studied by NFS [14, 16, 18–20, 48–50]. Interest in the transition between helical secondary structure and an unstructured coil conformation (or helix-coil transition) is important for both biological and technological applications. For example, during transcription of genomic DNA, RNA polymerase must unwind and dissociate the DNA double helix so that the RNA transcript can be synthesized [51]. Also, at the start of DNA replication, initiator proteins must dissociate the double helix at particular stretches along the genomic DNA referred to as replication origins. To facilitate the helix-coil transition, these replication origins often contain high fractions of A-T basepairs, which are less strongly bound than G-C basepairs [51].

Technological interest in the helix-coil transition is motivated by the development of advanced DNA sequencing methods. Inexpensive DNA sequencing methods promise to have an enormous impact on both personal health care and basic science [52, 53], and methods based on reading the DNA sequence as it passes through a nanopore have been actively developed [54–57]. The mechanics of DNA translocation through nanopores plays an important role in so-called nanopore sequencing methods and is a natural fit for NFS studies. Furthermore, research has shown potential for sequencing based on NFS itself. In a particularly dramatic demonstration of the power of NFS, Nakane et al. [48], collected hundreds of individual DNA duplex dissociation events, allowing the distribution of dissociation times under different applied forces to be determined and permitting statistical discrimination of sequences differing by single nucleotides. A number of other experimental protocols have shown the ability of NFS to discriminate DNA sequences by statistical analysis of the dissociation time of duplex DNA under force [17, 19].

As with enzyme–DNA complexes, experiments have shown a sharp dependence of the translocation probability of hairpin DNA on the transmembrane voltage [18, 20], which manifests itself as a relatively well defined threshold voltage for translocation of hairpin DNA through synthetic nanopores having diameters 1.0 nm $< d <$ 2.5 nm. However, these experiments arrived at a seemingly counter-intuitive result: pores having minimum diameters ≤ 1.2 nm showed much lower threshold voltages than pores having minimum diameters ≥ 1.4 nm [18, 20], i.e. more force was apparently required to drive the helix through larger pores than through smaller pores. This led to the suggestion that there was some fundamental change in the way by which translocation occurred between the larger and smaller pores. Here was a problem for which MD could give insight available by no other method: it was possible to visualize the translocation process for different pore sizes.

As shown in Fig. 14.4, a clear difference was observed in the mode of translocation between small pores and large pores in SMD simulations in which the coil of the hairpin DNA was displaced at a constant rate through the pore [18]. The largest constriction was 1.6 nm; thus, in all cases the double helix of the hairpin DNA (having a diameter of ~ 2.5 nm) had to be distorted to allow passage. For the pores having minimum diameters of 1.0 and 1.3 nm (Fig. 14.4a, b), the basepairs of the DNA were unzipped one by one and the DNA passed through the constriction as a single strand. However, for the pore having a minimum diameter of 1.6 nm, both strands passed through the pore simultaneously and some of the basepairs were

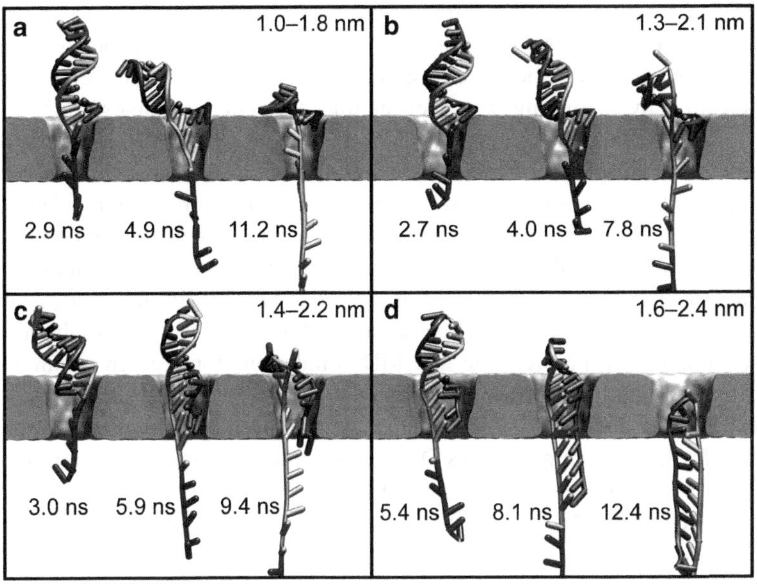

Fig. 14.4 Snapshots of MD simulations in which the phosphorus atom of the terminal base is pulled downward at a rate of 1 nm/ns. The pores have minimum diameters of (**a**) 1.0 nm, (**b**) 1.3 nm, (**c**) 1.4 nm, and (**d**) 1.6 nm. In all cases, the double helix, which has a diameter of $\sim$2.5 nm must be distorted to pass through the pore; however, the mode of passage is much different for the 1.6 nm pore than for the 1.0 and 1.3 nm pores. Figure adapted from [18] by permission of Oxford University Press

maintained, although the double helix was stretched and significantly distorted [18]. It is known that less force is required to unzip basepairs one-by-one (10–30 pN) than to stretch the DNA along its contour length (60–70 pN) [58]. Therefore, the reduction of the threshold voltage when moving from larger to smaller pores may be associated with the change in the translocation pathway from unzipping of the basepairs to stretching and distortion of the double helix.

Furthermore, there was a question as to the orientation of the hairpin DNA when it entered the pore. The DNA could pass through with the overhanging coil portion leading, as shown in Fig. 14.4a, b, as well as the with the double helix leading as shown in Fig. 14.4c. Due to the timescale limitations of MD, it was not possible to place the hairpin DNA in a random orientation far from the pore and observe which mode of translocation would eventually occur. Instead, Monte Carlo simulations were used to show that the probability that the hairpin DNA used in the simulations (having a 10 basepair double helix and a 50 basepair loop) arrived in the pore with the loop leading was significant [20]. We then began MD simulations in which either the loop of the helix was placed slightly outside the constriction of the pore or the coil was threaded through the constriction so that the translocation process could be observed in both orientations.

These MD simulations revealed that the DNA could pass through the pore in two distinct orientations – one in which the overhanging coil led translocation and another where the loop connecting the two strands of the double helix led translocation [20]. Although a transmembrane voltage was used to drive translocation rather than SMD pulling, we observed a similar dependence of the translocation mode on the pore size. Furthermore, for a 1.6 nm diameter pore, translocation by unzipping was seen at transmembrane voltages ≥ 6.5 V, where stretching/distortion was seen at ≤ 5 V. Figure 14.5a–c show snapshots from simulations of hairpin DNA translocation under an applied transmembrane voltage. Reduction of the constriction size resulted in a higher maximum electric field magnitude along the pore axis, while also reducing the depth at which the duplex DNA could move into the pore without unfolding. These competing factors resulted in dramatically different translocation behavior across different transmembrane voltages and pore geometries. The diagram in Fig. 14.5d illustrates which translocation modes were predicted by the MD simulations for a given minimum pore diameter and transmembrane voltage.

Because the duration of MD simulations is currently limited to a few microseconds for large systems such as DNA in a synthetic nanopore, certain translocation modes may not be sampled in the MD simulations. For example, spontaneous thermal unzipping of the hairpin DNA could occur on timescales inaccessible

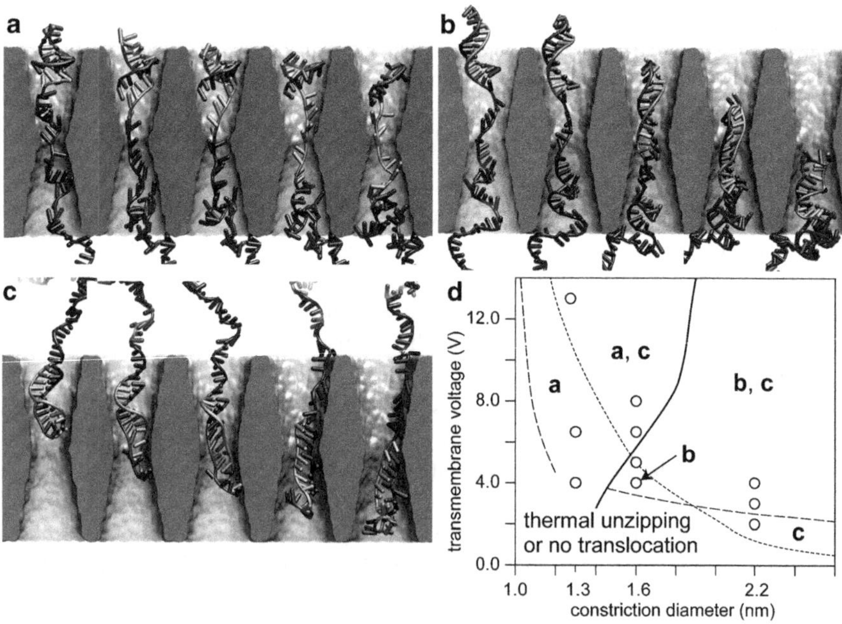

Fig. 14.5 Possible translocation modes for hairpin DNA in a synthetic nanopore having a diameter less than that of the double helix. (**a**) Unzipping of the basepairs one by one with the overhanging coil leading. (**b**) Stretching/distortion of the double helix with the overhanging coil leading. (**c**) Stretching/distortion of the double helix with the loop leading. (**d**) Diagram for predicting which of the translocation modes illustrated in the other *panels* might occur for a given pore diameter and transmembrane voltage. The *circles* show points that were probed by MD simulations. Figure adapted from [20] with permission from Elsevier

to MD, but might be a relevant mechanism for hairpin DNA translocation in experiments.

Another puzzling result of NFS experiments involving hairpin DNA was the wide range of ion current values measured. In experiments, ion currents ranging from $< 0.1I_0$ to $> 2I_0$, where I_0 was the current in the absence of DNA, were observed under identical bulk conditions and were attributed to the microscopic arrangement of the DNA within the pore [18, 20]. However, was the microscopic arrangement sufficient to produce such large swings in current? Did the explanation of such swings require interactions between multiple hairpin DNA molecules or the existence of impurities in the system? If indeed the range of current values could be associated with arrangement of the nanopore–DNA system, could the value of the current at a particular time be used to predict the arrangement at that time?

First, our MD simulations demonstrated that the range of current values observed in experiment could be explained by different arrangements of a single hairpin DNA molecules within the nanopore. Next, these simulations also suggested interpretations of the lowest and highest current values observed. For instance, the greatest reductions of the current were observed for the loop-first orientation (Fig. 14.5c) when the loop occupied the constriction of the pore [20]. However, the largest values of current ($2.17I_0$) were seen when the DNA had passed through the constriction and accumulated on the far side of the pore, just beyond the constriction. It also was possible to distinguish the translocation mode by the current values: because only one strand occupied the constriction during unzipping (Fig. 14.5a) rather than two during the translocation by the stretching/distortion pathway (Fig. 14.5b or 14.5c), the relative ion current (I/I_0) was significantly different.

While the simulations discussed above involved DNA translocation through synthetic nanopores, MD can also assist NFS experiments employing protein pores. Before synthetic nanopores were popular for nanopore experiments, the protein α-hemolysin was the nanopore of choice and is still widely used. Just as with synthetic nanopores, MD can be used to simulate α-hemolysin systems [59]; however, the time required for translocation can be much longer than in synthetic nanopores. This is principally because α-hemolysin has a very small minimum diameter (~ 1.4 nm [59]) and because the protein pore is suspended in a lipid bilayer, which is unable to withstand transmembrane voltages $> 1.0\,V$. Thus, experiments employing α-hemolysin typically involve smaller transmembrane voltages and consequently much lower dissociation rates than those employing synthetic nanopores. Therefore, we cannot simulate hairpin DNA translocation by simply applying the transmembrane voltage used in experiment, because hundreds of microseconds might be required to observe unfolding of the hairpin DNA. Using a larger transmembrane voltage would disrupt the bilayer and could lead to unrealistic conformational changes in the protein.

SMD presents a means to simulate translocation on timescales accessible to MD [44, 45]. However, one problem with conventional SMD simulations is that the force is applied to the molecule in an unrealistic way. Usually, the center of mass of some group of atoms is pulled at a constant velocity, resulting in unrealistic distributions of force along the molecule with respect to experiments in which an electric field drives the translocation [60]. Grid-steered MD [60] provides a means to apply a more realistic distribution of force to the molecule, which can result in a more realistic permeation trajectory. For example, Wells et al. [60] derived a three-dimensional electric field map of α-hemolysin from MD simulations. In grid-steered MD simulations, this electric field map was scaled by a factor greater than unity and applied to only the DNA. Thus, the force on DNA had a similar spatial distribution as it would under a larger transmembrane voltage without disrupting other portions of the system. The time required for simulation of translocation was reduced by many times, while the DNA adopted a much more realistic conformation during the trajectory than it would have using conventional SMD.

Much work remains to be done in understanding DNA duplex dissociation under force. While simulations revealed possible nanopore–DNA conformations that could lead to the extreme current values measured in hairpin DNA experiments [18, 20], more comprehensive studies of which arrangements give rise to which currents and to what extent the current value can be used to predict the arrangement are needed.

14.4 The Force on DNA in a Nanopore

In all of the NFS experiments discussed above, force was applied to the DNA by an external electric field. However, the force on the DNA as a function of the applied transmembrane voltage could not be directly measured. We can attempt to calculate the force on the DNA from a simple model. Suppose we have a system in which a long DNA molecule is threaded completely through the nanopore. Assuming that the electrolyte is a sufficiently good conductor, the electric field outside the pore is negligible compared to that inside; thus, we can consider the force on the DNA to be due to the electric field in the nanopore and, if necessary, in a small buffer region above and below the pore. Let the center of the nanopore be at the origin and the length of the nanopore and buffer be L. Assuming that the DNA between $-L/2$ and $L/2$ is positioned along the pore axis and has the shape of a B-form helix, we can approximate the linear charge density of the DNA as Q/a, where $Q = -2\,e$ is the charge of each basepair and $a = 0.34$ nm is the average rise of a basepair in random-sequence dsDNA. If we make the approximation that the system is cylindrically symmetric, we can attempt to estimate the force on the DNA by integration:

$$F_z = \int_{-L/2}^{L/2} dz\, E_z(z) Q/a = \Delta V Q/a \tag{14.1}$$

Here, F_z is the force on the DNA along the pore axis, E_z is the electric field in the z direction (along the pore axis) and ΔV is the transmembrane voltage. However, the force on the DNA in a nanopore is known to be substantially smaller than

$\Delta VQ/a$ [14, 15, 28], which was attributed to the effect of the DNA's counterions. In a hybrid setup, employing both a nanopore and optical tweezers, Keyser et al. [28] obtained an effective charge $Q_{\text{eff}} = -0.50 \pm 0.05\ e$, which is about 25% of the bare charge value, for DNA held fixed in pores having diameters from 6 to 11 nm.

What is the physical mechanism responsible for the reduction of the force from the bare electrostatic value of $\Delta VQ/a$? One could propose that a certain number of counterions bind to the DNA, which results in an effective reduction of the charge. Is this simple proposal correct? Keyser et al. [28] noted the "complex interplay between hydrodynamics and electrical charges in the screening layer of DNA." A theoretical description subsequently developed by Ghosal [61] highlighted the importance of electro-osmotic flow. If flow of the electrolyte solution is indeed important for determining the force on the DNA, how might this force depend on the properties of the nanopore?

Luan and Aksimentiev [62] used MD to calculate the force on DNA in a nanopore and determine the contribution of electro-osmotic flow to this force. Figure 14.6a illustrates the protocol used in the simulations. A virtual spring holds the DNA, while the DNA is subject to electrostatic force as well as that due to electro-osmotic flow; the force on the DNA can be calculated from the extension of the spring. Under an external field, K^+ and Cl^- ions in the nanochannel move in opposite directions; however, the K^+ counterions outnumber the Cl^- ions within 3.0 nm of the axis of the DNA, and therefore, induce a net flow of solution near the DNA. This flow moves in the direction opposite to the direction that the DNA would move if not restrained, exerting a viscous drag on the DNA and reducing the magnitude of the force on the DNA from its bare electrostatic value $\Delta VQ/a$.

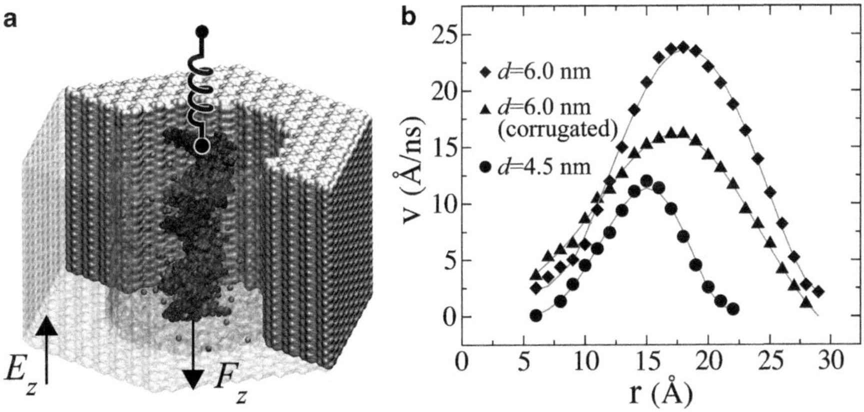

Fig. 14.6 Electro-osmotic flow in the vicinity of dsDNA. (**a**) Snapshot from MD simulation of effectively infinite DNA placed in an effectively infinite nanochannel. The center of mass of the DNA is attached to virtual spring to stall its motion under the force F_z, which results from external electric field E_z and the electro-osmotic flow due to ions in the pore. (**b**) Velocity of the electro-osmotic flow vs. distance from the DNA's axis in various nanochannels. *Diamonds*, *triangles*, and *circles* pertain to an atomically smooth channel having a diameter of 6.0 nm, a corrugated channel having a diameter of 6.0 nm, and an atomically smooth channel having a diameter of 4.5 nm, respectively. Figure adapted from [62]

Furthermore, the investigators have demonstrated that the electro-osmotic flow velocity and, consequently, the viscous force on the DNA depends on the geometry of the pore and the pore's surface properties [62]. Figure 14.6b shows the velocity profile of the electro-osmotic flow in several nanochannels. For 6.0 and 4.5 nm-diameter nanochannels with atomically smooth walls, the effective charge was shown to be $Q_{\mathrm{eff}} = -0.5\ e$, in agreement with experiment. However, in a nanochannel having a corrugated surface whose diameter varied as $d(z) = d_0 + A\cos(2\pi z/l)$ (where $d_0 = 3.0$ nm, $A = 0.2$ nm, and $l = 1.07$ nm), the effective charge was substantially higher. Finally, the study has shown that the stall force on DNA in a nanopore is $F_z = \xi\mu E_z$, where ξ is DNA's friction coefficient, which can be determined by measuring the time required for DNA to exit the pore in absence of electrostatic force, and μ is the electrophoretic mobility, which can be determined from DNA translocation experiments.

One might guess that a smaller-diamater pore would exhibit a smaller-magnitude electro-osmotic flow under the same bias, and hence the value of the effective charge would be closer to that of bare DNA. Indeed, van Dorp et al. [63], presented experimental evidence that Q_{eff} became closer to Q as the pore size decreased below 20 nm.

Simulations using 2.9 nm-diameter nanopores corroborated this relation between pore size and Q_{eff}. Figure 14.7a shows a system that consisted of initially unstrained DNA placed inside a nanopore having a minimum diameter of 2.9 nm in a membrane with a thickness of 10 nm. Figure 14.7b illustrates the second system in which the same molecule was placed in a pore of the same diameter, but here the membrane had a thickness of 5 nm. The system illustrated in Fig. 14.7c contained the same pore as the system shown in Fig. 14.7a except that the DNA was initially

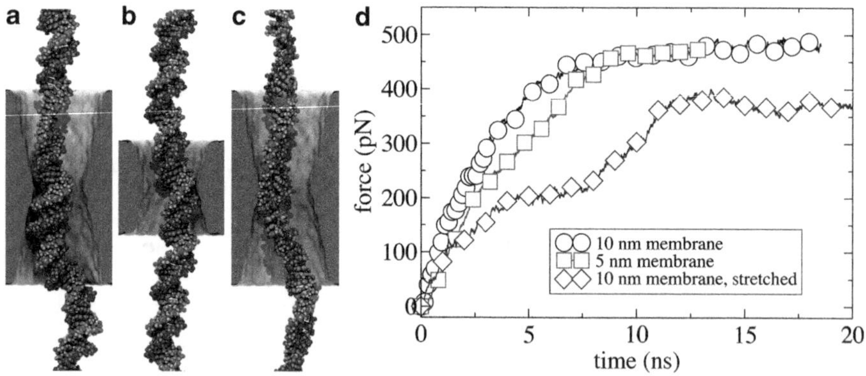

Fig. 14.7 Influence of the membrane thickness and DNA conformation on the stall force in a nanopore. In all cases, effectively infinite DNA was attached to a virtual spring, a transmembrane bias of 1.0 V was imposed. (**a**) Snapshot of unstrained DNA in a membrane having a thickness of 10 nm. (**b**) Snapshot of unstrained DNA in a membrane having a thickness of 5 nm. (**c**) Snapshot of strained DNA in a membrane having a thickness of 10 nm. (**d**) Force on the virtual spring attached to the center of mass of the DNA as a function of time after the 1.0 V transmembrane voltage was switched on for the three systems shown in *panels* **a**, **b**, and **c**. Figure adapted from [23] by permission of Oxford University Press

strained to 1.51 times its length at mechanical equilibrium. In each system, the DNA was bonded across the periodic system cell, making it effectively infinite. In the simulations discussed by Luan and Aksimentiev [62], the nanopores were also of effectively infinite length. However, simulations using nanopores of finite length should be directly comparable to those using nanochannels of effectively infinite length under the assumptions used in deriving (14.1). Each system was immersed in a solution of 0.1 M KCl, and a weak virtual harmonic spring restrained the center of mass of each molecule. Figure 14.7d shows the force exerted by the spring as a function of time after a 1.0 V transmembrane voltage was turned on. In all cases, the spring initially stretched rapidly. Eventually, the electro-osmotic force on the DNA was on average balanced by the force of the spring, allowing determination of the average force on the stationary DNA.

The average steady state force on the unstrained DNA in the 10 nm membrane was $<F_z> = -471$ pN – the negation of the average force on the spring, which is shown in Fig. 14.7d. Calculating the effective charge of the DNA by $Q_{eff} = <F_z/\Delta V>$, we obtain $Q_{\mathrm{eff}} = -0.97\ e$ for this 2.9 nm diameter pore – twice the value seen in larger pores – using $a = 0.33$ nm for the unstrained d(A_n) ·d(T_n) DNA. The force on the unstrained DNA in the 5 nm thick membrane was nearly identical to that on unstrained DNA in the 10 nm thick membrane. With $<F_z> = -365$ pN and $a = 0.50$ nm for the strained DNA, we obtain a significantly different value for the effective charge, $Q_{\mathrm{eff}} = -1.14\ e$, owing to different hydrodynamic properties of the strained molecule.

Analyses thus far have ignored the radial component of the force on DNA, which could also be probed in MD simulations. Simulations [20, 64, 65] and experiments [47] have highlighted the importance of the interaction between DNA and the surfaces of a synthetic nanopore during DNA translocation. Future work will be directed to determine the force on DNA held fixed in nanopores of various geometries, at different electrolyte conditions and varying strengths of DNA–surface interaction.

14.5 Conclusion

Through the examples of this chapter, we have shown how all-atom MD simulations can complement NFS experiments. First, the MD method allows the investigator to "observe" the microscopic behavior of the system, which provides invaluable insights into and a means to interpret puzzling results of NFS experiments. In some systems, the simulations can relate the experimentally measured blockage current to the microscopic conformation of the analyte in the pore [18, 20]. Furthermore, the simulations can partially compensate for one of the biggest deficiencies of the NFS method – the lack of direct measurement of the forces applied [22, 23, 62].

However, while MD can be a useful tool in many aspects of designing and interpreting NFS experiments, it has some limitations. The timescale accessible to the MD method, now ranging from a few microseconds to a millisecond, precludes exploration of many processes of interest. For example, rupture of an enzyme–DNA complex cannot be simulated by the MD method if the mean rupture time approaches the timescale of experiment. Among other processes difficult to model are the capture of analytes from solution by the nanopore and spontaneous dissociation of biomolecular assemblies, for example spontaneous unzipping of hairpin DNA, which could be relevant for nanopore translocation under some experimental conditions. Furthermore, the computational expense of large-scale MD runs can sometimes preclude the gathering of statistical data. We are sometimes left with making qualitative conclusions from a handful of trajectories, rather than quantitative results that could be obtainable by analyzing hundreds of trajectories.

Another factor that limits the predictive power of MD simulations is the accuracy of the molecular force field that describes the interactions between atoms in the simulation. One notable example of erroneous results due to force fields concerns the popular AMBER parm99 force field, with which DNA structures known to be stable in experiment made irreversible transitions to implausible structures in sufficiently long MD simulations [34]. Simulations of NSF experiments performed using this force field and its modified version [34] (that did not exhibit the spurious transitions) have shown some quantitative differences, while the qualitative behavior was preserved [20]. The force fields used to describe the surfaces of synthetic nanopores are particularly suspect as their interaction with DNA has not been calibrated. Furthermore, the composition and the structure of the synthetic nanopore surfaces themselves are not well known and depend on the fabrication methods [66].

Future advances in computer technology, simulation methods, and force fields will likely erase some problems that limit the application of the MD method to NFS today. As simulations in the microsecond range become routine, it will become possible to study NFS for processes in which phenomena such as spontaneous partial unfolding of proteins and nucleic acids are important. However, we strongly believe that MD simulation coupled with NFS experiments can even today form a single tool for the elucidation of biological phenomena and development of biotechnology that is more than a sum of its parts.

Acknowledgements We gratefully acknowledge contributions from Gregory Sigalov, Binquan Luan, and the group of Gregory Timp. This work is supported by grants from the National Institutes of Health (R01-HG003713 and PHS 5 P41-RR05969) and the National Science Foundation (PHY0822613). The authors gladly acknowledge supercomputer time provided through TeraGrid resources by a Large Resources Allocation grant (MCA05S028).

References

1. K. Neuman, A. Nagy, Nature Meth. **5**(6), 491 (2008)
2. R. Simmons, J. Finer, S. Chu, J. Spudich, Biophys. J. **70**(4), 1813 (1996)
3. J. Moffitt, Y. Chemla, S. Smith, C. Bustamante, Ann. Rev. Biochem **77**, 205 (2008)
4. S. Smith, L. Finzi, C. Bustamante, Science **258**(5085), 1122 (1992)
5. C. Gosse, V. Croquette, Biophys. J. **82**(6), 3314 (2002)

6. C. Lee, Y. Wang, L. Huang, S. Lin, Micron **38**(5), 446 (2007)
7. E. Evans, K. Ritchie, R. Merkel, Biophys. J. **68**, 2580 (1995)
8. W. Chen, E. Evans, R. McEver, C. Zhu, Biophys. J. **94**(2), 694 (2008)
9. S. Kim, P. Blainey, C. Schroeder, X. Xie, Nature Meth. **4**(5), 397 (2007)
10. I. Tinoco, Jr., C. Bustamante, Biophys. Chem. **101–102**, 513 (2002)
11. E. Evans, K. Ritchie, Biophys. J. **72**(4), 1541 (1997)
12. O. Dudko, G. Hummer, A. Szabo, Phys. Rev. Lett. **96**(10), 108101 (2006)
13. C. Ho, R. Qiao, A. Chatterjee, R.J. Timp, N.R. Aluru, G. Timp, Proc. Natl. Acad. Sci. USA **102**, 10445 (2005)
14. J. Mathé, H. Visram, V. Viasnoff, Y. Rabin, A. Meller, Biophys. J. **87**, 3205 (2004)
15. J. Mathé, A. Arinstein, Y. Rabin, A. Meller, Europhys. Lett. **73**, 128 (2006)
16. O.K. Dudko, J. Mathé, A. Szabo, A. Meller, G. Hummer, Biophys. J. **92**, 4188 (2007)
17. C. Tropini, A. Marziali, Biophys. J. **95**, 1632 (2007)
18. Q. Zhao, J. Comer, V. Dimitrov, A. Aksimentiev, G. Timp, Nucl. Acids Res. **36**(5), 1532 (2008). DOI 10.1093/nar/gkm1017
19. B. McNally, M. Wanunu, A. Meller, Nano Lett. **8**, 3418 (2008)
20. J. Comer, V. Dimitrov, Q. Zhao, G. Timp, A. Aksimentiev, Biophys. J. **96**(2), 593 (2009). DOI 10.1016/j.bpj.2008.09.023
21. B. Hornblower, A. Coombs, R.D. Whitaker, A. Kolomeisky, S.J. Picone, A. Meller, M. Akeson, Nature Mater. **4**, 315 (2007)
22. Q. Zhao, G. Sigalov, V. Dimitrov, B. Dorvel, U. Mirsaidov, S. Sligar, A. Aksimentiev, G. Timp, Nano Lett. **7**, 1680 (2007). DOI 10.1021/nl070668c
23. B. Dorvel, G. Sigalov, Q. Zhao, J. Comer, V. Dimitrov, U. Mirsaidov, A. Aksimentiev, G. Timp, Nucl. Acids Res. **37**, 4170 (2009). DOI 10.1093/nar/gkp317
24. C. Goodrich, S. Kirmizialtin, B. Huyghues-Despointes, A. Zhu, J. Scholtz, D. Makarov, L. Movileanu, J. Phys. Chem. B **111**(13), 3332 (2007)
25. Z. Siwy, L. Trofin, P. Kohli, L. Baker, C. Trautmann, C. Martin, J. Am. Chem. Soc. **127**(14), 5000 (2005)
26. G.V. Soni, A. Meller, Clinical Chemistry **53**, 1996 (2007)
27. G. Sigalov, J. Comer, G. Timp, A. Aksimentiev, Nano Lett. **8**, 56 (2008). DOI 10.1021/nl071890k
28. U. Keyser, B. Koeleman, S. Dorp, D. Krapf, R. Smeets, S. Lemay, N. Dekker, C. Dekker, Nature Phys. **2**, 473 (2006)
29. A. Aksimentiev, Nanoscale **2**, 468 (2010). DOI 10.1039/b9nr00275h
30. H. Chang, F. Kosari, G. Andreadakis, M.A. Alam, G. Vasmatzis, R. Bashir, Nano Lett. **4**, 1551 (2004)
31. R.M. Smeets, U. Keyser, D. Krapf, M.Y. Wu, N. Dekker, C. Dekker, Nano Lett. **6**, 89 (2006)
32. J.C. Phillips, R. Braun, W. Wang, J. Gumbart, E. Tajkhorshid, E. Villa, C. Chipot, R.D. Skeel, L. Kale, K. Schulten, J. Comp. Chem. **26**, 1781 (2005)
33. W.D. Cornell, P. Cieplak, C.I. Bayly, I.R. Gould, K.M. Merz, Jr., D.M. Ferguson, D.C. Spellmeyer, T. Fox, J.W. Caldwell, P.A. Kollman, J. Am. Chem. Soc. **117**, 5179 (1995)
34. A. Perez, I. Marchan, D. Svozil, J. Sponer, T.E. Cheatham, C.A. Laughton, M. Orozco, Biophys. J. **92**, 3817 (2007)
35. A.D. MacKerell, Jr., D. Bashford, M. Bellott, R.L. Dunbrack, Jr., J. Evanseck, M.J. Field, S. Fischer, J. Gao, H. Guo, S. Ha, D. Joseph, L. Kuchnir, K. Kuczera, F.T.K. Lau, C. Mattos, S. Michnick, T. Ngo, D.T. Nguyen, B. Prodhom, I.W.E. Reiher, B. Roux, M. Schlenkrich, J. Smith, R. Stote, J. Straub, M. Watanabe, J. Wiorkiewicz-Kuczera, D. Yin, M. Karplus, J. Phys. Chem. B **102**, 3586 (1998)
36. J. Comer, D.B. Wells, A. Aksimentiev, in *DNA Nanotechnology: Methods and Protocols*, Methods in Molecular Biology, vol. 749, DOI 10.1007/978-1-61779-142-0_22, Springer Science+Business Media, LLC 2011
37. M. Newman, T. Strzelecka, L. Dorner, I. Schildkraut, A. Aggarwal, Science **269**(5224), 656 (1995)

38. J. Watson, M. Gilman, J. Witkowski, M. Zoller, *Recombinant DNA: a short course* (WH Freeman, 1992)
39. F. Sanger, S. Nicklen, A. Coulson, Proc. Natl. Acad. Sci. USA **74**(12), 5463 (1977)
40. J. Shendure, G. Porreca, N. Reppas, X. Lin, J. McCutcheon, A. Rosenbaum, M. Wang, K. Zhang, R. Mitra, G. Church, Science **309**(5741), 1728 (2005)
41. R. Saiki, S. Scharf, F. Faloona, K. Mullis, G. Horn, H. Erlich, N. Arnheim, Science **230**(4732), 1350 (1985)
42. A. Jeltsch, J. Alves, H. Wolfes, G. Maass, A. Pingoud, Biochemistry **33**(34), 10215 (1994)
43. H. Viadiu, A. Aggarwal, Molecular cell **5**(5), 889 (2000)
44. H. Grubmüller, B. Heymann, P. Tavan, Science **271**, 997 (1996)
45. B. Isralewitz, S. Izrailev, K. Schulten, Biophys. J. **73**, 2972 (1997)
46. G.I. Bell, Science **200**, 618 (1978)
47. M. Wanunu, J. Sutin, B.M. B, A. Chow, A. Meller, Biophys. J. **95**, 4716 (2008)
48. J. Nakane, M. Wiggin, A. Marziali, Biophys. J. **87**, 615 (2004)
49. M. Wiggin, C. Tropini, V. Tabard-Cossa, N. Jetha, A. Marziali, Biophys. J. **95**, 5317 (2008)
50. J. Mathé, A. Aksimentiev, D.R. Nelson, K. Schulten, A. Meller, Proc. Natl. Acad. Sci. USA **102**, 12377 (2005). DOI 10.1073/pnas.0502947102
51. B. Alberts, A. Johnson, J. Lewis, M. Raff, K. Roberts, P. Walter, *Molecular Biology of The Cell*, 4th edn. (Garland Science, New York & London, 2002)
52. A. Guttmacher, F. Collins, New England J. Med. **347**(19), 1512 (2002)
53. M.L. Metzker, Genome Research **15**, 1767 (2005)
54. N. Ashkenasy, J. Sánchez-Quesada, H. Bayley, M.R. Ghadiri, Angew. Chem. Int. Ed. Engl. **44**, 1401 (2005)
55. D. Branton, D. Deamer, A. Marziali, H. Bayley, S. Benner, T. Butler, M. Di Ventra, S. Garaj, A. Hibbs, X. Huang, et al., Nature Biotech. **26**(10), 1146 (2008)
56. M. Rhee, M. Burns, Trends Biotech. **24**(12), 580 (2006)
57. J. Clarke, H. Wu, L. Jayasinghe, A. Patel, S. Reid, H. Bayley, Nature Nanotech. **4**(4), 265 (2009)
58. M. Rief, H. Clausen-Schaumann, H.E. Gaub, Nature Struct. Biol. **6**, 346 (1999)
59. A. Aksimentiev, K. Schulten, Biophys. J. **88**, 3745 (2005). DOI 10. 1529/biophysj.104.058727
60. D.B. Wells, V. Abramkina, A. Aksimentiev, J. Chem. Phys. **127**, 125101 (2007). DOI 10.1063/1.2770738
61. S. Ghosal, Phys. Rev. Lett. **98**, 238104 (2007)
62. B. Luan, A. Aksimentiev, Phys. Rev. E **78**, 021912 (2008). DOI 10.1103/PhysRevE.78. 021912
63. S. van Dorp, U.F. Keyser, N.H. Dekker, C. Dekker, S.G. Lemay, Nature Phys. **5**, 347 (2009)
64. A. Aksimentiev, J.B. Heng, G. Timp, K. Schulten, Biophys. J. **87**, 2086 (2004). DOI 10.1529/ biophysj.104.042960
65. J.B. Heng, A. Aksimentiev, C. Ho, P. Marks, Y.V. Grinkova, S. Sligar, K. Schulten, G. Timp, Biophys. J. **90**, 1098 (2006). DOI 10.1529/biophysj.105.070672
66. M. Wu, D. Krapf, M. Zandbergen, H. Zandbergen, P. Batson, Applied Physics Letters **87**, 113106 (2005)

Index

S.M. Iqbal and R. Bashir (eds.), *Nanopores: Sensing and Fundamental Biological Interactions*, DOI 10.1007/978-1-4419-8252-0,

O

P

R

S

9781441982513.